Lithics in the Scandinavian Late Bronze Age

Sociotechnical change and persistence

Anders Högberg

BAR International Series 1932
2009

Published in 2016 by
BAR Publishing, Oxford

BAR International Series 1932

Lithics in the Scandinavian Late Bronze Age

ISBN 978 1 4073 0414 4

Layout by Anders Gutehall
Translated by Alan Crozier

BAR Publishing is the trading name of British Archaeological Reports (Oxford) Ltd. British Archaeological Reports was first incorporated in 1974 to publish the BAR Series, International and British. In 1992 Hadrian Books Ltd became part of the BAR group. This volume was originally published by Archaeopress in conjunction with British Archaeological Reports (Oxford) Ltd / Hadrian Books Ltd, the Series principal publisher, in 2009. This present volume is published by BAR Publishing, 2016.

Printed in England

BAR
PUBLISHING

BAR titles are available from:

BAR Publishing
122 Banbury Rd, Oxford, OX2 7BP, UK
EMAIL info@barpublishing.com
PHONE +44 (0)1865 310431
FAX +44 (0)1865 316916
www.barpublishing.com

Acknowledgements

Deborah Olausson has given me inspirational guidance through the work of writing this dissertation. I owe her my sincere appreciation and gratitude. Many thanks to Mikkel Sørensen for his collaboration and assistance in the experimental studies. I am also extremely grateful to all the kind and knowledgeable people I have met at museums, universities, and archaeological institutions in Scandinavia, Germany and Poland. You all know what I mean when I say that a book like this would not have been possible without you.

Björn Nilsson, Helle Juel Jensen, Berit Eriksen, Peter Skoglund, and Lars Larsson have all given me valuable comments on earlier versions of the manuscript. Thanks to Björn Nilsson for the drawings, to Alan Crozier for the translation, and to Anders Gutehall for the layout. A great deal of this work has been done as part of a venture in contract archaeology, the Citytunnelprojektet. Thanks to all my colleagues at Malmö Museum who have assisted me in various ways.

My research and this publication have been financed through grants from the Swedish Research Council, Malmö Kulturhistoriska Förening, The Lejre Research Grant, Stiftelsen Konung Gustaf VI Adolfs fond for Swedish kultur, Ebbe Kocks stiftelse, Knut och Alice Wallenbergs stiftelse, Letterstedtska föreningen, Evers & Co:s fond for svenske videnskabsmænds studier i Danmark, Stiftelsen Fil. dr. Uno Otterstedts fond för främjande av vetenskaplig undervisning och forskning, Stiftelsen Bokelunds resestipendiefond, Stiftelsen Landshövding Per Westlings Minnesfond, and Stiftelsen Syskonen Willers Donationsfond. Generous contributions from Malmö Förskönings- och Planteringsförening and from Malmö Kulturhistoriska Förening have enabled my manuscript to be turned into this book. Thank you all very much.

Contents

Introduction **11**

Possible and impossible questions – about a study of sociotechnical change and persistence **12**

Part I: Requirements for a theory of technical development **17**

Interpretive archaeologies **17**

About interpretation 18

A crash 19

On knowing more about people than they do about themselves 21

Materiality **23**

The materiality of keys 25

Materiality and material 26

Engaged studies 27

Sociotechnical change and persistence **28**

Back to sociotechnical 30

On linear and non-linear processes 32

Technological frame 34

Action 34

Introduction and stabilization 35

Destabilization and restabilization 35

Participation 36

Summary thus far 36

Chaîne opératoire **37**

Chain of actions, production sequence, and procedures 38

Intentionality and the conceptual template of the maker 39

Some critical reflections on the chaîne opératoire 40

About classifying not according to what we know, but according to what we want to know **42**

Requirements for a theory of technical development – conclusion **45**

Part II: Large blade knives and other flint assemblages from the Late Bronze Age 47

A history of previous archaeological research 47

Large blade knives of flint in museum collections 49
Publications – southern Sweden and Denmark 51
The large settlement site excavations from the end of the 1970s until the middle of the 1990s 57
Major research excavations and publications from recent years 63
Some general surveys 64
A brief glance at central Sweden, Bornholm, northern Germany, Norway, and Poland 65
Summing up and formulating the questions 65

Methods 69

The onion problem 69
My source material 70
Flint as raw material 71
Attribute analysis 72
Use-wear analysis 74

Large blade knives of flint – morphology and manufacture 78

A morphological and technological attribute analysis of large blade knives of flint 79
Results – a morphological and technological attribute analysis of large blade knives of flint 94
The manufacture of large blade knives – experimental analyses 95
The experiments 96
Two reference large blade knives 97
Controlled experiments with specific questions 97
The first part of the experiment 98
The second part of the experiment 123
The third part of the experiment 130
Results of the manufacture of large blade knives – experimental analyses 134
Summing up – morphology and manufacture 134

Use-wear analysis of large blade knives 136

Previous functional analyses of large blade knives 136
Comparative analyses of published observations – polish from silicaceous plants 140
Comparative analyses of published observations – shafting of large blade knives 143
A comparative experimental analysis 145
Use-wear analyses, prehistoric large blade knives of flint 150
Summing up – functional analyses of large blade knives 164

Dating, contexts, and geographical distribution **165**

Summing up – dating, contexts, and geographical distribution 171

Assemblages of large blade knives of flint **172**

Analysis of assemblages of large blade knives 173
Results – assemblages of large blade knives 192

Special manufacturing sites for large blade knives of flint **196**

Previously known manufacturing sites 196
What should I look for in order to find manufacturing sites? 198
Result and discussion – special manufacturing sites for large blade knives 211

Discussion and conclusion – large blade knives of flint **215**

Large blade knives, a specialized artefact 218

Household production of flint tools in the Late Bronze Age **219**

Attribute analyses 220
The technological and morphological character of the debitage 229
Cores 231
A technological and methodological description of household production 231
What was made? 233
Retouched flakes and frost-cracked pieces with retouch 233
The appearance of the retouch and form 233
Formal tools 233
Use-wear analyses of unmodified flakes from household production 235
Comparisons with flint from earlier and later periods 237
Results of the analyses of household production 240

Two flint-knapping strategies and a hybrid **241**

Waste material from the production of large blade knives 243
Other production – an analysis of cores 243
Other production – function 246
Large blade knives of flint from the site 246
Discussion 250
Other places 251

Apalle and Hallunda – a previously unknown flint-knapping technology **253**

Results – Apalle and Hallunda, a previously unknown flint-knapping technology 256

Flint objects in cremation graves **256**

Analyses of cremation graves with finds of knapped flint 257
A burnt large blade knife of flint from a cremation grave 260
Conclusion – flint objects in cremation graves 260

Large blade knives and other flint assemblages from the Late Bronze Age – conclusions **261**

Part III: A sociotechnical construction of the large blade knife of flint 265

Actors, actants and networks 265

The construction of large blade knives 267

Flint as a substitute for bronze 267
From one metal to another 269
Iron production 271
Interpretations in contrast 271
Ritual specialists and flint craft 272
About places where specialists worked 272
Different ways of making large blade knives 273
The innovator 273
Local, regional, and supraregional 274
Changes in the landscape 275
Ritual depositions 277
Graves, burial places, and rock carvings 278
Bronze sickles 279
An age of flint, iron, and bronze 280
Change and persistence 281
From Bronze Age to something else and from something else to Iron Age 282

Conclusion – the sociotechnical construction of the large blade knife of flint 283

References 285

Unpublished references 303
Personal communications 303

Figure 1.1. This book is a study of knapped flint from the south Scandinavian Late Bronze Age. Here we see the only complete surviving tool with a shaft of wood and a knife blade of flint. The photograph, taken in 2007, shows the find from Stenild bog in Denmark which was discovered in 1898. Photo: National Museum, Copenhagen.

Introduction

Some time at the end of the Bronze Age a new tool was invented. It consisted of a wooden shaft and a knife blade of flint. Just one such tool survives today. It was found in 1898 in the Stenild bog in Denmark, figure 1.1. But there are many knife blades made of flint, as shown in figure 1.2. We know them from museums and private collections in southern Scandinavia and along the south coast of the Baltic Sea. Archaeologists call them by different terms. In Sweden they are known as lövknivar, literally "leaf-knives", in Denmark flækkeknivar, literally "blade knives", and in Poland "backed blades of Zele type", to mention just a few examples. I call them "large blade knives of flint". So, in a period which posterity has classified as the end of the south Scandinavian Bronze Age and the beginning if the Iron Age, figure 1.3, an implement was invented with knife blade made of flint. Why?

Much of the present book deals with this question why. But it is often difficult to find an answer to a word like why. Most readers are surely familiar with everyday situations when virtually impossible why-questions are asked. Why did that happen? Why did you do that? And we are expected to have a plausible reply. But what kind of answer can one give to such questions? I don't know? I didn't mean to?

But despite the difficulty of answering a why-question, I have posed it. Time after time I have asked myself – to the point of nagging – why? I have done this because I want to know. I want to know why large blade knives of flint were introduced at a time that is classified as a period of transition between bronze and iron technology, why they disappeared, and what happened in between.

I want to know how flint technology was a part of the society of the time, how it was perceived, what influence it had, how it was influenced, how it lived its own life, how it changed, perhaps was no longer appreciated and was then replaced by something else. In what way did it belong to social life, interaction, encounters, life, death, everyday existence, hard work, relaxation, festivities, rituals, seasons, world-views, cosmology, and the exercise of power? In other words, I want to know about the social shaping of a Late Bronze Age technology and the technical shaping of Late Bronze Age society.

Technology and society are both human constructs. Technology is created by technologists who invent, develop, and produce, as well as by people using what technologists have created – both "according to the manual" and in uses that no technologist could ever imagine. Technology is shaped by social structures, but also by the effort and commitment of individuals. But these individuals are also products of social shaping. How individual persons acted in particular events and how this happened on the basis of the contexts to which the persons belonged is also something I would like to know.

So this book is actually about something completely different from large blade knives of flint and the Late Bronze Age. It is about material culture and technology, about change and stability, and about sociotechnical change and persistence. It is also about how I view past and present, about how I handle the contrasts between the knowledge I want to achieve and the knowledge I believe I am able to produce. In that way the book is about asking questions, choosing methods to work with, analysing and interpreting, and creating a balance between the interpretations that are presented, the questions that are asked, and the methods that are selected.

The book is thus about the time long ago when this happened and the present day, now when I am talking about that time, and also about things that have happened since then and until now. That is what this book has been written to consider. What you as a reader get out of it is a completely different matter.

Figure 1.2. Knife blades, known as "large blade knives of flint", which were fitted into a wooden shaft to give a composite tool, survive in many examples in collections from southern Scandinavia and along the south coast of the Baltic Sea. These large blade knives are the main source material in this book. They are usually 11 to 15 centimetres long, weigh 60 to 120 grams, are three to four centimetres wide and one to one and a half centimetres thick. When in future I write "large blade knife" I mean the archaeological artefact represented by those in this picture. When I write "knife blade" I mean the edge part of the shafted tool of which the large blade knives were a part, as shown in figure 1.1. The photographs show front and back on knives from the Museum of Halmstad.

Possible and impossible questions – about a study of sociotechnical change and persistence

The study focuses on a period which in southern Scandinavia is called the latter part of the Late Bronze Age. This comprises the archaeological periods V and VI (Montelius 1885), in chronological terms c. 900–500 BC (Jensen 1997). On the continent this corresponds to Hallstadt Ha B2/3–Ha D or the latter part of the Lusatian culture, and locally in the core area for my case studies this period is called phase B of the Late Bronze Age (Björhem & Säfvestad 1993:51). This is the chronological framework constituted by the book's prehistoric source material, table 1.1, a framework that spans a time with a backward limit formed by the last centuries of the Urnfield culture and a forward limit consisting of the first century of the La Tène culture.

In geographical terms the study is confined to southern Scandinavia with present-day Sweden, Denmark, and Norway, together with parts of Germany and Poland, figure 1.4.

The archaeological material on which the studies are based has been collected in different ways. South-west Scania is the central area in my case study. More than 30 years of extensive development of a landscape which has been intensively used since the early Stone Age has produced a significant amount of source material here through archaeological excavations, and the area is one of the best-excavated in Europe (see, for example, Björhem & Säfvestad 1993; Billberg et al. 1996; Billberg & Magnusson Staaf 1999; Rudebeck & Ödman 2000; Lindhé, Sarnäs & Steineke 2001; Rudebeck et al. 2001). In my study I use parts of this source material, especially from the archaeological projects Öresundsförbindelsen, i.e. The Öresund Link Project (Billberg et al. 1996; Billberg & Magnusson Staaf 1999) and Citytunnelprojektet (Högberg 2001c; Lindhé, Sarnäs & Steineke 2001), projects in which I myself have been involved, figure 1.5.

There are previous studies of blade knives, and part of the material that I use has already been published. But by far the majority of the source material I have compiled simply by travelling around southern Scandinavia and searching in museum collections and archives. During these research trips and visits to museums I have met friendly and well-informed archaeologists and antiquarians who have helped me in my research by sharing their local knowledge, their familiarity with museum collections and exhibitions, and their own research. In this way I have gradually created and expanded the extensive archaeological source material on which this study is based, table 1.2.

Table 1.1. The chronological framework of the study. Within the Late Bronze Age it is the later parts that are the focus of this study, that is, c. 900–500 BC. The table is based on Jensen (1997:17, 2002:15) supplemented with Björhem & Säfvestad (1993:51). It shows the chronological framework for my case-study area (the Malmö area), for southern Scandinavia, and for central Europe.

BC	South Scandinavian Late Bronze Age	Central Europe	Malmö area
1000	IV		Phase A
		Ha B1	
900			
		Ha B2/3	
	V		Phase B
800			
700			
		Ha C	
	VI		
600			
		Ha D	
500	Early Iron Age		Early Iron Age

In this book I will navigate around this source material in different ways. I will move from the little to the large; from small introverted technological observations about individual flint flakes, observations which I and perhaps a dozen more people in the whole world care about, to an attempt to link this little aspect to the larger whole in a way that will be understood by anyone interested in material culture, technology, the Late Bronze Age, and sociotechnical change and persistence.

To be able to do this, I need not just a large amount of archaeological source material, but also a theoretical and methodological framework in which to work, a framework that discusses possibilities and limitations in my study of material culture and technology, a framework which consists of questions, theoretical stances, and a clear formulation of method, and which shows how I believe that the theories, questions, and methods with which I work are relevant tools for the job.

The first part of the book deals with this, how I think and how I have sought inspiration for my thinking in a study of flint objects and flint technology from the Late Bronze Age, that is, how I have built up my theoretical framework. In a study of contexts and the dilemma of working contextually, Ben-Ami Scharfstein, a philosopher at the University of Tel-Aviv, wrote that "no reasoning of action can be understood very well outside of its own context" (Scharfstein 1989:4). It is this "context" that is the subject of Part I, which deals with theories, methods, possibilities, and limitations. It is arranged in such a way that I begin with large generalizing questions and gradually work my way towards the more specific, from discussions about the possibility of interpreting the past, through material

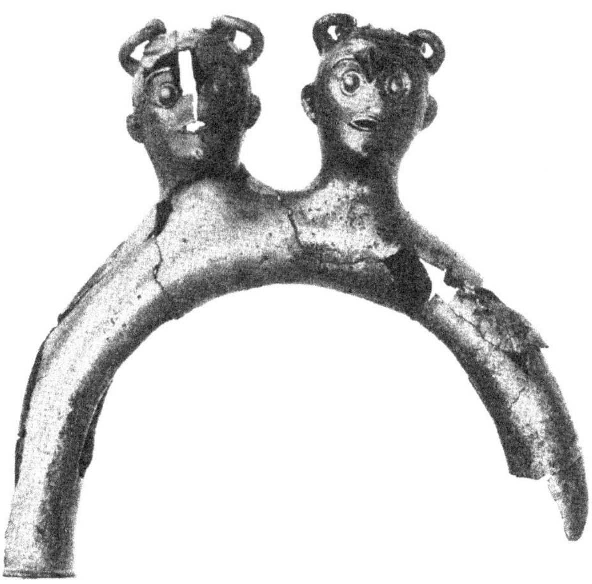

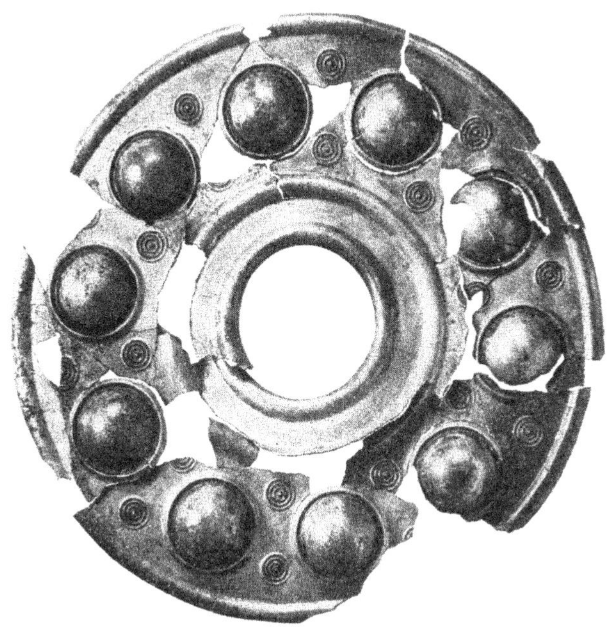

Figure 1.3. The South Scandinavian Late Bronze Age is usually associated with rich finds of metal. When shields of glistening bronze were drowned in the waters of lake Vänern, when lures of bronze were manufactured, when people used razors of bronze with intricate patterns and references to the contemporary cosmology of the sun's daily journey over the sky and its nightly transport under the earth, and when Etrurians in northern Italy were buried with a chariot as one of the rich array of grave goods – that was a time when a flint technology arose that is not comparable with any earlier or later flint technology. The figure shows, to the left, a semi-circular bronze tube from an horse's harness (Larsson 1975:175) and, to the right, a disc of a bronze lur (Larsson 1975:175), both from the famous Fogdarp find (Larsson 1975).

Figure 1.4. Geographic scope of the study. Case studies and detailed studies have been conducted with material from southern Sweden and eastern Denmark, especially from the Malmö area. This is the core area of the study (circle). The further away from this area I move, the more general the studies become.

culture and materiality, to technology and individual activities and acts. The focus is thus primarily on trying to work from premises and ambitions based on the possibilities and limitations inherent in archaeology, not on offering a theoretical framework into which to fit the archaeological material.

There is good reason here to emphasize a difference between ambitions and resources. What I want to do is one matter; what I can do is another. What I want to do is expressed in this study through my questions, my choice of study material, and theoretical discussions. What I can do with my questions is both enabled and restricted by the source material and the theoretical and methodological framework that I have selected. The resources do not always succeed in matching the ambitions; in other words, the questions are not always possible to answer. Sometimes they are impossible. But this gives us no grounds for changing the ambitions. When the resources are insufficient, I simply fall short of my ambitions. The consequence of this is that certain questions which I pose will be thoroughly examined, while others remain unexamined.

Part II of the book is an extensive study of archaeological material. It begins with a section about how a hundred years of Scandinavian research has handled and studied artefacts

Table 1.2. The collections from which my study material derives. From these museums I have studied collections of stray finds or material from excavations.

Sweden	Malmö Museums Lund University Historical Museum Museum of National Antiquities, Stockholm Österlen Museum, Simrishamn Helsingsborg Museum Kristianstad County Museum Blekinge County Museum, Karlskrona Halland County Museum, Halmstad Gothenburg City Museum Bohuslän Museum, Uddevalla
Denmark	National Museum, Copenhagen Roskilde Museum Sydsjælland Museum, Vordingborg Museum of Cultural History, Randers Svendborg Museum
Norway	Historical Museum, Oslo
Germany	Schloss Gottorf, Schleswig Landesamt für Boden-denkmalpflege in Mecklenburg-Vorpommern, Stralsund
Poland	Polish Academy of Sciences, Institute of Archaeology and Ethnology, Warsaw

of flint from the Late Bronze Age, a section that functions as a kind of encounter between the book's initial reasoning and reflection and the strictly steered analyses of the second part. This review of the history of archaeology leads to a great many questions specifically concerning the archaeological material. I then bring these questions into the other sections in Part II

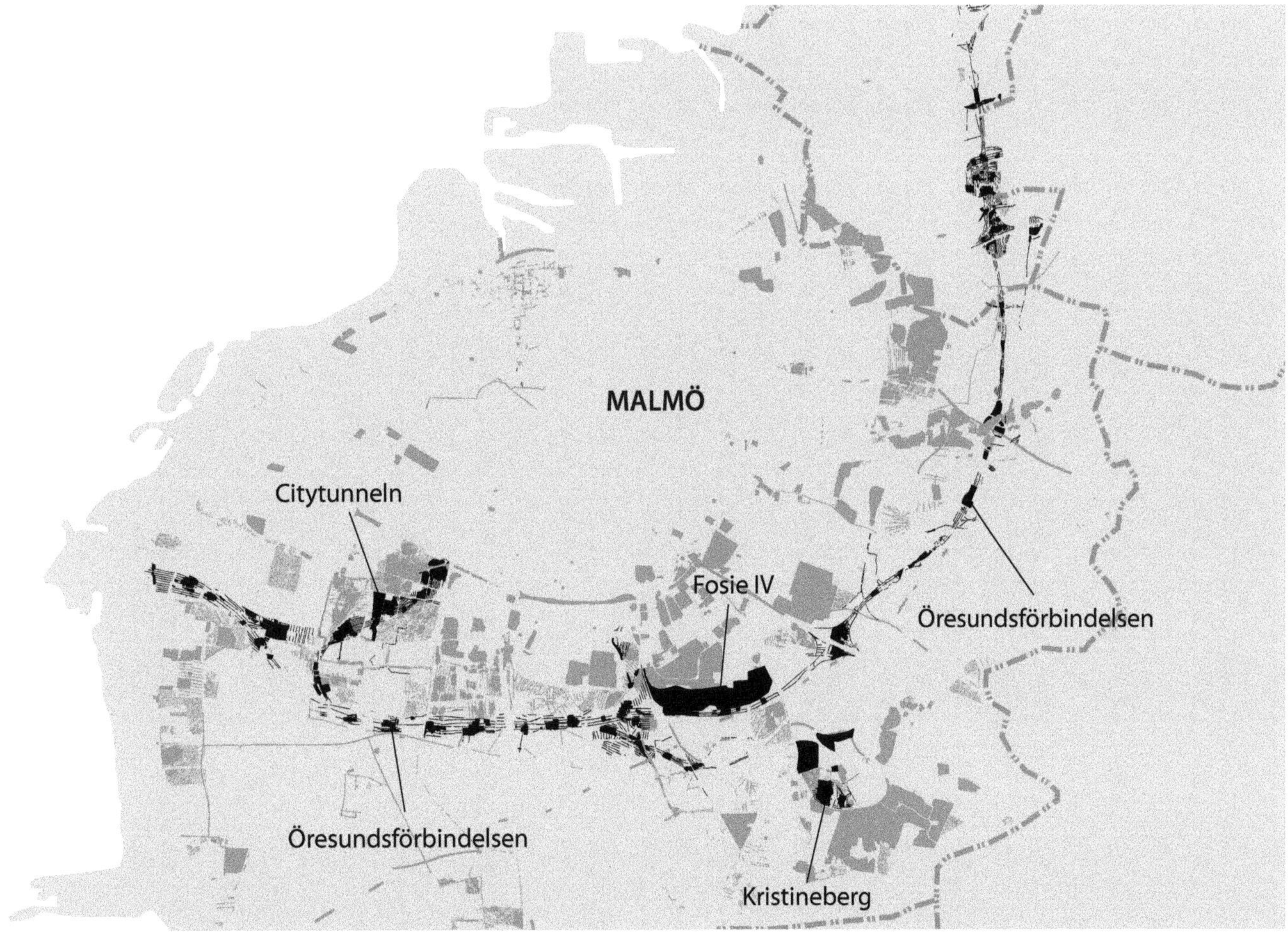

Figure 1.5. Commissioned archaeological investigations in the area around Malmö, the core area of my case studies, with the projects Öresundsförbindelsen, i.e. The Öresund Link Project, Citytunnelprojektet, Fosie IV, and Kristineberg marked.

of the book, which is a broad analysis of archaeological flint material from the Late Bronze Age.

In the last part of the book I link the results from Part I and Part II and interpret my analyses in the light of other scholars' archaeological knowledge of the Late Bronze Age. Focusing on sociotechnical change and persistence in a dynamic society in change, I discuss aspects of how what we call the Bronze Age became what we call the Iron Age.

Although the three parts of the book are connected, in that I could not have written any of the parts without the others, they can if one wishes be read separately. To put it simply, Part I is about archaeology, Part II about archaeological artefacts consisting of large blade knives and other flint assemblages from the Late Bronze Age, and Part III about interpretations of sociotechnical change and persistence at the end of the Bronze Age.

Part I: Requirements for a theory of technical development

The history of archaeological science and its epistemological and ontological discourse as part of Western world-views has been well discussed, both in Scandinavia (e.g. Jensen & Karlsson 2000; Olsen 2003a) and internationally (e.g. Hodder 1982, 1986, 2001; Shanks & Tilley 1987a, 1987b; Schiffer 1988; Barret 1994; Mackenzie 1994; Wylie 2002; Kristiansen 2004; Smith 2004). There is no reason to repeat what has already been said. Instead I will focus here on what I need to conduct this study, on different requirements for a theory of technical development. This is done in three sections, dealing in turn with interpretive archaeologies, materiality, and socio-technical change and persistence. Through a discussion of the chaîne opératoire and classification, I then focus on concrete methods for the studies of lithic material which I perform. The idea is to start with overall theoretical discussions and gradually approach more specific angles of approach and round off with an introduction to the methods I have selected.

Interpretive archaeologies

It is more than 45 years since Lewis Binford (1962) published his then pioneering text Archaeology as Anthropology, thereby establishing a critical and theoretical discussion within archaeology (Olsen 2003a). During the almost half-century that has passed since then, the theoretical discourse in archaeology has developed many facets and produced a wealth of ideas, while studies in the history of archaeology have established the subject's own intellectual history (Baudou 2004; Trigger 2006) – theoretical studies are a well-established sub-discipline with its own research agenda (for a critical discussion of this see Johnson 2006). But despite this, it still seems to be more the rule than the exception that newly written theoretical texts do not present new, self-sustaining perspectives, but new perspectives which are borne up by polemic against old ones (there are exceptions, such as Smith 2004). Why is this the case?

In part this can perhaps be explained by the pluralism that exists in today's archaeology, with a practice covering the entire range of theoretical stances, from extreme positivism to radical relativism (or from radical positivism to extreme relativism). Although the extreme examples are rare and the majority of what is published is in a broad central furrow, perhaps this diversity requires taking up a position. But above all it can probably be explained by the fact that the discussion of archaeological theory has developed into a separate genre, with a special plot and dynamic in the rhetoric of the narrative. To mark a theoretical stance, a text in this genre must say more about what the author does not want, that is, about previous studies in which weaknesses are discerned, than about what the author wants, that is, about the strength of the perspective he or she is advocating. And the most popular polemic that arises then, the real "bestseller", is processual versus post-processual archaeology. This polemic has become so popular that it has developed its own narrative tradition with pre-understandings and taken-for-granted assumptions that have acquired virtually mythical forms (for a discussion of archaeology's own narratives see e.g. Gustafsson 1996, 2001).

This polemic does not interest me. Without getting bogged down in self-analysis, I think there are two reasons for this. First, it gives me nothing. I have read and heard it often enough to know what it is about. The second reason is not so simple to express and more linked to my own research. As will be evident later on, my theoretical perception can be fitted into some form of post-structural or post-modern hermeneutic thought. Yet at the same time I work with lithic analyses, a sphere of archaeology that practically arose out of processual archaeology (see e.g. Swanson 1975). This means that I have been forced to take less interest in polemic and more in the possibilities and limitations that different perspectives give.

It is important to stress that I do not want to criticize a rhetoric that uses examples demonstrating the advantages and defects of earlier studies. Highlighting good and bad examples is an excellent way to clarify one's own and other people's stances, and this is something that I use frequently. What my critique concerns is that the discussion of archaeological theory has often got stuck in a rhetoric that is restricted to processual versus post-processual archaeology (see, for example, Shennan 2002; Rundkvist 2005), table 1.3.

But many have chosen to go beyond these polarizations to constructive attempts to deepen the theoretical reasoning and apply it to empirical studies (see e.g. Cornell & Fahlander 2002; Hagner & Rheinberger 2003; Nilsson 2003; Gustafsson & Karlsson 2004; Smith 2004; Andrén, Jennbert & Raudvere 2006; Carlsson 2007; Goldhahn 2007; Østigård 2007). And since the middle of the 1990s a multifaceted post-structural interpretative archaeology has emerged. The term "interpretive archaeologies" has been introduced for these archaeologies (Barret 1994:171; Mackenzie 1994:19ff; Shanks & Hodder 1995:5, see also Tilley 1993 for similar discussions but based on the concept of interpretative archaeology).

Interpretive archaeologies are archaeologies which aim to work through interpretation, in other words, they are about understanding and making sense. Interpretation is viewed as an ongoing process, always open to new perspectives and reinterpretations, and it can never be ended. Pluralism is a guiding star here, as a broad concept concerning different forms of interpretation – there are manifold ways of relating to the task of interpretation, and interpretations can be made pluralistic. The diversity comes from the fact that different questions are posed, that is, from the pluralism in the contexts where archaeological knowledge is produced. What has come to be called "the real past" thus exists as a historical, institutional, and social truth, rather than an inherent property of the archaeological artefacts that we dig up, in which prehistory has inscribed the truth about itself (Barret 1994:158; Shanks & Hodder 1995:5; Olsen 2003a:102).

These are perspectives discussed by Håkan Karlsson. He has listed a number of archaeology's post-structural theoretical premises (Karlsson 2000:148):

- There are no original meanings to discover in the past, since the meaning ascribed to the past depends in part on interpretations in the present.
- Archaeological interpretations of the past can never by nature be objective or free of values.
- As archaeologists we must decide what values we want our interpretations and our activities to have in the present.
- Archaeological knowledge does not accumulate since it is dependent on the socio-political contexts of the present.
- There is a dialectic between the archaeologist as interpreter and the material culture that is interpreted.

Table 1.3. Dichotomies exemplified with the conceptual opposites processual and post-processual archaeology and three of the many characteristics normally ascribed to these archaeologies.

Processual archaeology	Post-processual archaeology
Objective structures	Subjective constructs
Macroscopic overviews	Microscopic insights
Structures and laws	Contingencies

There are those who think that Karlsson's list of perspectives leads to arbitrariness and straight into the dead-end street of relativism, that they bring more confusion than clarity (see e.g. Nylén 1993; Moore 1994; Shennan 2002; Vandenberghe 2002; Wheen 2004; Nordin 2008). I do not agree. I find no reason for pessimism. These premises are not obstacles, they are challenges. They lead to the question: If the "real past" isn't there, what do I do?

This is in a way the essence of what I am trying to say here. If the past is a reflection of the present, how do I go about studying it? What knowledge do I want to produce, and how do I want to do it? To make progress with these questions I need to engage in a deeper discussion of interpretation, a discussion that concerns how I view the knowledge of the past that I believe I have arrived at.

About interpretation

To understand people in prehistory, to get close to them, to enter the heads of those long dead, to get at their thoughts, intentions, meanings, and purposes has long been a goal for theoretically oriented archaeology (Binford 1962; Hodder 1986; Renfrew & Zubrow 1994; Meskell & Preucel 2004). The following quotation is a textbook example.

> It is only when we make assumptions about the subjective meanings in the minds of people long dead that we can begin to do archaeology (Hodder 1986:82).

The ambition has been to explore what happened in the past, to understand people in prehistory on their own terms, to use methods that can capture their perception of the world. Reconstruction has been an important goal in this, the recreation of a vanished time (Olsen 2003a:243; Lucas 2005:54). In this perspective the past is viewed as something from which material culture is created and something that we as archaeologists can consequently recreate through theoretical and methodological analyses of the material culture that was once generated.

How this can and should be done has been discussed from many perspectives (see e.g. the variation in Hodder (ed.) 2001). The discussions have been permeated in different ways by the ambition that we as archaeologists, through good analyses, can bridge the gap between the fragments of a prehistoric culture in the form of the artefacts we find and the living social reality in which these objects were parts of a more complex whole than what, say, some potsherds or flint flakes might seem to be, figure 1.6.

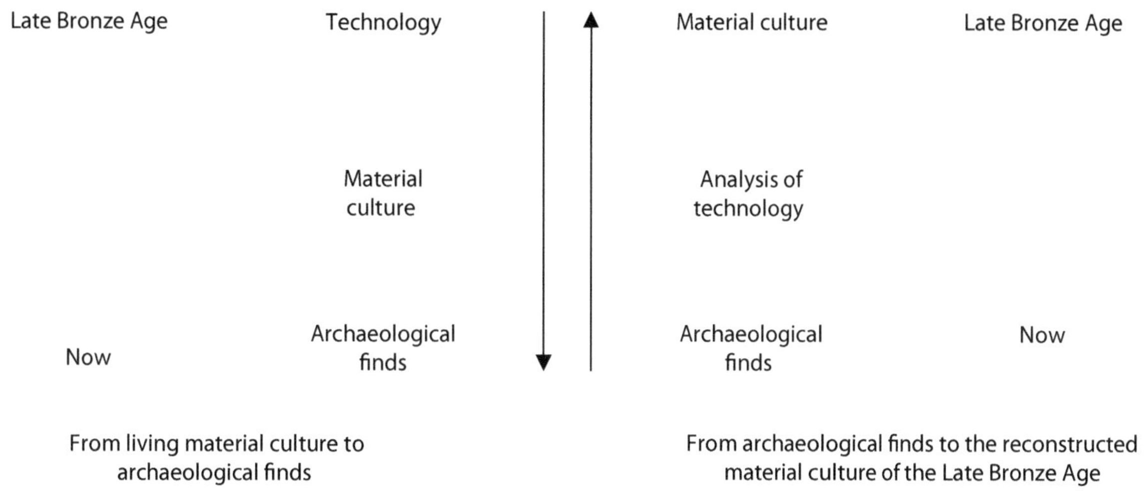

Figure 1.6. How a reconstruction could be structured. In the Late Bronze Age a technology generated a material culture which has become archaeological material, to the left in the figure. From this material, archaeology reconstructs the technology and through an analysis of the technology the material culture of the Late Bronze Age, to the right in the figure.

If we can just think of material culture as dynamic, as something that was once an active part of a living culture where social patterns and cosmology were worked into the artefacts, then we can also see through these fragments to the reality of the social and cultural lives of which the objects once were a part (Miller 2005:212ff).

This is problematic in two ways. First of all, I do not believe that we can get at a real prehistoric situation. It is, quite simply, inaccessible to us. It was so complex that it was not completely accessible even to the people who lived back then. Secondly, I believe that what we say about prehistory depends more on what we want to know than on what once happened (see Sundström 2003 for a similar discussion).

If we stop here for a moment and return to the question I asked at the beginning – why were large blade knives introduced at a time that is classified as a transition between bronze and iron, why did they disappear, and what happened in between? – then what I have just said might seem to be more than a little contradictory. I want to know why, yet simultaneously I say that prehistoric reality is not accessible to me. But I do not believe that I have built an internal contradiction into the study by highlighting this theoretical perspective. I believe that it is possible to work with questions about why, even though I do not believe in the possibility of recreating a prehistoric reality. Let me give an example, a fictitious one.

A crash

A traffic accident occurs on a street in the centre of a city. One car has driven into another. When this happens there are people who act in different ways or witness what goes on. The driver of the car that was hit perhaps did not notice the other car, and when he felt the impact he could not understand what had happened. The driver who collided with the car in front may have thought that it was going too slow, pushed the accelerator to overtake, and just then the driver in front braked and the crash happened without the driver behind really understanding why.

On the pavement, someone sees the whole course of events, from the cars driving one after the other until the one in front brakes and the one behind smashes into it. Someone else sees the cars driving close together and thinks that it does not look good, turns his back and goes round a corner, then hears the crash and turns back out of curiosity, sees the collided cars and thinks, “I knew that would happen.” Other people did not notice the cars on the street but hear the bang, look up, and see two cars that have just crashed. Someone else is a block away, hears the smash and thinks that there has been a collision. Yet another person hears or sees nothing of the actual accident but hears and sees the fire brigade, ambulance, and police who are called to the spot, thinks that something must have happened, and reads in the newspaper the following day that two cars have collided.

Everyone involved in this event observes and experiences different things. Subsequently everyone will agree that something happened and most of them will also agree that what happened was that two cars collided. Most people will agree that there was a lot of noise, and those who saw the event will probably agree that the cars were damaged by the crash.

But all those involved will later have different narratives about what happened. Depending on where they were and what situation they were in, they will tell different versions of the event. The person whose car was hit will say something like: “There was just this sudden bang.” The person who drove into that car might say: “I don’t understand what happened. Suddenly the car in front stopped moving.” The person who saw the entire chain of events may perhaps give a version indicating the cause: “I saw right from the beginning that it would happen. The car behind was driving far too fast and too close.” The person who just heard the emergency vehicles and read about the accident in the paper the next day might say: “Aha,

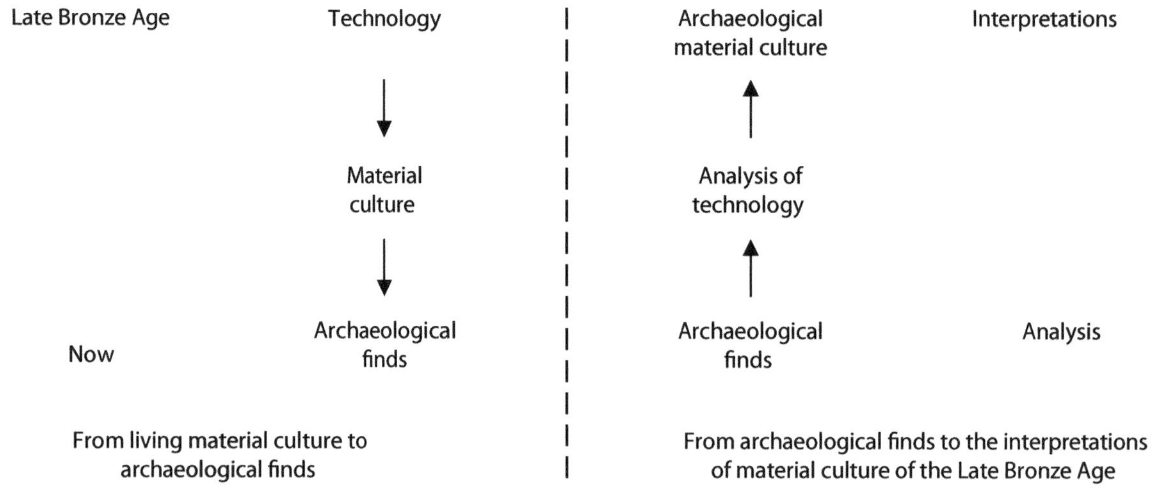

Figure 1.7. How the reconstruction perspective in figure 1.6 can be reversed. The left-hand side of the figure: Different technologies during the Late Bronze Age have generated a material culture from which I study the archaeological material in the form of flint tools and flint assemblages. This is a timeline from past to present. The right-hand part of the figure: I use the flint to analyse the technology and through the technology interpret the archaeological material culture. This is a timeline that stays in the present.

that's what I heard yesterday. I was sure something had happened."

When it comes to the interpretations of what happened, these too are based on different premises. The police officers who arrive on the scene are interested in finding out what happened. They interview the drivers involved and the witnesses, and they reconstruct the event on the basis of narratives and skid marks. The journalists who come are likewise interested in what happened. But they are also interested in how it felt to crash, what the witnesses felt when they saw the collision, how it sounded, what people thought, how people acted afterwards, and whether anyone was injured. The ambulance staff are also interested in any injuries, but more in order to provide help and relief and not so much to find out what the injured persons were thinking when the accident occurred.

Two years after the collision, one of the city's traffic planners visits the spot together with an expert in order to ascertain why the cars crashed just there. They analyse the distance between the pedestrian crossings, junctions, and traffic lights, and they categorize the collision in terms of traffic patterns, traffic flows, and previously known behaviours. Another two years later, representatives of one of the car makers analyse how the air bags in this car were activated and what effects this had.

None of these interpretations of the event is more right than another. No one is misinterpreting. Everyone agrees about the core of the events but creates different explanations depending on different conditions, purposes, and premises. That is how it is now. That is how it was in the distant past that archaeologists study.

That is why I do not believe it is possible to recreate a prehistoric reality. It was not even accessible for reconstruction back then, and is thus even less so for us now. And just as for the actors in the example, explanation and understanding are not primarily about what happened, but about what is relevant from any one of a great many angles, aims, ambitions, and questions. The traffic planner's interpretation of why the accident occurred is of little relevance to the police when they want to establish who caused the accident. The journalists' search for injured people is just as irrelevant to the ambulance staff. I therefore believe that the reconstruction is related more to what one wants to know than to what happened.

At the same time, it is true that the person whose car was hit had to buy a new one, the people who were injured suffered pain, were sick-listed, and had to undergo rehabilitation, and the journalist wrote an article about the event which was read by the person who heard but did not see the accident. The subsequent analysis of the accident in relation to known traffic patterns perhaps had the effect that the street was later rebuilt, and based on the analysis of the air bag in one of the cars, new air bags might have been developed. What happened happened, even if it can be understood from manifold angles.

To claim that prehistory can be interpreted pluralistically thus does not mean claiming that it did not happen. What we as archaeologists study and interpret is ambiguous but not arbitrary (Olsen 2003a:102). The ambiguity has a context. It is a matter of trying to understand the many different ways in which a lost world can be perceived (Scharfstein 1989; Thomas 2004:34). Michael Hagner and Hans-Jürg Rheinberger (2003:214), both working with science and technology studies (STS), have expressed this position as follows:

> It is an act of intellectual honesty rather than of revolutionary radicalism to maintain that a historicizing analysis of the role of the scientist, of basic epistemological categories, and even of the meaning of science itself is much more an attempt to liberate the concept of reason from its self-chosen chains rather than a flight from reason.

This makes demands on the way in which I handle my analyses in this study. How should I go about interpreting prehistory within this theoretical framework? How should I handle the opposition between my desire to ask a why-question and the impossibility of answering it? The consequence of what I have said hitherto is that what I do should be steered by an attitude that the past has happened, that when it happened it was pluralistic, that when it is studied today from pluralistic perspectives it is not arbitrary but ambiguous, and that it is possible to study the past, but not with the aim of establishing what really happened, but with the premise of interpreting a distant past on the basis of what we want to know today (Barrett 1994; Thomas 2004), figure 1.7.

In the next section I expand on this approach by discussing the claim that I know more about the people who lived then than they knew about themselves.

On knowing more about people than they do about themselves

Most archaeologists have probably at some time heard the critique of the subject according to which it is impossible to find out what happened in the past. Surely archaeology can't know very much about the people back then? Isn't the distant past far too distant for that? We don't know what people thought, felt, or talked about, and we find only fragments of their lives.

I would prefer to think differently. I believe that archaeologists know much more about the people who lived then than they did themselves. Or rather, archaeologists can know other things about those who lived then than they knew themselves (Miller 1985; Schiffer 2000). I will give an example by looking again at the story of the car crash.

The drivers involved in the accident understand what happened on their basis of their knowledge and experience, which are linked to the specific time in which they live and the places where they live. But based on an analysis extending over time and place, we can see in retrospect that the drivers were part of a larger context.

In the same year as the accident occurred, a new kind of tyre was introduced. These were marketed as "good and cheap". The marketing was effective, and soon many cars had these tyres, including the cars that crashed in my fictitious example. When the collision happened, no one considered the role of the tyres in the event. The tyres on both cars were new, in good order, and of the latest model, so no one reflected on them. But in an analysis performed 25 years later it is possible to see that the same type of accident happened at several different places, over a period of several years, and with exactly the same type of tyres that these cars had. An association over time and place between tyres and a specific type of accidents has thus been discovered. And it is precisely through this type of retrospective analyses that we as archaeologists can arrive at a knowledge that these tyres were part of a larger pattern which was only interrupted when a completely new type of tyre was introduced. This pattern was unknown to people at the time when the tyres were used, and it could not possibly have been known to the drivers or the people around them. When the accident happened, they did not know that it was part of what would become a whole series of future accidents. In this way we can know, with hindsight, more about them than they did about themselves.

I can take another example from my own research. At a number of places along the shores of southern Sweden and eastern Denmark, thousands of preforms of Neolithic flint axes have been found. These places are on beach ridges where flint nodules of high quality occur profusely. The preforms that have been found represent axe types that can be dated to the time from the Early Neolithic all the way to the Early Bronze Age. This means that preforms were manufactured at these places over a period of about two thousand years. And during this time an enormous amount of preforms were made and taken away from there to be shaped into axes, but many of these preforms were also left on the spot. This reveals a deep-rooted and very long tradition of making preforms for axes and leaving them there on these beach ridges, a tradition that began around 4000 BC and ended some time around 2000 BC (Högberg 2002, 2006).

This example is relevant here to show how continuity in the use of a place over a long time can help in an archaeological study applying a long-term perspective to interpret and describe something that the people who lived then could not possibly have known about themselves. When someone around 4000 BC made and deposited a preform for an axe at one of these places, this person could not know that he or she had just started a tradition that would last for two thousand years. When someone visited the place for the first time in 3500 BC to manufacture a preform, this person could not know how far back in time people had been doing the same thing, nor how far into the future this tradition would persist. But we know that as archaeologists (for a similar line of thought see Knutsson 2006).

The potential of archaeology to make comparisons over time and place thus gives us the possibility to see patterns, phenomena, effects, consequences, combinations, results, underlying structures, etc. that the people who lived then could not possibly have seen or understood. In this way we as archaeologists can know more things and different things than the people who lived then knew about themselves. This is a way for me to try to understand the archaeological record as a unique source of information in a long-term perspective (Schiffer & Skibo 1987:596; Witmore 2007:546ff).

Another way is through the testimony of material culture. William Rathje is well known for his studies in the late twentieth century of garbage from American households. He and his students have shown that the descriptions that people gave in interviews of what they had bought, eaten, and thrown away did not always agree with the remains in the garbage showing what they had actually bought, eaten, and thrown away. Sometimes this was due to lies; in other words, people would not say what they had actually done. But often it was a matter of unreflective and unconscious behaviour. Several people, for example, bought more food than they realized, many people threw away more than they believed they had done (Rathje & Murphy 1992). These are examples of how material culture can say more about people than they know themselves. In the

Figure 1.8. Paving stones. Note the smooth surface on the stone in the center of the picture.

same way, an investigation of a prehistoric context can tell us things about the people who acted in that context that they perhaps did not know about themselves. I will adduce yet another example, this time from my own experience. But first a clarification: when I use modern examples, it is to show how I think, not to claim that the way I think can be transferred to people who thought in the Late Bronze Age. So, here is the example:

A couple of years ago I bought a few cubic metres of paving stones to cobble a little patio in front of my house. The ones I bought were recycled Swedish granite and gneiss setts, figure 1.8.

In Sweden granite setts like these have been used to pave squares and roads since the start of the twentieth century, and town streets and country roads are full of them, although most of them now lie under several layers of asphalt. As the road network is renovated or rebuilt, these setts are lifted and recycled for sale (Olausson 2007). This was the kind of stones I bought. Many of them have surfaced squares and roads for many years, which means that the side of the stone that lay facing upwards has been worn to a smooth, well-polished surface.

Since I had never previously laid paving stones, I asked the men and women at the company where I bought the stones if they had any hints for me. They had. The most important thing was to place the stone with the thickest part downwards. The setts are cut in the shape of a slightly irregular cube. The sides are usually between 8 and 12 centimetres long. Since they are handmade, none of the six sides are alike; one of them is always bigger. The hint I received was thus to lay the stone with that side down. The reason was said to be that the stone is more stable. If the stone is placed with any of the other sides downwards, it risks becoming unstable. The stone then moves easily, with the consequence that the paving becomes uneven and easily damaged. The significance of the advice was reinforced with the comment: "that's how it's always been done". This advice is also supported in part by the text and illustrations in a paving handbook published by the leading organization in the business in Sweden (Nielsen 1994:44).

Armed with this advice, I started the job of laying the paving stones. Since I like stone and sometimes knap stone myself, I looked inquisitively at every stone I laid. I discovered quite soon that the advice I had been given was not compatible with the way the stones had formerly been laid. If the stones were supposed to be laid with the broadest side downwards, as I had been recommended to do by the people who sold them to me, this meant that the side opposite the thickest one must have faced upwards and consequently should show more wear and polish. But this was not the case. Several of the stones had worn surfaces which showed that they had not been laid with the broadest side down. This contradiction between the advice I was given and the testimony of the stones caught my interest, and I looked through all the stones I had bought. It turned out that, of the stones with worn surfaces, roughly half had been laid on the side opposite the broadest part. The other half had worn surfaces which showed that these stones had not been laid with the broadest part downwards.

This means that the company that sold the setts had given me two different "messages", a verbal one in the form of a hint for my work and a material one in the form of worn surfaces on the setts. The verbal message stated that the stones should be laid with the broad side downwards, because "that's how it's always been done". Otherwise the paving does not function properly. The material message stated that the stones had not always been consistently laid with the broad side downwards.

The material message shows that the advice I was given does not agree with what people in the past did. Those who gave me the advice do not know this. They believe their recommendations, as I know because I asked them. But through a simple observation of the material culture, that is to say, the setts and their worn surfaces, I could tell that their advice was not correct. This means that I, through the testimony of the material culture, know more about the people who gave me the advice than they know about themselves.

But what does all this mean for the study presented here? The consequence, as I see it, is that this book is not primarily about what people in the Late Bronze Age may have perceived or known about their own times. It is about what I know about the Late Bronze Age. Some of the stories that I present might perhaps have been recognizable to these people, while others concern things that the people who lived then did not know about themselves, figure 1.9.

This is therefore more about the process of making sense of things, rather than finding the meaning of artefacts in a prehistoric society. And then it is important what questions are asked, what methods are used to answer these questions, and how the results of this stand in relation to the questions asked and the methods selected.

In the next section I consider this in greater depth through a discussion of materiality, in order to further delimit and narrow the theoretical framework of the study towards artefacts and technology. I thus proceed from a discussion of the knowledge that I believe I can produce about the past, to a discussion of the study object of archaeology, material culture.

Materiality

Things are meaningful, they are central in our lives. They are our attributes, our coded signals to the world around us, reflecting who and what we are and our perception of ourselves. Things create an environment, help us as humans to navigate. They are our companions; we live our lives in a symbiotic integrated relationship with things – sometimes a happy one of joy and self-fulfilment, sometimes a bad one with constant quarrelling, envy, and unfulfilled desire, and sometimes things have just faded into memory or oblivion.

But an artefact is never anything in itself. Although objects are usually ascribed an independent meaning in everyday conversations, it is only in a specific situation that they exist with their meaning (Högberg 2000:13). In archaeology materiality has become an umbrella term for studies working with material culture based on ideas about artefacts as integrated actors in

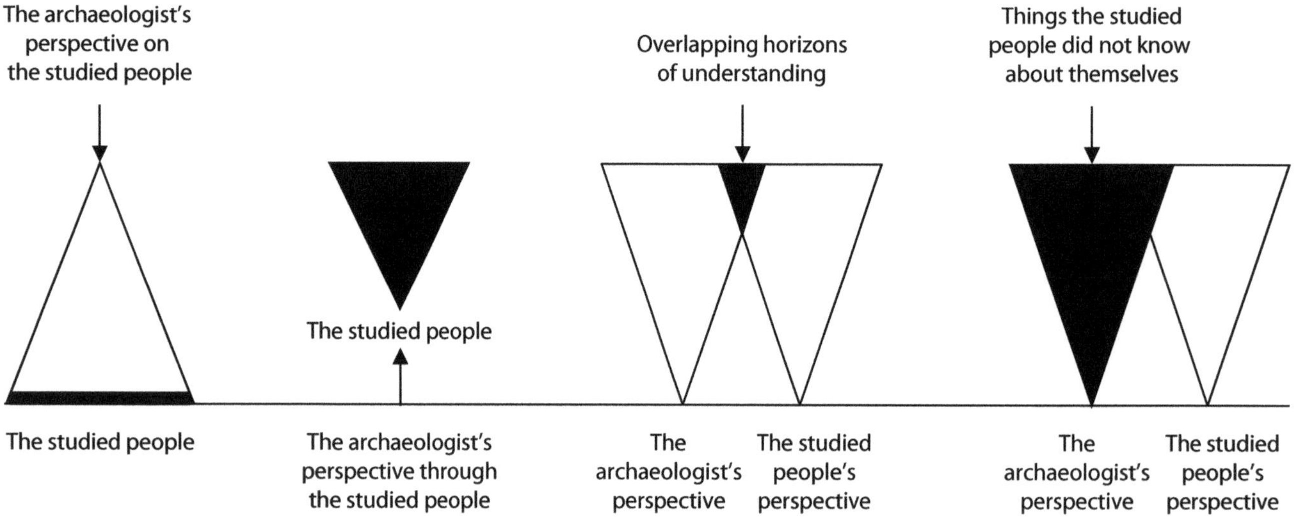

Figure 1.9. The Danish anthropologist Kirsten Hastrup at the University of Copenhagen (1999:166) has outlined the differences between ways of relating to the people or the society under study, in the figure called the studied people. Hastrup bases this description of differences on structuralism, ethnocentricity, and hermeneutics. According to her, the structuralist, to the left in the figure, looks down on the studied people, observes them from a suitable distance. In archaeology this perspective is close to positivism. The ethnocentrist, the second from the left in the figure, believes that every person is unique, that no generalizations are valid, and thus aims to see everything through the studied people. This perspective points towards relativism and is represented in archaeology in the most pronounced forms of indigenous archaeology. The hermeneutic, the third from the left in the figure, searches for overlapping horizons of understanding by trying to see things from the same angle as the studied people. This is in line with a great many studies in archaeology (see e.g. Baudou 2004, or Sørensen 2006a with references for the application of this perspective to lithic analyses). I take this deeper in the present study. I do not content myself with overlapping horizons, I think that I can investigate things that the people I study did not know about themselves, to the far right in the figure.

social life (see e.g. Graves-Brown 2000; Meskell 2005). Materiality studies have been described in different ways, table 1.4.

What these approaches have in common is an ambition to nuance artefact studies, to work one's way beyond simplified categorizations of objects based on dichotomies such as function versus symbolism. Materiality emphasizes that the artefacts are in the hands of the people who employ them; artefacts are created, used, and discarded. At the same time, artefacts are active, that is to say, they create the people who create, use, and discard them – people shape their objects while people are shaped by their objects (Lemonnier 1993; Hoskins 1998).

Materiality is also frequently described in terms of what it is not – it is not a conventional study of material culture in archaeology, it moves beyond simplistic readings of things (Meskell 2005:2), it is not about style or things as commodities and goods or as something only existing either in the heads of people or in their social relations with one another (Graves-Brown 2000:1). This shows how materiality, just as much as it is formulated as a starting point for analysis, has also been formulated as a move away from something else:

> Materiality signals both a disciplinary move away from empiricism and a new holistic concern with the understanding of the meaningful relationship between persons and things (Tilley 2007:19).

Material culture is thus regarded not just as evidence of something reflecting the conditions in which people act, but also as something that enables, governs, develops, and changes

Table 1.4. A selection of some perspectives from the last thirty years summing up how studies focusing on materiality view material culture.

Material culture not merely reflects culture but also actively constitutes it.	Hodder 1982
The study of materiality aims to elucidate the complex relationships in the interaction between social strategies, variability and material culture.	Miller 1985
Objects are the result of both the material and the immaterial world, and the material and immaterial world is a result of objects.	Miller 1998
Materiality is about how the very material character of the world around us is appropriated by humanity. Material culture is shaped by people; people are shaped by material culture.	Graves-Brown 2000
Consider things, and you will have humans. Consider humans, and you are by that very act interested in things.	Latour 2000
Material culture studies are the totality of approaches to all kinds of materiality and the interaction with, and influence on, humans, and vice versa. In other words, it is the study of the social world with an emphasis on the constraints and possibilities created by material factors; material culture is the medium through which human interaction and socialization take place.	Fahlander & Oestigaard 2004
Study of materiality is engaged in the dialectic of people, things and society, stresses the study of things in their own right.	Meskell 2005

(Latour 2000). Through this relationship between people and objects, materiality studies try to get behind the object, to see beyond what is directly obvious (Miller 2005:212).

The materiality of keys

A well-known example is the study of hotel keys by Bruno Latour (1992). Until the end of the 1990s, room keys in European hotels usually had a heavy, clumsy weight attached to the keyring. This was intended to remind the guests to hand in the key at reception every time they left the hotel. Otherwise there was a risk that the keys would be lost or forgotten in guests' pockets when they moved on. The weight on the keyring was evidence that a verbal reminder at check-in or a simple sign in reception to hand in the key every time you left the hotel was not sufficient to get the guests to do this. They were evidently focused on something less trivial than a room key. But the weight made the guests shift focus towards the key. It brought discomfort. It took up too much room in one's pocket or weighed down one's handbag. So guests went to reception to get rid of the key.

Latour uses this simple little artefact, a weight on a key, together with the social relations of which it is part, to show how original intentions change and how this happens in an intricate interplay between material and non-material, or if you wish between human and non-human actors.

The hotel owner's original wish was to get guests to hand in the key, a key which initially had no weight, at reception when they left the hotel. The owner therefore gave a verbal request when guests checked in: "Please hand in the key when you leave the hotel." The hotel owner discovered quite soon, however, that this was not sufficient. The keys disappeared and the cost of replacing them escalated. The owner then introduced an actor, a sign. It said: "Please hand in the key to reception when you go out." The sign was highly visible on the reception counter. This had a certain effect. Fewer keys disappeared. But the number of vanishing keys was still too large, So the hotel owner adopted an extraordinary measure. He introduced an object that bothered the guests. A weight was attached to the key, a weight so clumsy that it caused sufficiently great discomfort to the guest that he or she gladly handed over the key.

While the verbal and written instructions both appealed to the benevolence of the guest, the weight caused displeasure, which ought to be a hotel owner's last wish. Guests who feel discomfort perhaps do not stay or may never return to the same hotel. Yet the hotel owner evidently takes this risk to make his keys stay in the hotel. And it has the desired effect. Virtually all keys are handed in. Very few disappear. Gradually the hotel owner also discovered that neither verbal instructions nor the sign on the counter are needed. The weight does all the work that verbal and written communication had previously tried to do and only partially succeeded in.

What is interesting about this example is how it reveals a shift between original intentions and consequences and how material culture is part of this shift as an actor. The key with the weight reflects not just a social relationship, it is an active part of how such relationships are created and how people choose and are guided to act. The instruction followed by the hotel guests is not the same as the original instruction. In the process it has not been transferred from one instruction to another. Instead it has been translated. Material culture, that is to say, the key with the weight, does not reflect the original intentions. The keys with the weights reflect a renegotiation of the intentions, although closely associated with the original intentions. But they are not the same. What happens is not what we started with (Bijker 1995; Latour 1992, 1998, 2007).

The hotel owner no longer addresses the guests' benevolence by asking them to hand in the key when they go out. Instead the owner forces the guests to do this by causing them discomfort, which is supposed to make them feel such displeasure that they fulfil the owner's wishes. And the guests do not hand in the keys because the hotel owner wishes it; they do so in order to free themselves from a cumbersome object weighing down their pockets.

This change of intentions has taken place through a complex interplay between the hotel owner, the guests, and some objects in the form of keys, a sign, and weights. The meaning of material culture thus does not correspond to the idea that initiated the chain of events resulting in keys with weights. Material culture has changed the intention, helping to give it a new meaning.

But what does this mean for an archaeological study of flint objects from the Late Bronze Age? What Latour's study shows is that material culture changes, not due to any plan or original intention, but in intricate interaction between wishes, intentions, materials, negotiations, objects, ambitions, contexts, actions, events, consequences, etc. Objects are never fixed in their meanings and in the content ascribed to them. Innovations change. The example shows that a technological study cannot be performed with the assumption of following a given intention through social contexts in time and place. Instead one should follow intentions and consequences as a complex social interaction between objects and people (Latour 1992; Law 1992).

This makes special demands of an archaeological study. For what is intention and consequence in a corpus of archaeological material? What is change, what is constant? Let us consider the case of the key in Latour's example, which is repeatedly used to lock and unlock. And even now in the twenty-first century, when Latour's example has been outdated by new technology and the hotels' metal keys have been replaced by a plastic card with a magnetic strip, the lock with a reader, and the weight with a digital code that can be easily activated or demagnetized, the function of locking and unlocking still remains. But even this is fluid. I will cite another example.

In the mid-1990s I was working in southern Öland, the island just off the east coast of Sweden. Southern Öland was then and still is a sparsely populated rural area. Everyone knows everyone else. A woman from a small village invited me to her home. She said, "If you're passing, drop in. If I'm at home the door is open. If the door is locked and the key is on the outside, it means I'm in the byre, you'll find me there. If the door is locked and the key isn't in the lock, then I'm away. You'll find the key under the doormat, you can go in and put on the coffee till I get home. If the key isn't under the doormat, then I'm away for a long time." In this story, "locking and unlocking" is something completely different from what is associated with the room keys in Latour's study.

For an archaeologist, these examples are challenges. How am I to interpret material culture from a distant past? When I find an object during an excavation, when I transform fragments of a bygone material culture into a present-day body of archaeological source material, until I found it, the last physical thing that happened to the object was that it ended up in the ground. (And by this I do not rule out that non-physical things could have happened to the object – it could have been involved in thoughts and narratives since then.) I can assume that these finds were parts of a living material culture – that they had effects just as Latour's hotel keys did. But to state that this happened to these objects lacks interest and says nothing. It is a platitude. I need no archaeological analysis or interpretation to be able to state that an object I study was involved in a dialectical relationship between people and objects in the time or times when it was a part of the material culture. The interpretation must be deepened to include questions about what creates and carries meaning in the material culture I am analysing, and what the consequences were for the relationship between objects and people. But this is not easy. Let me provide another example, this time an invented one.

New lockers are installed for the pupils in a school. Half of them are green, the other half red. They are installed with no intention that the colours should reflect anything other than what happened to be in stock at the supplier. The difference in colour expressed by the actual arrangement is thus, initially, completely without meaning. After a period of use it turns out that an accepted terminology has developed among the pupils, so that they call each other greens and reds. It also turns out that the social patterns among the pupils to some extent coincide with whether the pupils are green or red. This terminology thus refers to the colour of the lockers. It is then a reasonable assumption that the lockers, although they did not have any meaning when installed, through use have become creators of meaning. Yet on closer analysis it turns out that the pupils were assigned their lockers by class, with some classes receiving green lockers and others receiving red ones. The terminology and the grouping that arose are thus based on a social pattern proceeding from the class to which pupils belong, not from the colour of the lockers. The colours of the lockers just happened to be handy terminology for the pupils, a simple way to express which class to which they belong.

If this example had not been invented, the differences in colour could have been held up as part of the material culture that created and carried meaning and resulted in social groupings of pupils. This could also be right. In the example there are groups who use a terminology referring to the colours of the lockers. But the reason for the groupings is not the colours. The colours are something completely different, a chance factor that merely reflects what the supplier happened to have in stock.

The examples of weights on hotel keys, locked doors, and lockers of different colours illustrate the problems that face me as an archaeologist. It is therefore of importance in a study of materiality to consider what it is in the parts of the material culture that are studied that I believe expresses differences and similarities that belonged to a system where they were regarded as carriers and creators of meaning. When translated to my study material, this leads to questions about what the flint objects I study can be thought to represent. How do I ask questions which can guide me in the work of interpreting meaning-bearing and meaning-creating aspects of materiality?

I will work more on this question a little later on in the text, but for the moment let me devote a few pages to a critical discussion focusing on materiality and material, a discussion that pinpoints a central problem in lithic analyses: how the gap between theoretical and practical knowledge can be bridged.

Materiality and material

Archaeology's use of materiality has not escaped criticism. With the support of Latour's studies and other works, Bjørnar Olsen has stressed that the archaeological discussion of materiality has forgotten material culture itself, missing the fact that this is not just an actor but also a manifest solid physical mass which both facilitates and resists (Olsen 2003b). A similar perspective has been put forward in a study by the Swedish ethnologist Åke Daun, showing that individuals' freedom of choice and manoeuvring space is limited in many more ways than we like to believe by the nature of the material world (Daun 1999:22ff).

This critique should be taken seriously. If the ambition is to study flint tools and flint assemblages as active parts in the dialectic social life of things and people during the Late Bronze Age, the things must have a substantial impact on the analysis, as is also clearly exemplified above with the aid of keys, locked doors, and the colour of pupils' lockers.

Another critic of materiality is Tim Ingold (2007). In a critical study of archaeology's interest in materiality, a critique that partly follows up both previous texts by Ingold (2000) and Olsen's text (2003b), he calls for a renewed interest in material rather than materiality. He argues that the growing pile of archaeological and anthropological literature dealing with materiality has had far too little to say about materials (Ingold 2007:1). Referring to Henry Hodges's (1964) book Artifacts, he lists all sorts of materials that people in prehistory used to make things, and which he claims studies of materiality have largely ignored: stone, flint, wood, bone, antler, hide, turf, soil, clay, plants, etc. This critique has also been put forward by others (Hurcombe 2007).

Ingold believes that, in an academic rhetoric where scholars compete to excel each other in theorizing material culture, the material itself is absent. Material culture does not seem to have been created from material, but from theoretical perspectives on the interaction between objects and people. In these perspectives, material is locked into the objects, not a malleable substance from which the objects are manufactured (Ingold 2007:11). This critique is worth dwelling on for a while.

Ingold has a point. Discussions of materiality seldom deal with materials. And this is a partiality that has created a weakness. But on the other hand, looking at the abundant archaeological literature dealing with materials, not least within lithic studies, it is obvious that most of them simply deal with the archaeological classification, description, and presentation of archaeological finds, i.e. they transform archaeological finds into present-day archaeological material culture in the form of archaeological source materials and museum collections (see Buchli 2004 for discussion). They are not interpreting prehistoric material culture. This critique of Ingold's perspective has also been delivered by others (Nilsson 2007; Tilley 2007). It leads to the question: Is it possible to go from materiality to material without falling back into a classificatory and descriptive handling of material culture?

A solution that Ingold suggests is that we can learn more about material if we ourselves work with the material that is a part of the material culture we want to understand:

> by sawing logs, building a wall, knapping a stone or rowing a boat ... could not such engagement – working practically *with* materials – offer a more powerful procedure of discovery than an approach bent on the abstract analysis *of* things already made? (Ingold 2007:3, original italics).

This is an interesting perspective and something archaeologists working with lithic technology and experimental archaeology have been discussing for many years. And in view of the enormous amount of literature that has been published with research based on sawing logs, building a wall, knapping a stone, rowing a boat, and many more practical experiments and interaction with materials, it is in fact rather odd, I think, that Ingold poses the question in this way. There are countless examples showing that this has already been done (see e.g. Coles 1979; Schiffer & Skibo 1987; Schiffer et al. 1994; Whittaker 1994; Meldgaard & Rasmussen 1996; Fansa 2002; Mathieu 2002; Hirth 2003, see also The Society of Primitive Technology and Exarc 2008).

But is it that simple, just knap a stone and you will understand the material behind the materiality? I don't think so. I know the material I am working with. I have collected flint nodules on beaches, knapped them into tools, and used these tools for different purposes. I have got flint splinters in my eyes, flint dust in my throat, bled from cuts from newly knapped flint edges, and endured stiff muscles from hard work with the tools. I have savoured the pleasure of working with a newly made flint tool and suffered distress when it has broken. But does this make me a better researcher in material culture? Ingold seems to thinks so. I don't.

Don't get me wrong here. I don't think there is anything wrong with the ambition in Ingold's critique. He carries on a theoretical discussion that is both interesting and necessary, urging people to think one step further than before. These are the perspectives of a cross-disciplinary researcher who has devoted a lot of time to studies of material culture and who is now testing new boundaries for materiality by putting the focus on practical interaction with material (see e.g. Molander 1996). But I do not believe the problem lies in how we as archaeologists interact with material or material culture.

I don't know how many modern flint knappers I have met who have drawn excessively quick conclusions by proceeding from personal experiences of practical work with the material to interpretations of how prehistoric flint craft was designed, organized, functioned, or did not function (for a similar critique see Schiffer & Skibo 1987:596f). (Probably as many as the number of archaeologists I have met who have drawn equally quick conclusions based on their experience of archaeological classification.) In North America, where modern flint knapping is a respected and appreciated craft and art in its own right, discussions of material versus materiality have been going on for several years, although using other words than material and materiality. When polarized it has been presented as a conflict between craftsmen and academics, but underneath the surface of these conflicts lies a deeper discussion of how we can make sense of the past based on present-day knowledge and know-how (Whittaker 2004; Kelterborn 2005a; Odell 2005; Patten 2005), i.e. a discussion very much on the same issues that Ingold raises, albeit from a practical more than a theoretical point of view.

The problem with Ingold's material versus materiality, as I see it, is not in how much we as archaeologists interact with the material. It lies in how we ask questions, select methods for working with these questions, and how relevant the interpretations we make out of this work are in relation to the questions we ask and the methods we select. Knapping flint doesn't make me a better interpreter of material culture. But it can make me ask questions, and make me use research methods and, consequently, make me come up with an interpretation which differs from those made by researchers who don't knap flint. But these results don't come out of my flint knapping, they come out of the questions I ask and the methods I use, which in turn, among other things, comes out of my flint knapping.

It is in the encounter between the researcher, the person conducting the study, and the artefact, the object of study, that the artefact acquires a meaning. Something happens there which gives meanings to material culture (Högberg 2000:15). So even if a perspective like materiality clearly shows the importance of understanding the dialectic between value-bearing and value-creating meanings of material culture, these meanings are not inherent in the objects that were once active parts of people's lives and that have been transformed into archaeological artefacts through an archaeological discourse (Olsen 2003b:100). The meanings must be created.

Engaged studies

I will return to the question I asked previously about how I can perform an interpretation that deals with meaning-bearing and meaning-creating aspects of materiality. To do this, I believe, requires more of a study than to show that materials and material culture play an active role and that this role is socially constructed. Questions about which active role and how it is socially constructed must be added (Reybrouck 2000:45). But this is not enough. It is not solely a question of which role material culture played and how it was constructed. It is a question of what I as a researcher am interested in analysing. In

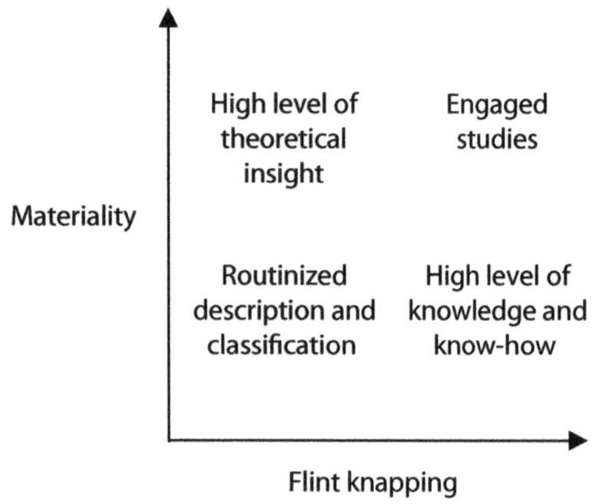

Figure 1.10. An illustration of how I view relations between theoretical and practical knowledge, or materials and materiality, to refer to Ingold's (2007) critique.

other words, it is important to understand that there is a which and a how, but it is the questions I ask and the methods I use that determine the which and how I will study, not the fact of acknowledging the presence of a which and how.

In the diagram in figure 1.10, theoretical knowledge in the form of materiality and practical knowledge in the form of flint knapping have been placed on separate axes. In the space between the axes, four different attitudes to working with lithic analyses are schematically defined. At the bottom left of the figure are studies showing low theoretical awareness and lack of practical experience. These are usually routine descriptive accounts of what the archaeological material looks like, what kind of tools it contains, and their dating. These studies have little or nothing to contribute to material culture studies. At the bottom right of the figure are studies characterized by low theoretical awareness and high practical knowledge and skill. These are usually practical examples of how tools are made or used. They also include studies investigating, for instance, fraction mechanics and patterns, post-deposition aspects and reduction sequences in lithic material. They mainly contribute to an understanding of the physical and practical creation and use of the archaeological material. At the top left of the figure are studies with high theoretical knowledge and little practical experience. These are normally characterized by a well-developed theoretical framework, but the finds tend to be merely classified and presented. They mainly contribute to a deeper understanding of the theoretical dimensions of material culture. The top right of the figure is the place I want to reach. To get there I need to ask other questions than those posed in other fields in the figure. With inspiration from Sergio Sismondo, a Canadian STS researcher, I have called this "engaged studies" (Sismondo 2008:13ff). These efforts have the ambition to contribute to high-level studies of both theoretical insight and knowledge and know-how. In that sense, the two-dimensional framework in the diagram allows us to see, not a conflict between Ingold's material and materiality, but a potential overlap (Sismondo 2008:21).

The concept of materiality means that the flint tools and flint flakes on which my analyses focus should be placed in a holistic and conceptual theoretical and interpretative framework considering social significance on a broad scale, addressing the social lives of things in relation to the social lives of persons (Tilley 2007:17f). This perspective is not that common in lithic analysis, where empiricism is strong (Torrence 1989:1ff; Odell 1996:4f). Therefore it is a challenge. How do I conduct an interpretative lithic analysis within the framework of materiality, using Late Bronze Age flint tools and assemblages as source material?

These are questions that are examined at greater depth in the next section, where I treat them as part of a discussion of sociotechnical change and persistence. I thus continue the discussion from above about materiality, but now with the focus on technology.

Sociotechnical change and persistence

That there was a technology, that it was used, that it was part of a sociotechnical framework and that the technology created material culture that was used, appreciated, rejected, and much more besides, all this is reasonably easy to observe. It is not even something that needs to be demonstrated in an archaeological study. It is if anything a matter of course as soon as a technology is noted in a body of prehistoric material. These are conditions and events we know existed as soon as we find archaeological artefacts that were part of this.

What is more difficult is questions about how, what, and why. Why was there a technology? What was it used for? Why was it considered important? How did it become important, and important in relation to what? How was it part of a sociotechnical framework? How was material culture created using the technology? How was the technology created on the basis of materiality? It becomes even more difficult when questions about consequences are brought up. What does this mean? What changes did a new technology entail? What did not change? And it becomes really tricky when questions are asked concerning the time when it happened. How did the people involved react? Did social patterns change? Were there options? Who acted? How did they react?

These are questions that interest me, and to work with them I must discuss some points of departure which concern what meanings I believe I can inject into the formulation "sociotechnical change and persistence".

Sociotechnical is a concept that has been explored by Wiebe E. Bijker, professor of technology and society in Maastricht. In a study of bicycles, bakelites, and fluorescent light bulbs that covers the time from the late nineteenth to the early twentieth centuries, he uses the term sociotechnical as a starting point for discussing the development of these different technologies as social processes – as a means to study both the social shaping of technology and the technical shaping of society (Bijker 1995:11).

Here it should be mentioned that the term sociotechnical should not be confused with the archaeological term sociotechnic – "artifacts where the material elements having their primary functional context in the social sub-systems of the total cultural system" (Binford 1962:219) – a term used in discussions of technological change (Schiffer & Skibo 1987:598). Although Bijker's "sociotechnical" and Binford's "socio-technic" are linguistically similar, the resemblance ends there.

Sociotechnical is a term which defines technology in a way that is in line with how studies of materiality define material culture, i.e. technology as deeply embedded in social life. People use technology to create things, and things created and the technology used create people. "Technological culture" has been highlighted as a key concept in studies of the social and technical as merged into one. It means studying technologies and societies from a cultural perspective:

> A focus on technological culture highlights how social interaction is mediated through technologies, and how technologies can only function when embedded in social institutions (Bijker 2006a:54).

Technological culture is a concept that is similar to the term material culture, just as sociotechnical resembles materiality. Although I have no ambition to replace one term with another, I think that "technological culture" is an interesting expression. Its semantic resemblance to material culture and materiality is worth thinking about since the similarity opens for the possibility of combining materiality and technological culture into one object of research in a sociotechnical study.

The first part of the term sociotechnical, the broad and polyvalent "socio", is used here as a reference to social aspects of interactions, connections, and associations between society, groups of people, and individuals. I have no ambition to give the concept a meaning that involves studying cultures or ethnic groups, nor society as the context in which everything is framed (Latour 2007). Instead the focus is on social formations, units or groupings in some period who lived in daily or recurrent contact of some kind and who were united or separated by actions and patterns of action (Cornell & Fahlander 2002:13ff; Vandkilde 2007:16). These actions are limited by the book's focus on technology in the Late Bronze Age and the study material in the form of objects of flint. I will develop this argument later on.

The second part of the term sociotechnical, the "technical" part, concerns the broad concept of technology – such a broad concept that it is actually defined more by its use as an object of analysis than by preformulated definitions (Pinch, Ashmore & Mulkay 1992; Greene 2004). But despite this, it is relevant here to give a short definition of the term, more with the ambition to fit it into my text and not so much to give a general definition.

Technology is about the use and knowledge of tools and crafts. It also has to do with systems, organization, and techniques. Different levels to which the concept of technology refers can be highlighted. It can concern a collection of objects, such as axes or knives of flint. It can also concern activities, the act of chopping with the axe or cutting with the knife. The term then refers to activities to do with manufacture, use, and handling. It can also concern knowledge, that is, what people know about and do with the technology in a broad social understanding of this (Lemonnier 1993; Bijker 2006b).

In line with this, Michael B. Schiffer and James M. Skibo have defined technology as "a corpus of artifacts, behaviors, and knowledge for creating and using products that is transmitted intergenerationally" (Schiffer & Skibo 1987:595). The first part of this definition agrees well with how I perceive technology. The second part about being "transmitted intergenerationally", however, is a definition that is more specific for other questions than the ones I ask. Technology as something that is handed down across generations is not a necessary part of a definition in relation to the questions with which I am working.

Technology is about objects, skills, activities, and actions. In archaeology technology has often concerned craftsmen who produced things through different activities and the use of these things (Kohring 2006). But it has also asked how, by whom, and why this has been done, in other words, studying technologies in use (Schiffer & Skibo 1987:595). Not least in lithic studies, technology has been examined in detail (Sinclair & Schlanger 1990; Inizan, Roche & Tixier 1992; Kooyman 2001; Holdaway & Stern 2004; Odell 2004). This is something I return to later in the book.

There is a difference in definition between technique and technology that is often pointed out (Roux 2003) and that I will consider briefly here. Technique refers to the capabilities of persons to do things. A flint knapper uses a particular technique when an artefact is manufactured, for example, strikes with a hammerstone or holds the core and the tools in a particular manner. Someone who reaps with a flint sickle uses a particular kind of technique to do so. Although techniques are commonly associated with tools or machines, they are not always necessarily connected to objects. Yet objects, techniques, and technologies are intimately interwoven.

Flamenco dance, for example, is usually performed together with objects and implements – dancing shoes with studs in the heels and under the toes, and a guitar are part of the technology of flamenco. But it is possible to dance flamenco without these objects, naked on a beach with only handclaps for rhythm. The technique of dancing flamenco then exists as embodied knowledge and an ability that is not connected to technologies in a tool-using activity (Ingold 1990:7).

On the other hand, it is not so simple. Even if one is not using any implements there on the beach, few dancers have learned flamenco naked on a beach. Flamenco is deeply rooted in its references to the objects. It is very difficult to envisage flamenco without the artefacts as aids. The dancing, the shoes, and the guitar are united in the image of flamenco. The objects shape flamenco, flamenco shapes the objects. So even if the dance is performed without shoes, few of the observers, and presumably not even the dancers themselves, do not imagine the rhythm of the shoes and the metallic sound of studs against a floor every time the naked foot stamps the sand on the beach.

The difference between technique and technology is thus more obvious in a terminological discussion than in practice. Techniques are embedded in and inseparable from technology

(Ingold 1990:7). In the future I will not ascribe any great importance to the differences between technique and technology, but it is important to mention the conceptual difference in order to avoid misunderstanding.

Back to sociotechnical

Having digressed on this discussion of "socio" and "technical", as I now return to the concept of sociotechnical it is important to stress that this concept is not a matter of a combination of, on the one hand, social aspects and, on the other hand, technical aspects. It is about something quite different. It is not two meanings united in one word, but a meaning that simultaneously comprises both parts of the structure of the word. The point is to try to ignore the "socio" on the one hand and the "technical" on the other:

> All relations should be seen as both social and technical ... Purely social relations are found only in the imaginations of sociologists ... and purely technical relations are found only in the wilder reaches of science fiction (Law & Bijker 1992:290).

I will elaborate somewhat on this perspective. Latour, along with others, such as Michel Callon and John Law, using what is called the Actor Network Theory (henceforth abbreviated as ANT), have emphasized objects, technologies, and people as collaborating actors in complex network relationships (see e.g. Latour 1992, 1998, 2000, 2007; Callon 1997; Law 1992). All parts of these networks, they claim, can be studied from the same point of departure. The differences between the parts of these network, for example, between humans and non-humans, is something that should be analysed, not presumed:

> the social is nothing other than patterned networks of heterogeneous materials. This is a radical claim because it says that these networks are composed not only of people, but also of machines, animals, texts, money, architectures – any material that you care to mention. So the argument is that the stuff of the social isn't simply human. It is all these other materials too. Indeed, the argument is that we wouldn't have a society at all if it weren't for the heterogeneity of the networks of the social (Law 1992:2f).

Latour thus does not talk about subject and object, about people and items, but about actants, in order to define both human and non-human actors. These are ascribed equal properties (Latour 2007:54f). Objects do not emerge from purely mental structures or materialist concerns (Martin 2005:285). Objects, technologies, and people participate in the sociotechnical, they shape it. If human beings form a social network it is not because they interact with other human beings. It is because they interact with human beings and endless other materials too (Law 1992:3). Actors and actants in a network are determined by relations to others in the network. The characteristics of individual actors, human as well as non-human, are the preconditions for how they can act. How they will act, however, is determined by the contexts to which they belong. Latour's (1992) study of hotel keys which I discussed above is a clear example of how objects, technologies, and people interact. I will provide another example.

Mobile phones were invented to ease communication. That is what they do. But at the same time they have also fundamentally changed our way of communicating with each other. They have created new spellings, new words, and new kinds of conflicts. These days, it seems as if all we need to do in order to communicate with each other is to type a few letters into our phones and text them to the phone number of the person we wish to communicate with. Our messages may say "cul8er". Recently, I read a newspaper article discussing the need for an SMS dictionary in order to help us understand this new language (Sydsvenskan 2008), and such dictionaries are available on the internet (SMS dictionary 2008). And, texting messages can be done, and is done, pretty much anywhere. This means that you can now take a shower in the middle of a "conversation" with a stranger. But using mobile phones during, for example, a theatre performance or at the cinema is annoying to many people and has also created all kinds of novel social conflicts. In one of the first episodes of the television series Californication from 2007, Hank Moody, the lead character, gets applause from a crowd when he beats up a young man who is repeatedly using his mobile phone at the cinema. This is a scene that would have made little sense only some fifteen to twenty years ago.

This example shows how things that have been shaped with certain characteristics are not developed on the basis of predefined needs, but from the users' creativity in combination with the objects' potential to satisfy these needs. Objects, people, and surroundings in this example are all actors in the development of new ways of expressing oneself and in the establishment of new conflicts. The ANT with its concept of actor/actant makes it possible to interpret what makes things happen in a way that dissociates it from people. Meanings can be traced to human and non-human actants, not to objects or subjects (Kärrholm 2004:122f).

This has consequences for my studies of flint technology from the Late Bronze Age. If my ambition is to investigate aspects of interactions between objects, technology, and people, the technological analyses must be performed on the assumption that the technologies belonged to heterogeneous networks.

ANT has been used before in archaeology (Greene 2004:158: Solli 2008:160ff). Per Cornell and Fredrik Fahlander, for example, have been inspired primarily by Latour's studies in what they call a microarchaeology, an archaeology that attempts to capture the possibility of translating a post-structural theoretical discussion into a practice, partly by focusing on people's actions (Cornell & Fahlander 2002).

Another example is studies performed by Valentine Roux. In an article about technological change, Roux has used the concept of "a dynamic systems framework" to discuss aspects of what causes technological change. Although Roux's study differs from ANT studies, her use of concepts is somewhat different from ANT, the inspiration from Latour above all is noticeable.

The clearest example of ANT in archaeology is a series of articles which introduced the theory under the expression "symmetrical archaeology" (Webmoor 2007; Witmore 2007). This multifaceted concept "symmetrical" comes from ANT and discussions of humans and non-humans role as actors (for a discussion of the concept of symmetrical, see Bijker 1995:275). Symmetrical archaeology, like interpretive archaeologies (see the earlier discussions) should not be regarded as yet another "-ism" in the theoretical polemic of archaeology. Instead it is an expression of an archaeological practice that seeks to put network relations between people and objects in the centre:

> At the outset it is important to highlight the most fundamental compass point for symmetrical archaeology, and that is how it attends to the relations of human and things (Webmoor 2007:563f).

> A symmetrical archaeology understands how human beings live *with* (to be distinguished from *in*) the world in terms of mixture and entanglements (Witmore 2007:559, original italics).

And although the discussions are about more than that (see e.g. Shanks 2007), these parts of symmetrical archaeology have in practice derived much of their inspiration directly from ANT (Olsen 2007).

These examples of ANT in archaeology are different, moving between present and past and between theory and practice. What they have in common is that the inspiration from ANT is about putting people, objects, and actions in the centre and viewing people, objects, and actions as parts of large-scale contexts – networks in which humans and non-humans interact.

It is important to emphasize here that ANT is an extensive theory which has been discussed for nearly thirty years. Latour's (2007) introduction to the subject is a book of about 300 pages, and many shelf-metres of books have been written on the topic since the start of the 1980s. Using the parts of ANT that deal with humans and non-humans as equal actors, actants, in integrative network is just scraping the surface, but it is scraping on the surface that has been done successfully in ANT studies (see e.g. Latour 2000).

There are some problems with ANT. It has been criticized for equating non-humans with humans (Vandenberghe 2002). This critique should be taken seriously. Objects do not have human properties such as intentions. Another problem is that it is a theory that has chiefly been applied to studies of modern industrial and post-industrial societies, and the theoretical discussion of ANT has also been carried on in polemic with social science. This has been an exacerbating circumstance here when I have been working with a prehistoric technology.

But despite critique and problems, the perspectives on non-humans as actors and thoughts about people, technologies, and objects as embedded in networks are inspiring, and I have found them very useful in this study. Yet this perspective should not be regarded as a definitive model for the description of everything. Just as little as any other theoretical framework, it is a guide to use for fitting interpretations into. It is an analytical perspective that is interesting to work with in an understanding of the complex social impact of technologies. It is a tool I use to unravel complications in the somewhat abstract perspective that permeates the discussions in this section, a perspective that is expressed, for example, in the formulation: "technology is embedded in the social, the social is embedded in technology". It is easy to say that technology and the social are one, and the word sociotechnical is excellent to use for that. But translating this into an analysis is more difficult. Here the perspectives developed in ANT are useful. It is important to point out, however, that network in the sense used here is a tool, not a description of something that exists. It is a tool for analysis, not something to be analysed (Latour 2007:131).

To sum up, then, the core of what has been said about the concept of sociotechnical is that it is not a matter of a dualism, with technology and objects on one hand and social relations on the other. It is a matter of chains of connections, networks of humans and non-humans. There are no pure social relations, just as little as there are pure technologies (Latour 1992, 2007).

Critics have argued that this perspective emphasizes the technology at the expense of other sections of society: "the economic, political and ideological constraints and influences on them, the broad historical changes affecting them and the more specific events leading them to the process under investigation" (Russell 1986:335). This is a relevant critique. A focus on technology risks overshadowing other aspects. But it is the point of departure chosen for this study. The ambition is to use analyses of the technology to illuminate other parts of people's lives. Obviously, a different point of departure, as I have already discussed, would have resulted in other questions, other methods, and hence other results.

The ambition in this book – to return to the questions from the introduction to this section – is to work with the questions why large blade knives of flint were introduced, used, and vanished during the last centuries of the south Scandinavian Late Bronze Age. The consequences of the discussion above about human and non-human actors belonging to networks is that analyses of why the technology arose or disappeared cannot be performed based on the technology itself, but proceeding from the networks in which the technology was active. The sociotechnical change and persistence is thus not a matter of technology a priori. It is about an interplay of change and persistence in the society of which the technology was a part. The ambition must therefore be to investigate how interaction between people, technology, and flint tools in the south Scandinavian Late Bronze Age created a sociotechnical network of change and persistence.

This brings me to the subject of the next section, which deals with a framework that makes it possible for me to work with both change and stability. Change and persistence are phenomena composed of different constituents. To work with this, I need to discuss two different aspects of how sociotechnical change and persistence can be described. This is done through a discussion of linear versus non-linear processes, and of the concept of technological frame.

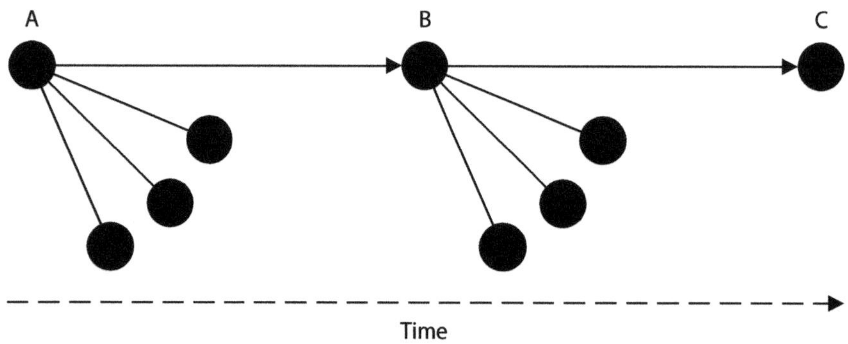

Figure 1.11. A traditional arrangement for a linear narrative. First came A, which led to B which then led to C. Both A and B have "sub-types", the three dots at A and at B. In a linear narrative like this these are described as sideline events, events that have not led to anything but a point alongside what is described as the linear development.

On linear and non-linear processes

In the discussions above I have held up the long-term perspective of archaeology as a way for archaeologists to work with interpretation of the past. The conclusion I drew was that I as an archaeologist, by looking at phenomena over a long time, can know more about people than they knew about themselves. Yet there is a problem with the long-term perspective that is important to consider here, a problem that can entice me as an archaeologist into describing what I here call "false linear processes" of change and persistence. By false linear process I do not mean that they are untrue, but that they are incorrect in relation to the claims they make to explain associations. I will develop what I mean by this through a critical discussion of interpretations presented in narratives about linear processes.

Development over time and place is often described as a linear process, from initiation to use and development – from A to B and from B to C in figure 1.11 (for a critical discussion see Lucas 2005:49ff).

The starting point for this way of describing development is a view of technology as linear: "technologies are linear in

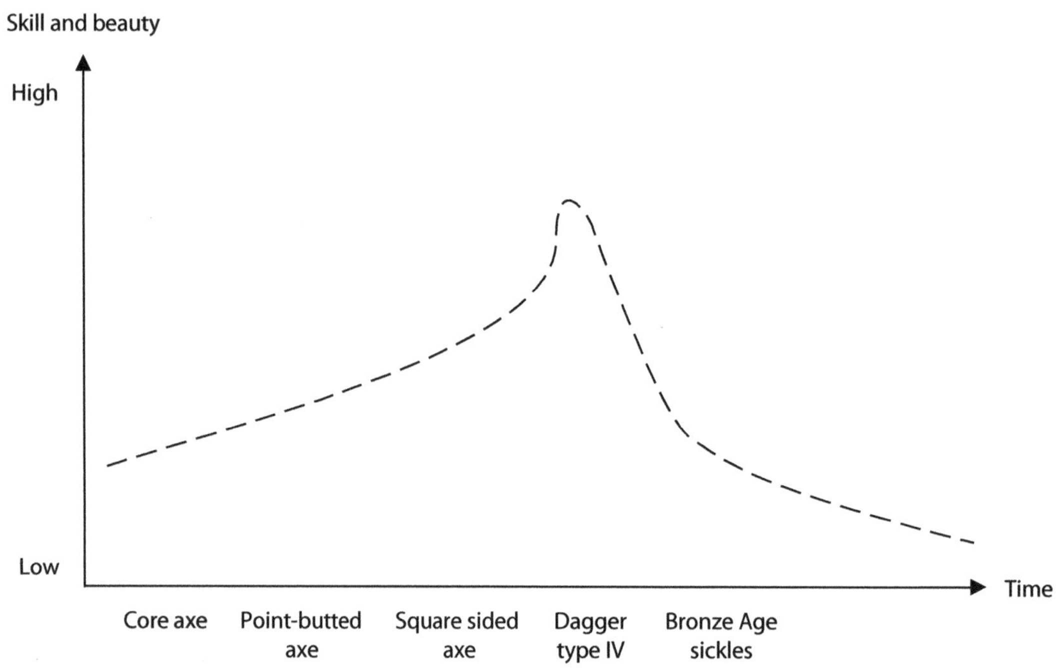

Figure 1.12. An example of the development of artefacts presented in a linear description. If we take a Late Mesolithic core axe, an Early Neolithic point-butted bifacial axe and a Middle Neolithic square-sided axe and put these into an evolutionary graph, with time on one axis and the skill and beauty of the artefacts we record on the other, it would look like an evolution over time and a linear development of technological skill and beauty. And if we add the Late Neolithic daggers and bifacial sickles as well, then the evolutionary graph will peak at the time of the Danish daggers type IV and then slowly degenerate into the Bronze Age. And this is of course not wrong. If we select these attributes for comparison, i.e. skill and beauty, and add time, we will get a rise and a fall in the graph. But if we want to interpret prehistory, it gives us a false impression of linearity. It suggests that nothing else happened. And it also suggests that people then were living in a rise or a fall. A person living in early Late Neolithic using a type I dagger didn't know anything about how that dagger technology would gradually develop into the production of type IV daggers, and that it then would peak the evolutionary graph of Scandinavian Neolithic flint-knapping technology, a graph comprising more than 2,000 years.

the sense that they necessitate a certain order of development" (Cresswell 1996:21, cited from a translation in Roux 2003:8). This way of viewing technology and development can be found in presentations of technological development from our own time, illustrated here with a narrative of the development of the car maker SAAB from 1945 to 1980 taken from a standard Swedish encyclopaedia:

> Construction of the first car, the test car Saab 92001, began in autumn 1945, and serial production of the Saab 92, with a transverse two-cylinder two-stroke engine between the driving front wheels, started in 1949. The Saab 93, with a longitudinal three-cylinder two-stroke engine, was launched in 1956, the same year as the prototype sports car Sonett. The estate car model Saab 95 came in 1959 and was manufactured until 1978. The Saab 96, with changes to details of the chassis, made its debut in February 1960. In the 1967 model the two-stroke engine was replaced by a Ford-manufactured four-stroke V4 engine. Manufacture of the Saab 96 ceased in 1979. The model was replaced by the Saab-Lancia 600. At the end of November 1967 the new medium-class model Saab 99 was presented, which from 1977 was also built with a turbocharged engine. May 1978 saw the presentation of the Saab 900, which was developed from the Saab 99 and differs from it, among other things, in that the nose has been extended. The sedan model of the Saab 900 was shown for the first time in March 1980 (Bra Böckers Lexikon 1980:102, s.v. SAAB).

Both the text and the illustration accompanying it give the illusion of a linear development. One model is developed out of another, and new types are introduced seemingly out of nothing. Neither text nor image say anything about the failures, the models that were developed but never went into serial production, or about all the suggestions, negotiations, and decisions that led to the different models. Nor do we see the consumers, the people who owned and drove the cars, in the picture or the text. Development seems to be powered solely by the manufacturer, not by an intricate interplay between manufacturer, motorists, mechanics, new laws and rules for cars, the change from driving on the left to driving on the right in Sweden in 1967, the expansive development of Sweden's road network during the time the text deals with, the oil crisis of the 1970s, the construction of motorways, and so on. These omissions leave a picture of a linear process, a narrative that is neatly arranged in retrospect.

We also find these linear accounts in narratives about prehistory. Several archaeologists have studied the archaeological material over time, from the birth of an artefact, through its development and up to its disappearance (see e.g. Steensberg 1943; Jensen 1987). Such studies result in an extract from prehistory which shows that the development resulted in C and that B and A preceded it, to refer to figure 1.11. Therefore the development of the artefacts is described as a linear process. Often this description also includes words charged with values, such as evolution and degeneration, for example, in descriptions of rise and fall over time, figure 1.12. This attitude to the past is sometimes called "the Whiggish perspective" (Martin 2005) and has been discussed, for instance, in studies in the history of archaeology (Gustafsson 2001).

It is far too simple to describe a technological evolution in this way as a linear process. It implies a non-disputable, enclosed, and predetermined process of development. This gives an impression that a long time ago, at the starting point at A in figure 1.11, it was already predetermined that A would become B and that B would become C, or that a Mesolithic core axe would become a Middle Neolithic square-sided axe, to use the examples from figure 1.12. It also gives the impression that C in figure 1.11 is an inevitable consequence of B and A, that it could not have been otherwise, and that the "sub-types" found at A and B were only parentheses in the grand linear narrative of development. This is a remarkable way to describe technological development (a similar critique has been put forward by others, e.g. Papousek 1989; Bijker 1995; Schiffer 2000; Roux 2003; Lucas 2005).

In a study of sociotechnical change and persistence like this, it is important to avoid regarding development as a linear process, from innovation to acceptance and rejection. But for an archaeologist it is an easy trap to fall into. We know how things turned out. The Bronze Age was followed by the Iron Age, and so it is easy to describe that process as a transition from bronze to iron. And it is also true; in a long-term perspective this has become the established narrative of the consequences. But it does not describe what happened, how it happened, why it happened, and who did it. The important thing about a study of blade knives is thus not to describe the process based on the knowledge of how it turned out: they

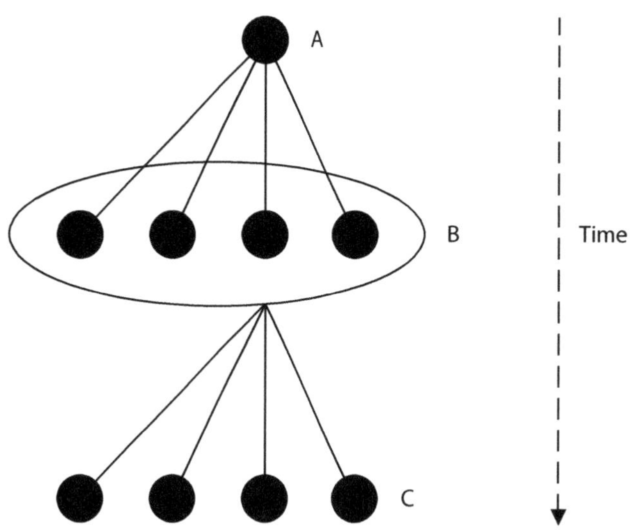

Figure 1.13. A non-linear presentation of technological development. The figure should be regarded as the opposite of the presentation of a linear narrative as illustrated in figure 1.11. In the figure here each part of the process is ascribed equal value in its time, without using the archaeological knowledge of which of these technologies would finally dominate. A leads to a great many different technological innovations. These are contemporaneous and thereby influence each other. The overall impression of this is given by B, which in turn is developed into a great many new technologies, one of which is C.

were introduced, used, and vanished. The important thing is to ask questions about how this happened, what was involved, and the meanings this may have had in different times (Beck et al. 2007:833ff).

To be able to do this, I must view technological development and change at any point in time as an unfinished topic of discussion, as a complex and non-linear process in which each period handles its specific technological development and stagnation. What the result of this will be over time is in no way predetermined, figure 1.13.

There is thus an essential difference between cause and effect which I find important to keep apart. As an archaeologist I view the effects of actions over a very long time. As I discussed above, this gives me the potential to know more about the past than the people who lived then did. But if it is my ambition to interpret the processes, that is to say, not just what happened but, for example, how and why, it is important to ignore the effects and focus on individual events as part of a process. How this can be done I will look at more closely through a discussion of the concept of technological frame.

Technological frame

The core of the discussions above is that it is difficult to find the meaning of a technological artefact or a technological system in the technology itself (Bijker 1995:6). In other words, large blade knives of flint and the technological system of which they were part cannot be investigated or discussed solely on the basis of the blade knives themselves. Instead I must study how the technology is shaped and how it is given meaning in ambiguous social interactions. To link up with what has been said before, this means that blade knives and flint tools from the Late Bronze Age must be studied in relation to the society – or the network, to use the term introduced earlier – in which they emerged, and are then viewed as meaning-bearing and meaning-creating in relation to several different aspects.

Studying a technological development therefore cannot be a matter of merely studying the properties of artefacts such as form or function. It must also involve studying the interaction between the meanings ascribed to these properties and how these properties create meaning. But one and the same artefact can have and give different meanings for different people and groups (Schiffer 2000). I will use examples from Bijker's (1995) study of nineteenth-century bicycles to illustrate this.

When the pneumatic tyre was invented and introduced at the end of the nineteenth century, it was with the express ambition of reducing vibrations in bicycles. The target group was broad: all cyclists. But the tyres did not become popular. The main reason for this was that the vibrations were not thought to be a sufficiently serious problem to justify replacing the cheap and durable solid rubber tyre with the more expensive and fragile pneumatic tyre. But when pneumatic tyres were later fitted to racing bicycles, it turned out that they had another advantage: they were much faster than solid rubber tyres. This success for the tyre in competitive sport changed the pneumatic tyre from just an anti-vibration tyre to a combined anti-vibration and high-speed tyre. As a result of this change the pneumatic tyre caught on generally, and in just a few years virtually all solid rubber tyres on newly produced bicycles were replaced with pneumatic tyres (Bijker 1995:77ff).

What this example shows is that it is not the observation that an artefact has a different meaning in relation to different social groups that is the important thing, in this case the group of ordinary cyclists and the group of racing cyclists. What is important is to try to analyse the meanings that different groups ascribe to the artefacts, how these meanings interact or compete for the preferential right of interpretation, and the consequences this has. In my analysis this means that I must work to define social groups (Bijker 1995; Schiffer 2000), social groups which in one way or another used the large blade knives of flint.

The concept of technological frame is useful for discussing this. A technological frame can be described as what structures interactions among the members of a relevant social group and shapes their thinking and acting:

> A technological frame comprises all elements that influence the interactions within relevant social groups and lead to the attribution of meanings to technical artefacts – and thus to constituting technology (Bijker 1995:123).

Technological frame is a theoretical concept that has great similarities to ANT's concept of network. In my studies these concepts, that is, technological frame and network, will be used alternately to enable analyses of the question why large blade knives were introduced, used, and vanished during the Late Bronze Age.

Action

This is something I will proceed with by focusing on action. It is thus time to put a little flesh and blood on the theoretical framework that I have built up. By focusing on action, people enter the picture, people who acted in different ways in relation to different purposes, emotions, and contexts. Cornell and Fahlander have emphasized action as a crucial starting point for interpretations:

> Human actions are performed by individuals with a certain degree of autonomy, but are also in large measure influenced by the structuring practices of the social group … the relationship between them is dialectic and the process structuring: individual actions are affected by structuring factors, which in turn consist of the result of actions (Cornell & Fahlander 2002:134).

In the same way, it is actions that I am in search of, actions translated into technology. To work with action, I need concepts to think with. These concepts must make it possible for me to follow and clarify different parts of a technological evolution (Bijker 1995; Beck et al. 2007). Through these concepts and the actions associated with them, I have the possibility to discuss the technological frame that I wish to examine in this study.

Some of these concepts are presented below. This is done through an in-depth discussion of the processes that can be part of the introduction, stabilization, destabilization, and restabi-

Concepts	Parts
Introduction	Innovation
	Acceptance
Stabilization	Negotiation
	Closure
Destabilization	Rejection
	Copying
Restabilization	Correlation
	Development

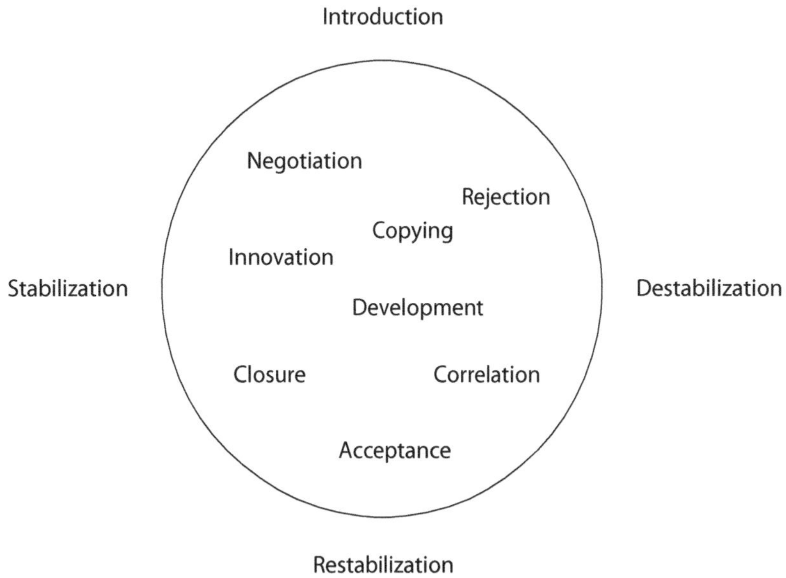

Figure 1.14. The figure shows the concepts I use to discuss the processes of introduction, stabilization, destabilization, and restabilization. These concepts are significant for my work of interpreting sociotechnical change and persistence. To the left in the figure the concepts are listed together with the parts of the different concepts that I discuss; to the right the concepts and their parts are set in a schematic figure to show that there is no hierarchy or predetermined order for the concepts, an impression that can easily be suggested by a list.

lization of an artefact, figure 1.14. These concepts should not be viewed as static or as a model into which to fit prehistory. I bring them up as conceptual tools which help me to think about what may have created change and persistence.

Introduction and stabilization

An innovation is something that has come into use. It is when something that has been invented has also been introduced and accepted for use. Many inventions are never introduced, that is to say, many objects and technologies which have been invented were never accepted in widespread use. Certain innovations create fundamental changes in activities and behaviours while others entail only minor modifications.

When a new object is introduced it is immediately a subject of negotiation. These negotiations are not bound to individual arguments or premises, such as efficiency or productivity. Instead the negotiations take place from a great many angles with a wide range of purposes and in a give-and-take between actors, human and non-human, within the network where the negotiation takes place. These negotiations have the result that the object is accepted or rejected by different groups.

It is also important to emphasize mistakes here. This is because it is not just an artefact's success that offers the possibility to investigate its development. Equally important are the mistakes, that is, the artefacts that were not accepted for use.

Closure is an important concept. It is defined as a process which means that a society's flexibility in the way of regarding an object disappears, consensus arises, and an accepted opinion of the state of affairs is constructed (Bijker 1995:85ff). Previous flexibility disappears. A state of stabilization has arisen, a stabilization which means that no discussions are needed as to what an artefact is, what it looks like, and what it is used for. This means that the pluralism of the artefact disappears (Bijker 1995:86). This stabilization can look different in different social groups. If a stabilization process leads to a closure that lasts for a very long time, that is, when a technology is used for generation after generation without any change to speak of, then this technology is persistent.

Destabilization and restabilization

Technology develops in interaction between all those involved in the network where the technology is used and acts. This interaction is not just about introduction and stabilization. It is a matter of technologies being rejected or copied, developed and changed; in other words, they acquire new and different meanings that lie outside the consensus that a closure represents.

Rejection means that a technology or an object is no longer considered worth bothering about in the same way as before. The meaning that was ascribed to the object by stabilization no longer applies. Copying is when a technology or an object is duplicated to imitate or resemble another object but the copying is not based on the processes that originally shaped the object or the technology that is copied; instead only fragments of the information contained in the object or the technology are copied. This copying upsets the stabilization of the object, gives it ambiguous new dimensions. Let me give an example.

The Fender Stratocaster, first launched in 1954, is one of the world's most famous guitar models. Musicians like Jimi Hendrix, Kurt Cobain (Nirvana), and John Frusciante (Red Hot Chili Peppers) have helped to make the "Strat" immortal. But it is expensive. Solid handiwork lies behind every guitar, with high-quality raw materials, technology, knowledge, and know-how. Because of its popularity and high price, the Stratocaster is one of the world's most copied guitar models.

Countless manufacturers have launched more or less credible imitations of the Stratocaster and, as a result, the term "Strat" is often used when referring to any guitar that has the same general features as the original, regardless of manufacturer. The success enjoyed on the market by Stratocaster copies has led Fender, the company that originally started making the Stratocaster, to release a copy of its own guitar, known as the Squier Stratocaster. This copy is cheaper and simpler than the original. It is produced with lower-quality raw materials and technology and with less know-how.

In little more than fifty years, the guitar once launched as the Fender Stratocaster has thus been transformed from a model to a concept that includes a great many different models, guitars, and associations with guitarists.

What this example shows is that copying is something that complicates how objects are perceived. What was perhaps clear during stabilization can be changed through copying. Innovations are developed beyond the control of intentions. A destabilization process has begun. This usually means that there are two alternatives for the future. Either the destabilization means that the object disappears, is not used any more, or a new closure process starts. I have chosen here to call the latter restabilization. This is in order to reveal the temporal perspective. But restabilization should actually be equated with a closure process. In a restabilization the object is correlated in a usage. New and old perceptions about the object are fitted into a world-view that increasingly moves towards stabilization. This can in turn lead to further development or change to get the object to suit the new demands entailed by destabilization.

Participation

It is also important here to touch on the question of the extent to which different actors belong to, participate in and interact with a technological frame. If an actor participates to a large extent in a technological frame, this actor will influence and be influenced by the network to which he, she, or it belongs. For instance, the large blade knife as an actor is probably highly involved in the network of large blade knife actors. Other actors perhaps play a smaller part. This can be investigated by adding different hypothetical activities, actions, and actors to the analysis, table 1.5.

These activities, actions, and actors all belong to different forms of relations. These can be simple, but also complicated. In figure 1.15 I have taken the four hypothetical actors listed in table 1.5 and put them into an interaction model. The aim of doing this is to clarify how complex the network to which different actors belong can be, and the extent to which they participate in different technological frames.

As figure 1.15 shows, an inventor can invent other inventions, a manufacturer can manufacture other things, the distributor can distribute other things, and the user can use large blade knives together with other things than blade knives. These actors can interact in several different ways. This shapes complicated network patterns. Several of the actions exemplified through the actors in table 1.5 and figure 1.15 can also in

Table 1.5. Examples of how sociotechnical change and stability can be studied in relation to several different aspects, here exemplified by activities, actions, and actors.

Activity	Action	Actor
Inventing	Invention	Inventor
Manufacturing	Manufacture	Manufacturer
Distributing	Distribution	Distributor
Using	Use	User

fact be one and the same action. For example, manufacture and use or invention may have been part of the same action, that is, something may have been made in connection with its use, and something may have been manufactured at the same time as it was invented. To be able to work with the question why large blade knives were introduced, used, and disappeared, these are aspects that must be worked into the analysis.

Summary thus far

Now that we have come a good way into the first part of this book, it is time to ask the question: what have I actually arrived at? In order not to get lost in the theoretical framework that I am in the middle of constructing, it is best to look back at what has been said hitherto. I began by asking a number of questions about large blade knives, technology, and the Late Bronze Age. These all deal with a specific question: Why were large blade knives of flint introduced and used and why did they then disappear at a time that has been classified as the transition between the Bronze Age and the Iron Age? To deal with this and other questions formulated here, I have devised a theoretical and methodological framework for this study. This work began with a discussion that highlighted ambiguity and pluralism as a point of departure. The result of the discussion is that what I do should be characterized by the following stances: the past has happened; when it happened it was pluralistic; when it is studied today from pluralistic perspectives it is not arbitrary but ambiguous; and it is possible to study the past, not from the premise that we can find out what really happened, but from the premise that we are interpreting a distant past in terms of what we want to know today. I also discussed aspects of the fact that I as an archaeologist can know more about the people who lived then than they did about themselves. This is partly thanks to the long-term archaeological perspective, and partly thanks to the testimony of material culture. These discussions have been on a general level and have, so to speak, staked out the direction for my studies.

The discussion was then taken further in a section about materiality. The focus was on how material culture can be perceived, still on a general level but with a gradual approach towards questions about materiality and technology. Here I noted that the objects are in the hands of the people who use them, that is, the objects are created, used, and discarded. But at the same time the objects are active, that is, they create those who create, use, and discard them. I highlighted the concept of engaged studies as a way for me to work with both theoretical and practical aspects of materiality.

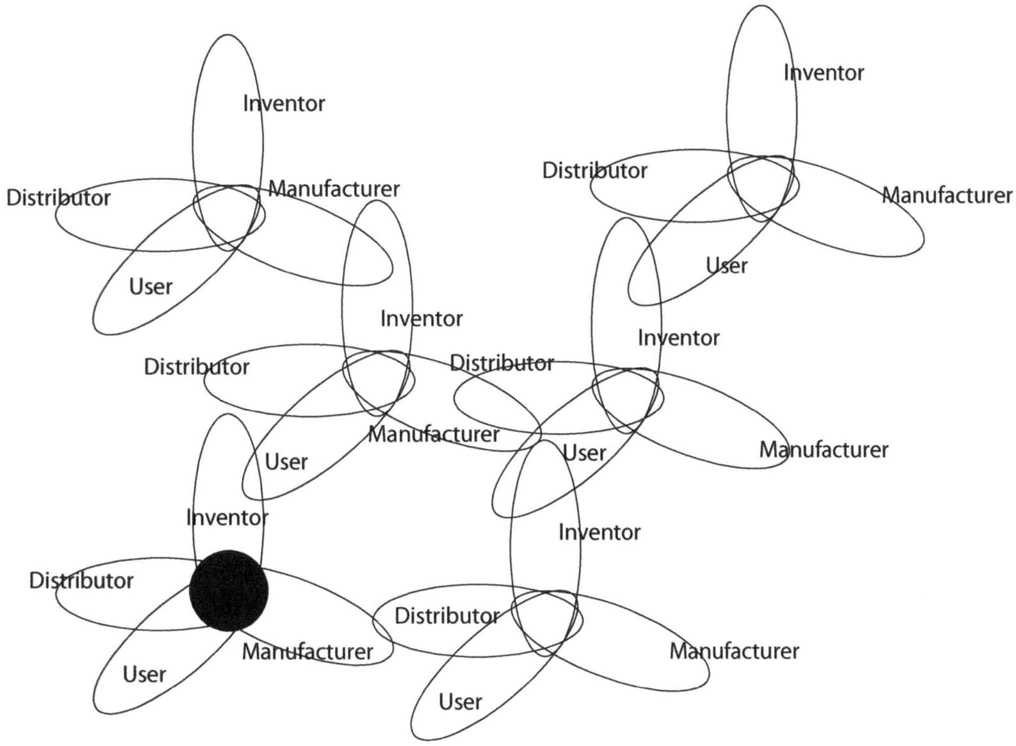

Figure 1.15. Here the actors listed in table 1.5 have been put into a sketch showing that parts of these actors' activities and actions can meet while other parts do not. Each oval corresponds to an actor's sphere of activity. The figure shows how complicated different networks of actors can be, in that actors belong to different networks through different actions. In this example the black lump in the network at the bottom left is the meeting that this book studies, with the actor "large blade knife of flint". The figure shows how a study must be conducted in relation to the network in which the knife was an actor, and in relation to the overall impression of other networks. The other networks in the figure could be, for example, once again in purely hypothetical terms, other crafts such as potter, bronze caster, house builder, wood carver, and stone cutter. Note that the figure is only a sketch allowing me to develop my ideas about what my analyses must comprise, not a template for analysis.

I then focused more narrowly on the core of this study, sociotechnical change and persistence. Here I emphasized the way ANT views humans and non-humans as actants in networks. I also brought in Bijker's concept of sociotechnical and noted that it is not a dualism, with technology and objects on the one hand and social relations on the other. It is about chains of connections and intricate interactions within networks of humans and non-humans. Here I also observed that there are no pure social relations, just as there are no pure technologies. I also emphasized the importance of avoiding narratives of false linear processes. Although we as archaeologists know the result of activities that were performed, enjoying the benefit of hindsight, it is important that these events are interpreted, not taken for granted.

Having come this far, I opened a discussion of technological frame. In this I came closer to a methodological approach by which I think I can study flint material from the south Scandinavian Late Bronze Age. Here I stressed the importance of studying actions. Actions are central in this study. A discussion was also conducted which exemplifies the importance of seeing different parts of the processes that can be included in change and persistence, and shows that these parts can belong in different ways to complicated networks.

At this stage in the discussions of the requirements for a theory of technological development, I proceed below to the last part of the theoretical and methodological framework that I intend to build with a discussion that further focuses on how I can work with actions. This is done by introducing the concept of chaîne opératoire. Then follows a discussion about the need to classify on the basis of the questions asked, and here I will gradually come closer to the study material, that is, knapped flint from the Late Bronze Age, which is the focus of part two of the book.

Chaîne opératoire

The chaîne opératoire is an approach that can be said, with some simplification, to aim at reconstructing the steps of actions that are expressed in a technology. As I discussed above on the subject of the technological frame, analyses of actions are central in my studies. To be able at all to discuss different parts of the networks to which the large blade knife may have belonged, and hence also to work with the question why it was introduced, was used, and disappeared, I must employ a method that puts action at the centre. This is the chaîne opératoire.

The chaîne opératoire has a solid foundation in European archaeological tradition. With its roots in French anthropology and archaeology, the method was established in the 1960s as a way to develop ideas about analyses of actions to comprise more than their material consequences (Audouze 2002). This was done by focusing on cultural and cognitive aspects of the action, that is, the technology, and not just on the result of the technology, that is, the objects (Leroi Gourhan 1993).

Since then the chaîne opératoire has been used for analyses of a wide range of materials and techniques (Lemonnier 1993; Audouze 2002; Roux 2003; Tostevin 2006, in press; Sørensen & Desrosiers 2008). In lithic analyses and research on flint knapping technologies, the chaîne opératoire has been discussed, developed, and applied in a great many studies and on material from highly diverse periods and places. Aspects of Palaeo-Inuit stone tool technology (Sørensen 2006a), Scandinavian Neolithic axes (Högberg 2006), French Upper Palaeolithic flint knapping apprentices (Karlin & Julien 1994), Scandinavian Mesolithic blade technology (Sørensen 2006b), Mousterian lithic assemblages from Lebanon (Copeland & Moloney 2003), Mesoamerican lithic technology (Hirth 2003), and stone-knapping evolution among early hominids (Roche 2005) are just a few of many examples of the variety of published studies within the framework of the chaîne opératoire.

The French term chaîne opératoire has been translated into several languages, for example, into Danish as den formative proces or "the formative process" (Madsen 1992:95), den operative proces or "the operative process" (Eriksen 2000:75), and den operative kæde or "the operative chain" (Sørensen 2006a:20), into Swedish as operativt schema or "operative schema" (Knarrström 2000:19) and into English as "operational sequence" (Bleed 2001:105). But since the French term is established and known in international publications (see e.g. Inizan, Roche & Tixier 1992; Shott 2003; Holdaway & Stern 2004; Odell 2004), I have chosen to use chaîne opératoire here without translation and henceforth without italics.

The chaîne opératoire has been well presented and is well rooted in lithic analyses focusing on technology (Inizan, Roche & Tixier 1992; Eriksen 2000; Holdaway & Stern 2004; Tostevin 2006, in press). I therefore do not intend to give any extensive presentation of the chaîne opératoire, referring interested readers to the studies mentioned here (see also Dobres 2000). What I will do is briefly present the concept within the framework of lithic studies and then discuss its use in relation to the questions that interest me.

The chaîne opératoire includes all the processes in a lithic technology, from the collection of raw material, through manufacture and use, to the discarding of the tool and the waste:

> The chaîne opératoire structures man's use of materials by placing each artefact in a technical context, and offers a methodological framework for each level of interpretation (Inizan, Roche & Tixier 1992:12).

In one of the many archaeological glossaries on the internet, the following definition can be found:

> chaîne opératoire: ordered chain of actions, gestures, and processes in a production sequence (e.g. of a stone tool) which led to the transformation of a given material towards the finished product. The concept is significant in allowing the archaeologist to infer back from the finished artefact to the procedures, the intentionality in the production sequence, and ultimately to the conceptual template of the maker (Tel el-Far'ah Dictionary/Pictionary 2008).

From this quotation we can pick out a number of terms that are usually considered essential in chaîne opératoire analyses. These are chain of actions, production sequence, procedures, intentionality, and the conceptual template of the maker. I use these terms in the discussion below.

Chain of actions, production sequence, and procedures

Chaîne opératoire analyses put action at the centre. This is done through a focus on the techniques, methods, and technologies employed in the manufacture or use of a tool. These techniques, methods, and technologies are often analysed in detail and combined into an interpretation of a production sequence.

Figure 1.16 shows an example of an illustration of one such analysis. The technology, which in this case concerns the manufacture of a square-sided axe, has been broken down into a number of steps. Each step has its special characteristics depending on the techniques and methods used.

Chaîne opératoire analyses thus do no focus solely on the end product of the manufacture, in this case an axe, but on the

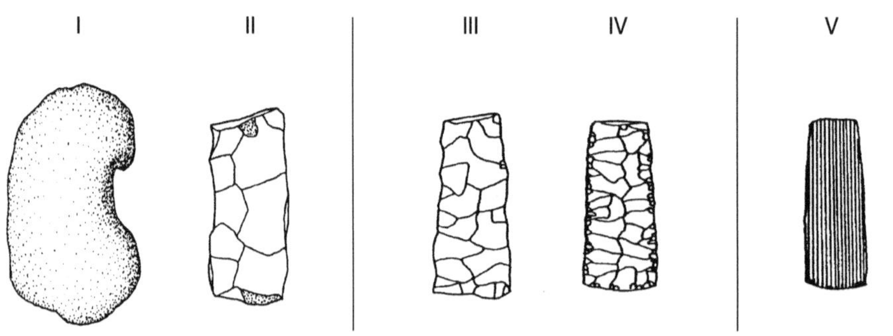

Figure 1.16. A simple chaîne opératoire for the manufacture of Neolithic square-sided flint axes. Square-sectioned axes are produced with what is known as the quadrifacial method. The basis of this method is a four-sided surface flaking that results in a blank with a squared or rectangular cross-section. The production of square-sectioned axes can be divided into various working stages, with the accomplishment of each stage being dependent upon the result of that preceding it, from stage one when suitable raw material is obtained, to stage five and the finishing polishing of the axe (Högberg 2006:196).

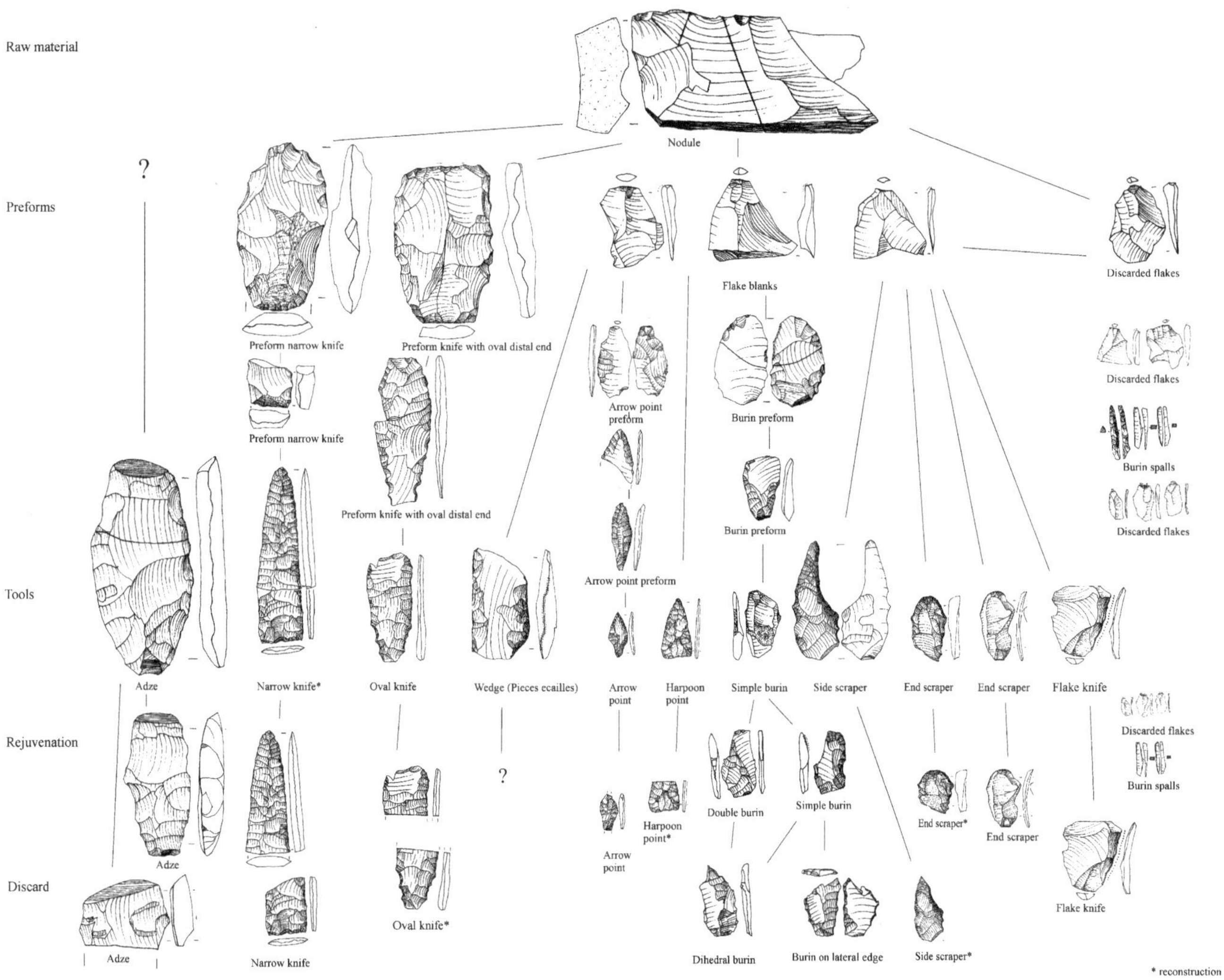

Figure 1.17. The figure shows an example of a chaîne opératoire taken from studies of Palaeo-Inuit technologies and traditions from the eastern Arctic, conducted by Mikkel Sørensen, National Museum in Copenhagen in Denmark. Sørensen's study is a fine example of how a chaîne opératoire approach can contribute new knowledge about the archaeological material and about the time and the contexts where this material was part of a living material culture. Through detailed analyses Sørensen, among other things, has charted the extraction and distribution of raw material, diagnosed different technologies and interpreted the processes by which different tools were manufactured, curated, and reknapped. The figure comes from Sørensen (2006a:141).

process. The aim of the analyses is to use the archaeological material to recreate the process that resulted in the axe and the flakes. In this way, everything is read backwards – from the end product and the waste material to the raw material and the various stages of production in between.

Each stage in a chaîne opératoire can in turn comprise broad and comprehensive analyses. The first step, collecting raw material, for example, can involve analyses of the availability, quality, and properties of different types of flint (Högberg & Olausson 2007). Step two in the example above, figure 1.16, the manufacture of a square-sided axe, can comprise different analyses of tools as regards their manufacture, different ways of using these tools, craft traditions, and an analysis of the place where they were manufactured (Sørensen 2006a). The aim of these analyses is to arrive at a broad understanding of the diverse variables that the technology and the manufacture comprise, figure 1.17.

Intentionality and the conceptual template of the maker

This part of chaîne opératoire analyses is an attempt to get at the thoughts and choices – both conscious and unconscious – of the knapper. By analysing flakes and tools in terms of different technologies and methods, and by refitting, it has been possible to perform studies of topics such as play, apprenticeship, skill, cognitive capacity, and division of labour (Fischer 1990; Karlin & Julien 1994; Schlanger 1994; Högberg 2008a).

Two concepts are important in this context: knowledge and know-how (Pelegrin 1990:118). Knowledge is defined as what can be learned by seeing, listening, and thinking, and what can be taught by showing, telling, and sharing experience. Knowledge is thus communicative, something that can be transferred from one person to another through conversation or actions. Know-how is muscular knowledge, that is, something that can only be learned by doing it yourself. It is a

Table 1.6. Knowledge and know-how, illustrated through a number of concepts. The figure comes from Apel (2001:28).

Knowledge	Know-how
Explaining	Acting
Explicit memory	Unconscious memory
Communicative	Intuitive
Theoretic memory	Muscle memory
Lost in case of lost memory	Not lost in case of lost memory
Words (two-dimensional)	Mental pictures (three-dimensional)
	Experience

tacit knowledge that is acquired through practical experience, not something that can be taught (Apel 2001:27), table 1.6.

Simple tool production, for example, making flakes from a core to use for cutting, requires little know-how. It is something one can learn almost completely by watching someone making flakes, thus acquiring knowledge, and then doing the same (Pelegrin 1990:123). Advanced tool production, for example, the manufacture of Scandinavian Neolithic bifacial daggers, requires a high degree of both knowledge and know-how. This manufacture can only be learned by gaining access to knowledge by seeing others manufacturing and hearing them tell about the knacks and methods for handling difficulties in flint knapping, along with many years of practice and gradual refinement of know-how (Apel 2001).

What makes knowledge and know-how such important concepts in chaîne opératoire analyses is the link to what a flint knapper knows and does not know (Pelegrin 1990). This "knowing" comprises the knapper's ideas about the technology – all the techniques, methods, forms, functions, and materials, the contexts and uses. What is permissible to manufacture, what is forbidden, what is in demand, what works, what does not work, what is needed, what is unnecessary, and so on? The flint knapper's work thus manifests society's knowledge, opinions, and convictions as to what a flint tool should look like, how it should be manufactured, where it should be manufactured, when, and in what contexts. How this "knowing" is used by the flint knapper is steered by intentions, either those of the knapper or those of the surrounding community (Pelegrin 1990; Eriksen 2000):

> The reconstruction of certain chaîne opératoire allows us ... to rediscover the process involved in techniques of production and, beyond that, the conceptual patterns from which they sprang (Karlin & Julien 1994:153).

The differences between knowledge and know-how that are represented in different ways of manufacturing flint tools make the concepts important in analyses of flint craft during the Late Bronze Age. These differences enable me, through the flint craft, to analyse different ways in which the flint tools were used and also give me a chance to discuss different dimensions of the flint craft in the Bronze Age.

Some critical reflections on the chaîne opératoire

There are, however, some problems with the chaîne opératoire that are important to consider. In the last fifteen years chaîne opératoire analyses have had a serious impact on North American archaeology, where they have encountered a well-developed and deep-rooted tradition of lithic analyses (see e.g. Hirth 2003). In the clash with what has been called the "reduction sequence approach" (see e.g. Bradley 1975), a method that has some features in common with the chaîne opératoire, a critical debate has arisen about similarities and differences between what many in North America perceive as something new, that is, the chaîne opératoire, and an established tradition, that is, the reduction sequence approach. This debate, despite some elements of ridicule (Shott 2003), has led to a much-needed critical scrutiny of the chaîne opératoire (Bleed 2001; Tostevin 2006, in press). I will not go into that debate here; for those who are interested I refer to Bleed (2001), Shott (2003), and Tostevin (2006, in press). I will instead use the debate to highlight some critical comments on the chaîne opératoire.

There is a tendency in chaîne opératoire analyses to overemphasize intentions and linear narratives. As discussed above, it is easy through archaeology to tell what I call narratives of false linear processes. We know how things turned out – a preform for an axe became an axe, to refer to the example above in figure 1.16 – and through this retrospective knowledge the narrative is described as linear, from preform to axe. But as I have shown in a different example, there are a number of manufacturing sites on beach ridges in southern Scandinavia where huge quantities of preforms have been found, cut to shape on these sites and then left there. If I hold a find from one of these sites in my hand and want to construct a chaîne opératoire, there is nothing to say that the intention of a preform was once that it should be manufactured into an axe. But – someone will perhaps object – the preform has been manufactured, and manufacture itself surely represents an intention? Yes, it has been manufactured. But there is nothing to say that the intention was to make a preform for an axe. In this example it is perfectly possible that the intention was that ten preforms would yield one or two axes and that each example of these axes actually was made of more than one preform and not in a linear series of manufacturing stages. Certain preforms would not become axes, they would be left on the site. Yet they still became axes in the form of the preform that was taken away and became an axe. This means that two or more preforms became one axe, figure 1.18 (Högberg 2006).

If I reconstruct the technology I am working with in a chaîne opératoire analysis, in this case the manufacture of square-sided Neolithic axes, by reconstructing the technology backwards from the finished axe and the flakes left by its manufacture, through different stages of production to the original raw material, the multifaceted action that the preforms for axes on the beach ridges represent will not be allowed to speak. Part of this critique has been discussed before (Schlanger 1994:144f).

Another problem with the linear narratives of chaîne opératoire analyses is that they give the illusion that nothing else has happened. As I mentioned before in the discussion of linear and non-linear narratives, this is a problem. For example, mistakes, inventions that never became innovations, or the interaction between the technology and other actors in the network to which the technology belonged – nothing of this finds room in a linear chaîne opératoire narrative. If we moreover go back to another of the earlier examples I considered, Latour's (1992) analysis of room keys, it was clearly shown that intention and consequence cannot be connected in a linear narrative. Intentions are changed through the actors' use of the objects. This is also the case with technology. What was initially an intention is perhaps not fully visible in the result of the manufacture. These are aspects of the technology that chaîne opératoire analyses often miss. In line with this, chaîne opératoire analyses also have a tendency to ignore non-humans as actors through the narrow focus of the method on addressing the human processes that bring artefacts from raw material through production steps to tools.

Another problem concerns the fact that chaîne opératoire analyses, in the practice where they are used, risk being transformed into a typology of technology, that is, an earlier descriptive classification of form and metrical attributes is replaced with a similar descriptive classification of technological attributes (for a critical discussion of this see Shott 2003; Dobres 2006; Nærøy 2007). The analyses are merely descriptive, not including the cognitive aspects of the method in the interpretation. This has had the result that several studies have been published where chaîne opératoire schemas have been used to describe the technology in different steps, not to interpret the technology in a context (for a critical discussion see Eriksen 2000; Bleed 2001). But one may wonder, does this critique have anything to do with the chaîne opératoire itself? Is it not just the fault of researchers who have not understood its full potential? Of course it is true that an approach cannot be blamed if it is not used to the full. But it is also a fact that the chaîne opératoire has a built-in trap.

Peter Bleed (2001) has performed a well-formulated analysis of different sequence models in which he has also discussed the chaîne opératoire. Bleed thinks that one of the grave defects of the method is that the cognitive basis for the chaîne opératoire takes for granted the underlying mental structures that are translated into actions. This means that individual steps in the manufacture and also the entire production process are interpreted as having been performed according to a plan (Bleed 2001:121).

> Chaîne opératoire studies rarely acknowledge a specific high-level theory perspective, apart from chaîne opératoire itself, for the origin of the questions they ask of the archaeological record (Tostevin, in press:7).

Since the chaîne opératoire emphasizes so strongly that cognitive aspects are implicitly built into the technology through the concepts of knowledge and know-how, it is easy to be misled into believing that a description of the technology is automatically a description of these cognitive aspects. But as I have stressed several times before, the important thing in this study is not to describe how something has certain properties, in this case cognitive aspects. The important thing here is to interpret what this means, how these aspects are expressed, when, what variation there is, who is involved, how the tools were developed and changed, and what consequences this had, in other words, why large blade knives of flint were introduced, used, and allowed to disappear. If questions like these are not posed in a chaîne opératoire analysis, there is a serious risk that the result will be a typological description of a technology performed using something that looks like a chaîne opératoire concept but which is in fact merely a description of the fact that there are cognitive aspects in a technology, which goes

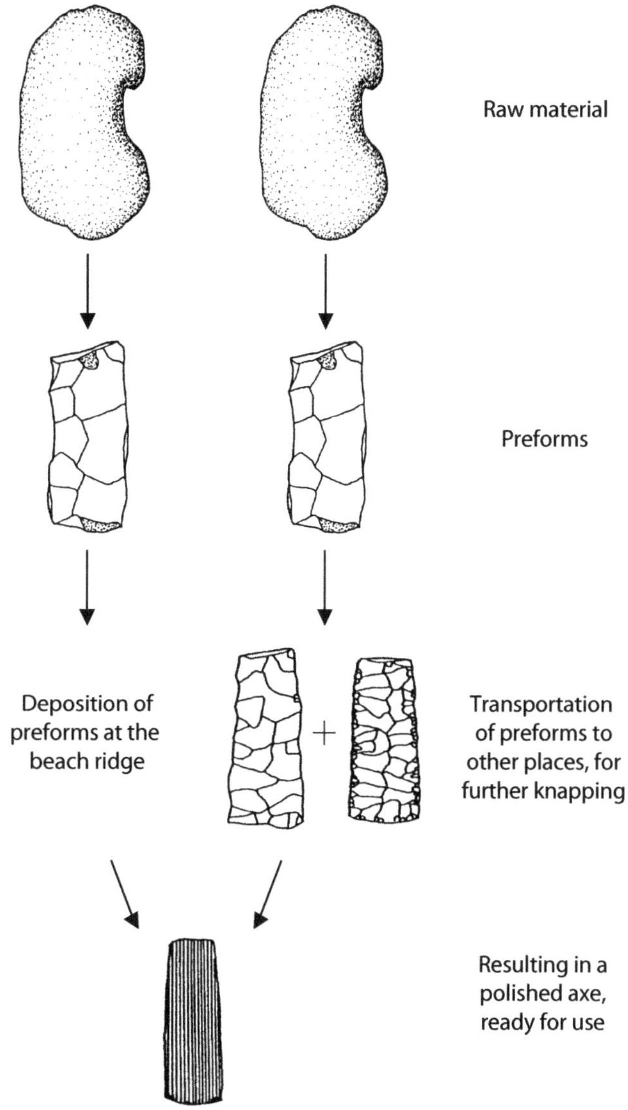

Figure 1.18. A chaîne opératoire for the production and life cycle for the preforms found on a beach ridge. A number of flint nodules are selected for the production of two or more preforms. One or more of the preforms are deposited on the beach ridge, while others are brought along to be made into finished axes at other locations. These actions, both the deposition of preforms on the beach ridge and the transportation of preforms to other locations, were part of the chaîne opératoire for axes (Högberg 2006:198).

without saying since we all know that this is the case. What one misses is the analysis of these cognitive aspects. This is not unique to the chaîne opératoire, however, but a shortcoming in all types of sequence models used in archaeology (Bleed 2001:101).

Yet despite this critique, the chaîne opératoire is a useful tool for my studies. Jacques Pelegrin is a respected scholar and tone-setting experimental flint knapper who has worked with the chaîne opératoire and lithic technology for many years. He has defined cognitive aspects of elaborate knapping activities, i.e. high levels of knowledge and know-how, in one of the most frequently cited articles in English on the subject. He writes (Pelegrin 1990:117):

> Such undertaking [flint knapping] – based on raw material which is never standard, and with gestures of percussion which are never perfectly delivered – cannot be reduced to an elementary repetition of gestures, or to the application of immutable sequences (as a machine would do). On the contrary, the realisation of elaborate knapping activities necessitates a critical monitoring of the situation and of the decisions adopted all through the process. If this is the case, then the capacity to mentally evoke the precise desired product is necessary for successful knapping, but it is not sufficient. The knapper has in mind successive goals, that is, a series of intermediary stages and geometric "cues". It is in respecting these, and with experience, that the anticipated result may be reached. These intermediary stages form a chain of intentions organised in a "conceptual schema opératoire". They are defined through certain geometric parameters, and they may represent the moment when a particular operation or technique changes to another. Between these stages, the actual and real situation is compared with the corresponding concept and diverse action modalities are evoked in order to correct a given state or to progress in the chaîne opératoire. Using experience, the knapper chooses the (most) adapted action modality – the one which is both possible and desirable.

This quotation sums up both the strengths and the weaknesses of the chaîne opératoire. Through the concepts of chain of actions, production sequence, and procedures, chaîne opératoire analyses put technology and action at the centre and through the conceptual template of the maker chaîne opératoire analyses put technology and cognitive aspects of manufacture at the centre. This enables me to work with systematic analyses of the interaction between technology, actions, and cognitive aspects (Schlanger 1994; Eriksen 2000). That is the strength of the chaîne opératoire method, what makes it possible to link the method to the discussions above about different actions involved in a technological frame.

Concepts such as intentionality and most adapted action show the defects in chaîne opératoire analyses. It is therefore important to balance the method against the discussions of materiality, ANT, and non-linear processes that I took up earlier, that is to say, to view actions as part of a sociotechnical network of actors and actants who both shape and are shaped by change and persistence. Here the perspective suggested by Gilbert B. Tostevin (in press), to regard each assemblage defined as the association of flint-knapping behaviours acted out at that particular spot in time and place, is a good starting point to bridge the possibilities and deficiencies in the chaîne opératoire approach.

I will now go on to elaborate some ideas about studies of lithic material and how I as an archaeologist think that the material should be handled. This is done in a section dealing with classification. This ends the first part of the book and also leads to a more concrete presentation of my source material and my methods in Part II of the book.

About classifying not according to what we know, but according to what we want to know

Flint tools have a form. They are shaped in one way or another to emphasize certain types of attributes. This may be an axe which needs a well-polished edge, a flake struck off to give a sharp flint edge, a dagger given all the latest fashionable attributes required of a dagger, or a preform for an axe that was shaped to be deposited on a manufacturing site (Olausson 1983; Apel 2001; Högberg 2006). Very many flint tools can also be characterized in terms of their form. Daggers and axes are clear examples. But just as many cannot. Tools which look different have been shown in analyses to be alike, and tools which look similar have been shown in analyses to be different – in both case as regards aspects of technology, function, and raw material (Briz et al. 2005).

The meaning of form and how archaeologists perceive and interpret it has long been a subject of discussion. This discussion can in principle be boiled down to two opposing views, those expressed in the Spaulding versus Ford debate (for discussion see e.g. Kooyman 2001:92; Wylie 2002:42; Holdaway & Stern 2004:219ff; Odell 2004:7):

- Types are inherent in the archaeological material. Types stand in relation to the intended form or mental template of the maker. We as archaeologists discover these types through our analyses. The types we identify were perceived as types during prehistory (Spaulding 1953, 1954).

- Types are invented by archaeologists. They are useful analytical tools but rarely have anything to do with prehistoric types (Ford 1952, 1954).

This polarization pinpoints essential issues for lithic studies. For even if these questions have been around for a long time and have been discussed by many scholars, with many alternative answers offered (e.g. Knutsson 1988; Hinnerson-Berglund 2000; Hirth 2003; Karsten & Knarrström 2003), they are still alive and far from settled (Kooyman 2001; Odell 2004; Briz et al. 2005). This is particularly true in Scandinavian archaeology, where by far the dominant way to handle flint material is to describe and classify what is found by an excavation in terms of accepted typological nomenclature (Helskog, Indrelid & Mikkelsen 1976; Andersson, Rex-Svensson & Wigforss 1978; Ballin 1996; Högberg et al. 2000). It should be stressed, however, that this is not unique to Scandinavia (see e.g. Bisson 2000; Tomášková 2005; Briz i Godino 2006). These nomenclatures are almost exclusively based on typological forms which are believed to have inherent properties that can be transferred to a prehistoric reality (Miller 1985; Bisson 2000). An artefact that is typologically classified as a scraper is interpreted as having been used to scrape with. An artefact that is typologically classified as a knife is interpreted as having been used to cut with. This is done without any discussion of the relations between typological observations (form) and use (function). Nor is there any consideration for different aspects of raw-material usage. This manner of classification has thus been heavily criticized (Miller 1985; Knutsson 1988; Bisson 2000; Vandkilde 2000; Knarrström 2001; Kooyman 2001; Cornell & Fahlander 2002; Karsten & Knarrström 2003; Tomášková 2005).

The consequences of archaeological flint material being sorted and classified in this way, based on predetermined typologies, are that it is sorted according to what we as archaeologists know, not according to what we want to know. We know that a specific form according to a predetermined nomenclature is defined as a particular tool, and that this tool should therefore be classified according to the same definition and sorted in the same pile as all other artefacts in an assemblage that have this form. We thus find out how many of this typological form there are in the assemblage. But we do not find out anything else. To learn anything more we must assume that there is something implicit, underlying, and inherent in this form and that the artefacts in the pile we have created through the classification correspond to this implicit, underlying, and inherent something. But we do not know this; it is just something we believe (for a discussion see Bisson 2000; Vandkilde 2000; Tomášková 2005). Cornell and Fahlander have criticized this type of classification of artefacts in the following, somewhat dramatic way:

> Classifying a bronze object from a grave as a razor means "killing" it. The classification not only locks the content of the object in one picture, it also actively prevents other descriptions of the information inherent in objects (Cornell & Fahlander 2002:3).

Instead of asking questions and classifying in terms of the needs raised by the questions, the material is routinely classified using nomenclature prepacked with what is considered essential knowledge (Bowker 2005:21ff; Witmore 2007:548). That will not suffice in this study. The question then is, what do I do instead? The answer is: to classify not according to what I know, but according to what I want to know. I will clarify this with yet another example from Bijker's (1995) studies of the development of the bicycle in the mid and late nineteenth century. He discusses the importance of classifying in terms of what you want to know and not of what you know. In the example Bijker compares the traditional high-wheeled Ordinary bicycle with the safety bicycle that was new at the time, figure 1.19.

The high-wheeled Ordinary or penny-farthing, with a big front wheel and a small rear wheel, and with the saddle almost directly on top of the front wheel, was an unsafe bike. If the front wheel bumped into something, such as a stone, it easily overturned. The cyclist was then thrown headlong over the handlebars towards the ground. This could cause injury. This type of bicycle was therefore perceived by many as an "unsafe bicycle". The safety bicycle, which resembles today's bicycle with two wheels of about equal size and the saddle placed roughly half-way between the wheels, was much safer. It was therefore regarded by many people as a safe bike.

For young men from the upper class, however, who wanted to show off their bravery to the ladies in the public parks each weekend, it would have been disastrous to ride around on a safety bicycle. For a display of skill and courage, the high-wheeled Ordinary was essential. Without the dangers associated with this bicycle, there was nothing to be brave with. This made the vehicle into a "Macho Bicycle" (Bijker 1995:74f).

What this example shows is that one and the same artefact can be perceived in diametrically opposite ways. There are no criteria for classifying the high-wheeled Ordinary which are not situation-bound. It was regarded as both downright unsuitable and as perfectly suitable. And these perceptions are related to social contexts, not to a predetermined classification. In this way, Bijker in the example presented here deconstructs the high-wheeled Ordinary into two artefacts: the Unsafe Bicycle and the Macho Bicycle (Bijker 1995:75). The reason he is able to do this is that he classifies the artefacts not according to what he knows, but according to what he wants to know.

For me it is essential to classify flint material in terms of what I want to know. This means that I cannot use classical morphological-typological or metrical-typological classifications such as the classification schema for flint that I myself once drew up (Högberg et al. 2000). I would then be classifying based on what I know.

Wolff-Michael Roth has devoted a comprehensive study to scientific classification, discussing it as a situated and socially distributed activity that does not eliminate ambiguity and discrepancy, but tends to minimize contradiction (Roth 2005). In line with this it is important to stress here that general categories are necessary so that archaeologists can talk to each other about the find material. For example, I myself have talked hitherto about large blade knives of flint as a general category, a category that I established at the beginning of the text by means of an overall description and some photographs. But it is important not to confuse these general categories of discourse with analytical tools that can be used for dealing with specific problems concerning interpretations. As will become clear later on, the concept in the form of a large blade knife that I use here to make myself understood will look completely different in the analyses I perform. The concept of large blade knife of flint will be deconstructed.

There is a big difference between, on the one hand, sorting flint objects into preset categories to create order and overview the material so as to be able to talk about it, and on the other hand, sorting flint objects into categories established on the basis of specific questions.

The Swedish archaeologist Knut Kjellmark (1866–1944), in his autumn years, criticized what he called "the terminology established by tradition" (Kjellmark 1944:273). He found it constrictive, forcing the material into compartments that were limiting rather than clarifying. In a posthumously published study of Stone Age settlement sites around the Mörrumsån river on the border between Blekinge and Småland in southern Sweden, he explicitly invokes this idea and suggests, apart from the traditional divisions, types such as *slingerflintor* "wind flints", *stötkantflintor* "impact-edge flints", *rivflintor* "grating flints", and *linjaler* "rulers" (Kjellmark 1944:272ff). Kjellmark's work is an example showing that terminology and types are under constant discussion and that the typology we now live with, even if it is "established by tradition", is also something that has always been a matter of discussion (Sjöström & Nilsson 2008).

A fundamental condition for theoretically based research on lithic material is to adapt the method to suit what you want to know, not what you already know (Odell 2004). I consider this critical if we are to succeed in combining the theoreti-

Figure 1.19. A safety bicycle, a traditional high-wheeled Ordinary bicycle and behind it a local home-made wooden bicycle. All these bicycles are from the late nineteenth century. The photograph was taken at the Johanna Museum at Skurup in Scania, Sweden.

cal with the empirical. It is a simple approach which, despite its simplicity, is crucial for the result of lithic studies. I will give two examples to further clarify the differences I see in the handling of flint material.

1. A body of material has been excavated. The flints are sorted and described according to an established classification system. Based on the types, groups, and sub-groups into which the material is sorted, analyses are performed: spatial, comparative, technological, functional, and so on. These analyses serve as a basis for interpretations of the material.

2. Before a body of material is analysed, the purpose of the analyses is defined. This is done by formulating hypotheses and questions. Based on this – the expectations of the knowledge that the study is to produce – a classification system is drawn up. The material is excavated and the material is analysed in terms of the hypotheses and questions. The analysis then serves as a basis for an interpretation of the material.

In the first example the interpretations of the material are based on a classification according to what is already known about the material. In the second example the material is interpreted according to a classification based on what you want to know about the material. The consequences are that in the first example the interpretations are governed by the classification, in the second by hypotheses and questions. This means that I must formulate specific questions, and establish categories for classification which are in relation to these questions.

Requirements for a theory of technical development – conclusion

This first part of the book has been about the formulation of questions about why large blade knives of flint were introduced, used, and then disappeared in a time that is classified as the transition between the Bronze Age and the Iron Age, and has contained discussions of a theoretical and methodological framework for working with this question. In a section about interpretive archaeologies I have held up ambiguity and pluralism as a starting point. One result of this is that the inquiry should follow an approach according to which the past has happened, when it happened it was pluralistic, that when it is studied today from pluralistic perspectives it is not arbitrary but ambiguous, and that it is possible to study the past, not with the premise of finding out what really happened, but with the premise of interpreting the distant past in terms of what I want to know today. In the light of the long-term perspective of archaeology and the testimony of material culture, I have discussed aspects of the fact that I as an archaeologist can know more about the people who lived then than they knew about themselves. These discussions have been held on a general level and have staked out the course of my studies.

The discussion was then taken deeper in a section on materiality. Here I observed that the objects are in the hands of the people who use them, that is to say, the objects are created, used, and discarded. Yet at the same time the objects are active, they create the people who create, use, and discard them. I highlighted the concept of engaged studies as a way for me to work with both theoretical and practical aspects of materiality.

I then focused more closely on the core of the study, sociotechnical change and persistence. To be able to analyse humans and non-humans as actants and actors in networks, I have held up Latour's ANT perspective as a point of departure. I also brought in Bijker's concept of sociotechnical. This was done in a discussion stressing that this is not a dualism, with technology and objects on one hand and social relations on the other. It is about chains of links and intricate interactions within networks of humans and non-humans. This discussion led to the observation that there are no pure social relations, just as there are no pure technologies.

I have also stressed the importance of avoiding narratives about false linear processes. Although we as archaeologists know from hindsight the results of activities that were performed, it is important that these events are interpreted, not taken for granted.

With this in my baggage I was ready for a discussion of a technological frame. In this I came closer to a methodology by which I think I can study flint material from the south Scandinavian Late Bronze Age. Here I underlined the importance of studying actions. Actions are central in this study. A discussion was also carried on exemplifying how essential it is to see different parts of the processes that may have been part of change and persistence, and that these parts may, in various ways, have constituted complicated networks.

This led to an introduction to the chaîne opératoire. Here I stressed the importance of the concepts of knowledge and know-how. The differences between knowledge and know-how that are represented in different ways of making flint tools enable me through the flint craft to analyse different ways in which flint tools were made and to discuss various dimensions of flint craft during the Bronze Age.

There are both strengths and weaknesses in chaîne opératoire analyses. Through concepts such as chain of actions, production sequence, and procedures, chaîne opératoire analyses put technology and action at the centre, and through the concepts of the conceptual template of the maker, chaîne opératoire analyses place technology and cognitive aspects of manufacture at the centre. This gives me the chance to work with systematic analyses of the interaction between technology, actions, and cognitive aspects. Concepts such as intentionality and most adapted action, however, show the deficiencies in chaîne opératoire analyses. I have therefore also stressed how necessary it is to balance the method against the discussions of materiality, ANT, and non-linear processes previously considered. This entails viewing actions as part of a sociotechnical network of actors and actants which both shape and are shaped by change and persistence.

The first part of the book ends with a section where I stress the importance of classifying more in terms of what I want to know than what I already know.

Figure 2.1. Photograph from Olof Christoffersson's collection of objects, Trelleborg Museum, Sweden. On the second shelf from the top is a large blade knife mounted together with Neolithic bifacial daggers and sickles. Published with the permission of Trelleborg Museum.

Part II: Large blade knives and other flint assemblages from the Late Bronze Age

To be able to work with the questions that I raised in the first part of the book, it is now time to change focus, from a general discussion of a theoretical framework to detailed analyses of the archaeological material on which this study is based. This is done by starting with a survey of what previous archaeological research has had to say about large blade knives and flint assemblages from the Late Bronze Age. The historical survey builds on extensive reading about what has been done and what has not been done, thus setting the agenda for continued studies by formulating a number of questions specific for the analysis of large blade knives and other flint assemblages from the Late Bronze Age.

This is followed by a number of chapters which tackle these questions by analysing the material in detail. The chapters begin with a description of the methods I use. I then analyse blade knives in terms of their appearance, manufacture, function, dating, contexts, geographical distribution, the occurrence of assemblages and special manufacturing sites. After that I turn to work with the parts of the Late Bronze Age flint material that are not large blade knives and then conclude the material analyses with a chapter about a hitherto unknown flint-knapping technology from Central Sweden and a brief chapter about the occurrence of flint objects in Late Bronze Age graves.

Since relatively little research has been done previously, it is necessary to present detailed and penetrating analyses of the material. At times this weighs the text down, and readers will gradually proceed from what has until now been a reasoning and reflective tone to chapters with hard science and extensive presentations of the results of analysis. This is particularly noticeable in sections giving the results of attribute analyses.

A history of previous archaeological research

Large blade knives have been included in archaeological collections as long as collecting has been going on. In the accession catalogues of museums in southern Scandinavia, blade knives were registered among the very first finds. In the earliest work of systematizing objects and periods, then, they were already a part of the museum collections. They are often found in the large private collections of archaeological finds built up by collectors and private archaeologists in southern Scandinavia around the last turn of the century. The large blade knife is thus an object and an archaeological artefact which has always been physically present in collections. It is neither unknown nor newly discovered, figure 2.1.

Despite this, it is not a generally known artefact. When I discuss large blade knives of flint with other archaeologists, they sometimes have difficulty assigning them to the correct time and place. Although many recollect having seen large blade knives at an excavation or in some collection, they find it difficult to talk about them. They are often uncertain about dating, geographic distribution, function, or context. It seems as if most of them know the facts, yet not entirely. They date from the Bronze Age, don't they, or is it the Late Neolithic? I suppose they were used for harvesting leaves, or can they really be used for that? (This question is asked since the artefact is called *lövkniv* or "leaf-knife" in Swedish, and it is usually interpreted as having been used for cutting foliage as animal fodder.) I have been asked questions like this many times. The large blade knife is, quite simply, a type of artefact which, although it has been included in collections the whole time archaeology has been a profession, seems to have lingered in the shadows. Why is this?

Figure 2.2. Large blade knives of flint from the exhibition at Roskilde Museum, Denmark. The exhibition was set up during the 1980s. Since the museum the previous year had excavated and published several settlement sites from the Bronze Age (Sørensen 1980, 1984–85, see also below) the blade knives were classified as belonging to the Late Bronze Age. The label in the showcase reads: "Blade sickle of flint, c. 800–500 BC".

In this chapter I will investigate what knowledge is available about this artefact, what archaeologists have written about large blade knives, how this has changed over time, what values are expressed when blade knives are discussed, and what interpretations of the past have been made based on studies of large blade knives of flint.

By examining how the large blade knife has been handled in museum collections and publications I want to try to capture the state of archaeological knowledge about this artefact. The aim is to present a history of archaeological research in order to show how large blade knives have been and are perceived in the archaeological discourse, and to establish a platform of knowledge on which to formulate and clarify the questions I am working with. I begin by describing briefly how blade knives have been handled in museum collections. This is followed by a longer presentation and examination of publications from southern Sweden and Denmark and then a quick look at central Sweden, Norway, Poland, and northern Germany. The ambition is to capture trends over time and thereby to highlight urgent problem complexes. I make no claim to present a complete account of publications about flint assemblages from the Late Bronze Age.

Large blade knives of flint in museum collections

In the 1830s the head antiquarian in Sweden, Bror Emil Hildebrand, began to arrange the collections at the Museum of National Antiquities in Stockholm according to the three-period system: Stone Age, Bronze Age, and Iron Age. His successors, Hans Hildebrand and Oscar Montelius, continued this arrangement and also increased the collections considerably (Bergström & Edman 2005). The ambition was to concentrate a selection of representative prehistoric artefacts from the whole country in one and the same place. The effect of this is that the Museum of National Antiquities has a very large collection of stray finds. It contains an estimated 30,000 or so objects, consisting of shelves with around 1,900 drawers with finds sorted according to province and parish. It takes roughly a week to look through the entire collection in a summary manner. This I know because I have done it. I have pulled out all the drawers one by one and gone through the contents in search of large blade knives.

Why, one may wonder? Parts of the collection have long since been available via the Internet. Thanks to the free service "Search the collections" it is possible to find things by key word, year, chronological period, etc. Why did I not content myself with that?

The story of accession number "Stora Hammar 8651a" can illustrate why. The find consists of 40 large blade knives of flint, preforms for blade knives, and blade-like flakes. Apart from a brief note in a find list in a text by Mark Wyszomirski (1979), the find has never been published. If I search for the accession number in the database I learn that it consists of a hoard from the Stone Age containing 40 knives (single-edged), found in the late 19th century on the lands of Stora Hammar in Scania, Sweden. Nothing is said about large blade knives, nothing about the Bronze Age. There are countless single-edged knives in the collection. Grouped together under this key word are Mesolithic blade knives, Neolithic flake knives, and Bronze Age large blade knives of flint. In a database search it is totally impossible to separate these knives from each other. If I had contented myself with a search of the database via the Internet, I would never have found the large blade knives, the preforms of blade knives, and the blade-like flakes hidden behind accession number 8651a from Stora Hammar. This is nothing unique for the Museum of National Antiquities. With the exception of some collections, see figure 2.2, I have encountered a similar situation at many of the museums I have visited.

At the Historical Museum in Oslo, for example, there are no large blade knives registered in the database. When I spoke to Norwegian archaeologists, this type of artefact was completely unknown to them, and when I visited the museum it was impossible to find any large blade knives of flint in the collections. Yet in the permanent exhibitions I found some knives, or rather preforms of large blade knives. In a showcase displaying Neolithic votive finds there are three preforms of blade knives grouped together with two Neolithic blades and a Neolithic axe, figure 2.3. With the aid of the accession number I was able to locate a further two preforms of large blade knives in the museum collections and archival material about the find. The five preforms of large blade knives are registered under the search term "Svære Flintflekker" (large flint blades).

Figure 2.3. A showcase in the Historical Museum in Oslo. Here three performs of large blade knives are displayed together with two Neolithic blades, the showcase on the far left. The preforms of large blade knives are the two upper objects and the one at the bottom right, above the big Neolithic axe preform. A further two preforms of large blade knives belonging to the same find as those in the showcase were discovered in the museum collections. The text in the showcase reads: "Gifts to sacred powers during the Stone Age." The five finds in the showcase, that is, two Neolithic blades and three preforms of large blade knives, are dated "possibly 2400–1700 BC". (The preforms of large blade knives have find number 14675.)

The examples above are good illustrations of the problems of working with large blade knives of flint from museum collections. They are, quite simply, hard to find. They are sorted under a very large variety of search terms, and extremely few museums have sorted them under the term large blade knife of flint (in Swedish lövkniv), table 2.1.

Large blade knives are dated to the Late Bronze Age, and one's first thought might be to search for them in material sorted as belonging to this period. But since large blade knives are made of flint, older classifications put them together with Stone Age material. This is something to which Henrik Thrane drew attention more than twenty years ago (Thrane 1985:146):

Table 2.1. Some designations under which large blade knives of flint have been sorted in the collections of Lund University Historical Museum, together with the year of acquisition. However, the collections in the museum have been reclassified in recent years according to modern terminology. That makes this collection one of the few where it has been possible to use the search term large blade knife (in Swedish lövkniv) and obtain a large number of hits.

Accession number	Year of acquisition	Designation
1-8378	1855–1869	Cutting knife of flint
8929-10143	1876	Single-edged curved knife of flint
15002-15589	1905–1906	Sickle of flint

Figure 2.4. Photograph of the tool from Stenild, Jutland in Denmark. It is the only known find of a complete tool where a large blade knife is used as a knife blade. Since it was first published in Blinkenberg's (1898) text, the photograph has been reproduced in several different publications, and the find today is one of the most frequently cited when large blade knives are discussed. A recent photograph of the find is figure 1.1 in this book. Photograph: National Museum, Copenhagen.

"LBA [Late Bronze Age] flint is traditionally included in the Stone Age sections of the museum store-rooms so that they tend to be forgotten or ignored." And since the artefact deviates from other Stone Age material, it has been sorted under a great many names. So the situation in most of the museum collections I visited is that, if I search in the stores according to chronology, that is to say, the Bronze Age, I will not find the knives there since they have been sorted out of this material. If I search in the collections by material or artefact type, that is, flint or knife, I will not find them there since they drown in the multitude of finds and search words under which they have been sorted. The only thing that then remains to do is talk to the staff who work with the collections and search through the collections by oneself.

Publications – southern Sweden and Denmark

Southern Sweden and Denmark are the core area of my studies. This area has yielded an extensive amount of knapped flint from the Late Bronze Age, published from the end of the nineteenth century until the present day. I have chosen below sometimes to present individual works separately, sometimes several authors together. The reason for this is that certain texts require detailed accounts while others deal with similar questions. The survey starts with a work from 1898 and ends with some texts from the end of the 1990s. The choice of the 1990s as the end point is due to the fact that this was when I started my first studies of flint assemblages from the Late Bronze Age (Högberg 2001a, 2001b). From then onwards, I myself am part of what I want to describe and hence also part of the work presented in this study. Several texts about flint assemblages in the Late Bronze Age have been published in the twenty-first century (Högberg 2005; Högberg & Olausson 2005; Jensen 2006; Knarrström 2007). These are discussed in other parts of the book. The focus in the survey of the texts in this chapter is on how the large blade knives of flint are treated, how they have been evaluated, what questions have been raised in research, and how the blade knives have been interpreted.

Chr. Blinkenberg, Oscar Montelius, and Otto Rydbeck – three early publications

When the shafted large blade knife mentioned above was discovered in 1898 during digging for peat in a bog at Stenild in Jutland in Denmark, it was a sensation. Finds of complete prehistoric tools are rare in southern Scandinavia. The find was published by Chr. Blinkenberg (1898) just a few months after it was found, and as we have seen, it is still today the only known find of a complete tool with a knife blade in its shaft, figure 2.4.

Blinkenberg's article presented finds of shafted Stone Age tools. He thus dated the large blade knife of flint from Stenild to the Stone Age (Blinkenberg 1898:145): "The flint blade that was found sitting in the shaft is of a well-known form: it is one of the Late Stone Age's common blade knives [*flækkeknive*] ..." Apart from describing the tool in detail, Blinkenberg considers the function and defines the tool as a sickle. He also raises the question whether "Stone Age people" were right-handed as we

Table 2.2. List of authors, with year of publication, whom I know to have published the same photograph of the tool from Stenild, a photograph from Blinkenberg's article (1898:143, fig. 9). Blinkenberg originally published three photographs of the tool, one on its broad side with the tool lying (1898:143, fig. 9), one on the narrow side with the tool standing (1898:143, fig. 10), and one showing a close-up of the shaft head without the blade (1898:145, fig. 11). The publications below reproduce only one of these, the first one. Of all the publications I know which have published photographs of the Stenild find, it is only Steensberg (1943) who did not use the same photograph as Blinkenberg. Lindälv (1964) has published a photograph showing a reconstruction of the Stenild find, but the caption does not state that the find in the picture in his publication is a reconstruction. The text rather gives the impression that the picture shows the original find. We can read the following: "Shafted flint blade from the Stone Age; find from Stenild near Hobro in Jutland" (Lindälv 1964:58).

Hatt	1937
Broholm	1953
Wyszomirski	1979
Ebbesen	1981
Jensen	1990, 1994, 2006
Vang Petersen	1993
Nielsen	2005

are now. He believes that the find from Stenild shows that this was the case (Blinkenberg 1898:155f).

Through time Blinkenberg's article has become one of the most frequently cited texts in studies of large blade knives. Even today there is hardly any publication about blade knives that does not mention the find from Stenild (see e.g. Jensen 2006), see table 2.2. The reason for this is obviously the unique character of the find itself, but also the usefulness of Blinkenberg's matter-of-fact and well-formulated description of the tool and the circumstances of the find. In his text some questions crystallize which, as we shall see below, have persisted into our own times. These questions concern dating, function, and the discussion of right- and left-handedness.

The find is now living a life of its own. I do not know how many more or less exact copies of the tool from Stenild I have seen when visiting museum collections, exhibitions, and reconstructed Bronze Age settings, figure 2.5. In this way the find has become the standard model for a shafted large blade knife.

When Oscar Montelius published his study *Minnen från vår forntid* in 1917, it was as a continuation of the work he had started more than 40 years earlier with the publication of the book *Svenska fornsaker* (Montelius 1917). *Minnen från vår forntid* is a large catalogue with elegant drawings of finds from the Stone Age and Bronze Age. A large blade knife is depicted (Montelius 1917:39). It is described together with Neolithic blades as an artefact from the Neolithic (Montelius 1917:22, catalogue numbers 570–572). It is thus clear that Montelius did not regard large blade knives of flint as a specific type of artefact but as a variant of Neolithic blades and blade knives. This is an outlook that, as we shall see later on, survived until quite recently (Lindälv 1964, 1967).

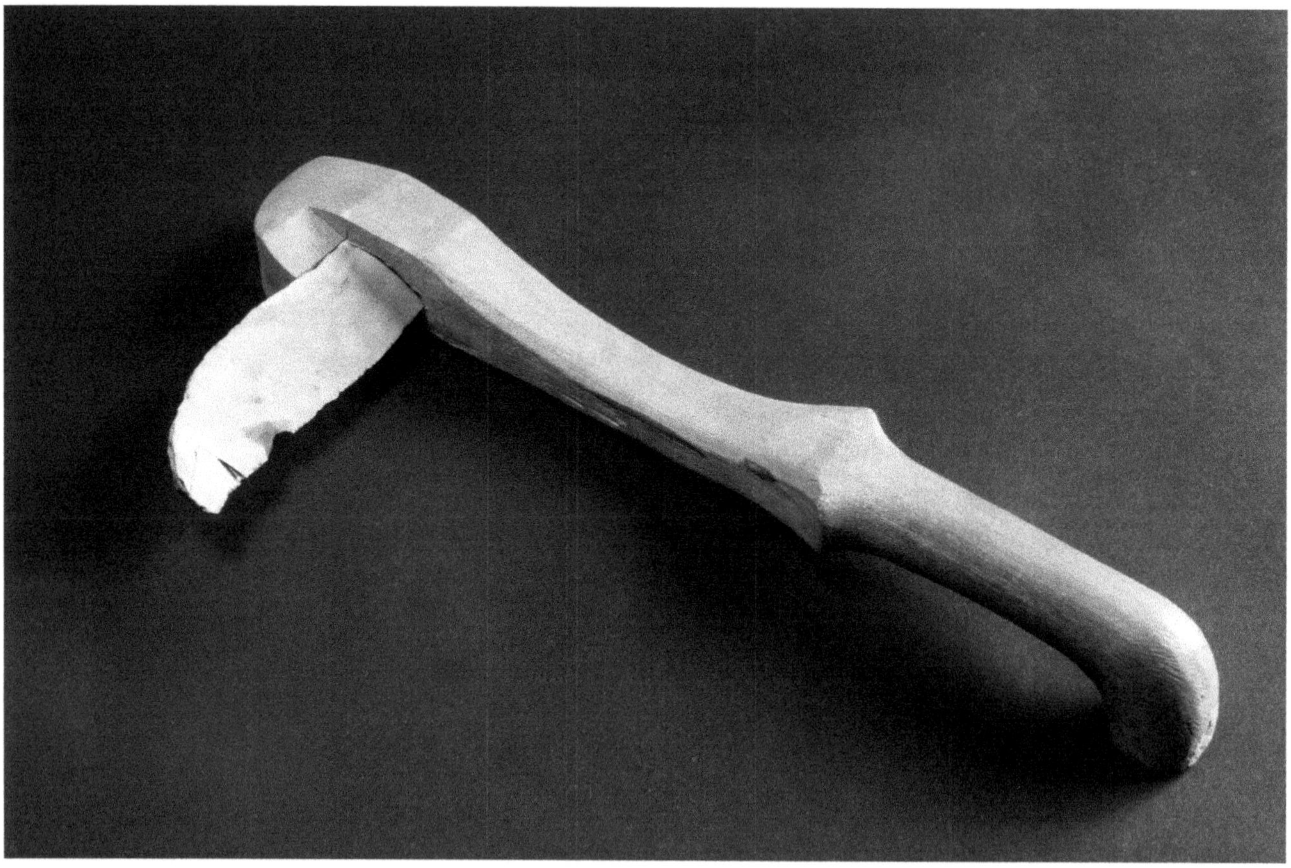

Figure 2.5. Reconstruction of the find from Stenild from the Lund University Historical Museum collections, catalogue number LUHM 9162. Photograph: Bengt Almgren.

In connection with the move to new premises of Lund University Historical Museum in Scania in southern Sweden, a book was published to mark the move. It included an article written by Otto Rydbeck (1918). The article presents closed Stone Age finds from dry ground and bogs in the museum's collections. In the text large blade knives are referred to as "sickles of flint" and described together with finds from the Neolithic (the Stone Cist period). The large blade knives are not discussed separately, but only in relation to other flint artefacts: "Since in the Stone Cist period the axe finds are in large measure succeeded by finds containing other tools, it seems probable that the sickles belong to this period" (Rydbeck 1918:32). It is obvious that Rydbeck, just like Blinkenberg twenty years previously, believed that flint artefacts of this type should automatically be placed in the Stone Age.

Müller – studies of flint tools from the Late Bronze Age are constituted

When Sophus Müller in 1919 published his study *Bopladsfund fra Bronzealderen* it was the first systematic comparative analysis of settlement site finds from the Late Bronze Age. It differs significantly from Blinkenberg's and Rydbeck's texts. This is partly because the authors were working with different find materials, and they had different levels of ambition. Whereas both Blinkenberg and Rydbeck discuss unique or closed finds, Müller's work focuses on settlement site material (Müller 1919:36):

> It could easily be supposed that this remarkable age, which to such a large extent utilized the foreign alloyed metal imported from far away, were not content with this but also had to use indigenous resources. But it is only the settlement site finds that demonstrate the extent to which and the way in which the use of flint and other stone was continued through time.

Müller bases his work on a number of Danish settlement site finds, see table 2.3. He notes that from all of these places there is a very large number of flint tools and flint waste. The starting point for his source-critical analysis is the question whether this flint material is from the Stone Age, which had previously – before his study – been taken for granted, or if the material came from the Late Bronze Age, and if so Müller thought that this had to be proved by argumentation, since flint tools had previously been assumed to date from the Stone Age.

Müller presented a thorough and comprehensive study. By means of comparative analyses of the morphology of the artefacts, the stratigraphy of the sites, and find contexts, he found that the vast majority of the flint material from these places should be dated to the Late Bronze Age. He also observed that the few finds of flint axes and flint daggers in the material should be perceived as older artefacts in younger contexts.

Large blade knives of flint, designated as *flækkeknive*, are defined as an artefact from the Late Bronze Age. They are described as being big and coarsely shaped, with a long edge, a

Table 2.3. Settlement sites from the Late Bronze Age with finds of large blade knives which Müller (1919) presented in his study, here together with data on the year of excavation and the number of large blade knives and fragments of blade knives found.

Excavation site	Year of excavation	Large blade knives, whole	Large blade knives, fragments
Voldtofte	1909–11 & 1915–16	3	2
Hasmark	1913–14	7	Many
Troldting near Bulbjerg	1913	26	17
Abbetved	1915–18	20	17
Dyrehavegaard, Nyborg	No data	13	Some

flat rounded back, rounding off towards the tip, and curving inwards at the opposite end to enable shafting. The length is 10–15 cm. Some knives are shaped all the way along the back and the tip and have shafting retouch, while others have less shaping and lighter retouch. He also mentions that in the find from Abbetved, see table 2.3 above, there are blades that are not modified, that is, not retouched along the back or for shafting, but show a damaged and worn edge and therefore deserve to be regarded as tools.

Müller set his conclusion that the large blade knives are from the Late Bronze Age in relation to Blinkenberg's (1898) dating of the find from Stenild and thought that the dating needed to be revised (Müller 1919:83): "Since the finds treated here have shown that the tool was also regularly used in the Bronze Age, the Stenild sickle can no longer be assigned with certainty to an earlier time." The formulation "no longer ... with certainty" nevertheless hints at some ambiguity. Müller does not seem to want to date the large blade knives of flint categorically to the Late Bronze Age alone. The knives he analysed were in his opinion clearly from the Late Bronze Age. Yet he seems to believe that the type – that is, the blade sickle or knife – also existed during the Stone Age. Müller is somewhat contradictory, however, since he writes a few pages later in his publication that one result of his study is that large blade knives of flint should be regarded as typical of Bronze Age settlement site finds (Muller 1919:89). This uncertainty about dating, as we shall se later on, is a running theme of publications right up to the present day (Becker 1990; Björhem & Säfvestad 1993).

Müller's study is well done and, just like Blinkenberg's (1898), frequently cited in later Danish research on the Bronze Age (see e.g. Ethelberg 1982). Müller formulated and treated important questions about chronology, function, find contexts, and spatial distribution. He described the appearance of large blade knives and the contexts in which they were found. He observed that flint tools seem to have been of great significance for everyday life in the Late Bronze Age (Müller 1919:81).

This is interesting. As early as 1919 it was clearly established that flint was used as a raw material in the Late Bronze Age. Yet, as we shall see below, this is not reflected in any general awareness among archaeologists during the twentieth century. It is also interesting that only one year elapsed between the article by Rydbeck presented above (1918) and Müller's article. They must have been writing their texts at virtually the same time but arrived at such different conclusions about the dating of the large blade knives.

Müller's study was pioneering and set a standard, so much so that Thrane, just over sixty years later, noted that not very much had happened as regards new research on flint assemblages in the Late Bronze Age since Müller had published his work in 1919 (Thrane 1985:148).

Oldeberg, Sandklef, and Hatt – early remarks about function

In the 1930s a number of texts were published, all discussing the functions of blade knives and bifacial sickles. The texts focus on prehistoric agriculture and discuss the reaping of cereal crops and the harvesting of leaves as animal fodder.

In 1932 Andreas Oldeberg published a study of bifacial flint sickles. The aim of the article was to argue that these artefacts were sickles and not saws, as they were believed at that time to be. In the study Oldeberg mentions the Stenild find together with Neolithic blades and knives (Oldeberg 1932:214f). Oldeberg gives a rather confused description of the find, stating incorrect dimensions and describing the edge as being retouched. He also refers to a discussion claiming that the knife blade in the Stenild find is perhaps not contemporary with the shaft (Oldeberg 1932:214): "In recent times some doubt has been expressed as to whether the flint flake is actually contemporary with the handle." I have not encountered this doubt anywhere else, nor have I seen it cited in other publications. Oldeberg does not say where this claim comes from.

As regards the function, Oldeberg dismisses the idea that large blade knives were used as sickles. Instead he thinks that they would have been suitable as meat knives (Oldeberg 1932:214): "they are much too heavy and clumsy to have performed efficient service as sickles ... but they would be extremely handy as meat knives." This statement is not based on any analysis, but solely on an opinion about the function of the knives in relation to their appearance. Oldeberg also has a rather long discussion of flint knives in hoard finds, suggesting that they may have been sacrificed in honour of "a corn-mother" (Oldeberg 1932:215). In this discussion too, he regards the large blade knives and Neolithic blades as one and the same type of artefact. Oldeberg also discusses bifacial sickles, arguing that they were not saws but were used as sickles.

Two years later, Albert Sandklef (1934) published a response to Oldeberg's article. In the text Sandklef argues that the sickles were not used as sickles at all, but as knives for harvesting leaves for animal fodder, as indicated by the Swedish name of the artefact – *lövknivar* (Sandklef 1934:285). Just like

Oldeberg, Sandklef regards the large blade knives and Neolithic blades as one and the same type of artefact, and just like Oldeberg he mentions the Stenild find as an example. Sandklef dismisses Oldeberg's theory of a harvest sacrifice as "totally improbable" (Sandklef 1934:289).

Five years after Oldeberg's text (1932), Gudmund Hatt (1937) published the book on prehistoric Danish agriculture, *Landbrug i Danmarks oldtid*. In it he cites Blinkenberg's (1898) interpretation and dating of the Stenild find. He also brings up Oldeberg's discussion of harvest offerings to the gods (Hatt 1937:68). In this publication, just as in Sandklef's (1934) text, the Swedish term *lövkniv* is used, not as an identification of the function of the large blade knives of flint but in a discussion of Neolithic bifacial sickles (Hatt 1937:69).

None of these studies connect with what Müller (1919) wrote, and all of these authors accept that large blade knives and Neolithic blades knives are one and the same artefact. Although the texts focus on bifacial sickles, they also mention large blade knives of flint and the Stenild find. It is interesting to note that the term *lövkniv* and the function of harvesting leaves are emphatically asserted in the texts, but in connection with descriptions of bifacial sickles, not of blade knives. It is clear that there was a discussion of terms and function in the 1930s and that previous classifications were no longer considered suitable. It is also clear that there was no consensus in this discussion. For example, P. V. Glob, in a publication from 1951, states that bifacial sickles were used as tools for harvesting grain and foliage (Glob 1951:32). The discussion of harvest sacrifices that the texts bring up is something I have not seen published anywhere else.

The question is whether these articles together have given rise to the Swedish term *lövkniv*, at first in a discussion of bifacial sickles but subsequently as a general term for knife blades which were interpreted as having been used for harvesting leaves. As we shall see later on, the term *lövkniv* was to come back in a publication by Brøndsted (1939, 1958) a few years later, then as a designation for both bifacial sickles and large blade knives of flint.

Steensberg, Glob, Broholm, and Brøndsted – comparative assessments enter the picture

In 1943 Axel Steensberg published a study of prehistoric harvesting tools. It is based on extensive experiments and artefact studies. In his text Steensberg argues that Neolithic blade knives and large blade knives of flint are one and the same type of artefact and that they should mainly be dated to the Neolithic. He discusses the Stenild find and dates it to the Early Bronze Age. The argument for this dating invokes analogies with sickles from southern Europe and Egypt (Steensberg 1943:74):

> That this implement cannot date from the Late Stone Age is to be seen on comparing with other sickles found in Europe and Egypt, as corn sickles proper, with an angle between edge and handle, do not occur at the transition to the Metal Age, whereas the characteristic upward bend of the handle from the plane of the blade occurs first in Southern Europe in the Copper Age and is not definitively demonstrated until the Early Bronze Age.

Steensberg also believes that the fact that the Stenild find has a knife blade made of flint in the shaft and not a knife of bronze should be interpreted in economic terms; flint was used when bronze was not available (Steensberg 1943:74). This reasoning is clarified in a description of a trial involving shafting and reaping (Steensberg 1943:160):

> It then occurred to me to make … the handle … [and] … the head on the model of the Stenild sickle. Actually it was natural to think that this implement, datable as it is to the Bronze Age, had been created in the likeness of the bronze sickle – that it was merely a simpler and cheaper process to place a flint flake in the haft than the usual knob sickle of bronze.

Two assessments of flint are worth noting here. Flint is a surrogate for bronze, and the Stenild shaft is believed to have been made for bronze sickles, not flint sickles. This is despite the glaringly obvious fact that it was found with a flint blade which was held firmly in the shaft with the aid of wooden wedges. Steensberg does not refer to Müller's (1919) study and discussion of the dating.

Over twenty years after Müller, Peter V. Glob (1938) published a study of stone objects from the Bronze Age. The focus is on axes and clubs of groundstone, but one section is devoted to flint assemblages of the Bronze Age. Glob gives a detailed description of how he regards this find material (Glob 1938:61):

> In the Late Bronze Age flint technology was very primitive. Only things made from knapped discs and flakes were used now. The large settlement site finds give detailed information about the extent to which flint was still used. The careful flaking technique of the Stone Age was almost totally lost. Discs and flakes are coarse and almost always have remains of the flint's natural cortex on the surface. The lack of blade cores in the finds clearly shows that the technique was very different from that of the Stone Age. Retouching no longer involves fine flaking, but coarse crushing and chopping. Flint is now used only for rather simple tools such as knives, scrapers, and awls. Bearing in mind the very different flint technology of the Stone Age and the Late Bronze Age in examining the flint sickle from Stenild, it seems that it can best be assigned to the Late Bronze Age. In the Stone Age they would scarcely have used such a poor flake as that of the Stenild sickle; in the Late Bronze Age, however, they could not do any better.

Glob adds nothing new as regards the description of large blade knives and flint assemblages from the Late Bronze Age. He says what Müller had already said. What is new is the comparative evaluations of flint technology in relation to Stone Age flint material. In Glob's text we get not only a description of the large blade knives themselves but also a description and assessment of them in relation to something else.

This is an outlook that was to be great significance for how flint material from the Late Bronze Age would be assessed by archaeologists, and Glob's text can be viewed as a norm for the view of flint assemblages at this time and for a few decades to come. Gone is Müller's nuanced and descriptive attitude, replaced by an evaluative and polarizing description, the essence of which is that the flint tools and the technology are primitive, that it is not just coarse but coarse in relation to the refinement of the Stone Age, and that something was lost during the Late Bronze Age. Glob even thinks that they would not have used such bad flint tools in the Stone Age, but in the Late Bronze Age they knew no better.

This value judgement recurs in the survey of Danish prehistory, *Danmarks Oldtid*, published by Johannes Brøndsted in 1939 (1958), and the survey of the Bronze Age, *Danmarks Bronzealder*, published by H. C. Broholm in 1949. Both refer to Müller's study and, just like Glob (1938), do not contribute any new knowledge about flint assemblages from the Late Bronze Age. Broholm, however, presents some new settlement site material from Kornerup in Zealand in Denmark (Broholm 1949:180). Both publications have graphic descriptions of flint assemblages from the Late Bronze Age:

> one is amazed to see that Bronze Age people adopted only a few types, and simple ones at that, of the diverse beautiful and practical flint tools that Neolithic people had developed ... but it is even odder that the highly developed flaking technique that the Stone Age mastered had fallen into oblivion. ... Nor was any care expended on the coarse blade knives; they seem to have only made the effort to obtain a sufficiently long and powerful edge, and the similarity between these blade knives and the one in the grain sickle from Stenild suggests that the coarse blade knives occurring on Bronze Age settlement sites must have been used as reaping tools (Broholm 1949:196).

> Settlement sites from the Late Bronze Age provide the important information for cultural history that the old flint technology still survived in simplified, primitive form and was used for certain tools necessary for everyday life, ... not least of course where poverty or remoteness obstructed access to metal on any scale. People had to have cutting and scraping tools whether metal was available or not, and so they continued with flint. ... Bronze Age flint scrapers are as a rule large and coarse, made of round or oblong discs or broad flakes. The knife is the old flake knife, it too often thick and coarse, and undoubtedly still used, when nothing better was available, as a leaf knife and grain sickle (Brøndsted 1958:257).

Several reflections can be made on the basis of the quotations above. In the texts the use of flint in the Late Bronze Age is regarded as an old flint technology that survived in simplified and primitive form. The large blade knife from the Late Bronze Age was the old blade knife. These formulations reveal a perspective of evolution and recession. The technology is old and has degenerated into a simplified and primitive form, and it is "the old flint knife" that lives on.

Flint craft in the Late Bronze Age is thus defined as a lingering echo of the quality craft of an earlier age. It is clear that the opinion is that the thorough skill in handling flint displayed by flintsmiths in the Stone Age has been replaced by something coarser which can be compared to a neglected and almost forgotten craft, albeit one that has not totally ceased. Here we also find the assessment that large blade knives of flint are not standardized. This specific observation about the manufacture of large blade knives has not been mentioned in any previous literature. The observation is not founded on any analysis, however, but on an emotional opinion. Flint tools from the Late Bronze Age are perceived as something used in poor circumstances and in the absence of metal, simply because nothing better could be had.

These four publications (Glob 1938; Steensberg 1943; Broholm 1949; Brøndsted 1958) express an evaluative classification of the flint finds from the Late Bronze Age which is not derived from earlier publications. It is new for its time. As we shall see, these assessments were to have consequences in that they live on to some extent today.

Strömberg, Lindälv, and Gregersen – some brief notices

In 1952 Märta Strömberg published an article about the objects in the private collection that the farmer Per Knutsson had built up on his farm in Trolle-Ljungby Parish in north-east Scania, Sweden. In the collection there were (I use the past tense because the collection was destroyed in a fire a few years after Strömberg studied it) several large blade knives of flint which Strömberg dated to the Late Neolithic (Strömberg 1952:88). This is a somewhat remarkable dating in view of the fact that the texts by Müller (1919), Glob (1938), and Broholm (1949) were all available at the time, texts to which Strömberg referred in other contexts (Strömberg 1954). Perhaps Strömberg was inspired more by Rydbeck's (1918) article and took the information about the dating from there.

The dating to the Stone Age recurs again in a publication by Elof Lindälv from 1964. Lindälv, who was a passionate private archaeologist and later honorary doctor at Gothenburg University, presented a small study of hoard finds from the Varberg district in Halland in south-west Sweden. In the study he discusses a find of 24 Middle Neolithic blades. Lindälv links these blades to the find from Stenild which Blinkenberg (1898) had published, even though the blades had been discovered in a hoard with four axes and two chisels dated typologically to the Middle Neolithic (Lindälv 1964:58). Judging by the photograph in the article, the blades should also be regarded as Middle Neolithic. They are not Late Bronze Age large blade knives of flint. This is yet another distinct example of the same confusion of two different types of blade knives which can be followed from Blinkenberg (1898) and Montelius (1917), and which was expressed by scholars like Oldeberg (1932), Sandklef (1934), and Hatt (1937) thirty years before Lindälv. A big hoard of large blade knives of flint is dated by Lindälv to the Late Neolithic, with the addition that such tools of flint were no doubt also used during the Bronze Age (Lindälv 1964:60). Three years later, Lindälv published a book in which he repeated these datings (Lindälv 1967).

Another private archaeologist who was active at this time was Søren Gregersen from Denmark. In a brief note in the Danish journal *Skalk* in 1960 Gregersen appeals to other collectors and hobby surveyors for information about finds of large blade knives. Gregersen mentions that he has collected more than 1,700 large blade knives within a limited area near Tommestrup at Stevns, Denmark. This is a huge number of knives and, as Gregersen also notes, it is more than the total number of finds of similar knives from all the then known settlement sites. He suggests that they might not only be settlement site remains, wondering whether this could be a factory (Gregersen 1960:29). Gregersen regards the large blade knives as a key artefact for Late Bronze Age settlement sites.

Gregersen gives an indication, for the first time in published studies, of the occurrence of production places for large blade knives of flint. This is something new, and despite the fact that the text is only a brief note, with no analysis, it is regularly referred to in later publications (see e.g. Björhem & Säfvestad 1993:77). Despite this, no one until now has analysed and published the material described by Gregersen; see the chapter "Special manufacturing sites for large blade knives" later on.

On the subject of these three brief notes, it is interesting to observe that in neither Strömberg's (1952) nor Lindälv's (1964) texts have the Danish studies of flint assemblages from the Late Bronze Age published since Blinkenberg (1898) been used as references, whereas Gregersen (1960) uses them. This points up a difference between Danish and Swedish publications. Danish archaeologists since the start of the twentieth century have dated large blade knives of flint to the Late Bronze Age, with a more or less hesitant reservation that they might also have occurred in earlier periods (Müller 1919; Glob 1938; Broholm 1949; Brøndsted 1958). Since the start of the twentieth century Swedish archaeologists have dated these large blade knives to the Late Neolithic (Rydbeck 1918; Strömberg 1952; Lindälv 1964). In the south Scandinavian archaeological view of large blade knives in the 1960s, then, there are two different opinions about the dating: the Danish view based on detailed studies of settlement site material and the Swedish view based on studies of finds from other contexts than settlement sites. There are exceptions. As mentioned above, for example, Hatt (1937:66) in his book *Landbrug i Danmarks Oldtid* refers to Blinkenberg's (1898) examination of the find from Stenild and its dating to the Stone Age without mentioning Müller's (1919) published results, and Strömberg in 1954 published a comprehensive study of Scanian settlement site material dated to the Bronze Age, emphasizing knapped flint as important find material.

Stjernquist and Thrane – settlement site material

At the start of the 1960s, extensive remains of settlement sites dated to the Late Bronze Age were investigated at Hököpinge in south-west Scania, Sweden. These have been published by Berta Stjernquist (1969). Questions on which the study focused included charting and dating, and the work reflects the efforts at the time to establish chronologies and typologies for Bronze Age settlement site remains (see e.g. Strömberg 1954). The finds are rich, consisting of pottery, flint objects, groundstone tools, organic material, bronzes, and residues of bronze casting. The flint assemblages are meticulously examined in text and image, and much effort is devoted to an attempt at typological classification. Stjernquist stresses the importance of not classifying flint objects from the Late Bronze Age in terms of typologies applied to Stone Age flint (Stjernquist 1969:84ff). She observes that there is a great difference between flints from the Early and the Late Bronze Age. These differences are formulated on the basis of typological observations (Stjernquist 1969:114ff). The type of evaluative comparisons previously published by Glob (1938), for example, is conspicuous by its absence in the study. Stjernquist does not mention large blade knives of flint. She seems to think that they do not exist in the finds from Hötofta. Judging by the many photographs published in the book, however, there are a number of broken large blade knives in the material (see e.g. Stjernquist 1969:105 & 109).

In 1971 Henrik Thrane (1971) published the results of excavations at Jyderup Skov in Zealand performed by the National Museum in Copenhagen. Since the middle of the 1960s, extensive remains from the Late Bronze Age have been examined here, and the site has been excavated with the ambition of a total investigation. The material is rich, with pottery, flint tools, fragment of bronzes, and finds from bronze casting. The flints are highlighted as a significant part of the material, and large blade knives of flint, designated as *knive* (knives) and *segle* (sickles), are defined as reaping tools (Thrane 1971:159). The flint assemblage is not evaluated in any way.

The interesting thing about these two publications is that they focus on chronology, typology, and overall settlement site structure, and that flint assemblages are part of this work. Another interesting detail is that the comparative assessments found in literature from previous decades have vanished.

Wyszomirski and Ebbesen – hoards in focus

With the exception of earlier brief notes (see e.g. Blinkenberg 1898 and Lindälv 1964), it was not until the years around 1980 that large blade knives in hoards were subjected to detailed discussion and analysis.

In 1979 Mark Wyszomirski published a study of a number of Swedish hoards of large blade knives. The study focused on typology and function. Wyszomirski divided the large blade knife as a type into two groups depending on whether the retouching was on the left or right side. This revives the discussion started by Blinkenberg (1898) as to whether the blade knives were retouched in relation to right-handedness and left-handedness. Based on use-wear analyses in a low-magnification microscope, Wyszomirski examined the function of large blade knives and came to the conclusion that they should be regarded as multifunctional tools, that is, they were used for many different tasks on many different kinds of material (Wyszomirski 1979:148). With reference to Müller (1919), he dates the knives no more precisely than to the Late Bronze Age, but with the reservation that they could also have existed in the Early Bronze Age (Wyszomirski 1979:149). Wyszomirski does not express any opinion on how the hoard finds should be interpreted.

What is new in Wyszomirski's study is his investigations of the function of large blade knives of flint. Whereas previous authors reasoned about the function in terms of the morphology of the knives, practical experiments, or shafting, Wyszomirski discusses the function based on the results of use-wear analyses. This was a new approach for the time in Scandinavian archaeology, as a consequence of which the article is frequently referred to in later publications (see e.g. Björhem & Säfvestad 1993 and Jensen 2006). What is interesting to note is that Wyszomirski's article is the first to publish analyses of a Swedish assemblage and with explicit references to Danish research.

In an article from 1981, Klaus Ebbesen discusses a number of Danish hoards. He dates the large blade knives in terms of their morphology, and by analogies with previously published settlement site finds. What makes his description and dating somewhat out of the ordinary is his belief that the finds belong to the Bronze Age because they so obviously do not come from the Stone Age. No Stone Age technology has such coarse platforms in his opinion (Ebbesen 1981:12). This echoes a similar opinion in Steensberg's (1943) study from forty years earlier. It is also in line with Glob's (1938) assessment of the flint technology from the Late Bronze Age, but it differs in that Ebbesen merely makes an observation, not an evaluative comparison. He states that this coarse technology can be found in settlement site material dated to the Late Bronze Age, that is, 1000–500 BC or periods IV, V, and VI.

He interprets the hoards as votive deposits and notes that the large blade knives of flint were not included in the material analysed in contexts where Bronze Age hoards of bronzes were discussed. This is an important observation which, as we shall see below, has not been followed up by any subsequent scholar. As regards the function of large blade knives, he finds this hard to specify; he thinks that they could possibly have been used for harvesting leaves in the late summer (Ebbesen 1981:17).

What was new about these two publications is that they specifically dealt with hoards of large blade knives. Both clearly demonstrate the existence of the phenomenon. None of them enters a discussion about possible interpretations of the functions and meanings of these hoards in the Late Bronze Age. Both studies regard the hoards in the light of the known Stone Age custom of depositing flint tools in bogs and wetlands.

The large settlement site excavations from the end of the 1970s until the middle of the 1990s

In the 1970s and the early 1980s a large number of archaeological excavations were conducted in Denmark and Sweden. They were occasioned by extensive infrastructure investments in combination with an increased awareness of the importance of investigating archaeological remains before they were destroyed by construction work (Rudebeck et al. 2001). As a result, many remains of settlement from the Late Bronze Age were discovered. Some of the excavations from this time have not been published in their entirety (for exceptions see e.g. Björhem & Säfvestad 1993 and Rudebeck & Ödman 2000), but a good number have been published in shorter articles and compilations. Many of these present and discuss finds of flint tools from the Late Bronze Age (see e.g. Ethelberg 1982; Sørensen 1980, 1984–85; Ödman 1992; Bertelsen et al. 1996). What these small articles and compilations have in common is that they present large and complex assemblages, in comparison to earlier studies with the exception of Müller (1919). The excavations from the 1970s onwards have generated a very large amount of finds. This has led to a more in-depth discussion of flint assemblages and large blade knives of flint from the Late Bronze Age.

Ethelberg – remains of settlement in south-west Denmark

At the end of the 1970s three sites with remains of settlement were excavated in southern Jutland and the island of Als, close the present-day German border. Several pits containing pottery and flint assemblages from the Late Bronze Age were found. The sites are called Grundtvigs Allé 1–4, Stensgård Vest, and Dybbøl Øst. Per Ethelberg (1982) presents and discusses these finds in an article. He begins the study by formulating a hypothesis (Ethelberg 1982:108): "Throughout most of the Bronze Age, bronze was primarily a status metal and secondarily a utility metal." The hypothesis is based on the assumption that, if bronze was appreciated more for its status than for its utility, this should be reflected in the use of other materials for tool making. Ethelberg thought that these materials could be organic, but also flint and groundstone. In the study Ethelberg referred to Müller (1919) (although he calls the Troldting site near Bulbjerg by the name Troldbjerg) and noted, just as Thrane (1985) would do a few years later, that not much had been investigated since Müller's study was published (Ethelberg 1982:108f).

The archaeological material from the excavations is dated with the aid of ceramic typology to Montelius periods V and VI. The flints from the different pits are presented under standard designations: cores, scrapers, edge flakes, blade knives, flake knives, flake awls, retouched pieces, blades, flakes, and debitage. The cores are defined as bipolar.

From Grundtvigs Allé 1–4 three large blade knives of flint are described, all of them broken, and from Stensgård Vest two blade knives, made from large, coarse blades. The large blade knives from both sites have clear traces of use wear, in this case macrodamage, on the edges (Ethelberg 1982:118ff). From Dybbøl Øst there are no large blade knives of flint. Ethelberg instead describes two small blade knives (Ethelberg 1982:121ff):

> One of them is made from a thin a-blade. The back retouch is fine and one-sided. There are traces of wear [in this case macrodamage] on the edge. The other knife is made from a thin flake and given oblique cross-retouch. The two knives are 6 and 5 mm thick respectively. … There are no large, coarse blade knives with alternating back retouch.

This is something new. Judging by the description and drawings, it is clear that there were no large blade knives of flint in the material from Dybbøl Øst; instead Ethelberg regards the smaller knives as being comparable to the Late Bronze Age large blade knives of flint. I have not encountered any such knives in previous studies. As we shall see later on, however, this is a recurrent feature at other settlement sites in Jutland (Draiby 1985).

Ethelberg characterizes the flint technology as coarse, although more in line with Müller's descriptive approach (1919) than with Glob's (1938) evaluative stance (Ethelberg 1982:118): "Most pieces have immense bulbs of percussion. Bulbar scars are very common. Occasional flakes show that bifacial knapping was not an unknown technique. However, there are no bifacial tools." Ethelberg notes that the majority of the flint tools are made of coarse flakes and blades. He thinks that flakes and blades should therefore be viewed as the most important units in tool manufacture (Ethelberg 1982:128). Yet he notes that people in the Late Bronze Age also produced smaller flakes and that occasional tools made of smaller flakes have been found. He himself qualifies the description he gives of flint assemblages as coarse (Ethelberg 1982:130): "Even if the technology is coarse, this does not mean that it is primitive ... people used the technique that was required for what they wanted to make." This attitude to flint tools from the Late Bronze Age is completely different from the one expressed by Glob (1938), Brøndsted (1958), and Broholm (1949) forty years earlier. Ethelberg also defines a reduction process by registering the occurrence of cortex on tools and flakes (Ethelberg 1982:129f): "On the majority of the scrapers, bits of cortex of varying size are still preserved. This rarely occurs on blade knives. This indicates that people first made discoid flakes, then blades, and finally the small fine flakes."

This is the first time that anyone has attempted to define a process of manufacture and reduction on the basis of an analysis of the flint material. Previous authors' views of manufacturing processes were based on opinion, not analysis. That all artefacts derive from one and the same manufacturing process, as Ethelberg believes, means that all objects were produced on the same site and from the same flint nodule. As we shall see below, it is not quite that simple, and interpretations of the reduction process are something that needs further investigation. This is evident not least from Gregersen's brief note in *Skalk*, see above, about a possible production site at Stevns in eastern Denmark (Gregersen 1960), to which Ethelberg does not refer.

As regards the function of large blade knives of flint, Ethelberg did not observe any gloss on the knives he analysed and therefore concluded that they were not used as reaping tools. He thought that reaping tools were made of bronze (Ethelberg 1982:130). This opinion is in line with the discussion of function in the 1930s that I summarized above (Oldeberg 1932; Sandklef 1934; Hatt 1937; Steensberg 1943).

What is interesting and new about Ethelberg's study is that he takes it for granted that flint assemblages can come from the Late Bronze Age, and that he bases his study on a hypothesis, that is, specific questions linked to the flint material. Although Ethelberg reaches the conclusion that his hypothesis did not pass the test, since he thinks that it was too simply formulated, the study is one of the few examples since Müller (1919) of anyone proposing a hypothesis and a question concerning flint material from excavations of settlement site remains from the Late Bronze Age. Also interesting is his nuanced view of flint assemblages from the Late Bronze Age which completely abandons the vocabulary introduced in the 1930s.

Yet another point worth noting is the consistent use of terminology and classification of the flints by key words. This is an example of the standard terminology which emerged at several museums of contract archaeology in Scandinavia, above all devised to classify material from the Stone Age (Helskog, Indrelid & Mikkelsen 1976; Andersson, Rex-Svensson & Wigforss 1978), a standard terminology that is still used today in a few variants (see e.g. Ballin 1996; Högberg et al. 2000). Defining flint assemblages from the Late Bronze Age with classification systems intended for flint tools from the Stone Age is, as we shall see below, not without complications (see also Stjernquist 1969).

What is also interesting and new is the descriptions of the flint objects, which differ from previous examples. This applies to the smaller knives from Dybbøl Øst, which Ethelberg regards as the same type of artefact as large blade knives of flint but not resembling them morphologically, and it applies to traces of bifacial and pressure retouching in the flint assemblage. The question is whether this represents a knapping tradition that is geographically confined to south-west Scandinavia.

Sørensen – excavations in Zealand

At the end of the 1970s and the start of the 1980s a number of excavations were conducted in the area around Roskilde on the island of Zealand in Denmark. At several sites there were pits with copious finds, dated by ceramic typology to the Late Bronze Age. Parts of this material are presented and discussed in two articles by Søren A. Sørensen (1980, 1984–85). Sørensen describes and dates the flints, and in his description of flint assemblages and find contexts he follows selected parts of previously published Danish studies of settlement site material, from Müller (1919) onwards. However, he does not refer to the article by Ethelberg (1982), despite the obvious similarity as regards the find circumstances and the material.

What is new in Sørensen's studies is the ambition to narrow the dating of the large blade knives of flint within the Late Bronze Age, and his suggestion for a chronological division of large blade knives. In a thorough analysis of find contexts, pottery, and the large blade knives from the many pits, together with comparisons with previously published settlement site finds, Sørensen clearly shows that large blade knives of flint occur together with pottery that is usually dated to the Late Bronze Age periods V and VI, see figure 2.2 (Sørensen 1980:27f). He also shows that large blade knives do not occur in period IV or in the Early Iron Age, and that bifacially knapped sickles and pressure flaked tools occur together with pottery that is dated to period IV of the Late Bronze Age (Sørensen 1984–85:60f). These are important findings. Few previous studies had so clearly placed the large blade knife of flint in the last two periods of the Late Bronze Age and so clearly demarcated the large blade knife chronologically from earlier bifacial tools. It is interesting to note the difference from the findings of Ethelberg (1982), who thought that bifacial technology and pressure flaking occur together with large blade knives of flint. Sørensen studied material from eastern Denmark whereas Ethelberg's material was from western Denmark, and the differences between their results show the importance of investigating flint assemblages from the Late Bronze Age in terms of hypotheses about possible geographical differences and similarities.

Sørensen proposes a typological evolution of the large blade knife from a shaft-retouched blade-like knife in period V to a back-retouched blade-like knife in period VI (Sørensen 1984–85:61). This proposed typological division has not gained general acceptance and, as we shall see later on, it has not been possible either to confirm it or refute it when it has been followed up in other studies (Björhem & Säfvestad 1993:77).

Björhem and Säfvestad – the study of large blade knives is stabilized

On the outskirts of Malmö in south-west Scania, Sweden, there were large-scale archaeological excavations in the late 1970s and early 1980s. The results were published in 1993 by Nils Björhem and Ulf Säfvestad in the book *Fosie IV, bebyggelse under brons- och järnålder*. This publication has become a standard reference work in southern Scandinavia settlement site archaeology. In it the authors present and analyse the huge amount of settlement site material from the Bronze Age and Iron Age yielded by the excavations. In 1983, just after the excavations had been completed, Björhem and Säfvestad (1983) published a summary description of the find material and their preliminary thoughts and hypotheses before they began the work that would result in the publication from 1993 (Björhem & Säfvestad 1983:17f):

> The flints mainly consist of coarse flakes but there are also plenty of tools such as awls and scrapers and the characteristic blade (leaf) knives (in Swedish *spån(löv)knivar*). The blade knives have a distinct tendency to appear only together with the later pottery. This could possibly suggest a greater need for leaves and grass for fodder when people were forced to stable certain animals during the winter after ecological/climatic changes at the end of the Bronze Age. Naturally, we know very little about this, and at present it can be no more than a bold guess. The find combination of large blade knives of flint and later pottery is nevertheless interesting to note.

Although this was written more as a hypothesis than as a result presentation, it clearly demonstrates how the authors at this time thought about large blade knives. They are blades. They occur together with later pottery. They were used to cut fodder for the animals in the form of leaves and grass.

This opinion seems to have made an impact on Swedish research in the 1980s. Two small notes from contract excavations in southern Scania refer unquestioningly to large blade knives of flint as having been used to cut leaves: "Finds of about twenty so-called leaf knives ... show that winter fodder was cut for domesticated animals" (Jonsson 1986:3); "At the end of the Bronze Age leaf knives of flint became common. They were used for gathering leaves as winter fodder for the animals" (*Arkeologi och naturgas i Skåne* 1985:19).

In the publication from 1993, Björhem and Säfvestad follow up the hypotheses from 1983. The archaeological material on which the studies are based is enormous. In total Björhem and Säfvestad present finds of more than 90 kilos of flint objects, of which some sixty objects are identified as being large blade knives of flint (Björhem & Säfvestad 1993:66ff). This is the highest single number of large blade knives analysed from one and the same excavation. The study has a solid foundation in previous research, and Björhem and Säfvestad (1993:72) observe that large blade knives are the only flint object hitherto identified as diagnostic for the Late Bronze Age in southern Scandinavia.

The focus is on typology, chronology, and manufacture. The aim is to demarcate types or variants and any chronological differences in the large number of blade knives. They date the large blade knives to phase B of the Late Bronze Age, see table 1.1 (Björhem & Säfvestad 1993:77). Through analyses of pottery from 59 features they were able to sort the material chronologically. Of 54 dated large blade knives, 53 were found in features with phase B pottery. One occurs in a feature with phase A pottery. They analyse previously published studies and note that the chronological pattern observed in the material from Fosie IV is generally valid for south-west Scania, Sweden: "To sum up, the large blade knives seem to appear mainly in periods V and VI of the Late Bronze Age, but the boundary backward in time is as yet difficult to ascertain" (Björhem & Säfvestad 1993:77).

They also write that the difference in typology presented by Sørensen (1984–85), see above, which he believes to be chronologically determined, cannot be demonstrated in the material from Fosie IV. The type of knife that Sørensen assigns to period V, the one with retouch only for shafting, is not found in the material that Björhem and Säfvestad examined. They ask whether this knife is a local Zealand variant or if its absence in the Malmö material might after all be due to chronological differences (Björhem & Säfvestad 1993:77).

With the ambition of investigating whether large blade knives of flint were standardized or not, the authors performed extensive metric and morphological analyses. They measured length, maximum width, and thickness, they described the shape of the distal part of the large blade knife and registered whether the retouch was on the right or the left side. This results in a division of large blade knives of flint into two types, type I and type II, figure 2.6.

In this context they mention Wyszomirski's (1979) typological division based on right or left retouch – 35% of the large blade knives from Fosie IV have the retouch on the right side – but they reject it having observed that right or left retouch is a variable that is independent of the other variables that interact in the classification into type I and type II.

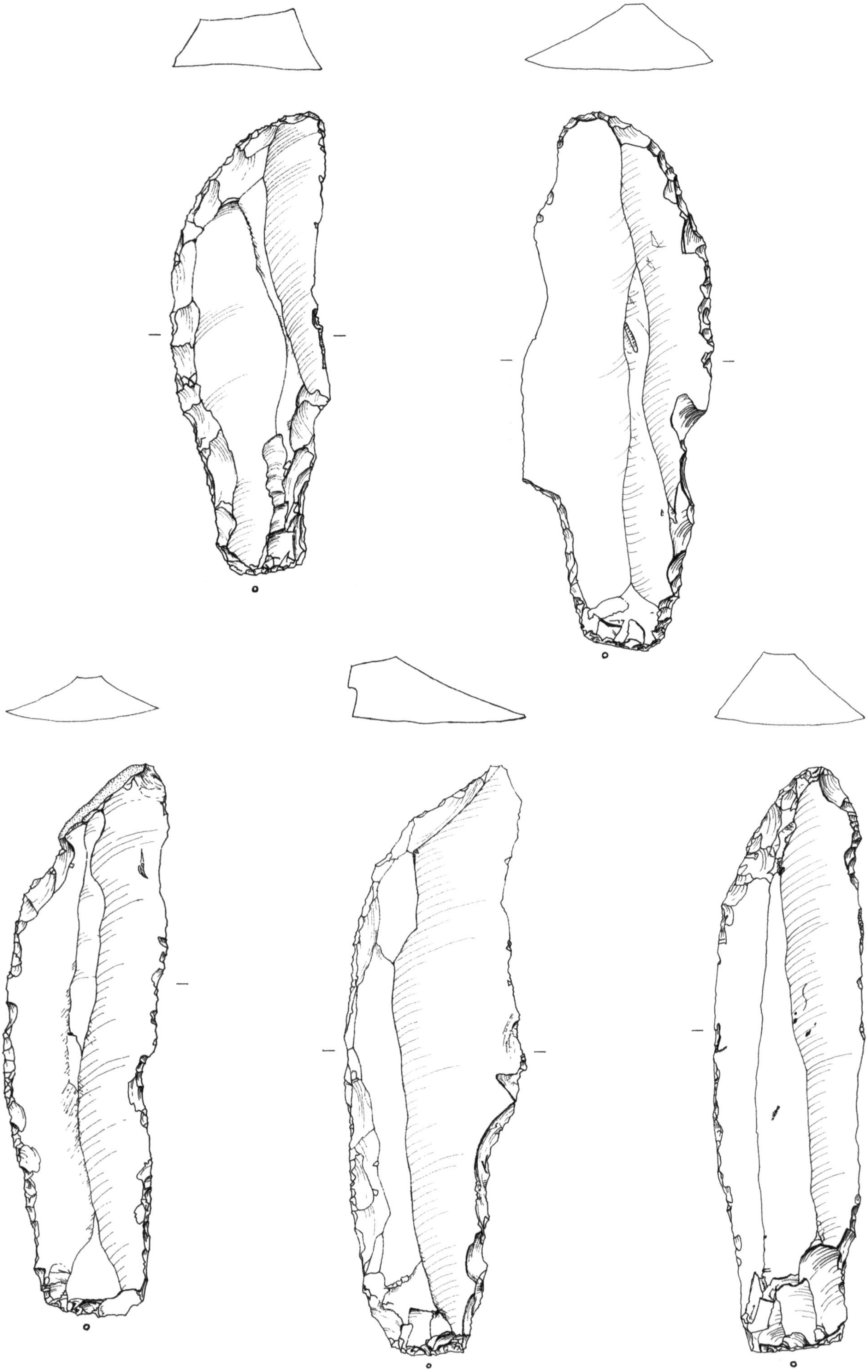

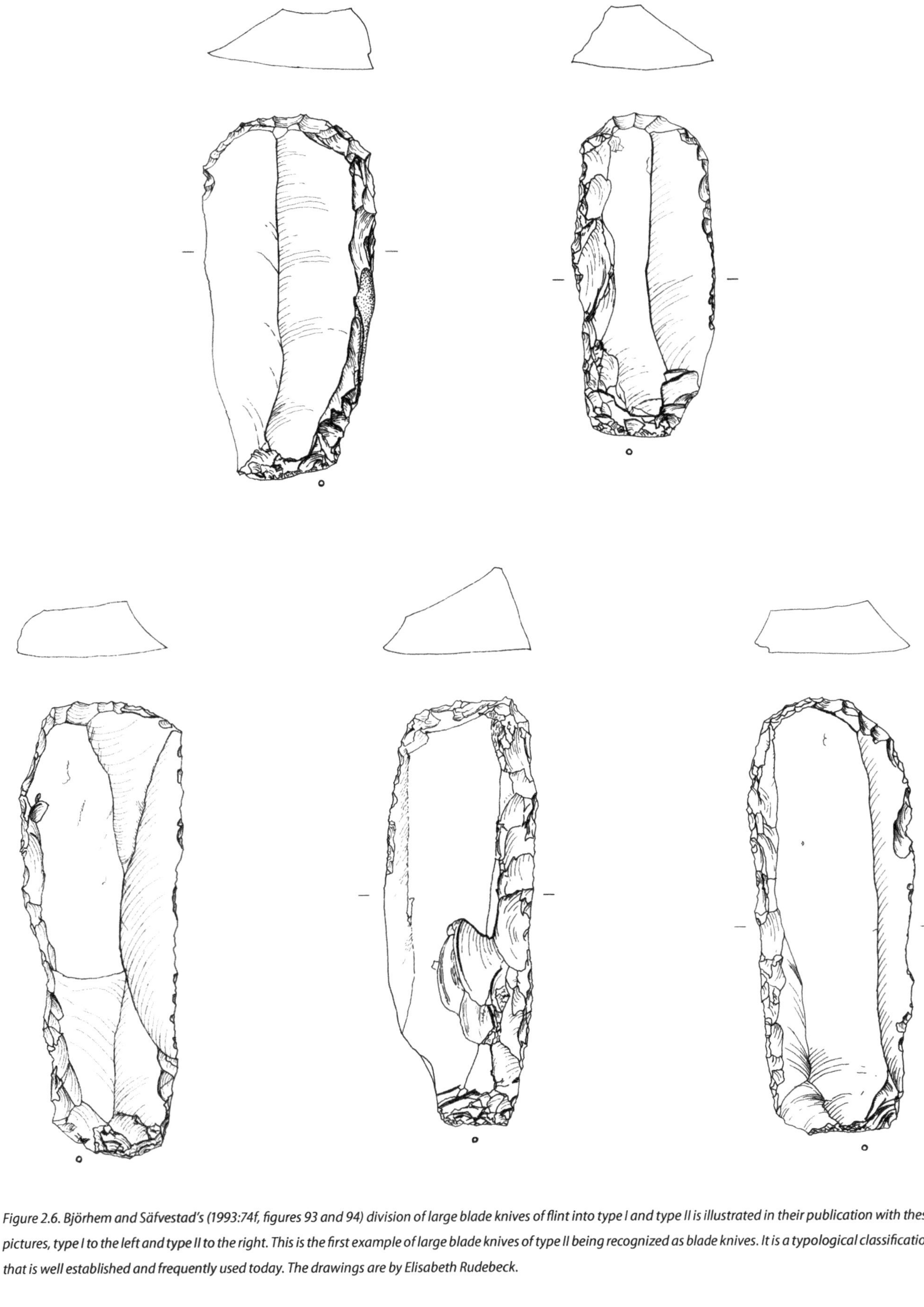

Figure 2.6. Björhem and Säfvestad's (1993:74f, figures 93 and 94) division of large blade knives of flint into type I and type II is illustrated in their publication with these pictures, type I to the left and type II to the right. This is the first example of large blade knives of type II being recognized as blade knives. It is a typological classification that is well established and frequently used today. The drawings are by Elisabeth Rudebeck.

Björhem and Säfvestad also test, with references to Ebbesen (1981), Wyszomirski (1979), and Blinkenberg (1898), whether right and left retouch may be connected to right- and left-handedness. They conclude that this is not the case, but they leave the question dangling by putting forward a number of hypotheses (Björhem & Säfvestad 1993:76): "One possibility is that the different placing of the retouch has no real practical significance, but was instead governed by various social/cultural factors that are difficult for us to assess today."

They also consider whether the large blade knives of flint were multifunctional tools, with reference to Wyszomirski (1979), and propound the hypothesis that right and left retouch perhaps should be connected to different functions. They also discuss the possibility that the placing of the retouch reflects different sets of knives that each user had.

As regards the production of large blade knives, Björhem and Säfvestad (1993:77) note that blade cores are lacking in the settlement site material which they studied and which includes large blade knives. The lack of blade cores, in their opinion, is evidence that large blade knives were not produced on the settlement sites. This view is diametrically opposed to Ethelberg's (1982:129f) interpretation of the reduction process as a continuous series of flaking whereby all the tools were manufactured at the settlement site. There is no reference to Ethelberg's study. One hypothesis put forward by Björhem and Säfvestad, with reference to Gregersen (1960), is that the large blade knives were produced centrally by specialized large blade knife knappers on special workshop sites.

Björhem and Säfvestad's study is systematic. It builds on well-formulated hypotheses, and the methods used to test these hypotheses are clearly described. With the hitherto largest number of large blade knives of flint from a single excavation, they had the opportunity to perform realistic comparative analyses. They present several new findings. The division of the large blade knives into types I and II was not previously noticed. They show that these two types not only look different, but were also manufactured from the beginning as two different types. This typological classification is firmly established and frequently used today (e.g. Eliasson & Kishonti 2003; Jensen 2006), although, as we shall see below, it is not heeded universally (Christensen 1996).

Through extensive comparative analyses, chiefly based on ceramic typology, they clearly demonstrate that large blade knives occur in what they define as phase B of the Late Bronze Age. This agrees with Sørensen's (1984–85) dating of large blade knives to periods V and VI of the Bronze Age. They ask important questions about manufacture, production sites, and function, questions which lead to well-formulated hypotheses for further work rather than to concrete results. What Björhem and Säfvestad do not investigate are interpretations of the Late Bronze Age based on the large blade knives. Their analyses stop at the examination of blade knives as an archaeological artefact. They do not proceed to interpretations of the large blade knife as a meaning-bearing object in the Bronze Age. Nor is that the aim of their studies (Björhem & Säfvestad 1983:25, 1993:1).

Björhem and Säfvestad's work can be defined as the work that has stabilized the study of large blade knives. By linking analyses of the extensive archaeological material to previous research, they set a clear framework for the typology and chronology of large blade knives of flint.

Bertelsen et al. 1996 – settlement sites in central and north-west Jutland

The end of the 1980s saw the start of the project "Bronze Age Settlement Sites in Viborg County". The aim of the project was to present the many, mostly unpublished, settlement site excavations performed in central and north-west Jutland, Denmark (Bertelsen et al. 1996:10). In the final publication from the project (Bertelsen et al. 1996) two chapters are devoted to flint assemblages. These sum up previously published thoughts about flint tools from the Late Bronze Age. In a chapter dealing with typological aspects of the flint material there is a discussion of large blade knives. These are defined as being made from long, broad, and often fairly thick blades. The retouch along one edge is emphasized as being characteristic and linked to the way the blade knife was used (Mikkelsen 1996:67): "the flat retouched section gave a place for the finger to rest on." This was the first time anyone published the opinion that large blade knives were held in the hand unshafted during use, although Steensberg had tested this experimentally in his studies from 1943. Other large blade knives, according to Mikkelsen, were knapped to shape at the proximal part in order to be shafted. He mentions that there are blades without retouch in the material but claims that they could very well have been used as knives. Several of the large blade knives of flint in the study have been subjected to use-wear analysis by high-magnification microscope, leading to the interpretation of the blade knives as multifunctional tools.

In the study several large blade knives of type II, as Björhem and Säfvestad (1993) defined this type, are classified as scrapers (Mikkelsen 1996:70). This is because these knives, according to the standard nomenclature for Stone Age assemblages, are morphologically defined as scrapers. The large blade knives of flint, which are thus defined in the study as scrapers, were analysed for use wear. The result shows that they were used as knives, which led to surprise that scrapers were used as cutting tools. This results in the interpretation that people in the Bronze Age had a more flexible attitude to the use of tools than people in the Stone Age (Christensen 1996:89).

The study is an example of how restrictive it can be to classify flint tools from the Late Bronze Age with the aid of nomenclature suited to flint tools from the Stone Age. In the light of Björhem and Säfvestad's (1993) study, to which no reference is made, the interpretation that there was a more flexible attitude to the use of tools in the Bronze Age than in the Stone Age is astounding. The study is also an example of how Danish and Swedish research has sometimes followed parallel trails without ever meeting; compare what we saw above regarding the publications by Rydbeck (1918), Strömberg (1952), and Lindälv (1964).

Major research excavations and publications from recent years

In the same period as these large-scale settlement site excavations there were also a number of research excavations and projects.

Draiby – Fragtrup in Vesthimmerland, northern Jutland, Denmark

An area at Fragtrup in northern Jutland, Denmark, was excavated in stages in the 1950s and 1960s. The remains consisted of two very well-preserved long-houses with clay floors and occupation layers, a couple of other houses, features with settlement site finds between the houses, and traces of bronze casting and pottery making. "The Fragtrup houses", as they have become known, are among the first Bronze Age houses to have been found in Scandinavia (Draiby 1985:127). The remains are complex and well investigated. After the excavations a great deal of the material was unpublished until Bente Draiby's article from 1985.

The remains are dated by ceramic typology and radiocarbon to an interval in the Late Bronze Age ranging from the start of period V to the start of period VI (Draiby 1985:201ff). Draiby devotes the major part of her text to an account and interpretation of the house remains and the pottery finds. Although flint assemblages were given low priority during the excavations, they were documented (Draiby 1985:205) and are treated as an important part of the archaeological material, with potential to be used for interpretations of activities on the excavated house sites. The flint assemblages are assessed in a nuanced way comparable to Müller's (1919) publication, to which Draiby also refers. The knapped flint is described as coarse, with a great deal of cortex preserved and with the minimum of working necessary for the purpose (Draiby 1985:205).

There are several knives in the material, but no large blade knife of flint (Draiby 1985:205ff). The knives are somewhat similar in form to large blade knives, but, as far as I can judge from the drawings in the publication, they are not like large blade knives as regards the technology. There are also a few bifacial knapped tools in the material. Here it is interesting to compare with Ethelberg's (1982) description of flint assemblages from a number of excavated settlement sites dated to the Late Bronze Age. Ethelberg also presents blade knives which do not look like large blade knives of flint but which he thinks were used in the same way. He also describes finds of bifacial technology and pressure flaking in the material. Similar observations have been made by Berit Valentin Eriksen in analyses of a large body of settlement site material from excavations at Bjerre in Thy, north-west Jutland, Denmark. Eriksen also describes knives which were used as large blade knives but do not look like them (Eriksen in press). As stated above, then, there is good reason to investigate whether there is a difference between the western and eastern parts of southern Scandinavia.

Becker and Jensen – a unique hoard find from Spjald

At Spjald in western Jutland in Denmark an area with remains of houses from the Late Bronze Age was excavated in 1969–1971. In the remains of a house foundation, more specifically in the hole for a roof-bearing post from house XII, there were six large blade knives of flint together with a flint scraper, the remains of two moulded splint boxes, a wooden vessel, pottery, and lumps of resin (Becker 1972, 1990). This is the only known excavated and published assembled find of large blade knives. (In his study from 1979 Wyszomirski mentions two hoards found by excavations at Löddeköpinge in south-west Scania, Sweden, but these are until now unpublished.)

The results from the excavations at Spjald have been published by Carl Johan Becker (1972), who in collaboration with a number of other scholars also published a special article dealing with the finds in the specific post-hole from house XII (Becker 1990). The remains from the site are broadly dated to the Late Bronze Age, 1000–500 BC (Becker 1990:195). The special article presents and discusses the material, the circumstances of the find, and the analyses of the objects.

Becker calls the tool a "schweres Klingenmesser aus Flint" (Becker 1990:196). With references to Müller (1919), Glob (1938), and Ebbesen (1981), he dates these blade knives to the Late Bronze Age, with no further specification. He also makes one reservation: that the form, i.e. the large blade knife of flint, could also appear in the archaeological material from the very rare settlement sites from period III (Becker 1990:200). He also draws attention to the confusion in the literature, where the large blade knives are assigned a range of datings from the Neolithic to the Late Bronze Age.

He describes the appearance of the large blade knives from Spjald in a way that agrees with earlier descriptions. He includes no value judgements in his description as Glob (1938) does, confining himself to a description in line with Müller's (1919). Becker observes that the find from Spjald is unusual in that he does not know of any similar find of large blade knives being deposited in a post-hole (Becker 1990:208). He discusses left and right retouch, with reference to Wyszomirski (1979) and Ebbesen (1981), and states that, if these authors are correct (a matter on which he does not take a stance), then there are blade knives for both left- and right-handed people, that is, for more than one person, in the find from Spjald (Becker 1990:201). He also considers the raw material, concluding that two of the blade knives are made of beach flint and one of Jutland flint with provenance about 100 km north of Spjald (Becker 1990:202).

As part of the special study, Helle Juel Jensen performed use-wear analyses in a high-magnification microscope (Jensen 1990). This was the first time the function of a well-documented find was analysed, and the first time large blade knives were analysed with high magnification. It is also the only study of this kind hitherto, but see Christensen (1996) above, and it is often cited (see e.g. Björhem & Säfvestad 1993; Jensen 2006). Jensen concludes that the large blade knives of flint from Spjald were used for cutting reeds (Jensen 1990:209f), and Becker interprets this to mean that they were used for collecting reeds for thatching (Becker 1990:201). This interpretation is repeated in later publications (Jensen 1994:81ff; Jensen 2002:349), and it is also followed up in a brief note in the journal *Skalk* by Christian Adamsen in 2002. He writes there about a private collection of at least twelve large blade knives, all collected in the south Jutland bog Køng/Lundby Mose (Adamsen 2002). Adamsen interprets these as coming from occasions when people in the Bronze Age cut reeds for thatching and dropped the knife blades by mistake.

Becker's and Jensen's studies have yielded new knowledge in several ways. The find in itself was new, of course, a hoard in a post-hole in the remains of a long-house. Also new is the fact that this was the first time large blade knives of flint were subjected to functional analysis with a high-magnification microscope. The results show that the large blade knives from Spjald were used for working with reeds, and Becker interprets this specifically to mean that the knives were used for collecting reeds for thatching houses. This was the first time that anyone so narrowly defined a function for large blade knives of flint. A third aspect that no one had considered before is the question of which type of raw material the large blade knives were made of. Becker had studied flint as raw material since the early 1950s (see e.g. Becker 1959, 1980a, 1980b, 1980c) and when the article was published he was thoroughly familiar with the different types of flint available in the studied area. His analysis shows that the large blade knives from Spjald were made of flint types taken from at least two different places.

Olausson – the Ystad Project

The start of the 1980s saw the launch of a large-scale project on landscape archaeology in Scania, southern Sweden, "The Cultural Landscape during 6000 Years in Southern Sweden – the Ystad Project" (Larsson, Callmer & Stjernquist 1992). In a special publication from the project Deborah Olausson has written about parts of her studies of the Bronze Age cultural landscape (Olausson 1992:251ff).

Large blade knives of flint are here called "pruning knives" and are dated to the Late Bronze Age (Olausson 1992:268). In an account of archaeological activities and observations in the area covered by the project, Olausson mentions details of an excavation conducted by Lund University Historical Museum in 1922 on lands belonging to the Ruuthsbo estate. The finds consist of pottery dated to periods V–VI and one large blade knife of flint (Olausson 1992:257). This excavation remained unpublished, apart from a brief note by Strömberg (1954). An interesting question is whether the previously mentioned difference between early Danish (e.g. Müller 1919) and Swedish (e.g. Strömberg 1952) studies as regards datings of large blade knives would have arisen if this excavation had been published.

Without adopting any standpoint on the function, Olausson mentions that large blade knives may have been used for reaping (Olausson 1992:268). She also refers to authors who think that the large blade knife was used for harvesting leaves (Olausson 1992:271). This is a good illustration of the uncertainty about the function of large blade knives of flint that prevailed at the time when the article was written. In a discussion of raw materials and trade, Olausson mentions bronze but not flint (Olausson 1992:274). This is something she shares with other authors (see e.g. Thrane 1975; Jensen 1981).

Some general surveys

Throughout the twentieth century, general surveys and collections of articles were published on the subject of prehistoric Sweden and Denmark. These publications changed over time. From having been geared to presenting finds (see e.g. Montelius 1917), to using the finds to tell about the prehistory of a country or region (Broholm 1949), to describing the past without finds (Stenberger 1964; Hagen 1967) and telling about the past and illustrating these narratives with finds. Some of the find-based surveys have been discussed above (Broholm 1949; Brøndsted 1958). But what is the situation in surveys published in recent years? I will discuss this briefly here using three books from Sweden (Sjöberg 1998; Welinder 1998; Burenhult 1999) and one from Denmark (Jensen 2002).

Stig Welinder, in his survey of Neolithic and Bronze Age agriculture, *Jordbrukets första femtusen år* (Welinder 1998), describes large blade knives of flint in words and pictures. He dates them to the Late Bronze Age. In functional terms they are said to be reaping sickles (Welinder 1998:146) and a drawing illustrating leaf cutting in the Late Bronze Age renders a copy of the Stenild find (Welinder 1998:204f). Large blade knives are thus viewed as tools for harvesting crops and leaves.

Göran Burenhult's books *Arkeologi i Norden* are surveys of Swedish archaeology (Burenhult 1999). In the part dealing with the Late Bronze Age he states that flint assemblages are part of the find material from settlement sites. With reference to Björhem and Säfvestad (1993) he mentions large blade knives, using a term meaning *blade-leaf-knives* of flint, and associates them with cutting leaves as winter fodder for stabled livestock (Burenhult 1999:26f).

A popular work from Gothenburg City Museum published in 1998 presents a large number of finds from the museum's collections and exhibitions (Sjöberg 1998:62f). A couple of pages are devoted to a hoard of large blade knives of flint found in Halland, Sweden. The hoard is described as a votive find. The large blade knives are characterized as a variant of the Neolithic bifacial sickles used for reaping or cutting leaves. They are dated to the Neolithic or the Bronze Age. This description of large blade knives of flint is familiar from Lindälv's (1964, 1967) publications from the 1960s, see above. Lindälv described the same find in the same way (Lindälv 1964:60), and it is obvious that no more recent research has been included in the narratives about this find.

In the comprehensive survey *Danmarks Oldtid*, the volume about the Early Bronze Age, Jørgen Jensen mentions large blade knives of flint in one sentence (Jensen 2002:349). He refers to Becker's (1990) study from Spjald and cites his interpretation of the large blade knives as tools for cutting reeds for thatching. The large blade knives are called *flintflækker* or flint blades.

A brief glance at central Sweden, Bornholm, northern Germany, Norway, and Poland

In the survey above I have confined myself geographically to southern Sweden and Denmark, the core area of my studies. Here I take a quick look both northwards and south across the Baltic Sea. I look briefly at how large blade knives of flint are designated in literature dealing with central Sweden, Bornholm, the Baltic coast of Germany, Norway and Poland.

Central Sweden

A number of large-scale excavations have been conducted in the Mälaren region. Several of these uncovered extensive finds from the Late Bronze Age, and sites like Hallunda (Jaanusson 1981), Apalle (Ullén 2003), and Pryssgården (Borna-Ahlkvist, Lindgren-Hertz & Stålbom 1998; Borna-Ahlkvist 2002) are well known in Bronze Age research. Although flint does not occur naturally in the ground in these parts of Sweden, several finds of knapped flint and flint objects have been discovered on these sites. No large blade knives, however, have been registered among the finds.

Bornholm

Finn Ole Nielsen has published a survey of the state of research on the Late Bronze Age in Bornholm. He thinks that the period has been neglected by scholars. There are no well-excavated settlement sites from this time (Nielsen 2005:115ff). In a context where hoard finds are discussed, he considers two finds of large blade knives of flint. These have previously been discussed by Ebbesen (1981). What is interesting about Nielsen's article is that the large blade knives are presented together with hoards of bronzes. This is new. The hoards are interpreted as "votive deposits", but this interpretation is not further elaborated (Nielsen 2005:126). The use of flint during the Late Bronze Age is explained in terms of the lack of bronze, since deliveries were heavily reduced around 700 BC (Nielsen 2005:122f). This is a view of flint as a substitute material that echoes publications from the 1930s.

Germany

From Mecklenburg-Vorpommern in northern Germany, material from a number of settlement sites dated to the latter part of the Late Bronze Age has been published. Some publications mention flint assemblages as significant parts of the find material (Lampe 1982; Auler 1999; Lütjens 2001), but few studies of flint assemblages with large blade knives from the Late Bronze Age have been published (Auler 1997). I have not found any publication dealing with large blade knives of flint from other parts of northern Germany.

Norway

As mentioned in the introduction to this chapter, the large blade knife of flint is an unknown artefact in Norwegian archaeology. No Norwegian archaeologist that I have spoken to knows anything about large blade knives from the Bronze Age. I have not found any text dealing with this artefact, although the texts treat both periods and archaeological contexts where large blade knives of flint could hypothetically be part of the find material. In a book by Øystein Kock Johansen (1993), for instance, the author examines hoards from the Bronze Age in Norway. A small chapter is devoted to tools of stone and flint. The study covers all the periods of the Bronze Age, but the only flint tools considered are daggers of type VI, that is, finds from the Early Bronze Age. Large blade knives of flint do not exist in the "awareness" of the study (Johansen 1993:67ff). There is no mention of large blade knives in Norwegian surveys (Hagen 1967), nor in works on nomenclature (Ballin 1996).

Poland

The large blade knife of flint is a known artefact in Poland. Examples have been published together with settlement site material (Moskwa 1976:277), and the objects have also been subjected to functional analysis (Kruk 1994). The distinctive thing about Poland is that flint mines from the Late Bronze Age have been investigated, demonstrating extensive production of large blade knives, known as Zele knives, resembling Scandinavian large blade knives (Lech & Lech 1997). This is the only known example of an excavated site of raw material extraction and production of large blade knives of flint from the Late Bronze Age.

Summing up and formulating the questions

This chapter has painted a picture of earlier opinions and studies of large blade knives of flint, together with a consideration of the questions asked about the large blade knives. The aim has been to show how large blade knives have been and still are perceived in the archaeological discourse, and to establish a platform of knowledge to use in order to formulate questions for my work.

The survey shows that studies of large blade knives of flint have been conducted for more than a hundred years. Dating, function, typology, technology, and find contexts have been discussed in terms of many different questions and examined using a multitude of methods. Sophus Müller laid the foundation for these studies in 1919 through his publication of a number of Danish settlement site finds. Archaeological knowledge of large blade knives and flint assemblages from the Late Bronze Age has then grown successively over the years. Yet this has not taken place as a cumulative build-up of knowledge. Instead we have seen repetitions, take-for-granted assumptions, indifference, *ad hoc* solutions, and unfounded claims going hand in hand with careful research, analysis, and presentation of new knowledge. Sometimes the research has built on earlier findings, sometimes not. Sometimes research has been done by scholars in parallel, with no cross-fertilization, sometimes not. In certain periods the differences between different authors have been large, often associated with national borders, and sometimes the differences have been minor, figure 2.7.

Research has chiefly presented new finds but also to some extent raised new questions and reported the results of analyses. Above all, the extensive rescue excavations since the end of the 1970s have generated new find material, and in analyses

	1898	1918	1919	1932	1934	1937	1938	1943	1949	1952	1958	1960	1964	1979	1981	1982	1984–85	1985	1986	1990	1992	1993	1996	1998	1998	1999
Dating																										
Stone Age	x			x	x	x																				
Late Neolithic		x								x																
Stone Age and Early Bronze Age								x					x											x		
Bronze Age																										
Late Bronze Age, maybe also Stone Age			x				x		x		x															
Late Bronze Age, maybe also Early Bronze Age														x						x						
Late Bronze Age												x			x						x		x			
Late Bronze Age, periods V–VI																x	x	x	x			x			x	x
Function																										
Sickle	x	x	x		x	x	x	x	x	x	x	x	x													
Meat knife				x																						
Multi-tool														x								x	x			
Knife for harvesting leaves														x	x			x	x		x	x		x	x	x
Harvesting tool																	x			x	x			x	x	
Not a harvesting tool																x										

Figure 2.7. Schematic presentation of assessments and research findings concerning large blade knives of flint in different times, exemplified here with a century of published ideas about dating and function. As the figure shows, there is no cumulative build-up of knowledge in archaeological research about these aspects of large blade knives. Opinions and assessments come and go at different times. What the figure also shows is that, although knowledge has not been accumulated, there has been a change over time in the overall impression of opinions about dating and function. As regards dating, this change is not clear and cannot be pinpointed. As regards function, the change is clearer and can mainly be associated with new methods in functional analysis which were applied for the first time to large blade knives in Wyszomirski's study published in 1979. It should be pointed out that this methodological change applies generally to lithic studies, not specifically to large blade knives of flint. For an overall discussion of the consequences of this methodological development see Knutsson (1988), Odell (2004), and Holdaway and Stern (2004).

and publications of the results new and more profound questions have been considered and studies have been performed, yielding both good results and well-formulated hypotheses for future studies. Through new methods, research from the 1990s onwards has generated new knowledge about large blade knives.

In early studies the Late Bronze Age large blade knives of flint were classified simply as blade knives and sorted under the same terminology as Neolithic blades and blade knives. This classification entailed a problem which has dogged research for decades, and which several studies have tackled in different ways. Since the Late Bronze Age large blade knives of flint were classified as blades and placed in the collections together with Neolithic blades, this has led to problems since the Late Bronze Age large blade knives of flint do not really look like Neolithic blades. In several studies it is obvious that this is a disturbing difficulty, a source of dissatisfaction. Yet even though Müller (1919) demonstrated early on that the large blade knife is an artefact from the Late Bronze Age, it was not really until after the large-scale excavation projects and studies from the early 1980s that the classificational confusion was ended and the Late Bronze Age large blade knife of flint began to be perceived as an artefact category of its own which should not be linked to Neolithic blades and blade knives.

I have not been able to ascertain where the Swedish term *lövkniv* comes from, table 2.4. The survey has not given any unambiguous answer, just hints. The term appears in a number of works from the 1930s where it is used as a designation indicating the function, but in a discussion of bifacial sickles, not Late Bronze Age large blade knives of flint. These works confuse terminological and functional discussions as regards different types of tools. Terms and names such as blade, knife, blade knife, leaf knife, sickle, harvest and leaf-fodder gathering are used interchangeably. The effect is that it is difficult to work out what is actually meant (Oldeberg 1932; Sandklef 1934; Hatt 1937). A few years later the term *lövkniv* is found in the literature, but this time as a designation for both bifacial sickles and blade knives (Brøndsted 1958). At the end of the 1970s Wyszomirski (1979) uses a phrase meaning "so-called leaf knife". At the start of the 1980s Björhem and Säfvestad first employ the term *spån(löv)kniv* ("blade (leaf) knife") and then in the same work switch to *lövkniv*, which is also used in other Swedish publications from the 1980s (*Arkeologi och naturgas i Skåne* 1985; Jonsson 1986). In a publication by Björhem and Säfvestad from 1993 the term *lövkniv* is the taken-for-granted designation for this type of artefact (Björhem & Säfvestad 1993:66ff), and in nomenclature from the twenty-first century the term is firmly established (Högberg et al. 2000).

Table 2.4. A selection of the names for the artefact that I call a large blade knife of flint which I have found in publications and in museums' accession catalogues. LUHM = Lund University Historical Museum, SHM = Museum of National Antiquities, Stockholm.

Designation	Translation	Reference
Skärknivar af flinta	Cutting knives of flint	LUHM 1855–1869
Eneggad krokknif af flinta	One-edged curved knife of flint	LUHM 1876
Flækkeknive, Segl	Blade knife, sickle	Blinkenberg 1898; Müller 1919; Glob 1938; Broholm 1949; Brøndsted 1958; Thrane 1971; Jensen 2002; Jensen 2006
Skäror av flinta	Sickles of flint	LUHM 1905–1906; Rydbeck 1918
Blade, Flake, Meat knife		Oldeberg 1932
Kornsegl	Grain sickle	Brøndsted 1958
Schweres Klingenmesser aus Flint	Heavy blade knife of flint	Becker 1990
Pruning knife		Olausson 1992
Spånkniv	Blade knife	Sandklef 1934; Lindälv 1964
Kniv, eneggad	Knife, one-edged	SHM 2007
Spån(löv)kniv	Blade (leaf) knife	Björhem & Säfvestad 1983; Burenhult 1999
Lövkniv	Leaf knife	Brøndsted 1958; Björhem & Säfvestad 1983, 1993; Jonsson 1986; Knarrström 2007; Högberg et al. 2000; Högberg 2005; SHM 2007
Large blade knife of flint		Högberg 2004a
Large retouched blade knives		Jensen 2006
Store flintknive	Large flint knives	Jensen 2006

It is obvious that the Swedish term *lövkniv* was not established on one specific occasion. Instead it seems to be a terminology that has emerged through different publications and which evidently had different phases of use which are not always visible in the texts. From having been a designation for all tools used for harvesting leaves (Oldeberg 1932), to becoming a defined type of artefact (Björhem & Säfvestad 1993) and a key term in a nomenclature (Högberg et al. 2000).

How have Late Bronze Age large blade knives of flint been assessed as archaeological artefacts? Although the large blade knife was held up early in the twentieth century as an important artefact for understanding society in the Late Bronze Age, few studies have been performed with the ambition of using this artefact type as a starting point for this. The emotionally charged and comparative opinions about flint assemblages from the Late Bronze Age that made such an impact in the years around 1930 set an agenda for the study of flint tools from the Late Bronze Age which influenced interpretations of this material for several decades to come. This still survives to a certain extent today – not specifically in studies focusing on flint material from this time but more in studies of other aspects of the Late Bronze Age – often unreflective but present and clad in modern language. In a text from 1999 discussing Bronze Age agriculture, for example, we read (Lagerås & Regnell 1999:273):

> Available data do not indicate any radical technical improvements of agricultural tools during the Bronze Age, and agrarian changes during the period cannot at present be interpreted in terms of technology.

The large-scale archaeological excavations at Fosie IV in Malmö, as we saw above, unearthed about sixty whole and fragmentary large blade knives of flint (Björhem & Säfvestad 1993:72). One may wonder whether the conclusions in this quotation, that "agrarian changes during the period cannot at present be interpreted in terms of technology" would have been formulated in the same way if these large blade knives of flint from Fosie IV had instead been bronze sickles?

Another example of the same outlook on flint assemblages in the Late Bronze Age is expressed in the survey of agrarian history in southern Jutland, *Det Sønderjyska Landbrugs Historie* (Ethelberg et al. 2000). Under the heading "technological progress" we read that flint as a raw material underwent a "renaissance" during the Late Bronze Age, yet not, in the author's opinion, as a technological innovation but as a reaction to the lack of metal (Ethelberg et al. 2000:236ff). The possibility is not considered that people in the Late Bronze Age actively made a choice in certain contexts to use flint instead of metal for specific purposes.

In this view of flint assemblages with large blade knives we can detect the comparative assessments from the publications by Glob (1938), Steensberg (1943), Broholm (1949), and Brøndsted (1958). These propagate the myth of the superiority of metal, in contrast to flint which is regarded as the poor man's raw material (for a critical discussion of this see Magnusson Staaf 2000). It is obvious that the value judgements about flint assemblages from the Late Bronze Age that were first expressed in the 1930s had a powerful impact and still affect research today (Högberg & Olausson 2005).

Another point worth highlighting is that earlier research consisted of small case studies of material linked to specific excavations (Ethelberg 1982; Sørensen 1984–85; Becker 1990; Björhem & Säfvestad 1993) or studies based on narrowly focused questions (Wyszomirski 1979; Ebbesen 1981). There has been no general study comprising several different aspects of flint assemblages and large blade knives of flint in the Late Bronze Age. In this connection it should also be noticed that flint assemblages and large blade knives are absent in works about the Bronze Age which deal with trade, contacts, economy, ritual, power, social aspects, technological innovations, everyday life etc. Large blade knives of flint have not been considered worth incorporating in general studies of cultural history in the Late Bronze Age.

The above survey and discussion is a good point of departure for defining a number of interesting research areas. These include things that have been studied previously and given interesting results and hypotheses for continued work, and areas not covered by previous research. The survey has provided a basis for formulating questions dealing with aspects of typology, chronology, function, hoard finds, context, technology, manufacture, geographical distribution, variation, raw materials, trade, exchange, contacts, influences, and the absence of interpretations of large blade knives of flint in other fields of Late Bronze Age study.

Typology

Several different ways of classifying and typologizing large blade knives of flint have been presented, tested, and used. Some authors have discussed types based on whether the blade knife is retouched on the right or left side. This typology can be traced back to the first published studies of large blade knives (Blinkenberg 1898; Müller 1919). It has also survived into our own times (Wyszomirski 1979; Björhem & Säfvestad 1993). The placing of the retouch has been set in relation to right- and left-handedness. Is such an interpretation sustainable? Or could there be other reasons for left and right retouch?

Sørensen defines two types depending on whether the large blade knife has shafting retouch or backing. He thinks that these types have a chronological basis, the former from period V and the latter from period VI (Sørensen 1984–85:61). Björhem and Säfvestad also define two types, type I and type II. This is based on metric and morphological values (Björhem & Säfvestad 1993:72ff). Sørensen's typology has been tested by Björhem and Säfvestad (1993:77), who were unable to prove or disprove it.

Both Sørensen and Björhem & Säfvestad base their typological observations and typologies on metric and morphological attributes. This raises some questions for further work. Are there functional and technological types that Sørensen or Björhem & Säfvestad have not defined because their methods have paid no heed to function and technology?

Chronology

Despite extensive and thorough studies of settlement site material there is still uncertainty about the dating of large blade knives of flint. It is perfectly clear that they occur in the latter part of the Late Bronze Age, periods V and VI. Yet the doubt in earlier studies about the chronological placing of large blade knives lives on. We have everything from publications that still date large blade knives of flint to the Stone Age (Sjöberg 1998), to studies expressing uncertainty as to whether large blade knives existed in Bronze Age periods III and IV or not (Becker 1990). The dating of large blade knives is therefore something that requires investigation. From what periods are there large blade knives in the archaeological material and which periods have none? Another problem concerns the difficulty of specifying exactly when within the time span of periods V–VI large blade knives of flint were used. When were they introduced? How long were they used? When did they cease to be used?

Function

The studies of the function of large blade knives of flint which have been published were performed on limited amounts of material. The results of these studies are uniform inasmuch as they observe that the large blade knife was used as a sickle, but disagree when it come to how the large blade knife of flint was used and for what. As regards the function and meaning, studies of hoards of large blade knives show that there are different perceptions of the social contexts to which the large blade knife belonged. So even if a great deal has been done about the function and meaning of large blade knives, a great deal still remains to be investigated: How were large blade knives of flint used, for what, and in what contexts?

Hoards

Several authors regard assemblages of large blade knives of flint as something to be associated with the tradition established in the Stone Age of depositing one or more objects in the ground as an offering (Ebbesen 1981:17f; Nielsen 2005:216ff; Jensen 2006:108f). Becker's analysis of the find from Spjald, however, shows that there was variation in the patterns of deposition. Becker thinks that this find is unusual and virtually unique and therefore cannot be interpreted within the framework of available comparisons (Becker 1990:208). This raises several questions. Are all assembled finds of large blade knives the same? Are there different kinds of deposits? Is it correct to place hoards of Late Bronze Age large blade knives within the same interpretative framework as Stone Age hoards?

Contexts

The studies that have been published have concerned large blade knives of flint in settlement site contexts or hoards (see e.g. Müller 1919; Ebbesen 1979; Björhem & Säfvestad 1993). The question is whether large blade knives are found in other contexts. In what connection do we find the large blade knife of flint and where do we not find it?

Geographical distribution and variation

It is obvious that large blade knives of flint have been found in southern Sweden, Denmark, Poland, and northern Germany. What remains to investigate is where the geographical boundaries run. Are large blade knives to be found in Norway, in central Sweden, or in other parts of northern Europe?

There is also uncertainty as to whether the flint objects found in south-west Denmark look the same as those from eastern Denmark and Sweden (Ethelberg 1982; Draiby 1985). That there are other differences between these areas has been demonstrated by Glob (1938) in an analysis of Late Bronze Age groundstone axes and shaft-hole axes. This raises some interesting questions. What do flint assemblages look like in different parts of the areas where the large blade knife of flint occurs? Are they similar, different, or varying?

Technology and manufacture

Few studies have been done about the technology and production of large blade knives of flint. There are no studies examining production processes and few examples of attempts to find out where large blade knives were manufactured. Broholm (1949:196) states that blades and blade cores are not found in settlement site material from Late Bronze Age. This is rather strange in view of the fact that Müller (1919:75) noted both blades and blade cores in the material, but it is also interesting in the light of Björhem & Säfvestad's (1993:77) observations about the absence of blades and cores in the material they looked at. This gives rise to a number of questions. How were large blade knives of flint manufactured? Where were they manufactured? Is it right to search for blade cores in order to demonstrate production? Is there other evidence that could indicate the production of large blade knives?

To be able to analyse the technology with which large blade knives of flint were made, I also have to study other flint knapping from the Late Bronze Age. What knapping technologies existed simultaneously with large blade knives?

Raw material

Becker (1990:202) is the only scholar who has discussed the provenance of the flint from which large blade knives were made. Aspects of availability and shortage of raw material in relation to the occurrence and use of large blade knives of flint have not been explored, nor questions concerning trade, exchange, and contacts from the perspective of raw material. Where was the flint that was used for the production of large blade knives collected? Was the flint handled as a commodity for exchange or trade?

Trade, distribution, exchange, contacts, and influences

Although many studies have investigated Bronze Age contact routes and distribution systems for bronzes (Thrane 1975; Jensen 1981; Persson 2001), no such studies have included large blade knives of flint. There are no published studies of how and where large blade knives were distributed and what contact networks and channels this distribution might have comprised. How were large blade knives distributed? What channels or networks of contact could the blade knife have been part of? Who took part in the handling of large blade knives? Is it possible to trace influences between different parts of the studied area?

The absence of interpretations of large blade knives in other fields of Late Bronze Age study

It is a simple observation that, even though settlement site finds and hoards of large blade knives have been known for more than a hundred years, no studies (see e.g. Thrane 1975; Jensen 1981) have included flint assemblages and large blade knives of flint as part of the archaeological material on which to base research into general regional or boundary-crossing aspects of topics such as contacts, economy, social conditions, power, trade, or, as Ebbesen (1979) has pointed out, rituals and sacrifices. Can studies of large blade knives contribute anything to answer these questions?

Methods

Through this survey of the history of archaeology I have highlighted different aspects of my study material. This has raised interesting questions for further work. With this as a basis, it is now time to introduce the methods I use and to discuss how I plan to work with the archaeological source material. The analyses of the extensive archaeological material have been performed in different ways, depending on the questions under study. My methods will therefore be specifically discussed in connection with the questions that the methods concern and the analyses performed. However, there are three ways of working which permeate virtually all parts of the material studies: raw material analysis, attribute analysis, and use-wear analysis. These are well-known components of lithic analyses (see e.g. Swanson 1975; Inizan, Roche & Tixier 1992; Andrefsky 1998, 2001; Whittaker 1994; Eriksen 2000; Holdaway & Stern 2004; Odell 2004) and have previously been used in analyses of flint assemblages from the south Scandinavian Bronze Age (Högberg 2001a; Knarrström 2000, 2001). What they have in common is that they are excellent methods to use in a study of actions based on the chaîne opératoire (Sinclair & Schlanger 1990). But before I describe these methods I will briefly touch on some questions to do with lithic analyses and my study material.

The onion problem

There is a trend in archaeological material studies to focus increasingly on parts of the whole rather than general descriptions (Kristiansen 2004; Kristiansen, Cornell & Larsson 2004). For studies of lithic material this means that, while there is a need for methods and theories that can be applied to large amounts of material and can offer a potential for generalizations, methodological development has progressed towards more refined methods and analyses that give detailed knowledge of limited and specially selected parts of the material (Torrence 1989; Odell 1996).

These detailed studies of functions and technologies have contributed to the deconstruction of overall analyses and shown their inadequacy in the generalizations that have been made, and they have yielded detailed knowledge of individual tools and techniques (van Gijn 1989; Jensen 1994; Fullagar 1998). But the problem is that the details have become so specific that it is difficult to link the detailed knowledge to a larger context, figure 2.8.

This is nothing unique for archaeology. In an analogy trying to describe the multiple experiences of comfort, Witold Rybczynski, architect and professor of urbanism at Wharton University of Pennsylvania (1986:230), wrote:

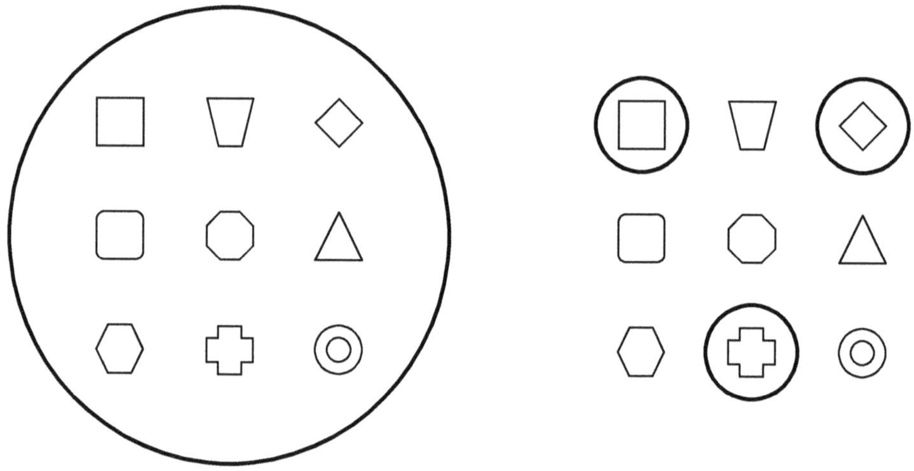

Overall analysis, the total impression
rather than the specific

Detailed analysis, the specific
rather than the totality

Figure 2.8. Left: an analysis of the total impression of all activities and expressions that yield generalized knowledge about the totality but say nothing about the details. The overall impression is not representative of any of the individual details. Right: An analysis of selected parts of the material that yields detailed knowledge about little but says nothing about the overall impression of the material. Each individual detail is not representative of the total impression of all the details.

> It is like trying to describe an onion. It appears simple on the outside, just a spheroidal shape. But this is deceptive, for an onion also has many layers. If we cut it apart, we are left with a pile of onion skins, but the original form has disappeared; if we describe each layer separately, we lose sight of the whole. To complicate matters further, the layers are transparent, so that when we look at the whole onion we see not just the surface but also something of the interior.

In the same way, as a researcher in lithic analysis I risk ending up with an analysis of a fragment of a layer, or even worse, of the contact surfaces between two layers. Consequently I will never see the onion, no matter whether that onion is the overall impression of a single flint assemblage or the south Scandinavian Late Bronze Age.

Getting round the onion problem is essential to be able to work with the questions that I bring up in this study. In an investigation like this it is a challenge to try to go from specific details to interpretations of broader perspectives. This is done by studying the material at different levels. Some parts of the source material I analyse in detail, while other parts are analysed in a more general way. By discussing details in relation to the totality and the totality in relation to details, I can partially get round the onion problem, figure 2.9.

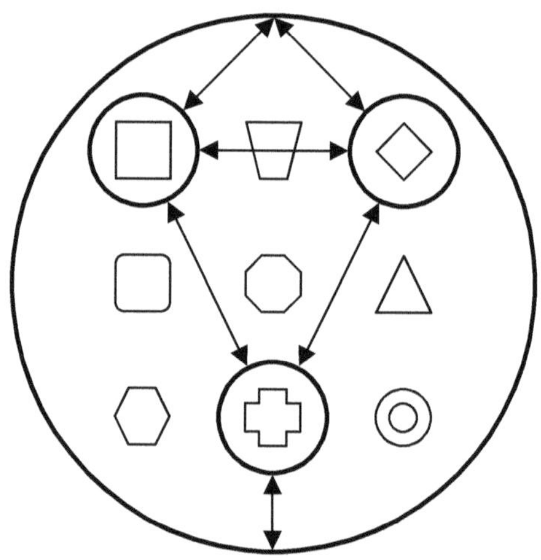

Figure 2.9. To get round the problems that lithic studies have with, on one hand, the need for methods and theories that can be applied to large bodies of material and that can offer the potential for generalizations, and on the other hand a methodological development towards detailed analyses which generate knowledge about limited and specifically selected parts of the material, I have chosen to do both. Detailed studies are performed on parts of the material, while material from the entire study area is examined more generally. By then analysing details against details, details against the totality, and the totality against details, I can partially escape these problems. The figure should be seen as an evolution of figure 2.8.

My source material

Sophus Müller suggested as long ago as 1884 that the context in the form of similarities and differences must be the foundation for the way in which an object is interpreted (Müller 1884:196). In concrete terms, this means analysing which objects are found together in time and place, how, and in what contexts (Hodder 1986; Olsen 2003a). The contexts I have chosen are ones I know where flint from the Late Bronze Age was used as raw material for the manufacture of objects. This means that the context basically consists of flint assemblages, and parts of this context consists of different situations in which flint assemblages are found archaeologically or where flint can be said through archaeological analysis to have been used.

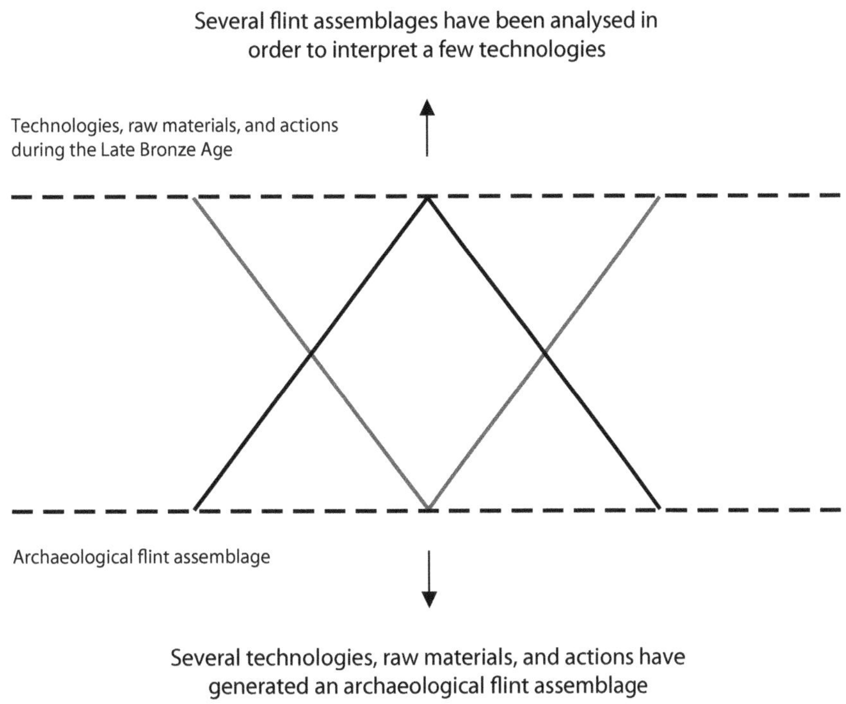

Figure 2.10. Sketch illustrating the dynamics in the study between technologies during the Late Bronze Age and analysed archaeological material. The base of each triangle corresponds at the top to a multitude of technologies, raw materials, and actions during the Late Bronze Age and at the bottom to a multitude of analysed archaeological material. Several technologies have resulted in an assemblage of archaeological material. Several archaeological assemblages have been used to analyse a few technologies.

It is appropriate to reflect here briefly on the selection of study material. First it must be stressed that a great many different raw materials and technologies were used in the south Scandinavian Late Bronze Age for a wide variety of actions. Wood, hides, leather, clay, cotton, linen, peat, reeds, nettles, copper, tin, lead, iron, bronze, bone, and antler were used by people to weave, spin, cast, carve, forge, build, chop, and scrape; to manufacture tools, ornaments, clothes, tables, chairs, and benches; to build houses, roads, and monuments, and to sow, reap, process, and cook food, to mention just some examples from an extremely complex everyday life. Of all these technologies, raw materials, and actions, and the many others not mentioned, my studies deal with just a few technologies where I focus on a single raw material, flint. This of course leads to an extremely limited study, but it is at the same time focused, figure 2.10.

Secondly, it must be stressed that the amount of study material – that is, knapped flint – from different parts of my study area, varies greatly. An excavation in eastern Denmark can yield finds of hundreds of kilos of knapped flint, while an excavation in eastern central Sweden might find only about twenty pieces of knapped flint weighing a total of 150 grams. This difference is due to variations in the availability of the raw material, flint, which I consider in the next section.

Flint as raw material

Flint as a raw material occurs profusely in south-west Scandinavia and along parts of the Baltic coast of Germany. I have previously presented a study of flint as a raw material together with Deborah Olausson in the book *Scandinavian Flint: An Archaeological Perspective* (Högberg & Olausson 2007). The study discusses different types of flint, their appearance, properties, and availability, with a detailed survey of occurrences and variations. In the work presented here I use the results from this earlier study and when doing so cite references to it. There will therefore be no discussion here of flint as a raw material. Readers who are interested in the topic are referred to the publication by Olausson and myself and the other works cited there.

What I will mention, however, is the great variation within my study area as regards the availability of flint as a raw material. In south-west Sweden, eastern and north-west Denmark, and along Germany's Baltic coast there is a great deal of flint. At several places there is exposed limestone and chalk, and high-quality flint is accessible in geological formations. At other places the upheaval in the landscape caused by the Ice Age has deposited large amounts of flint in till layers. This flint is of varying quality.

In south-west Denmark the supply of flint is much smaller. There is some till flint, but none in limestone or chalk layers. Along the west coast of Sweden and up along the southern coast of Norway there is likewise no flint in limestone chalk layers. Along the shores, however, one can find till-transported flint which was used for making tools all through prehistory. This is of highly fluctuating quality. In the rest of Sweden and Norway, with few exceptions, no flint is naturally available.

This great difference in the supply of flint as raw material between different parts of these areas is of great significance for this study. This is a topic to which I will return later.

Attribute analysis

Flakes, i.e. the product or by-product of stone tool production or core reduction, show numerous traces of the flint-knapping that created them. These traces consist of various attributes, all with their special characters. The composition of the attributes is directly related to the technique and method employed in the flint-knapping, and the variation as well as the distinctiveness of the diagnostic attributes increases with the complexity of the flint-knapping (Inizan, Roche & Tixier 1992:39ff.; Andrefsky 1998:29; Holdaway & Stern 2004:37). Both the attributes and the variations in the morphology of the attributes, what I call variables, thus show more or less distinct and diagnosable traces of the technique and method used when knapping. For example, the association between the length and thickness of a flake, the thickness of the platform and the platform angle has been clearly demonstrated (see e.g. Dibble & Whitaker 1981). Another example is how the degree of fragmentation of the flakes relates to the raw material, techniques, and methods employed by the knapper (see e.g. Sullivan & Rozen 1985; Amick & Mauldin 1997). By analysing the character of different attributes, separately or together, it is thus possible to reconstruct flint-knapping techniques and methods (Holm 1985; Knutsson 1988; Madsen 1992; Ballin 1995).

There are a great many different ways in which this can be done (see e.g. Kooyman 2001:77ff). I have chosen to base my studies on a comprehensive analysis of technological attributes and an experimental study based on the analyses of the technological attributes. This has been done within the framework of a chaîne opératoire analysis, to enable me to discuss the actions in which Late Bronze Age large blade knives of flint and other knapped flint from the Late Bronze Age played a part as actants, and the actions that actors performed with these flint objects.

Table 2.5. Terms and definitions used in the attribute analyses of large blade knives and flakes (from Inizan, Roche & Tixier 1992; Whitaker 1994; Rosen 1997; Andrefsky 1998; Kooyman 2001; Holdaway & Stern 2004; Odell 2004). Note that the definitions are adapted to the analyses performed here. No claim is made to provide general definitions valid for all types of analyses of lithic material.

Term	Definition
Ventral surface, ventral side, or lower face	The side of the flake that has broken away from a core, nodule or worked piece. It shows no evidence of previous flake removals. It is opposite the dorsal surface.
Dorsal surface, dorsal side, or upper face	External surface of the flake, showing flake scars from previous removals, cortex or a weathered surface. It is opposite the ventral surface.
Lateral margin	Left and right side of a flake defined when looking at the dorsal side of the flake with the proximal end upwards. Note that this is a definition which differs from publication to publication. Some define the left and right lateral margins looking at the dorsal side of the flake, but with the proximal end downwards and some looking at the ventral side of the flake with the proximal end downwards.
Proximal end or proximal fragment	The end of a flake containing the striking platform. Opposite the distal end.
Medial part of a flake or medial fragment	The mid-section part of a flake, defined by the absence of the proximal and distal ends.
Distal end or distal fragment	The end of a flake where the force of the original point of impact terminates. Opposite the proximal end.
Platform, striking platform, or butt	The striking platform, or merely the platform, is the surface on the end of the flake struck to cause a flake to detach. It is also the surface of a core or a worked piece on which the force has been applied at the removal of a flake.
Platform preparation	Abrasion, trimming of removals on the edge of the core, from the platform down on to the dorsal surface, made before the knapping of a flake.
Impact point, point of impact, point of percussion, or point of force application	The point or area where the force is applied to fracture a piece of flint. It is often visible on the butt as a white crush mark, sometimes inside a ring crack.
Ring crack	A small ring or crescent around the impact point.
Bulb of percussion, bulb of force, or bulb	A more of less conchoidal shape which has developed on the proximal end of the ventral surface of a flake, radiating from the point where the force was applied.
Bulbar scar, bulb scar, or eraillure scar	A small flaking scar on the bulb of percussion that results from an extra small flake being detached when the main flake is detached. The small flake removed is called an eraillure flake
Fissures, radial fissures, or hackles	Fracture marks on the ventral surface of a flake that radiate out from the point of impact. They are often seen on the bulb of force (bulbar fissures) or at the lateral margins (lateral fissures).
Compression waves, compression rings, ripple marks, ripples, or conchoidal ripples	Concentric waves on the ventral surface on the flake which radiate out from the platform and bulb of force.
Dorsal flake scar, negative removal or negative flake scar	The mark, i.e. the complementary surface, left by a flake removed from a piece of flint, visible on cores, the dorsal side of a flake, and on retouches.
Dorsal ridge, arris, or flake scar ridge	Ridge on the dorsal side formed by the intersection of two or more flake scars from previously removed flakes.
Cortex	Remains on the dorsal surface or the butt of the chalk or limestone surface of the nodule from which the flake was removed.
Weathered surface	Remains on the dorsal surface or the butt of an altered crust on the surface of the nodule from which the flake was removed.

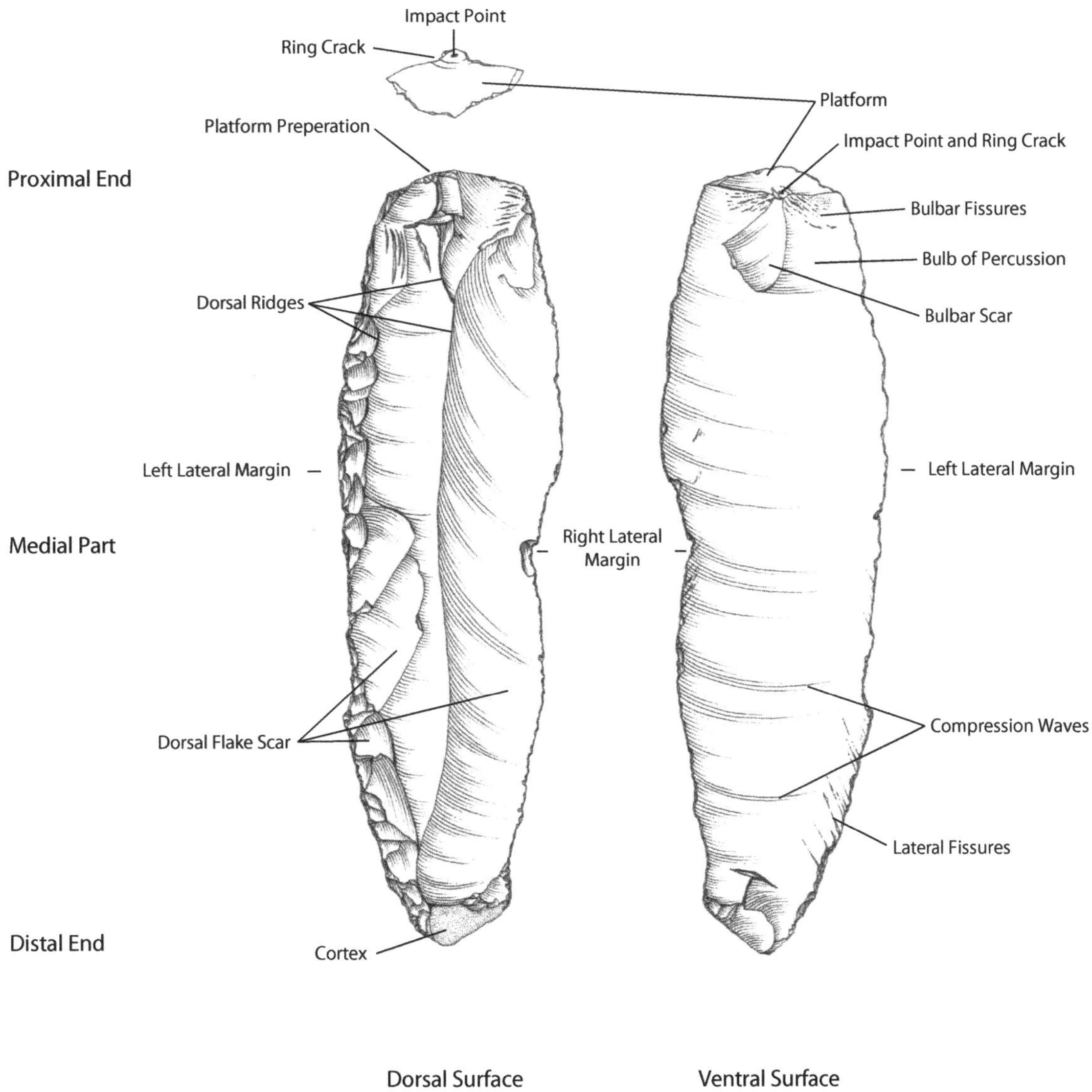

Figure 2.11. Illustrations of terms used in the attribute analyses.

There has been some criticism of the attribute analysis of flakes and debris. Michael J. Shott (1994) has highlighted three weaknesses in a well-formulated study. In several analyses the attributes are viewed as given a priori instead of being something to be analysed, For example, different platform attributes are usually taken as an indication that soft or hard technique was used, without this being questioned, analysed, or interpreted (Shott 1994:77). In my study I avoid this in two ways. Firstly, I handle the attributes in my registration as characteristics of the flakes which serve as a basis for an interpretation of the overall impression of the attributes, I do not interpret the attributes separately. Secondly, I analyse the attributes in comprehensive manufacturing experiments and thus interpret the material based on the results of several studies, not merely the attribute analyses. The other weakness of attribute analyses that Shott discusses is the lack of uniformity between different studies, which makes them hard to compare (Shott 1994:77). Here I can only agree. This study is no exception. But to facilitate any future comparisons I clarify my assumptions and premises, an approach emphasized by others (Andrefsky 1998:64ff). The third point of criticism that has been voiced concerns the focus of attribute analyses on form at the expense of size. In other words, several studies have focused on for example the morphology of the platform without considering its size (Shott 1994:78). In this study I use both form and size as attributes.

All of the attributes and variables in the following technological and metric analyses have been chosen due to the belief that, separately or together, they can give a picture of the character of large blade knives and other flint material from the Late Bronze Age, and that they can serve as a foundation for interpreting the technology of which the Late Bronze Age flint craft was a part. The terms and definitions I use are shown in table 2.5 and figure 2.11.

In the registration a flake is defined as a piece with a ventral surface, i.e. a lower face displaying one or more traces of knapping in the form of a bulb of percussion, bulbar scar, compression waves, or fissures (Högberg 2001a). If the flint shows one or more of these traces of knapping, it is defined as a flake. Debris, i.e. pieces of flint with a dimension of ≤ 10 millimetres

Table 2.6. Attributes and variables used in different variants in the analysis.

Attribute	Variables	Use
Appearance and shape	Length, width, thickness, weight, degree of fragmentation, fracture patterns, curvature, and morphology of ventral surface	Size and form are vital components in lithic debitage and debris analysis (Shott 1994:77f). The length, width, and thickness of a flake are associated with how the core has been shaped and how the flake has been detached from it. The degree of fragmentation is a variable that can be used, among other things, to describe and distinguish different reduction strategies (Amick & Mauldin 1997:25). Degree of fragmentation and fracture patterns can also be used to interpret which type of force detached the flake. The variables are also important for the interpretation of other flake attributes. To define differences between, for example, length and width ratios from different reduction strategies, it is essential to determine the degree of fragmentation and fracture patterns (Andrefsky 1998:86f). The curvature of the flake shows what the front of the core looked like.
Dorsal surface	Number and type of dorsal ridges, negative removals, and the amount of cortex	The occurrence of cortex and negative removals on the dorsal side of the flake tells us about the character of the flint-knapping method and can thus distinguish different reduction strategies (Andrefsky 1998:104ff.; Inizan, Roche & Tixier 1992:47; Clarkson, Vinicius & Mirazón Lahr 2006). The number of negative removals on the dorsal side indicates the frequency of previous removals from the part of the core from which the flake was split. It is the surfaces between the ridges on the dorsal side of the flake that are counted as negative removals.
Platform	Form, appearance, length, width, status, and occurrence of ring crack	The platform is the surface on the proximal part of the flake that received the force to detach it (Inizan, Roche & Tixier 1992:79f). Different flint-knapping techniques and methods give different types of platforms (Andrefsky 1998:88ff). The platform is therefore an important attribute to analyse in order to characterize and distinguish different reduction strategies. Ring crack together with other attributes can be used to distinguish different flint-knapping techniques (Kooyman 2001:12f).
Platform preparation	Occurrence, morphology, and type	Platform preparation is the knapping done at the platform edge on the dorsal side of the flake at the part of the platform which is to receive the force of the blow. The occurrence of platform preparation is usually associated with a well-thought-out knapping method (Inizan, Roche & Tixier 1992:46f). Platform preparation can be done in different ways. Specific manufacturing techniques often involve specific ways of preparing the platforms (Inizan, Roche & Tixier 1992).
Bulb of percussion	Appearance and form	The morphology of the bulb of percussion depends on the way in which the knapping was done. Different techniques and methods give different forms of bulb of percussion, and attributes such as bulbar scars, fissures, and hertz cone in the bulb of percussion can be used to distinguish different reduction strategies (Kooyman 2001:20ff).
Retouches	Appearance, form, and position	Retouches have been studied from several aspects, for example, the type of tool used for the retouching (Patterson 1998:27ff), what was shaped with the aid of the retouch (Nishiaki 2000:37ff), reduction strategies (Dibble 1995:311ff), where on the flake the retouch is placed, what the actual retouch looks like and how it was applied (Inizan, Roche & Tixier 1992:75ff). The look, form, and position of the retouch can be used to analyse different types of manufacture.

(Högberg et al. 2000; Larson & Kornfeld 1997:7), have not been registered.

Table 2.6 defines the attributes and variables of these attributes that I use. Also presented in the figure is a summary of what the attributes and variables can be used for in an analysis. In connection with the presentation of the results of the analysis later on, the use of each attribute with its variables is described in detail.

Use-wear analysis

Use-wear analysis is the study of distinctive patterns of wear on the edges of tools. Prehistoric artefacts made out of lithics, bone, antler, wood, metal, shells, and glass have been the subject of research (e.g. Odell 2004). In 1964 the English translation of the Russian researcher S. A. Semenov's work *Prehistoric Technology* was published (Semenov 1957, 1964). This event is generally regarded as the starting point for the western research development of microscopic use-wear analysis of lithic artefacts. Since then research has developed and the method has improved. Years of extensive experimental work combined with carefully considered studies of prehistoric artefacts from different periods and places, thoroughly conducted blind tests, source-critical studies of research methods, and mechanical and chemical studies of post-depositional aspects have developed use-wear analysis of lithic tools into a now well-established method with its own research agenda (see Ibáñez & González 2003; Odell 2004; Holdaway & Stern 2004; Longo & Skakun 2005, and references cited in these texts).

Although the development of the method has enhanced the possibilities to get good results out of many types of prehistoric lithic assemblages, there are still quite a few common problems shared by practically all researchers concerning the potential of interpretation of the function of tools using microscope analysis.

On a general level this has to do with the questions of worked materials versus working tasks. Use-wear analysis can be employed to interpret the kind of material the tools have been in contact with, for instance different plants, meat, hide, shells, bone, and antler. However, the working tasks performed, i.e. the production and the product produced, cannot be interpreted from use-wear analysis. There is no direct relationship between the analysis of worked material and the task that was performed in prehistory. For example, if an analysis comes up with the result that a flint knife was used for cutting wood, it is impossible to know what was manufactured out of those thousands and thousands of things which wood can be used for.

On a more detailed level the problem has to do with the process of development of wear traces on specific tools used on specific worked materials (van Gijn 1989). Many lithic tools were used for such a short time that only faint traces of wear have developed. The identification of this is called generic polish (see for example Jensen 1994). Tools were also used on worked materials which do not give specific wear traces, i.e. different worked materials give traces which cannot be distinguished from each other. A classic example of this is the problems – still not solved in spite of a great deal of effort – in characterizing and classifying wear traces which make it possible to clearly separate tools used on bone from tools used on antler in a prehistoric assemblage (Kelley 1979; Vaughan 1985; van Gijn 1989).

Use-wear analysis with high-power approach

Use-wear analysis can be conducted microscopically under low-power magnification, i.e. magnification up to approximately 100× (e.g. Odell & Odell-Vereecken 1980), and under high-power magnification, i.e. magnification from 100–500× (e.g. Keeley 1980; Jensen 1994). Although these different approaches are two ends of a continuum (Holdaway & Stern 2004:41), they both have their advantages and disadvantages (Odell 2004:139ff). High-power magnification, known as the high-power approach (henceforth abbreviated HPA), is considered as the more appropriate use-wear method for distinguishing different worked materials (Odell 2004:148f.). Since a deeper understanding of the analysis of worked material is one of the purposes of this study, HPA has been chosen for use-wear analysis.

The method of use-wear analysis with HPA have been thoroughly discussed and challenged in several publications (see Knutsson 1988; Jensen 1994; Levi Sala 1996 and Odell 2004 for a review). It usually involves the identification, analysis and interpretation of three things: the used areas on the tool, the mode of use, and the worked material (Bamforth 1988). In this study I alternate between these perspectives depending on the questions that I am investigating. In a section about use-wear analysis of large blade knives of flint there is a detailed study of traces of use on the large blade knives. This study focuses on several different aspects of use-wear analysis. In other parts of the book I employ use-wear analyses as a general method for answering simple questions such as the presence or not of traces of use on tools and flakes.

It is important to emphasize here that use-wear analysis is a matter of interpretation. It is what I see in my microscope that I interpret, and these interpretations are based on the experience I have. Just as in other archaeology, then, this is something based on questions, choice of method, analysis, and interpretation. I will therefore discuss source-critical aspects (see e.g. Levi Sala 1996) in connection with the analyses performed.

HPA is carried out with a binocular metallurgical microscope with reflected light. In this study the tools were analysed with an Olympus BX41 microscope with a magnification range of 50–500× and using light-field illumination as described by Jensen (1994). Photographs were taken with a digital camera mounted on the microscope and the photos were computer-manipulated using the programs analySIS 3.1®, with an EFI module, and Photoshop 6.0.

On analySIS® with EFI application as a method for analysis and image rendering

A problem with function analyses in high-magnification microscopes concerns the documentation and rendering of images. Since the analysis is done at high magnification, the depth of focus is minimal. At 200× magnification it is 2.15 μ, using my equipment. The edge on a flint knife may seem smooth to the eye, but when it is magnified 200 times it is anything but smooth. This means that the whole field of vision is rarely in focus in the microscope, see figure 2.12.

One can compensate for this in the analysis by shifting the area of focus within the field of vision by switching back and forth between different areas on the flint edge that is being analysed. This can be different areas alongside each other, different areas on a crest or in a trough, or a combination of both, figure 2.13. By shifting back and forth several times one can gain an all-round impression of the polish that is observed within the field of vision, which can thus be described and analysed. This method has also been described by other researchers (see e.g. Jensen 1994:19).

Yet this overall impression is difficult to recreate in an image. A photograph shows only a horizontal plane of the polish. Different topographical variations thus cannot be rendered

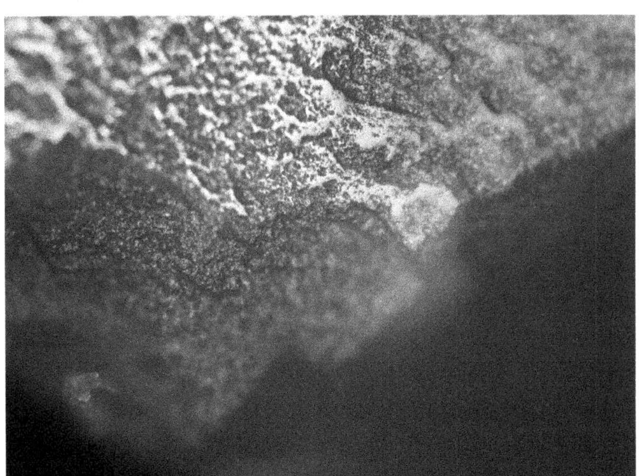

Figure 2.12. Picture of the field of vision at 200× magnification. Roughly 20% of the polish is in focus, the rest is not. The photograph shows the edge of a large blade knife of flint. Find number MK20:201876.

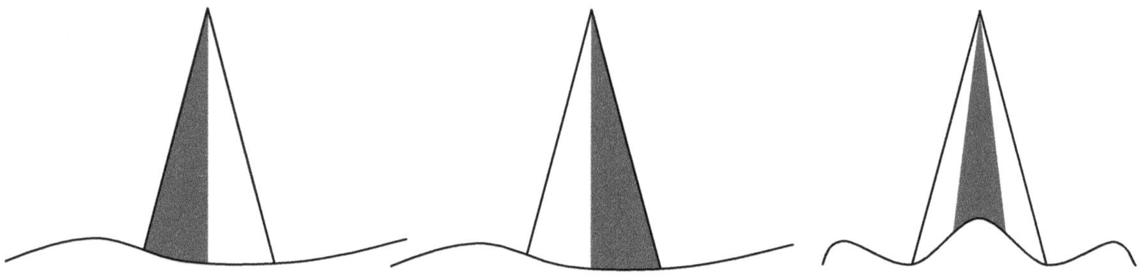

Figure 2.13. By quickly shifting focus in the microscope's field of vision one can get an all-round picture of the polish that is to be described and analysed. The problem is that this holistic view can never be rendered on an ordinary photograph, since this reproduces only one of all the focus areas one alternates between. An isosceles triangle indicates the extent of the field of vision in the microscope, a grey triangle shows the area of focus within the field of vision.

since they are out of focus. Moreover, the extent of the polish and the transition between polish and edge is impossible to render on one and the same photograph (van Gijn 1989:12).

Photographic documentation of the polish is nevertheless important in use-wear analyses (van Gijn 1989; Jensen 1994). There are various reasons for this. It is not possible to have more than one tool under the microscope at a time. When it is necessary to compare polish on different tools, good photographic documentation and the researcher's memory are the only alternatives. Another reason is publication. Good photographic documentation is essential so that other specialists can critically assess the evidence on which the analyses are based.

In the past these problems have sometimes been resolved by pasting together photographs of different areas of focus in the same field of vision and publishing them as a montage (see e.g. Owen & Unrath 1986, plate 36). This is a time-consuming method, both during the analysis and in the subsequent work, and it only functions at low magnification. Usually the problem has not been solved at all. Instead the appearance and extent of the polish and its relation to the edge have been illustrated as well as can be done with the best picture available.

Today a great many different solutions are available for computer control of image rendering in microscope analyses. I have chosen a program called analySIS® to which I have added an EFI application (EFI stands for Extended Focal Imaging). The program was produced for efficient image analysis in microscopy. Through the program I can control the digital camera which is mounted on the microscope, and all photography is done via the computer.[1]

The great advantage of this program is that it gives an opportunity to handle the problems of photographic documentation and image rendering. Through the program I can create an image with a focus over the whole field of vision that I have in the microscope. The approach is first to take several photographs within the same field of vision but with different depth of focus. The result of this is a great many pictures on which the polish is in focus in different parts of the field of vision. Through a simple manoeuvre in analySIS®, all these different images can be combined into one picture showing the whole field of vision in focus. When this is done the parts of each picture that are out of focus are removed. What is left on each picture is then only the part of the field of vision that is in focus. When all these images are combined into one, it creates a picture showing the overall impression of all the pictures' focal areas within a field of vision. This image is computer-manipulated. The program's present selection function determines what should be included or removed on each picture, figure 2.14.

The quality of the pictures depends on how carefully the different individual photographs are taken. One absolute condition is that the object to be photographed is perfectly still. It is the focus of the microscope that should be changed, not the object itself. When I photograph heavy large blade knives

Figure 2.14. Example of the use of analySIS® with an EFI application. The picture shows the appearance and extent of the polish and its relation to the edge of a large blade knife of flint. The analysed area has a topography that makes it impossible for the whole field of vision to be in focus simultaneously. The area has therefore been photographed in eight different images, to the right. Each of these has a different focal area within the same field of vision in the microscope. When these pictures are combined the result is a computer-manipulated image in which virtually the whole field of vision is in focus, above. The photograph, originally at 200× magnification, shows the edge of a large blade knife. Find number MK20:201876.

1 When I started to use analySIS® in 2003 it was a fresh, new program. As I write these lines after four years' work, the program is so old that it is no longer obtainable. There are, however, other programs with the same functions which can be obtained through microscope suppliers.

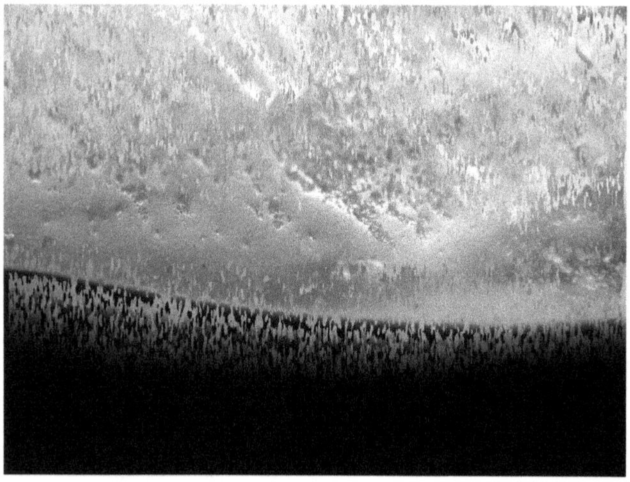

Figure 2.15. Example of an image manipulated in analySIS® where the object moved during the photography. The picture is full of lines that spoil the rendering of the polish. The photograph, originally at 200× magnification, shows the edge of a large blade knife. Find number MK20:201876.

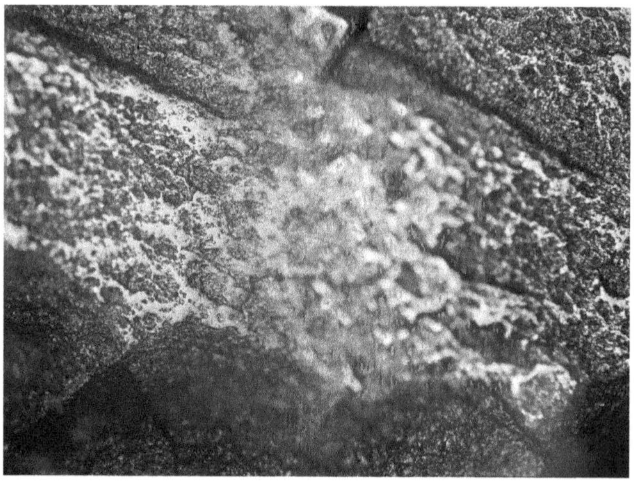

Figure 2.16. Example of an image manipulated in analySIS® where the number of photographs – three – taken as a basis for manipulation is too small. Parts of the picture are therefore blurred. The photograph, originally at 200× magnification, shows the edge of a large blade knife. Find number MK20:201876.

of flint, for example, it is sometimes difficult to get them to lie still on the microscope. The fixative I use is quite simply not strong enough. The large blade knives move, not so much as to be visible to the naked eye, but over a period of a few minutes this is visible in the photo documentation. The effect of this is that when the images are combined into one picture, there are sometimes small lines on the manipulated image. When these lines affect the analysis or the clarity of the image rendering, the picture has to be redone, figure 2.15.

Another necessary procedure is to take enough photographs within the field of vision so that all parts of the flint edge are photographed in focus. If this is not done, parts of manipulated image become blurred. If this affects the clarity, the picture must be retaken, just as when the object moved, figure 2.16.

The advantage of using this program is that polish which is observed on a flint edge can be rendered in a way that an ordinary photograph cannot manage. The problems described above with photographic documentation and picture rendering thereby disappear. When I use analySIS® in this study, this is indicated in the caption. I also state the numbers of photographs employed as a basis for the manipulated image that is published.

Large blade knives of flint – morphology and manufacture

After this account of the methodology, it is time to tackle the find material. As we saw above, there have been few studies of how large blade knives of flint were manufactured. In a work from the 1980s Ethelberg voiced the opinion that large blade knives and other flint objects from the Late Bronze Age were made in one and the same reduction process and from one and the same core. First large flakes were produced, which were then retouched to scrapers, then blades were made which were used for large blade knives, and finally smaller flakes which were used for smaller retouched tools (Ethelberg 1982:129f).

The result of the analyses of the comprehensive material excavated at Fosie IV in Scania, Sweden, shows that large blade knives of flint should be regarded as standardized. We have seen that Björhem and Säfvestad (1993:73) divided large blade knives into two types, type I and type II. Type I has a curved (asymmetrical) tapering distal part. It is described as being between 110 and 145 millimetres long with a straight concave edge, often with severe wear damage. Both types have a width of 35–45 millimetres and a thickness mostly in the interval 10–20 millimetres. Type II has a rounded to straight (symmetrical) distal part. It is between 80 and 110 millimetres long, has a straight edge which often does not show any great wear damage, and there is no shafting retouch. Through a study of fracture patterns on fragmentary large blade knives of type I, Björhem and Säfvestad clearly demonstrated that type II blade knives are not type I large blade knives that were broken and reknapped as type II large blade knives. This study was performed to show that the blade knives are two different types, not one type at different stages of curation (cf. Dibble 1987; 1995). This raises the question whether this observation could be confirmed by a technological analysis, that is, if type II differs from type I not just morphologically but also in manufacturing technology. Björhem and Säfvestad also searched for remains of production but could not find any traces of manufacture in the material (Björhem & Säfvestad 1993:77).

The studies by Ethelberg (1982) and Björhem and Säfvestad (1993) thus raise a number of questions for further work. Apart from these, there have been no studies focusing on aspects of the manufacture of Late Bronze Age large blade knives of flint. This gives a weak foundation for me to base my studies on and few previous studies to relate my findings to. At the same time, it gives an opportunity to work with a research field that has hardly been touched before.

Table 2.7. Sources of the large blade knives of flint examined in the study.

	Number of large blade knives of flint
Museum collections	295
Excavations	159
Total	**454**

Table 2.8. The number of whole and partial large blade knives of flint included in the analysis.

Whole large blade knives	357
Parts of large blade knives	97
Total	**454**

The work was done in two parts, a detailed attribute analysis and an experimental analysis. The aim of the attribute analysis was to investigate what the large blade knives look like, what different types of technological attributes are found on them, and how the morphology and attributes can be interpreted. The aim of the experimental analysis is to examine aspects of the manufacture of large blade knives and the waste resulting from the manufacture. These analyses are important in providing a basis for discussing similarities and variations in the morphology and manufacture of large blade knives, in the appearance of the debitage from the manufacture, and to serve as a foundation for analyses of function, dating, assembled finds, and manufacturing sites in the subsequent chapters. The attribute analyses and the experimental manufacturing analysis are thus in many ways the core of the material studies, constituting results in themselves and providing evidence for other studies.

A morphological and technological attribute analysis of large blade knives of flint

The morphological and technological attribute analysis of large blade knives is based on the attributes and variables of these attributes presented above. A total of 454 objects are included in the analysis. About a third of the material on which the study is based come from various excavations, the rest from museum collections, chiefly of stray finds, table 2.7.

The hitherto most comprehensive analysis of large blade knives is the work done by Björhem and Säfvestad (1993) in connection with studies of the material excavated at Fosie IV in south-western Scania, Sweden. Their analysis covered 82 whole and partial blade knives, 68 from the Fosie IV excavations and a further 14 from other excavations (Björhem & Säfvestad 1993:72f). I have chosen not to include these blade knives in my analyses, so that I can compare my findings with theirs, without comparing the same knives. However, nine of the large blade knives from Fosie IV are included in my analysis as reference material.

The attributes I have registered are those that I was able to observe on the large blade knives. Since I registered both whole blade knives and fragments, table 2.8, this means that the total number of large blade knives of flint included in the analyses of each attribute varies. For example, a large blade knife with a crushed platform may have a clearly preserved bulb of percussion while another which also has a crushed platform has no bulb of percussion left. Another example is large blade knives which lack a distal part. Some are broken in such a way that, despite the break, I was able to measure the width and thickness in the distal part, while others are broken in such a way that these dimensions could not be measured.

The large blade knives have been registered according to twelve attributes, with a total of thirty-three variables, table 2.9. Detailed explanations as to how each attribute is used in the analysis are given below in connection with the presentation of the results from the analyses of each attribute.

Of these attributes it is only the appearance of the large blade knife that has previously been examined by others (e.g. Wyszomirski 1979; Björhem & Säfvestad 1993). However, I myself have published a small study in which some of the attributes were considered (Högberg 2004a).

Table 2.9. The attributes and variables used in the analyses of the large blade knives of flint.

Attribute	Variables
Morphology of the large blade knives	Type Placing of the retouch Weight Length Width Thickness
Dorsal side	Number of negative removals Bidirectional dorsal flake scars Number of ridges Constructed ridge Amount of cortex
Edge outline	Curvature of the edge
Exterior platform angle	"Angle de chasse"
Morphology of the platform	Flat Faceted Cortical
Platform type	Rhombic or not
Platform preparation	Extending from the edge and down over the dorsal side in seven different variants Preparation flake with feathered termination Preparation flake with hinge termination
Size of the platform	Length Width
Ring crack on the platform	Ring crack around the point of impact Ring crack beside the point of impact
Morphology of the bulb of percussion	Well-formed or not Marked Diffuse Occurrence of bulbar scar Occurrence of moustache Occurrence of fissures
Retouch	Appearance Application
Raw material	Type of flint

Table 2.10. The number of type I and type II large blade knives of flint in the study, whole and fragmentary.

Type	Number	%
I	433	95
II	21	5
Total	**454**	**100**

The morphology of the large blade knives

The analysis of the morphology of the large blade knives of flint includes assigning the large blade knives to type I or II, examining the placing of the retouch, and comparing length, weight, width, and thickness.

A total 454 large blade knives were assigned to either type I or type II, table 2.10. The types are identified according to criteria from Björhem & Säfvestad (1993:72ff), see also the discussion in the previous chapter. The majority, about 95%, of the large blade knives in my analyses are type I large blade knives. Roughly 5% of the large blade knives are of type II. Of these, nine come from museum collections and twelve from excavations. This is a slightly smaller proportion of type II large blade knives than in the material from Fosie IV, where 12% of the knives were of type II.

The question of the placing of the retouch on the left or right side of the large blade knife is, as we saw above, something that has followed research on large blade knives from the very start, associated with the question of whether people in the past were left- or right-handed to the same extent as we are today (Blinkenberg 1898; Wyszomirski 1979; Ebbesen 1981; Björhem & Säfvestad 1993). This has also been used to typologize large blade knives of flint (Wyszomirski 1979:145f).

Before we look at the results, it is important to clarify that the definition of this variable has been handled in different ways by different researchers. Wyszomirski (1979:145) defined left and right side by looking at the dorsal side of the large blade knife with the proximal end downwards. Ebbesen (1981:7f) and Björhem and Säfvestad (1993:76) looked at the ventral side with the proximal end downwards. This means that Wyszomirski's left side is the right side for Ebbesen and for Björhem and Säfvestad, a difference that Björhem and Säfvestad did not notice in the comparison of their own results with Wyszomirski's (Björhem & Säfvestad 1993:76). To further add to the confusion, I have chosen to determine right and left side by looking at the dorsal side of the large blade knives with the proximal end upwards. This means that my right side is the same as for Ebbesen and for Björhem and Säfvestad.

Table 2.11. The placing of the retouch on large blade knives of flint, number and percentage.

	Number	%
Right	239	55
Left	193	45
Total	**432**	**100**

Table 2.12. The length of whole large blade knives of flint, stated in centimetres and percentages.

Length	Number	%
9	20	6
10	22	6
11	43	12.5
12	76	22
13	66	19
14	56	16
15	35	10
16	21	6
17	8	2
18	0	0
19	0	0
20	1	0.5
Total	**348**	**100**

A total of 432 large blade knives are included in the analysis of the placing of the retouch. Table 2.11 shows that roughly half of these large blade knives are right-retouched and the other half are left-retouched.

This differs from Björhem and Säfvestad's finding that roughly a third of the large blade knives were right-retouched, and from Wyszomirski's results, where about a third of the large blade knives were left-retouched. I will return to the discussion of the placing of the retouch later on in the section about my manufacturing experiments.

Length

The length is measured in centimetres on intact large blade knives, from the platform to the distal part along the direction of percussion (Andrefsky 1998:97f, fig 5.8a & c). Altogether 348 whole large blade knives are included in the analysis of the length of blade knives.

Table 2.13. The length of the 16 whole large blade knives of type II in the study, stated in centimetres.

Length	Number
8.0	1
8.5	4
9.0	0
9.5	1
10.0	4
10.5	0
11.0	2
11.5	2
12.0	0
12.5	1
13.0	1
Total	**16**

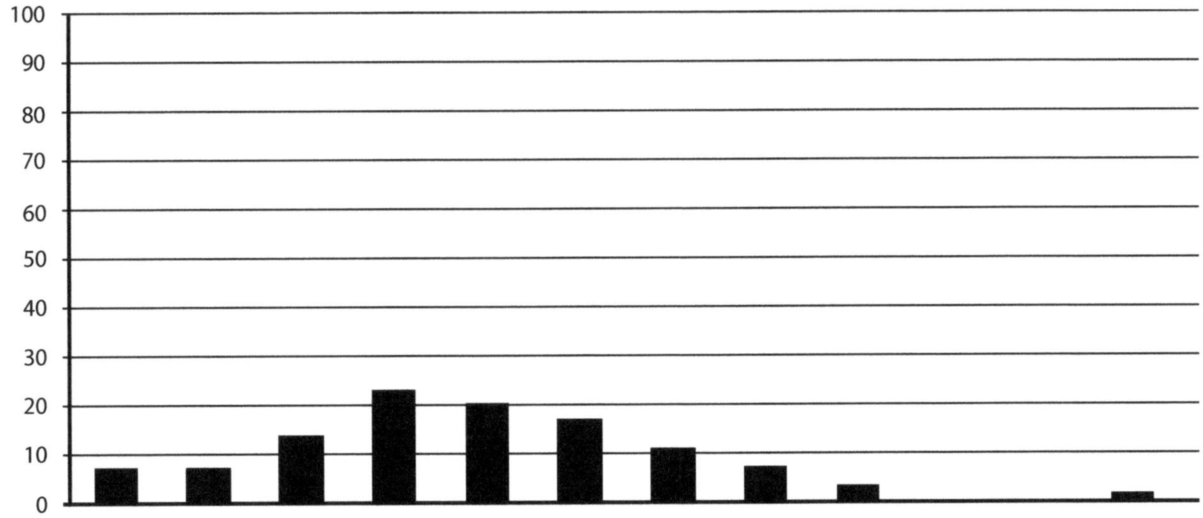

Figure 2.17. The length of large blade knives of flint in percentages. The length on the horizontal axis starts from the left with 9, then 9.1–10, 10.1–11, and so on up to 19.1–20 to the far right, stated in centimetres.

Table 2.12 and figure 2.17 show that more than half of the large blade knives are between 12 and 14 centimetres long and that 80 per cent of the blade knives are between 11 and 15 centimetres long. The question is which large blade knives are shorter and which are longer than the average. Type II large blade knives, as defined by Björhem and Säfvestad (1993:72ff) are generally shorter than type I according to their study. Table 2.13 sums up the length of the 16 whole large blade knives of type II in my material. This shows that the majority of the large blade knives of type II are shorter than the average of all large blade knives. It may be noted that the large blade knives of type II in Björhem and Säfvestad's (1993:73) study are between 8 and 11 centimetres long. In my material there are four type II large blade knives that are longer than 11 centimetres.

I have also investigated whether there are other large blade knives than type II blade knives which are shorter than 12 centimetres. I did so by searching out blade knives shorter than 12 centimetres in my database and then studying what these blade knives have in common and what they do not. The result shows that there is an association between length and find spot.

Table 2.14 clearly shows that large blade knives of flint from find spots on Rügen are shorter than other type I large blade knives. The majority of the blade knives from find spots on Rügen are between 8 and 10 centimetres long. It is only from length 10 centimetres and upwards that other large blade knives of type I enter the picture. An interesting observation in this connection is that the large blade knives which I have analysed from Schleswig-Holstein are not like those from Rügen but do resemble other large blade knives of type I.

As regards large blade knives which are longer than the average, I have not found any others that deviate. Those which are more than 15 centimetres long are large blade knives of type I from different places in Denmark and southern Sweden which look like any other type I large blade knives, apart from being longer.

Table 2.14. The length of large blade knives of flint in intervals up to 12 centimetres long, all large blade knives from find spots on Rügen in my material and all other large blade knives of type I in my material.

Length	Blade knives from Rügen, Germany	Other type I blade knives
-8.5	8	0
8.6–9.0	4	1
9.1–9.5	4	0
9.6–10.0	6	2
10.1–10.5	4	11
10.6–11.0	7	18
11.1–11.5	1	31
11.6–12.0	1	10

Table 2.15. The weight of whole large blade knives of flint stated in intervals of 20 grams and percentages.

Grams	Number	%
21–40	16	5
41–60	48	13
61–80	96	27
81–100	78	22
101–120	56	16
121–140	31	9
141–160	18	5
161–180	8	2
181–200	2	0.5
201–220	1	0
221–240	0	0
241–260	2	0.5
Total	**356**	**100**

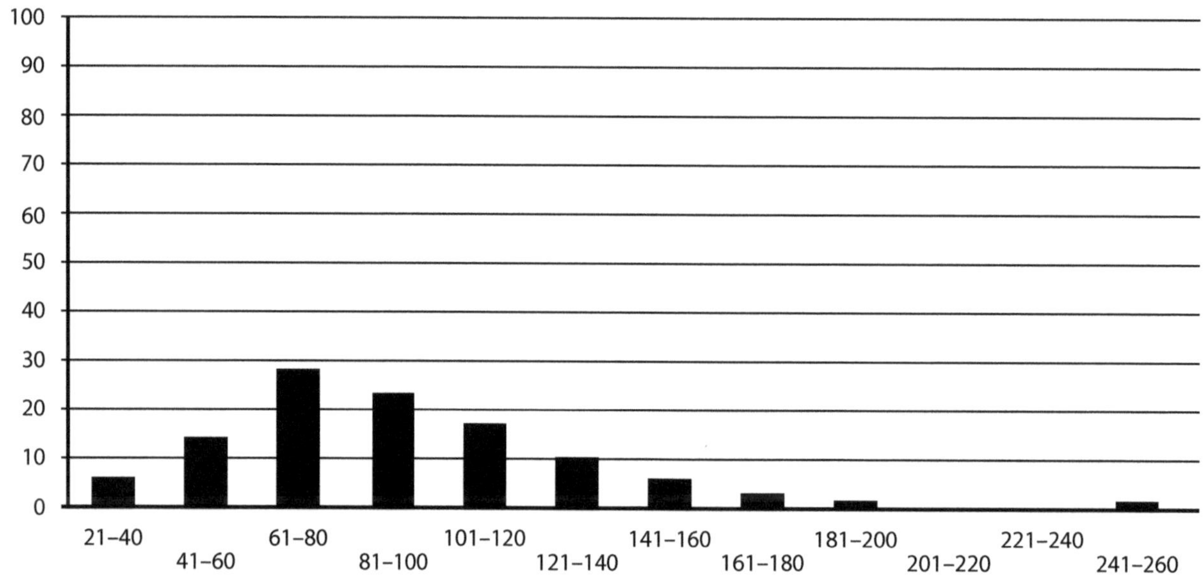

Figure 2.18. The relationship between the weight of large blade knives of flint, stated as a percentage on the vertical axis and in grams on the horizontal axis.

The result of the analysis of the length of large blade knives shows that the size intervals illustrated in the bar chart in figure 2.17 can be explained by the fact that the large blade knives in the interval 8 to 10 centimetres mainly consist of the majority of type II large blade knives and large blade knives from Rügen, Germany. Most of the other large blade knives are between 11 and 15 centimetres long, and more than half of these are 12 to 14 centimetres long. Although this result is in line with that of Björhem and Säfvestad, who found that all of the large blade knives included in their study were between 8 and 14.5 centimetres long (Björhem & Säfvestad 1993:73), it also differs somewhat in that the material analysed here also includes large blade knives of flint that are longer.

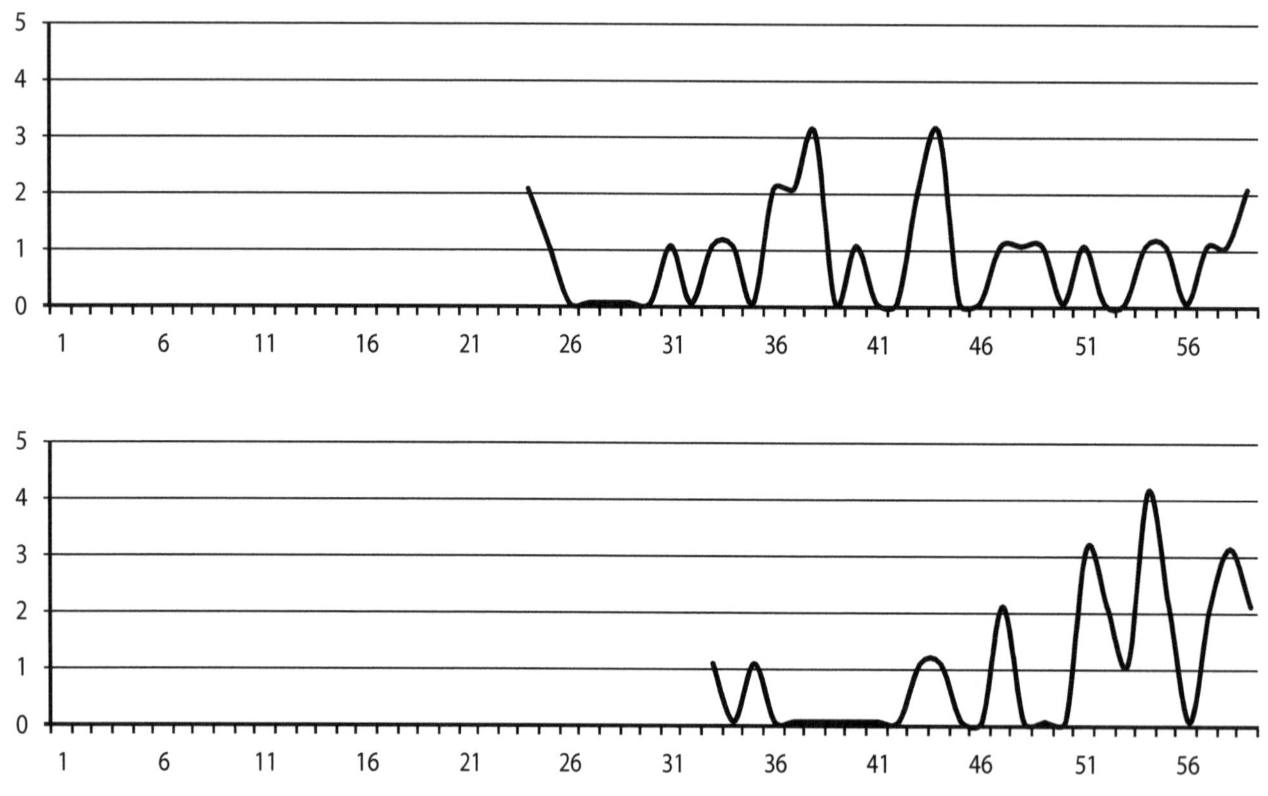

Figure 2.19. Comparison of the weight of large blade knives of flint from find spots on Rügen, Germany, top, with other type I large blade knives, bottom. The horizontal axis shows the weight in grams and the vertical axis the number of blade knives of each weight. To bring out differences and similarities in the material with greater visual clarity, I have joined the individual values in a line.

Table 2.16. The weight of the 16 whole large blade knives of type II, stated in intervals of 20 grams.

Length	Number, type II
41–60	3
61–80	7
81–100	1
101–120	3
121–140	2
Total	**16**

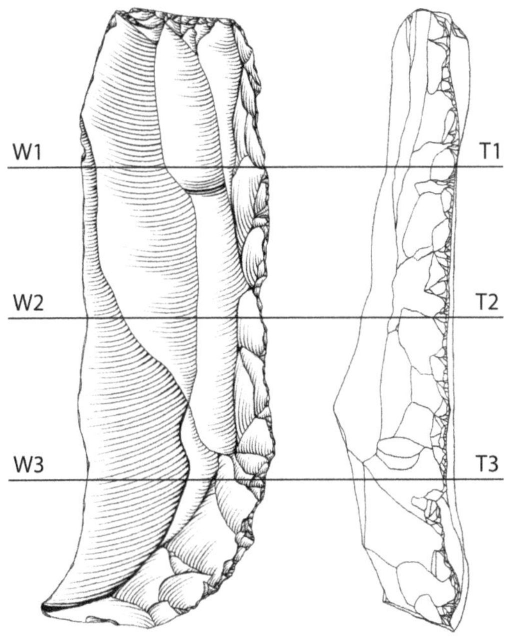

Figure 2.20. The width and thickness of large blade knives of flint was measured at three places per blade knife, at a quarter of the maximum length, at half of the maximum length, and at three quarters of the maximum length. Since both whole blade knives and fragments of blade knives are included in the analysis, the number of measured large blade knives for each dimension differs.

Weight

A total of 356 whole large blade knives were analysed for weight. Table 2.15 and figure 2.18 show that half of the whole large blade knives weigh between 60 and 100 grams and that roughly 80 per cent of the large blade knives weigh between 41 and 120 grams. Also, there are large blade knives that weigh slightly less or slightly more than the vast majority of the large blade knives.

I have studied whether the differences indicated by the analyses of the length of large blade knives presented above are also reflected in the weight of the large blade knives. Figure 2.19 shows that there is, once again, a difference between large blade knives in my material from find spots on Rügen and other type I large blade knives. The ones from find spots on Rügen weigh less. It is only from around 50 grams per large blade knife that other type I blade knives enter the picture. Again it is the case that the large blade knives from Schleswig-Holstein are like the other large blade knives of type I, not like those from Rügen.

Type II large blade knives do not differ in weight from type I large blade knives, as is evident from a comparison of tables 2.15 and 2.16. This is interesting in view of the fact that type II blade knives are generally shorter than type I. This means that type II blade knives have a greater mass than type I blade knives. I have also examined large blade knives which weigh more than average. These do not otherwise differ from the rest of the type I large blade knives.

The result of the analysis of the weight of large blade knives shows that the majority of large blade knives of types I and II weigh between 60 and 120 grams and that the majority of the large blade knives from find spots on Rügen weigh between 20 and 60 grams.

Width and thickness

The width and thickness of large blade knives has been measured with three dimensions per blade knife, at a quarter of the maximum length, at half the maximum length, and at three quarters of the maximum length (Andrefsky 1998:99), figure 2.20. The number of blade knives belonging to each group is shown in table 2.17.

Data on the actual width and thickness of large blade knives are combined in table 2.18. The relations between the actual measurements of large blade knives are shown in figure 2.21.

The analysis of the width and thickness of large blade knives shows that the blade knives are generally very similar. The majority of the measurements are within limited intervals. Most large blade knives are between three and four centimetres wide and one to one and a half centimetres thick. This result agrees well with that of Björhem and Säfvestad (1993:73), but figure 2.21 also shows that, even though the agreement is large, there are differences.

The question is whether the difference in length and weight that the analyses above have found between the large blade knives from find spots on Rügen and other type I blade knives also applies to the width and thickness of large blade knives. In figure 2.22 I have added the values for large blade knives from Rügen together with the values for other type I blade knives.

It is clear that the majority of the large blade knives from find spots on Rügen are among the low values. These blade knives are thus both narrower and thinner than the average for large blade knives of flint. The differences presented above bet-

Table 2.17. The number of large blade knives of flint in each group of dimensions.

	Width 1	Width 2	Width 3	Thickness 1	Thickness 2	Thickness 3
Number	413	407	400	414	407	401

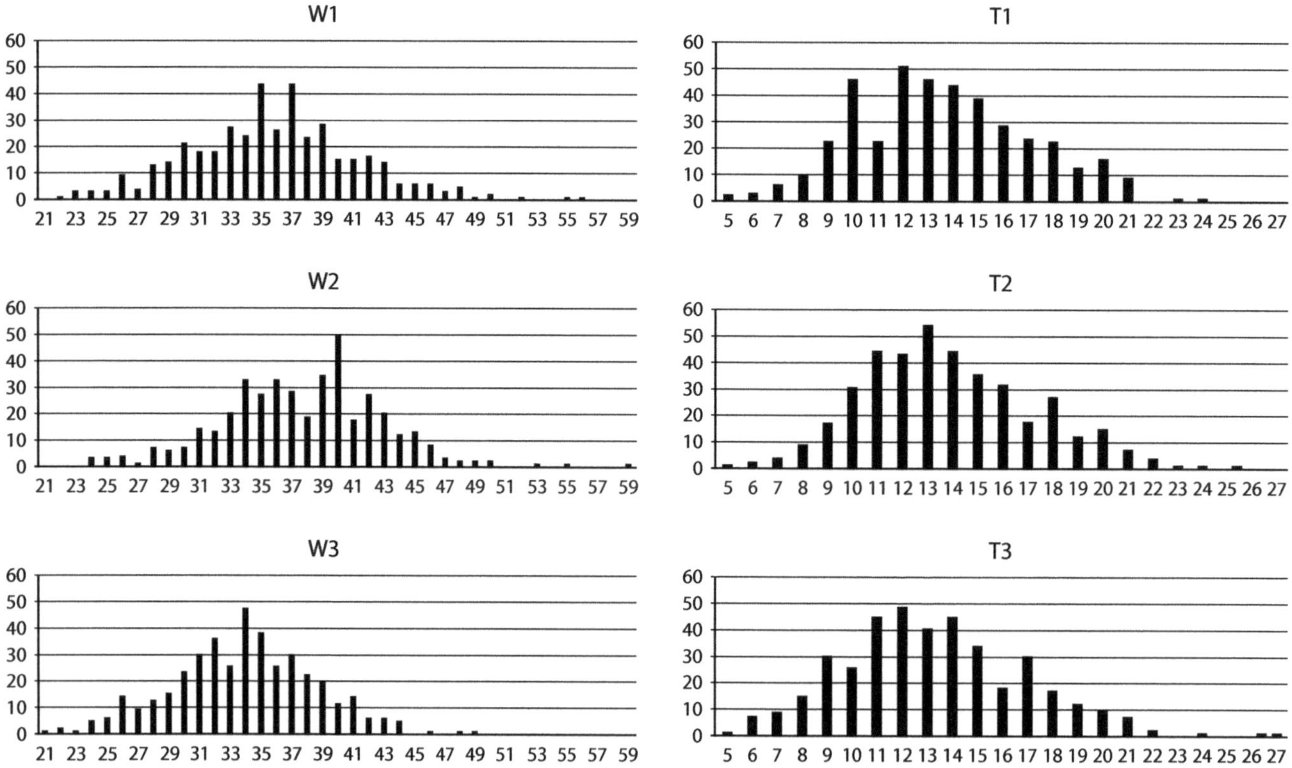

Figure 2.21. The width of large blade knives of flint, to the left, and the thickness, to the right, measured at three places on each blade knife, stated in millimetres on the horizontal axis and number of blade knives on the vertical axis. Note that the scale for millimetres on the horizontal axis differs between the charts for width and those for thickness.

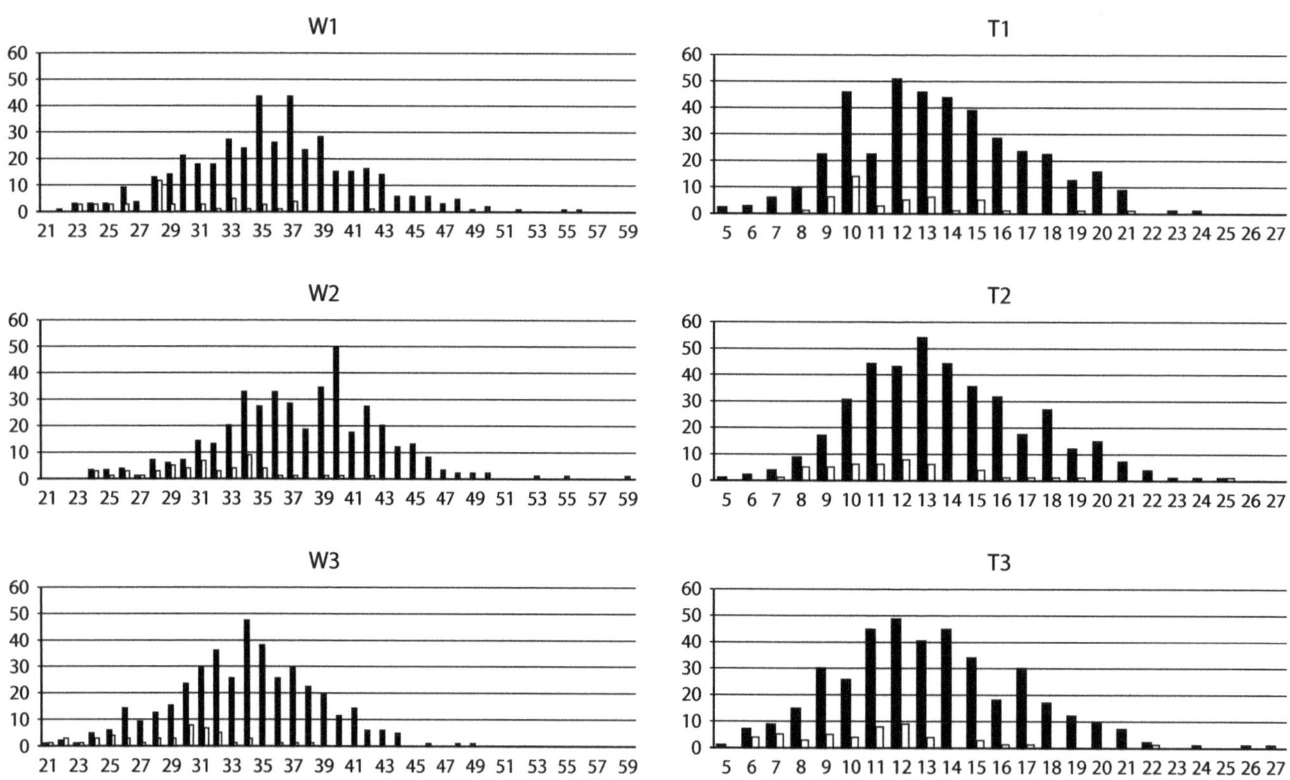

Figure 2.22. Comparison of the values for the width and thickness of large blade knives of flint from find spots on Rügen, Germany, white bars, and all large blade knives of type I, black bars. The values are stated in millimetres on the horizontal axis and number of blade knives on the vertical axis. Note that the scale for millimetres on the horizontal axis differs between the charts for width and those for thickness.

Table 2.18. The actual width and thickness of large blade knives of flint, measured in three places and stated in millimetres and number of blade knives per measurement. W = width, T = thickness.

Millimetres	Number of large blade knives per dimension					
	W1	W2	W3	T1	T2	T3
5				2	1	1
6				3	2	7
7				6	4	9
8				10	9	15
9				23	17	30
10				47	31	26
11				23	45	45
12				52	44	49
13				47	55	41
14				45	45	45
15				40	36	34
16				29	32	18
17				24	18	30
18				23	27	17
19				13	12	12
20				16	15	10
21	0	0	1	9	7	7
22	1	0	2	0	4	2
23	3	0	1	1	1	0
24	3	3	5	1	1	1
25	3	3	6	0	1	0
26	9	4	14	0	0	1
27	4	1	9	0	0	1
28	13	7	12			
29	14	6	15			
30	21	7	23			
31	18	14	29			
32	18	13	35			
33	27	20	25			
34	24	32	46			
35	43	27	37			
36	26	32	25			
37	43	28	29			
38	23	18	22			
39	28	34	19			
40	15	49	11			
41	15	17	14			
42	16	27	6			
43	14	20	6			
44	6	12	5			
45	6	13	0			
46	6	8	1			
47	3	3	0			
48	5	2	1			
49	1	2	1			
50	2	2	0			
51	0	0	0			
52	1	0	0			
53	0	1	0			
54	0	0	0			
55	1	1	0			
56	1	0	0			
57	0	0	0			
58	0	0	0			
59	0	1	0			

ween large blade knives from Rügen and Schleswig-Holstein also apply to width and thickness; the large blade knives from these two parts of northern Germany are different. The few other large blade knives that are wider or thicker than the average do not otherwise differ from other type I blade knives.

The result of the analyses of the morphology of the large blade knives of flint shows that the majority of the large blade knives in the material are of type I. Only a few type II large blade knives have been registered. Roughly half of the large blade knives have the retouch on the left side, the other half on the right side. The analyses of the length, weight, width, and thickness of large blade knives shows that there are both similarities and differences in the material. The differences concern the distinction between type I and type II blade knives, and the discrepancy between type I blade knives from find spots on Rügen and other type I blade knives. The majority of type II blade knives and the type I blade knives from find spots on Rügen are 8–10 centimetres long. The majority of other type I blade knives are 11–15 centimetres long and more than half of these are between 12 and 14 centimetres long. The result of the analysis of the weight of large blade knives shows that the majority of the blade knives of type I and II weigh between 60 and 120 grams and that the majority of the blade knives from find spots on Rügen weigh between 20 and 60 grams.

Type I and type II large blade knives are often between three and four centimetres in width and one to one and a half centimetres in thickness. Once again there is a difference in that blade knives from find spots on Rügen are narrower and thinner. As regards large blade knives which are longer, heavier, thicker, or wider than the average there are otherwise no differences. These large blade knives are type I knives which are slightly bigger than average. There is a difference between large blade knives from German find spots on Rügen and from Schleswig-Holstein. The latter resemble other type I knives from Scandinavia.

The dorsal side

The morphology of the dorsal side has been analysed in terms of the following variables: the number of negative removals, bidirectional dorsal flake scars, the number of ridges, constructed ridge, and the occurrence of cortex, figure 2.23. Since different ways of manufacturing flakes and blades, and hence also large blade knives, cause differences in the look of the dorsal side, important attributes to analyse are negative removals and ridges. These attributes are in direct relation to the appearance of the core that was used. What is the dorsal side on a large blade knife was part of the front of the core just before the blade knife was detached (Andrefsky 1998:101ff). The appearance of the negative removals and the ridges and where they occur on the dorsal side are therefore important variables in analyses of any similarities and differences in the way in which the large blade knives of flint were manufactured.

If a negative removal is to be registered it must run in the direction of percussion of the large blade knife over a quarter or more of the surface of the dorsal side. Flakes from platform preparation are not counted. What is registered as bidirectional

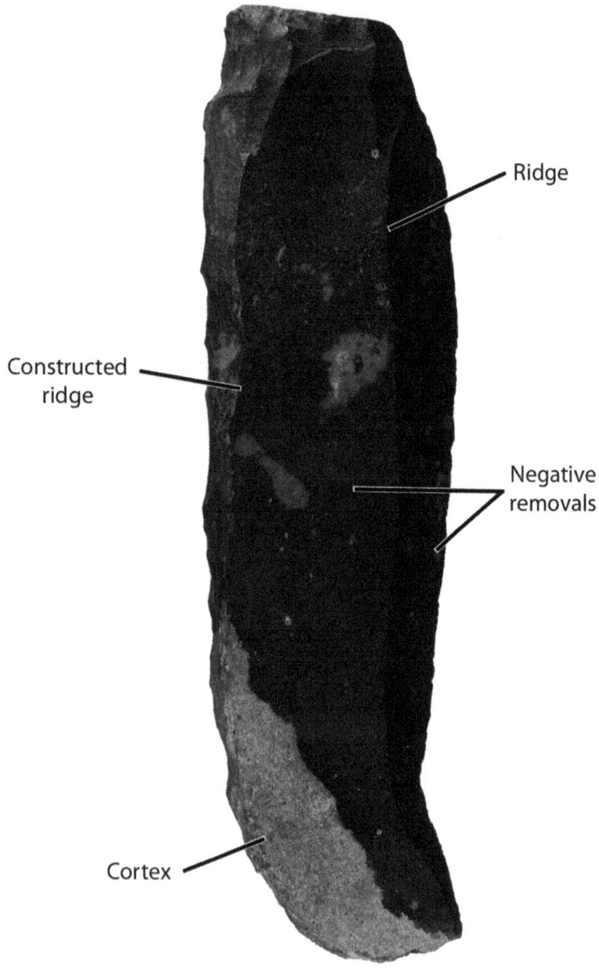

Figure 2.23. Examples of variables: the number of negative removals, number of ridges, constructed ridges, and the occurrence of cortex.

dorsal flake scars are those going against or at right angles to the direction in which the blade knife was detached. Negative removals from retouching are not counted.

The number of ridges has been registered. To be defined as such, a ridge must be demarcated by two negative removals and run over more than a quarter of the surface of the dorsal side. Ridges from platform preparation or retouching are not counted. The occurrence of constructed ridges has been noted. A constructed ridge is one that has formed on the dorsal side of large blade knives through the retouching (backing), that is to say, a ridge consisting of a line formed by the distal end of the negative removals from the retouch.

The occurrence of cortex has been noted. This was done on a sketch of each large blade knife that was registered.

A total of 357 large blade knives were included in the analysis of negative removals on the dorsal side. Table 2.19 shows that the majority of the large blade knives have two or three negative removals on the dorsal side.

A total of 454 large blade knives were included in the analysis of the number of ridges on the dorsal side. Table 2.20 shows that the vast majority of the large blade knives have one or two ridges on the dorsal side.

Table 2.19. The number of negative removals on the dorsal side.

Negatives	Number
1	30
2	216
3	109
4	1
5	1
Total	**357**

If the number of ridges on each large blade knife is classified according to type I and II blade knives, a difference is found, table 2.21.

The majority of type I blade knives have one ridge while the majority of type II blade knives have two ridges on the dorsal side. Of 357 large blade knives, 72 have a constructed ridge. Otherwise there is no difference between large blade knives with and without a constructed ridge.

Only six of 454 large blade knives have a bidirectional dorsal flake scar on the dorsal side. This means that virtually all large blade knives lack bidirectional dorsal flake scars. It is interesting to note that two of the six large blade knives that have bidirectional dorsal flake scars are type II blade knives. As mentioned above, about 5% of my study material consists of type II blade knives, yet this type accounts for some 33% of those with bidirectional dorsal flake scars.

Of 454 large blade knives, 104 have cortex on the dorsal side. All are type I blade knives. Few of these have more than 10% of the dorsal side covered with cortex. Often the cortex occurs in small areas either on the distal part of the blade knife or along the back of the knife.

The result of the analysis of the morphology of the dorsal side shows that large blade knives of flint usually have one or two ridges, two or three negative removals, and hardly ever any bidirectional dorsal flake scars on the dorsal side. About a fifth of the large blade knives have a constructed ridge and roughly as many have occurrences of cortex on the dorsal side. When cortex does occur it normally covers only small areas on the dorsal side.

The result also reveals a difference between type I and II large blade knives. None of the type II blade knives has very much cortex, a much larger share of type II than type I blade knives have bidirectional dorsal flake scars, and there is a difference between the number of ridges, as a larger proportion

Table 2.20. The number of ridges on the dorsal side of large blade knives of flint.

Ridges	Number
0	12
1	290
2	148
3	3
4	1
Total	**454**

Table 2.21. The number of ridges on the dorsal side of large blade knives of flint, classified as type I and II blade knives, number and percentage.

	Type I		Type II	
Number of ridges	**Number**	**%**	**Number**	**%**
0 and 1	294	68	8	38
2	137	31.5	11	52
3 and 4	2	0.5	2	10
Total	**433**	**100**	**21**	**100**

of type II blade knives have two ridges while the majority of type I blade knives have one ridge. Björhem and Säfvestad (1993:73ff) investigated the possibility that type II large blade knives could be reknapped type I blade knives. By comparing the two types for features such as length and width they found that "several factors argue firmly against the reknapping theory" (Björhem & Säfvestad 1993:76). The results of my analyses agree well with those of Björhem and Säfvestad. There is nothing to suggest that type II large blade knives are reknapped type I large blade knives.

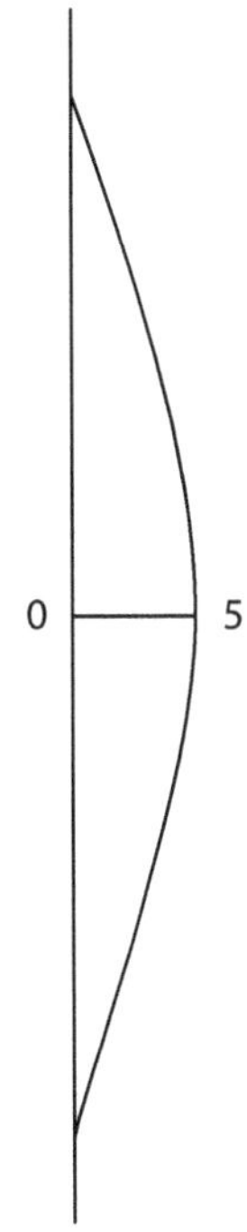

Figure 2.24. The edge outline of the large blade knives of flint has been measured on about 80 per cent of the medial part of the blade knives. The proximal and distal parts have not been considered in this measurement. The curvature is graded as 0 to 5 depending on how much the edge curves in relation to a flat surface: 0 = straight edge, 5 = edge curving one centimetre, and 1–4 = edge curving 2, 4, 6, and 8 millimetres respectively. Note that the value 5 is given to all blade knives curving more than those graded as 4. It should be emphasized here that curvature is a difficult attribute to measure. A flint edge can curve in thousands of different ways. The values are therefore not absolute. An edge that I graded as 1 one day could perhaps be graded as 2 the next day. What is important to stress, however, is that the differences are distinct between values 0 and 1 on the one hand and values 4 and 5 on the other hand. The differences expressed by these values are real.

Edge outline

A large blade knife of flint usually has a long continuous edge running along virtually the whole lateral side. The appearance of this edge is crucial for how the blade knife can be used. A straight and well-formed edge cuts in one way and a curved edge in a different way. The appearance of the edge also gives an indication of what the core looked like. A conical core gives a large blade knife with a curved edge. A core with a straight front gives a large blade knife with a straight edge (Inizan, Roche & Tixier 1992:65). The form of the edge is thus important for ascertaining what the edges looked like and how the large blade knives were made.

The curvature of the edge has been measured against a level surface (Andrefsky 1998:107f). Roughly 80 per cent of the edge on the medial part of the large blade knife has been analysed. The parts of the edge at the extreme proximal and distal ends have not been included in the measurement. The curvature has been given different values, with 0 meaning a straight edge and 5 meaning an edge that curves one centimetre or more, figure 2.24. Large blade knives with an uneven edge that curves in different directions are also classed as 5. A total of 357 large blade knives were included in the analysis of edge curvature.

The result of the analyses of edge curvature shows that the majority, almost four fifths, of the large blade knives have a straight edge, table 2.22. The differences found in the material, that is, the fifth of the large blade knives with slight curvature, cannot be explained through comparisons with other attributes. A large blade knife which has a more curved edge than others does not differ in other attributes.

Table 2.22. Values for edge outline.

	Number
0	156
1	124
2	45
3	16
4	10
5	6
Total	**357**

Exterior platform angle

Platform angle is an attribute that has been held up in several studies as being important in order to define, analyse, and distinguish different manufacturers (Dibble & Whittaker 1981; Dibble & Pelcin 1995; Davis & Shea 1998; Shott et al. 2000). The platform angle can be measured either as the interior flaking angle, that is, the angle found at the transition between the platform of the flake and its ventral side, or as the exterior platform angle (also called "angle de chasse"), that is, the angle found at the transition between the platform of the flake and

Table 2.23. Exterior platform angle, stated in degrees, number, and percentage.

Degrees	Number	%
65	6	1.5
70	55	16
75	214	61
80	69	19.5
85	6	1.5
90	3	0.5
Total	**353**	**100**

Figure 2.25. The morphology of the platform: flat, faceted, and cortical.

Figure 2.26. A rhombic platform.

its dorsal side (Inizan, Roche & Tixier 1992:37; Holdaway & Stern 2004:120ff). I have measured the exterior platform angle on a total of 353 large blade knives.

The result of the analyses of the exterior platform angle shows that more than half of the large blade knives have a platform angle of 75 degrees and that virtually all blade knives have a platform angle of 70–80 degrees, table 2.23. It is interesting to note that the three large blade knives of Zele type from Poland (Lech & Lech 1997) included in the material have platform angles of 80, 85, and 90 degrees. They thus differ from the majority of the other blade knives. The large blade knives from the find spots on Rügen do not differ as regards the platform angle attribute, nor do type II blade knives.

Morphology and type of the platform

The morphology and type of the platform are important attributes for distinguishing different reduction strategies and also one of the most frequently used characteristics in attribute analyses (Inizan, Roche & Tixier 1992:81; Madsen 1992:105f; Shott 1994:80f). The appearance of the platform has been analysed according to whether it is flat, faceted, or cortical, figure 2.25.

A total of 387 large blade knives were analysed for the morphology of the platform. Table 2.24 shows that the predominant platform is flat. Among the eight blade knives that do not have flat platforms, there is no other attribute to distinguish them from other large blade knives.

The platform type has been analysed according to whether it is rhombic or not, figure 2.26. Of the total 413 large blade knives with preserved platform, it is rhombic on all of them. No other shapes of platform occur.

The result of the analyses of the morphology and type of the platform shows that virtually all of the large blade knives of flint have a platform that is flat and rhombic.

Table 2.24. Morphology of the platform, with the different types stated in numbers.

Morphology	Number
Flat	379
Faceted	6
Cortical	2
Total	**387**

Platform preparation

Just like the attribute morphology and type of the platform, platform preparation is frequently used in attribute analyses to distinguish different reduction strategies (Inizan, Roche & Tixier 1992:81; Madsen 1992:105f; Shott 1994:80f). I have analysed platform preparation based on the extent of the preparation, figure 2.27 and the occurrence of negative removals from special preparation flakes which I call preparation flakes with feathered termination and preparation flakes with hinge termination, figure 2.28. A feathered termination flake is an overhang removal with a smooth termination at the distal end of the flake. Hinge termination is a termination caused when the force of the impact used to remove the flake turns or rolls away from the core (Andrefsky 1998:18, 86). These different kinds of terminations on preparation flakes are both deliberately made; they are not accidentally caused by mistakes.

A total of 402 large blade knives are included in the analysis of the extent of platform preparation. Table 2.25 shows that the majority of the blade knives are prepared along the platform. It also shows that there is variation in the appearance of the extent of the platform preparation from the platform edge down over the dorsal side.

In the analysis of the platform preparation of large blade knives I discovered a difference between the blade knives

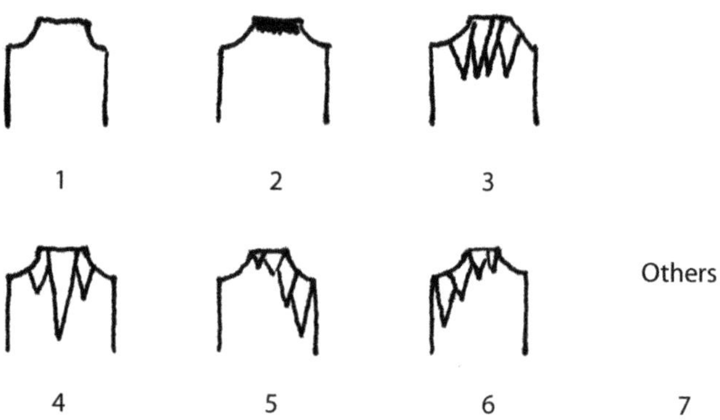

Figure 2.27. The extent of the preparation on the dorsal side of large blade knives of flint has been registered in seven variants.

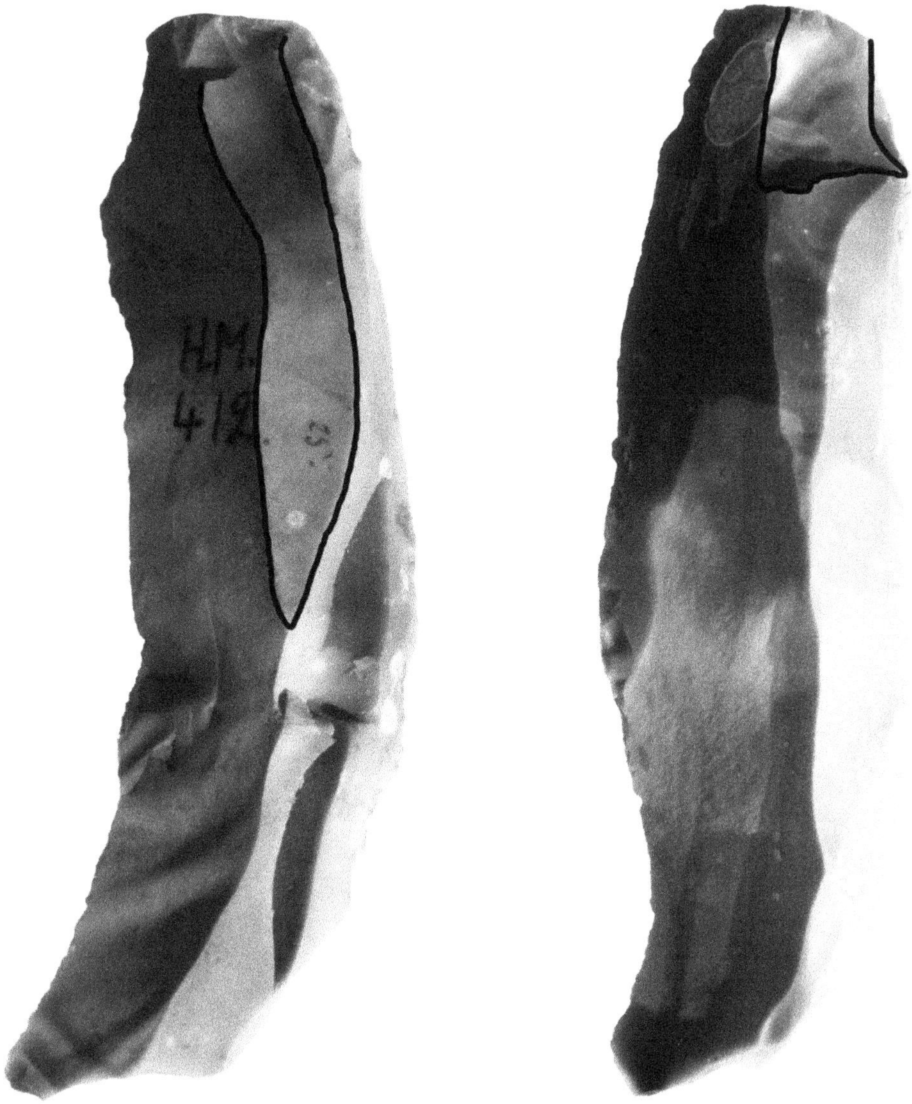

Figure 2.28. Large blade knives of flint with negative removals from preparation flakes with feathered termination and preparation flakes with hinge termination.

which is not fully visible in the attribute analysis as I defined it. The majority of the blade knives have platform preparation running down past the shoulders of the blade knives. The edge between the platform and the dorsal side is then usually shaped in a continuous line, see figure 2.29. What I discovered in the analysis was that some of the large blade knives do not have a smooth line here. Instead the edge between the platform and the dorsal side consists of a ridge formed by two negative removals. Often these negatives run from the platform down over large parts of the surface on the dorsal side, see figure 2.29.

Figure 2.29. Right: a large blade knife where the edge between the platform and the dorsal side is smoothly shaped in a continuous line. Left: a large blade knife where the edge between the platform and the dorsal side consists of a ridge.

Table 2.25. The extent of platform preparation, stated in number and percentage for each variant.

Extent	Number	%
1	63	15.5
2	103	25.5
3	42	10.5
4	150	37.5
5	20	5
6	20	5
7	4	1
Total	**402**	**100**

Table 2.26. The extent of platform preparation divided into large blade knives where the platform edge is a ridge formed by two negative removals and other large blade knives, stated in number and percentage for each variant.

Extent	With ridge		Other	
	Number	%	Number	%
1	14	47	49	13.5
2	12	40	91	24.5
3	0	0	42	11.5
4	2	6.5	148	40
5	0	0	20	5
6	0	0	20	5
7	2	6.5	2	0.5
Total	**30**	**100**	**372**	**100**

Table 2.27. The number of large blade knives of flint with preparation flakes with hinge termination, divided according to the number of preparation flakes present on each blade knife.

Number of preparation flakes on each blade knife	Preparation flakes with hinge termination
1	77
2	56
3	25
4	4
5	0
6	1
7	1
Total	**164**

Since I realized that this difference between different types of platform edges would not be visible in the attribute analysis, what I did when I found large blade knives on which the platform edge is a ridge, as described in figure 2.29, was to give these blade knives a special marking in my registration.

When the extent of the platform preparation on these blade knives is compared with the result of the analysis above, a significant difference is seen between large blade knives on which the platform edge is a ridge and blade knives where the edge between the platform and the dorsal side is smoothly shaped in a continuous line, table 2.26 and figure 2.30.

Of the large blade knives where the platform edge is a ridge, virtually all either have no platform preparation or else preparation that does not run down past the blade knife's shoulders, i.e., variant 1 or 2 in figure 2.27. Of the blade knives where the edge between the platform and the dorsal side is smoothly shaped in a continuous line, three fifths have preparation running down past the blade knife's shoulders.

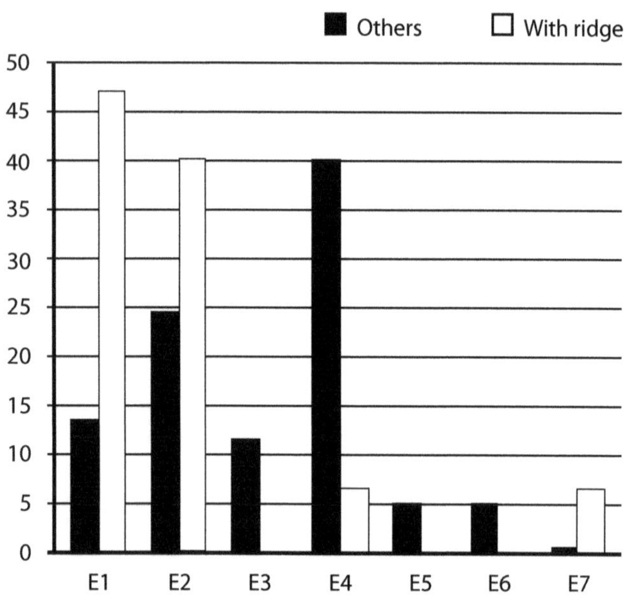

Figure 2.30. Bar chart showing the differences presented in table 2.26, stated as percentages. E = extent of preparation according to the variants in figure 2.27.

Of the total 402 large blade knives included in the analysis of platform preparation, 164 have preparation flake with hinge termination and 136 have preparation flake with feathered termination, table 2.27 and 2.28. Many of the blade knives, 66 in all, have both types of preparation flakes, and a total 237 of the large blade knives have preparation flakes with either hinge or feathered termination. Table 2.27 and 2.28 shows that it is most common to have one, two, or three preparation flakes with hinge or feathered termination.

If we go back to the earlier analyses of differences in the appearance of the preparation, we note that none of the 30 large blade knives where the platform edge is a ridge has preparation flakes with hinged termination, and that only three of them have preparation flake with feathered termination. The difference in the extent of the platform preparation is thus found again as regards the occurrence of hinge and feathered preparation flakes.

The result of the analyses of the platform preparation of large blade knives of flint shows that the majority of the blade knives have platform preparation in the form of trimming of the platform edge and small preparation flakes running down over the surface of the dorsal side. The look of this preparation varies somewhat. Many large blade knives also have preparation flakes with hinged and feathered termination as part of the preparation. The result also shows that there are blade knives with a different preparation or none at all. These are large blade knives where the platform edge is a ridge formed by

Table 2.28. The number of large blade knives of flint with preparation flakes with feathered termination, divided according to the number of preparation flakes present on each blade knife.

Number of preparation flakes on each blade knife	Preparation flakes with feathered termination
1	117
2	15
3	3
4	1
Total	**136**

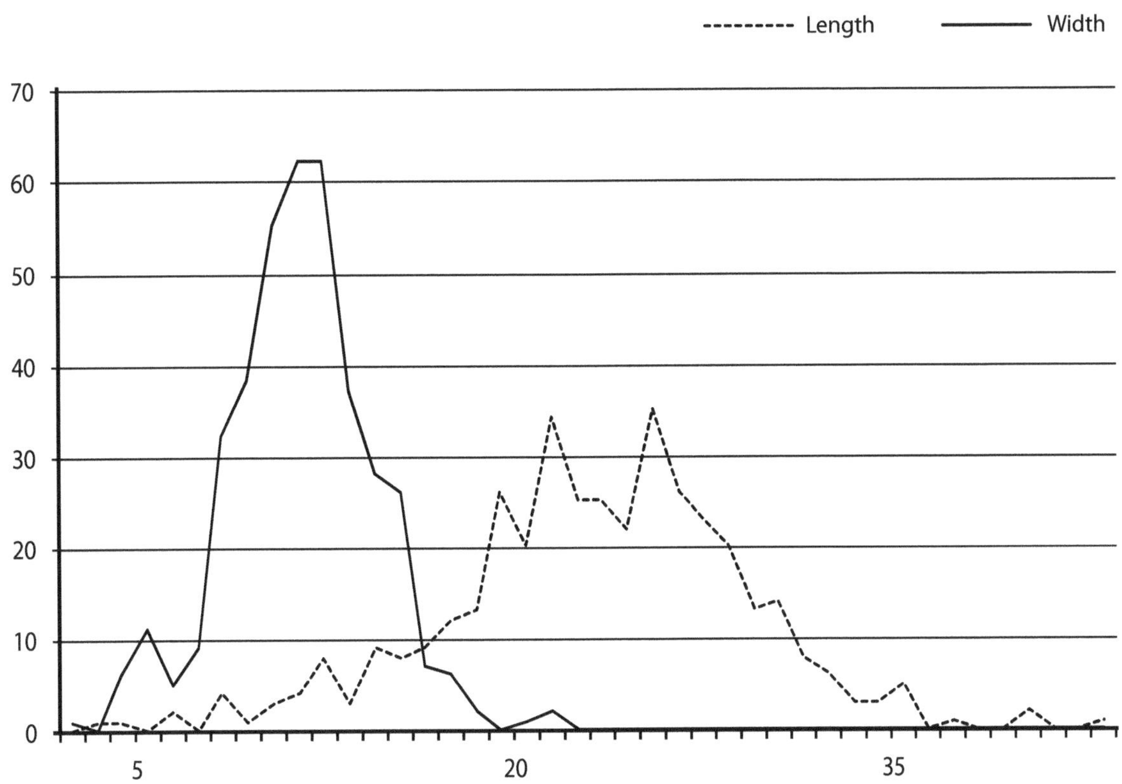

Figure 2.31. The size of the platform, stated in numbers and width and thickness in millimetres. To highlight differences and similarities I have once again joined the values to give continuous lines.

two negative removals running from the platform down over the surface of the dorsal side.

The way in which platforms are prepared is crucial for the way in which the knapping is done (Inizan, Roche & Tixier 1992:46f). The difference in the manner of platform preparation is thus an indication of different ways of manufacturing large blade knives. This will be discussed below.

The size of the platform

The size of the platform is a metric variant of the appearance of the platform (Inizan, Roche & Tixier 1992:81; Madsen 1992:105f; Shott 1994:80f). The size has been measured in width, the distance between the platform edges as defined by the two lateral sides, and thickness, the distance between the platform edges as defined by the dorsal and ventral side (Andrefsky 1998:93, figure 5.5). Altogether 390 large blade knives are included in the analysis of the size of the platform. Table 2.29 and figure 2.31 show that the majority of the platforms are 1.5 to 3.0 centimetres in width and 0.6 to 1.5 centimetres in thickness.

I have investigated whether any of the differences in the material presented above for other attributes also exist as regards the attribute the size of the platform. I have not found any such differences.

Ring crack on the platform

Ring crack on the platform can be used to identify different types of tools used in the manufacture. Soft tools rarely cause a ring crack, while hard tools often cause a ring crack (Kooyman 2001:13). The occurrence of ring crack on the platform has been registered here when it occurs around the impact point and when it occurs beside the impact point, figure 2.32.

A total of 390 large blade knives are included in the analysis of ring crack on the platform, table 2.30. Of these 371, or 95%, have ring crack around the point of impact. This is thus a very common attribute. About a quarter of these, 103 in number, also have one or more ring cracks on the platform beside the point of impact, table 2.31. No large blade knife has ring crack only beside the impact point. The table shows that

Table 2.29. Width and thickness of the platforms, grouped in intervals of five millimetres, stated in centimetres and number of large blade knives of flint per interval.

Size	Width	Thickness
−0.5	2	18
0.6–1.0	10	139
1.1–1.5	32	215
1.6–2.0	80	17
2.1–2.5	141	1
2.6–3.0	96	0
3.1–3.5	25	0
3.6–4.0	3	0
4.1–4.5	1	0
Total	**390**	**390**

Figure 2.32. A platform on a large blade knife of flint with ring crack around the impact point, and beside the impact point, to the left and the right.

there are usually one or two ring cracks on the platform beside the point of impact.

I have investigated whether any of the differences in the material presented above for other attributes also exist as regards the attribute ring crack on the platform. I have not found any such differences.

The result of the analysis of ring crack on the platform shows that almost all of the large blade knives have ring crack on the platform around the point of impact. Of these, a quarter also have ring crack beside the point of impact.

The morphology of the bulb of percussion

How the bulb of percussion on a flake developed depends on the type of production in which the knapping was done. Bifacial manufacture of a dagger, for example, gives more flakes with a diffuse bulb of percussion, while manufacture of square-sided axes gives flakes with a marked bulb of percussion (Högberg 1999a:85f). The morphology of the bulb of percussion has been registered in terms of the variables well-formed bulb of percussion or not, whether the bulb of percussion is marked or diffuse, and whether the bulb of percussion shows a bulbar scar, moustache, and bulbar fissures. Moustache is a variant of the attribute bulb of percussion and consists of an incipient hertzian cone, linking the butt to the bulb (Holdaway & Stern 2004:34), figure 2.33.

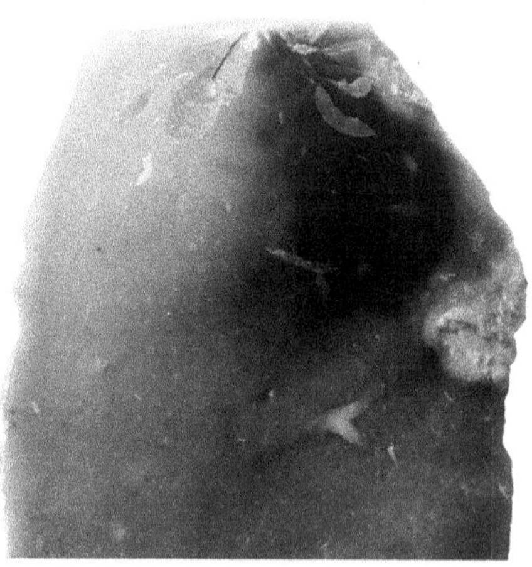

Figure 2.33. Variables of the attribute the morphology of the bulb of percussion which I have used in the analyses. A flake with a well-formed and marked bulb of percussion with bulbar scar, moustache, and bulbar fissures.

Table 2.30. The number of large blade knives of flint with and without ring crack around the point of impact.

	Number
With ring crack	371
Without ring crack	19
Total	**390**

Table 2.31. The number of large blade knives of flint with ring crack on the platform beside the point of impact.

Number of ring cracks beside the point of impact	Number of blade knives
1	77
2	20
3	2
4	2
5	1
6	0
7	0
8	1
Total	**103**

A total of 390 large blade knives are included in the analysis of the morphology of the bulb of percussion. The result shows that most large blade knives have a marked and well-formed bulb of percussion with bulbar scar, moustache, and bulbar fissures, tables 2.32, 2.33, 2.34, 2.35 and 2.36.

There is a difference in the material as regards the variable diffuse bulb of percussion. Roughly a fifth of the large blade knives from find spots on Rügen, Germany, have a diffuse bulb of percussion, as opposed to around 5 per cent of other blade knives, table 2.37. This difference is small, however. Once again, the large blade knives from Schleswig-Holstein are like those from Denmark and Sweden but unlike those from Rügen. For other variables I have not found any difference between different parts of the material.

Retouch

The retouch on large blade knives has been studied through a description of the appearance and form of the retouch in two parts: the morphology and position of the retouch. The analysis of the retouch morphology describes the look and character of the negatives of the retouch and has been classified in three groups, see figure 2.34.

Retouch position states whether the retouch was done on the dorsal or ventral side of the flake, classified here in four groups, see figure 2.35.

Table 2.32. The number of large blade knives of flint with and without a well-formed bulb of percussion.

Well-formed bulb of percussion	Number
Yes	370
No	20
Total	**390**

Table 2.33. The number of large blade knives of flint with marked and diffuse bulb of percussion.

	Number
Marked	363
Diffuse	27
Total	**390**

Table 2.34. The number of large blade knives of flint with and without bulbar scar.

Bulbar scar	Number
Yes	245
No	145
Total	**390**

Table 2.35. The number of large blade knives of flint with and without moustache.

Moustache	Number
Yes	345
No	45
Total	**390**

Table 2.36. The number of large blade knives of flint with and without bulbar fissures.

Fissures	Number
Yes	364
No	26
Total	**390**

Table 2.37. Comparison of large blade knives of flint with diffuse bulb of percussion, showing blade knives from find spots on Rügen in relation to other large blade knives, in number and percentage.

	Number	%
Large blade knives from find spots on Rügen, Germany	52	
Proportion with diffuse bulb of percussion	10	19
Other large blade knives	338	
Proportion with diffuse bulb of percussion	17	5

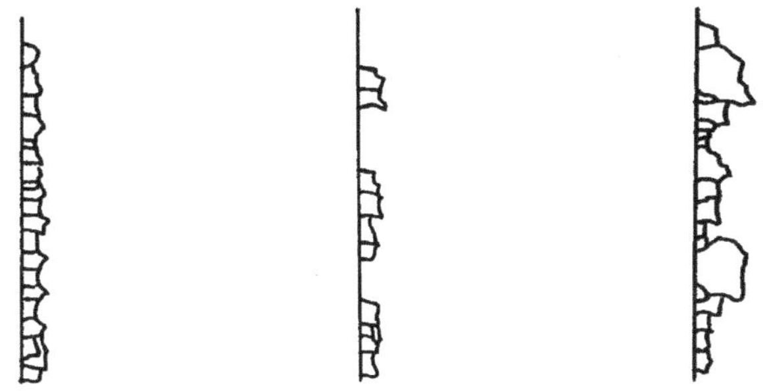

1 2 3

Figure 2.34. 1: Continuous and regular retouch. 2: Partial and regular retouch. 3: Continuous and irregular retouch (from Inizan, Roche & Tixier 1992:85f). Continuous retouch consists of an uninterrupted sequence of negative removals in the retouched area. Partial retouch is a sequence of negative removals interrupted by unretouched areas between the retouched areas (Inizan, Roche & Tixier 1992:85f). Regular retouch is a smooth retouch consisting of similar negative removals of homogeneous width and height. Irregular retouch is an uneven retouch consisting of negative removals of varying height and width (Andrefsky 1998:77ff).

A total of 454 whole and fragmentary large blade knives are included in the analysis of retouch. The result shows that 440 large blade knives are retouched, with only 14 blade knives lacking retouch. Tables 2.38 and 2.39 show that continuous, regular, and direct retouch are by far the most common. The number of large blade knives with other retouches is small.

I have investigated whether any of the differences presented above for other attributes also exist as regards the morphology and placing of the retouch, but no such differences have been found.

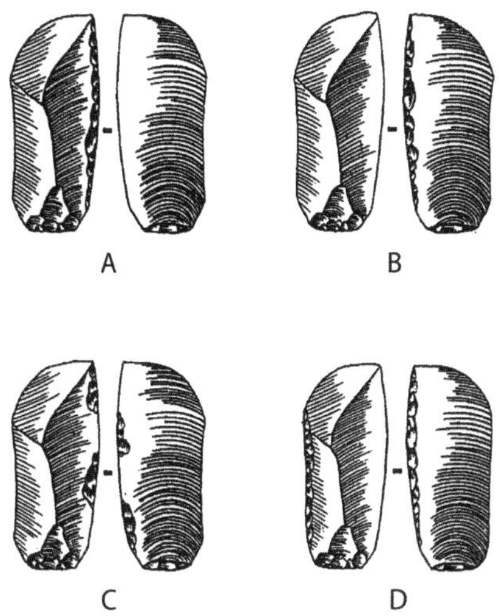

Figure 2.35. A: Direct retouch. B: Inverted retouch. C: Alternating retouch. D: Alternate retouch. Direct retouch: the retouch is seen on the dorsal side of the flake; inverted retouch: the retouch is seen on the ventral side of the flake; alternating retouch: the retouch is seen on both the dorsal and the ventral side on the same retouched edge; and alternate retouch: the retouch is seen on both the dorsal and the ventral side but not on the same edge (Inizan, Roche & Tixier 1992:94).

Table 2.38. The morphology and position of the retouch stated as the number of large blade knives of flint.

Morphology		Position	
1	421	A	415
2	19	B	10
3	0	C	14
		D	1
Total	**440**		**440**

Table 2.39. Frequency of different combinations of morphology and position of retouch on large blade knives of flint.

Combination		
Morphology	Position	Number
1	A	396
1	B	10
1	C	14
2	A	19
1	D	1
		440

Type of flint

The type of flint from which each large blade knife is made has been registered, in general terms for all large blade knives and specifically for the assemblages presented later in the chapter "Assemblages of large blade knives of flint". The aim of the general registration was to give an idea as to whether the large blade knives were made of a great many different types of flint or if only certain types were used. For the registration I therefore noted whether the flint from which each large blade knife was made is Senonian or Danian. The reason for my choice of such general type designations as Senonian and Danian flint is that a large proportion of the large blade knives have a heavy patina and lack cortex. This makes it difficult to arrive at a more precise definition. When it has been possible, however, I have noted specific varieties of these flint types. As is evident from table 2.40, the majority of the large blade knives are made of Senonian flint.

Table 2.40. Flint types of the 454 large blade knives. Note that I have primarily registered a difference between Senonian and Danian flint. When it has been possible, however, I have specified the different varieties of these types (Högberg & Olausson 2007).

Type		Number	%
Senonian		423	93
	Falster flint 6		
	Common Kristianstad flint 4		
Danian		25	6
	Brown Bryozoan flint 8		
	Bryozoan flint, Funen variety 9		
Till flint		6	1
Total		**454**	**100**

An interesting difference in the type of flint which is connected to the differences discussed above between type I large blade knives with and without platform preparation is that blade knives without platform preparation have a somewhat larger share of cortex on the dorsal surface and are made from a Scandinavian Senonian flint that usually occurs as tabular flint. This tabular flint is found, for example, in the flint-mining areas at Södra Sallerup in Scania, Sweden, or at Bjerre in north-west Jutland, Denmark (Högberg & Olausson 2007:88).

Results – a morphological and technological attribute analysis of large blade knives of flint

The result of the attribute analyses of large blade knives of flint reveals both similarities and differences. The majority of the large blade knives are type I. These are between 11 and 15 centimetres long, weigh 60–120 grams, are three to four centimetres wide and one to one and a half centimetres thick. They usually have one or two ridges and two or three negative removals on the dorsal side and lack cortex. The majority of the blade knives have a straight and well-formed edge and an exterior platform angle of 70–80 degrees. The platform is flat and rhombic with a ring crack around the impact point. The platform is usually 1.5 to 3.0 centimetres in width and 0.6 to 1.5 centimetres in thickness. Platform preparation in the form of trimming along the platform edge and thin preparation flakes running down over the dorsal side commonly occur. Many large blade knives also have preparation flakes with hinged and feathered termination as part of the preparation. Most blade knives have a marked and well-formed bulb of percussion with a bulbar scar, moustache, and bulbar fissures. Virtually all large blade knives have continuous, regular, and direct retouch. Roughly half of the blade knives have the retouch on left side, the other half on the right side. The majority of the large knives are made of Senonian flint.

The differences are that a small proportion of the large blade knives do not look like the type I blade knives that have just been described. The majority of type II blade knives and type I blade knives from find spots on Rügen, Germany, are between 8 and 10 centimetres long. Type I and type II blade knives weigh the same, while the majority of the large blade knives from find spots on Rügen weigh between 20 and 60 grams. Type II large blade knives thus have a greater mass than type I blade knives. Type II large blade knives are as thick and wide as type I large blade knives, whereas blade knives from find spots on Rügen are narrower and thinner. It is interesting to note that the large blade knives of flint in my material from the Schleswig-Holstein area in Germany look like blade knives from Denmark and Sweden, not like those from Rügen. Also interesting is that the large blade knives from south-western Denmark which have been discussed by Ethelberg (1982) and Draiby (1985), see the chapter "A history of previous archaeological research", look like blade knives from Rügen. There is thus a difference in the appearance of large blade knives in

Figure 2.36. From the left, a type I large blade knife with a well-prepared platform, a type I large blade knife without platform preparation, a type II large blade knife, and a large blade knife from Rügen, Germany.

northern Germany and south-western Denmark. The result also reveals other differences between type I and II large blade knives. Few of the type II blade knives have any cortex, a much larger share of type II than type I blade knives have bidirectional dorsal flake scars, and there is a difference between the number of ridges, with a larger share of type II blade knives having two ridges while the majority of type I blade knives have one ridge. It is interesting that the three large blade knives of Zele type from Poland (Lech & Lech 1997) included in the material have platform angles of 80, 85, and 90 degrees. They thus differ from the majority of the other large blade knives. The result also shows that there are blade knives of type I which lack platform preparation. These are large blade knives where the platform edge is a ridge formed by two negative removals running from the platform down over the dorsal surface.

Based on this finding, I have divided the large blade knives of flint into four groups, figure 2.36. The result shows that large blade knives of type I with a well prepared platform are similar in appearance and form, and they are by far the most common type in the material. There is also a type I large blade knife without platform preparation which is not as common but distinct and easy to recognize. Type II large blade knives are not common either but they are also distinct and easy to recognize. Large blade knives from Rügen have several technological attributes which resemble those of type I blade knives, but they differ in size. They are smaller. Type I and II large blade knives have been described before (Björhem & Säfvestad 1993:74f). Large blade knives of type I without platform preparation and large blade knives from Rügen have not been described previously.

The large blade knives in each of these groups resemble each other, with a great many shared attributes. Between the groups there are both similarities and differences. This makes the boundaries between the groups vague.

The morphological and technological attribute analyses have shown what the large blade knives of flint look like and the attributes and variables of these attributes found on the blade knives. I will now go on to tackle the question of how the large blade knives were manufactured. This is done in an experimental study of the production of large blade knives.

The manufacture of large blade knives – experimental analyses

How is a large blade knife of flint made? Is it a sophisticated craft or something that anyone can do? What does the reduction process look like? These seemingly simple questions are not exactly easy to answer. For even if several scholars have had opinions about the technology represented by the large blade knives and others have formulated hypotheses about manufacturing and production processes, as we saw in the historical survey of archaeological research above, no research has been done about how the large blade knives were actually made. To be able to work with questions about the manufacture and technology of large blade knives of flint, I have performed a number of experiments at manufacture.

Experimental archaeology has been a vital part of lithic research for a long time (Johnsson 1978). The pioneering experiments by Skertchly (1879) and Sehested (1884) in the late nineteenth century are often held up as early examples, even though instances of previous research can be found (Johnson 1978; Whittaker 2004). Since then the topic has developed and today's researcher can lean on a well-established international sub-field of archaeological research (Odell 2004). There are many experimental studies of raw material, technology, and function, providing a solid base for further research (Swanson 1975; Inizan, Roche & Tixier 1992; Whittaker 1994; Andrefsky 1998; Eriksen 2000; Kooyman 2001; Holdaway & Stern 2004; Odell 2004). And, as Douglas B. Bamforth and Nyree Finlay (2008:3) recently put it: "We know what it takes to make flaked stone tools."

How this knowledge is translated into interpretations of archaeological material and events in the past varies, however. As we saw above, research in lithic analysis has been developing its methods for a long time, whereas theoretical discussions have not been developed to the same extent (Odell 1996:4f), and it is not uncommon that excessively hasty conclusions are drawn from practical experiences rather than from interpretations of the archaeological material. This has been harshly criticized by others (see e.g. Whittaker 1994:282f) and out of this critique there have come methodical and systematic studies with well-formulated research agendas (see e.g. Callahan 1987; Kelterborn 2002, 2005b; Hirth 2003). What they have in common is the principle of building the experiments on a solid knowledge of the archaeological material, well-formulated questions, and clearly expressed methods, all with the ambition to create a foundation for interpretations through analogies (Mathieu 2002). It is important here to point out that the analogies must be interpreted in relation to research questions and contexts (Finley 2008:72). Practically gained experiences are not the same as interpretation. Experimental manufacture shows how something might have been made, not how it actually was made.

In Scandinavian research, experimental archaeology follows these general trends, and it has long been an important part of lithic studies (Kragh 1964; Olausson 1983; Knutsson 1988; Madsen 1992, Jensen 1994; Högberg 1999; Eriksen 2000; Apel 2001; Knarrström 2001; Sørensen 2006a, 2006b). In particular, the Lejre Research Centre in Denmark has for more than forty years encouraged and financed extensive experimental research, always with a special focus on lithic technology (Meldgaard & Rasmussen 1996; Lejre Forsøgscenter 2008). Experimental archaeology is a vital part of my studies, and a lot of the thinking presented in this book is based on experiments. And, for a Scandinavian researcher, it is obvious that there are previous experiences to lean on.

The experiments

I have conducted a number of experimental studies in order to ascertain how Late Bronze Age large blade knives of flint were produced. Together with the archaeologist and flint knapper Mikkel Sørensen I did this at the Lejre Research Centre in Denmark in 2000 and 2005 (reference number HAF 6/00 and HAF 04/05; experiment reports: Högberg 2000b, 2005b). The results of these experiments have been supplemented with further series of trials performed at Malmö Heritage, Sweden, in 2003 and 2007. Mikkel and I have several years' experience of flint knapping and experimental archaeology. Mikkel is a skilled knapper with more than twenty years' experience (Sørensen 2006a, 2006b).

The point of departure for the manufacturing experiments are the results of the attribute analyses of prehistoric large blade knives of flint presented in the previous chapter. In these analyses four groups of large blade knives were defined. A type I blade knife with a well-prepared platform, a type I blade knife without platform preparation, blade knives of type II, and blade knives from Rügen. These groups represent both variation in the morphology of the tool and uniformity as regards aspects of manufacturing technology.

The uniformity is seen in the fact that raw material of high quality was used for the production. Technological attributes such as platform angle (angle de chasse) and the size and form of the platform are the same. Flat platforms with a ring crack around the point of impact, a well-formed and marked bulb of percussion with moustache and fissures on the bulb of percus-

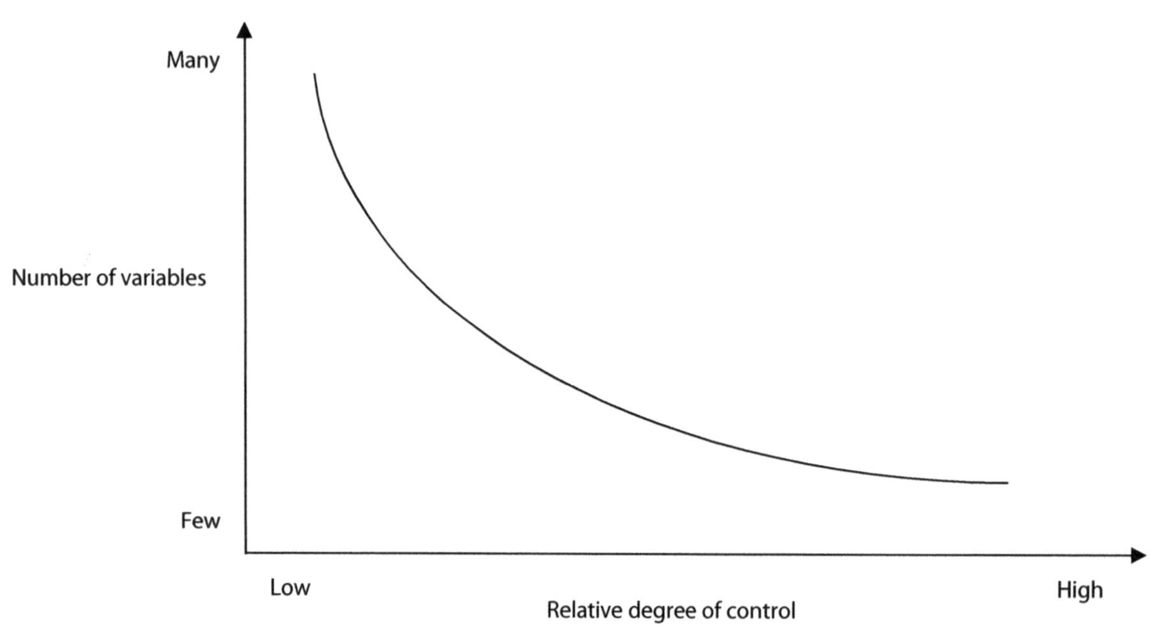

Figure 2.37. By discussing different approaches, asking questions, testing these in controlled experiments, discussing again, asking new questions, testing again, and so on, we created situations where we constantly learned new things about how large blade knives of flint can be made. In this way we were able, figuratively speaking, to move gradually right along the curve in this graph by reducing the number of variables in the experiment step by step and thereby also increasing the degree of control in the work. The figure has been redrawn from Mathieu (2002:11).

sion commonly occur. The edge is characterized by showing virtually no curvature.

The variation is most clearly seen in the obvious difference in size between blade knives of type I, those of type II, and blade knives from find spots on Rügen. A more subtle variation is that, while the majority of the blade knives have a well-prepared platform with preparation flakes running down over the dorsal side, some blade knives lack platform preparation. The variation also concerns the occurrence of cortex on the dorsal side. On the majority of the large blade knives this is rare, and when it does occur the cortex often covers less than 10% of the dorsal surface. But some large blade knives have much more cortex, sometimes with up to 50% of the dorsal surface covered.

Two reference large blade knives

These observations of morphological and technological attributes serve as a foundation for the manufacturing experiments. Based on these attributes, two fictitious "reference large blade knives" of type I were selected, one with and one without platform preparation. These reference blade knives were used as yardsticks and templates for the experimental production. They were chosen for the experimental manufacture because they are the most common types. No manufacture of type II large blade knives and blade knives from Rügen has been attempted.

Both the reference blade knives are 12.5 centimetres long, three to four centimetres wide, and one to one and a half centimetres thick. They are made of high-quality Senonian flint. They have straight edges, a marked bulb of percussion with moustache and fissures, ring crack around the point of impact on the flat rhombic platform, and a platform angle of 75 degrees.

The reference blade knives differ as regards platform preparation. One of them has a smooth and well-prepared platform edge and negative scar left by preparation flakes with hinged and feathered termination running from the platform edge down over the dorsal side. The other blade knife lacks this and has instead a platform edge consisting of a ridge formed by two negative removals running from the platform down over the dorsal surface. Although both reference blade knives show an average of all the attributes found on the respective types of analysed prehistoric large blade knives, they are not representative of any of these. It is rare to find all attributes on one and the same large blade knife of flint. However, the morphology of an individual attribute on a reference large blade knife is a representative mean value for each of the attributes analysed on the prehistoric large blade knives.

Controlled experiments with specific questions

Since the experiments performed by Sørensen and myself were the first ever attempted, there was no specific knowledge about the manufacture of large blade knives of flint to draw on for the planning of the work. Instead we had to proceed from what was known about prehistoric large blade knives and our previous general knowledge of both modern and prehistoric lithic technology (see e.g. Högberg 1999, 2001a, 2002, 2004a; Sørensen 2006a, 2006b). This required us to design the experiments in such a way that we learned something new, not just reproduced what we already knew. The method we used was to create a learning situation where we sought to put ourselves in situations that forced us to alternate between previous experience, questions, new knowledge, and new questions. By discussing different approaches, asking questions, testing these in controlled experiments, discussing again, asking new questions, testing again, and so on, we created situations where we constantly learned new things about how large blade knives of flint can be made. In this way we gradually reduced the

Figure 2.38. The learning in the experiment was a constant alternation between good and bad ideas and successful and unsuccessful series of experiments. Here is a photograph from the fourth round of experiments in making large blade knives with Kristianstad flint. Extract from the experiment diary: "Mikkel knapped the flint nodule clean by trying to get it to split at its 'natural' cracks. With this block Mikkel has shaped the core in a different way from what he tried earlier. He previously shaped the block while simultaneously removing preforms for large blade knives. With this core he has shaped it properly, all around, and then started to strike preforms for blade knives. He did not succeed. No preform was produced from this core." (Högberg 2005b).

number of variables in the experiment while simultaneously increasing the degree of control in the work, figures 2.37 and 2.38. This way of conducting experiments has been defined as the emic approach:

> The emic approach leads to the identification of stages of production and diagnostic flakes related to these stages and defined by a polythetic attribute set. The dynamics of true production situations are here accounted for. This research rests on exclusively archaeological premises; it is based on continuous feedback between the archaeological and experimental products in a never-ending chain of corrected simulations (Knutsson 1988:14).

The experiments were performed as controlled trials to answer specific questions, table 2.41.

The study comprises three parts:

- an attempt to determine which technologies and methods created the technological attributes observed in the archaeological material and from this to define reduction processes and one or more chaîne opératoire for the experimental manufacture
- to create and describe reference material for what the waste material from the manufacture of large blade knives can look like and from this to define diagnostic flakes from the production of large blade knives
- to fragment parts of this waste material in order to illustrate what it looks like when fragmented and determine whether it is possible to recognize diagnostic flakes in fragmented flakes resulting from the production of blade knives

The latter two aims are dependent on the success of the first. A total of 40 manufacturing series were performed during eight days' work, table 2.42.

The attribute analysis performed above shows that large blade knives were made of different types of flint with different properties. It was therefore important to conduct manufacturing experiments with different kinds of flint. For the experiment I selected three different types of block flint – Falster flint, Brown Bryozoan flint, and Kristianstad flint – and one tabular flint, Scandinavian Senonian flint, figure 2.39. These flint types are represented in the archaeological material.

Table 2.41. There is no self-evident link between the two columns in the table. The link is a construction based on hypotheses that must be analysed using adequate methods, leading to results that must be interpreted. What I do with experimental archaeology, the column to the left, is to acquire experience and knowledge so that I can move my premises forward and ask relevant questions, and adjust and fine-tune methods so I can work with a body of archaeological material to answer the questions in the column to the right. So when I write in this chapter about how large blade knives are made, I really mean "are" made, not "were" made.

The present	The Late Bronze Age
How are large blade knives made?	How were large blade knives made?
What does the reduction process look like?	What did the reduction process look like?
What does the production waste look like?	What did the production waste look like?

The first part of the experiment

The aim of the first part of the experiment was to clarify the techniques and methods that generated the technological attributes observed in the archaeological material. The overall question was:

- How are large blade knives of flint made?

This question was broken down into a number of subsidiary questions about technique, manufacturing tools, and the morphology of the core:

- Are large blade knives knapped with direct or indirect technique?
- What kinds of tools give the special technological attributes found on large blade knives, and what material should the tools be made of: stone, antler, or metal in the form of bronze, copper, or iron?
- How should these tools be used?
- How should they be held during manufacture?
- What should the core look like?
- Is it a platform core, and if so what does it look like?

The work on these questions led in turn to questions about reduction strategies and details in the manufacture:

- What reduction strategies are needed to make large blade knives?

Table 2.42. Manufacturing series in the experiments. HAF refers to the experiment number assigned by Lejre Research Centre, MK to manufacturing series performed at Malmö Heritage.

Series number	Manufacture
HAF 6/00, 1–16	Tests of different tools
HAF 6/00, 17–18	Manufacture of preforms for large blade knives
HAF 6/00, 19	Retouching of preforms for large blade knives
HAF 6/00, 20–21	Manufacture of preforms for large blade knives and analyses of diagnostic flakes
HAF 6/00, 22	Manufacture of preforms for large blade knives and analyses of preparation flakes
HAF 6/00, 23–25	Manufacture of complete series of preforms for large blade knives and large blade knives of flint
HAF 04/05, 26–27	Tests of previous experience of Common Kristianstad flint
HAF 04/05, 28–30	Manufacture of complete series of preforms for large blade knives and large blade knives of flint
MK 03, 31–35	Manufacture of preforms for large blade knives from tabular flint
MK 03 & 07, 36–40	Fragmentation and working of diagnostic flakes

Figure 2.39. Three different types of block flint – Falster flint, Brown Bryozoan flint, and Kristianstad flint – and one tabular flint, Scandinavian Senonian flint. The introductory experiments were performed with block flints of Falster flint and Brown Bryozoan flint, and the majority of the results presented below come from manufacture with these flint types. Additional experimental series were performed with Kristianstad flint. The experience from these experiments was translated into an attempt to manufacture large blade knives of tabular flint, i.e., of Scandinavian Senonian flint.

- How should the platform, front, and base of the core be shaped?
- How should the platform be prepared?
- How should the core front be prepared?
- What is the chaîne opératoire for the manufacture of large blade knives of flint?

Tools and techniques

The first part of the experiment began with a comprehensive series of trials in which several different tools and techniques were used to remove flakes resembling large blade knives from platform cores. The aim was to become familiar with flint-knapping techniques and methods suitable for manufacturing such large flint objects as large blade knives, and also to systematically test different techniques for recreating the technological attributes that have been observed in the archaeological material and exemplified in the two reference large blade knives of flint, figure 2.40. The flint used comprised different types of Falster Flint and Brown Bryozoan flint.

It is generally known that different tools used for flint knapping give different types of technological traces on flakes and cores. Yet few studies have been devoted to examining the differences between different tools used in one and the same type of technique, for example, different hammerstones used with direct technique (but see Wenban-Smith 1989). It was therefore important in the first part of the experiment to test different types of tools in several different ways.

Hammerstones, antler punches, and clubs of antler, bone, and wood are known through archaeological finds from several prehistoric periods (Holdaway & Stern 2004). Studies of Late Neolithic Danish daggers have demonstrated that metal tools were used for flint knapping in prehistoric times (Stafford 1998; Barrowclough 2004). From historical times there are several examples of flintsmiths using metal tools (Kragh 1964; Mason 1978; Gebel 1980; Högberg 1999b), and many modern flintsmiths use metal in their craft (Whittaker 2004). In a Late Bronze Age context, then, it is hypothetically possible that tools of stone, antler, bone, wood, copper, bronze, and iron were used in a great many variants of indirect and direct techniques. Iron is included since iron production was introduced to Scandinavia in the Late Bronze Age (Hjärthner-Holdar 1993). The tools and combinations of tools that were tested are shown in figure 2.41 and table 2.43.

Figure 2.40. Mikkel Sørensen tests different techniques during the introductory part of the experiment. Extract from the experiment diary: "The aim was to try to find out what type of tool was used. Mikkel and I agreed that it ought to be a hard technique. Also, we both thought that a hammerstone was the most likely. We tested many different direct and indirect techniques, both to recreate the right attributes and to exclude different techniques. While we were testing different techniques we also discovered some mechanisms in flint knapping associated with the manufacture of such big objects as large blade knives. We both learned a great deal" (Högberg 2000b).

Since the technological attributes of the platform and the morphology of the bulb of percussion are not solely due to the type of tool that is used, but also to what the core looks like and which type of flint the core is made of (Högberg 1997), all the parts of the first series of experiments were performed with identical cores made from the same type of flint.

We quickly learned that the occurrence or not of the attributes on the reference large blade knives that we aimed to achieve was not the only essential factor. Also important is how these attributes appear on the blade knives, that is, what they look like in detail, figure 2.42a–d.

The result of the first part of the experiment is a series of eighteen flakes differing in the appearance of platforms and bulbs of percussion, figure 2.43.

When these flakes are compared with the registered technological attributes of the large blade knives of flint in the archaeological material, as manifested in the reference large blade knives, it is observed that the flake from test 11 is the one that best agrees with the prehistoric large blade knives. In test

Table 2.43. Techniques, tools, and combinations of tools tested in the initial series of experiments.

Test 1	Direct technique with forged iron lump.
Test 2	Indirect technique with antler punch and wooden club.
Test 3	Indirect technique with bronze punch and wooden club.
Test 4	Indirect technique with copper punch and wooden club.
Test 5	Indirect technique with hammerstone as punch and wooden club.
Test 6	Indirect technique with hammerstone as punch and stone as club.
Test 7	Indirect technique with hammerstone as punch and wooden club.
Test 8	Indirect technique with iron punch and wooden club.
Test 9	Direct technique with antler club.
Test 10	Direct technique with hammerstone, large, hard but elastic, flat, pointed.
Test 11	Direct technique with hammerstone, medium-sized, hard but elastic, long and narrow, somewhat rounded.
Test 12	Direct technique with hammerstone where the stone is held like a club. The stone the same as in test 10.
Test 13a	Direct technique with hammerstone where the stone is held as a club. The stone the same as in test 11.
Test 13b	Direct technique with hammerstone of flint, large and rounded.
Test 14a	Direct technique with hammerstone, large, soft, flat, and rounded.
Test 14b	Direct technique with hammerstone, large, very hard, flat and rounded.
Test 15	Direct technique with hammerstone, large, soft, long and narrow, and rounded.
Test 16	Direct technique with hammerstone, large, hard, oblong, and flat impact surface.

11 we used a medium-sized, long and narrow, and somewhat rounded hammerstone with direct technique, figure 2.44. It is oblong and virtually cylindrical with rounded but not pointed or blunt ends. These ends were used as percussion surfaces. The rock is hard but elastic.

In this introductory part of the experiment we also discovered that the platform preparation can be done with the same hammerstone as used for removing the flakes and the straight edge is most easily achieved if the block is supported against something. Sørensen supports it against the ground or his thigh. These were important observations for the continued experiments.

After these initial series of trials focusing on tools and techniques, we proceeded with the experiment, now with the aim of learning how to make large blade knives. We agreed that a hypothetical manufacturing chain for blade knives should consist of three stages:

- Preparation of a suitable flint nodule into a core
- Further working of the core including preparation for flaking off a preform for a large blade knife and then flaking off a preform
- Concluding retouching of the perform for a blade knife

Figure 2.41. The tools used in the experiment.

With this as our point of departure, we tested different variants of cores and ways of holding them, and we gradually built up our knowledge of what is needed to make something that looks like a preform for a large blade knife. We were severe in our judgement of what was manufactured. It was not just supposed to look like a preform for a blade knife, it also had to be a preform for a large blade knife in terms of technology and morphology, so that it could be transformed into a large blade knife which looks in every detail like a prehistoric large blade knife of flint.

The work on this is presented below in two sections, one for type I blade knives with a well-prepared platform and one for type I blade knives without platform preparation.

Figure 2.42a. We aimed for a distinct ring crack on the platform which lies more within the line between the platform and the dorsal side than on it.

Figure 2.42c. We aimed for a broken instead of a continuous line at the transition between the platform and the dorsal side.

Figure 2.42b. We aimed for a distinct moustache on the bulb of percussion, not in the form of a hertz cone that "jumped" out of the bulb of percussion but more a cone that is visible in the bulb of percussion.

Figure 2.42d. We aimed to make radial fissures running from the moustache fracture in the bulb of percussion.

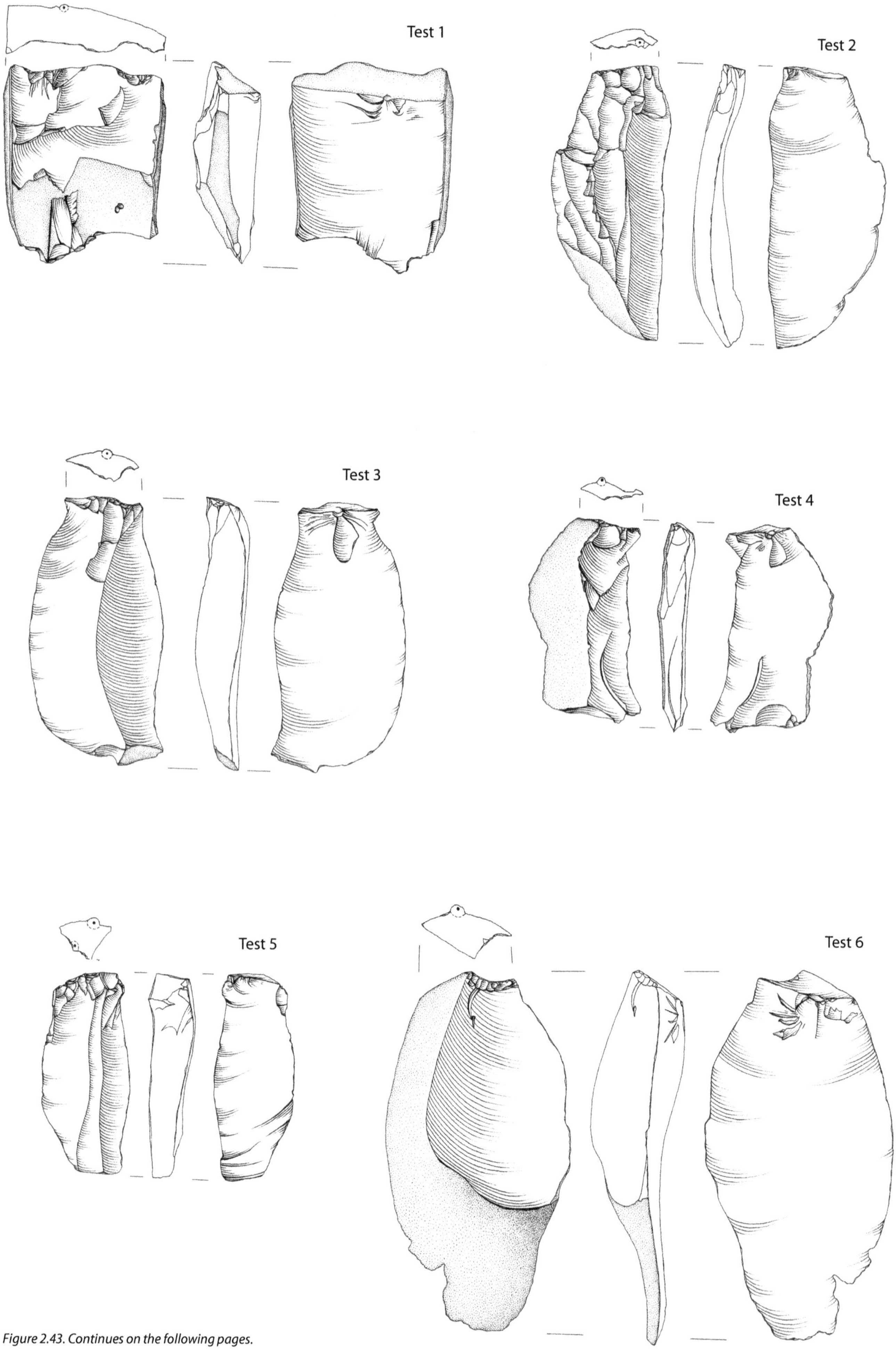

Figure 2.43. Continues on the following pages.

Test 7

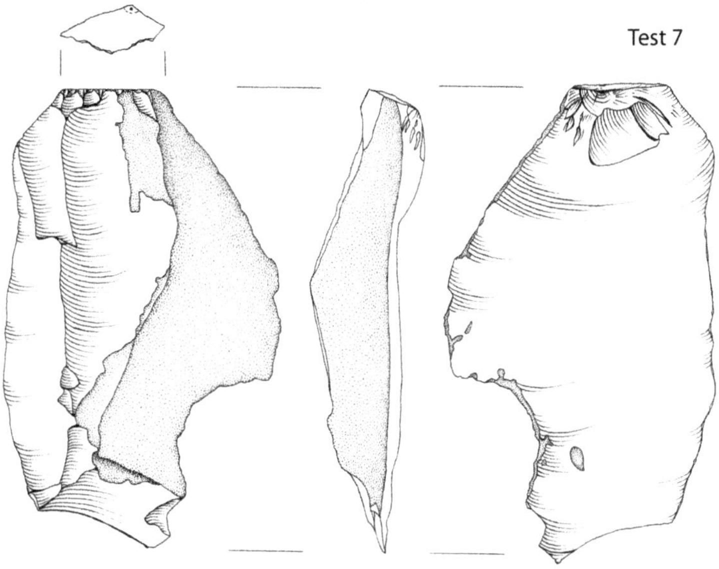

Test 8

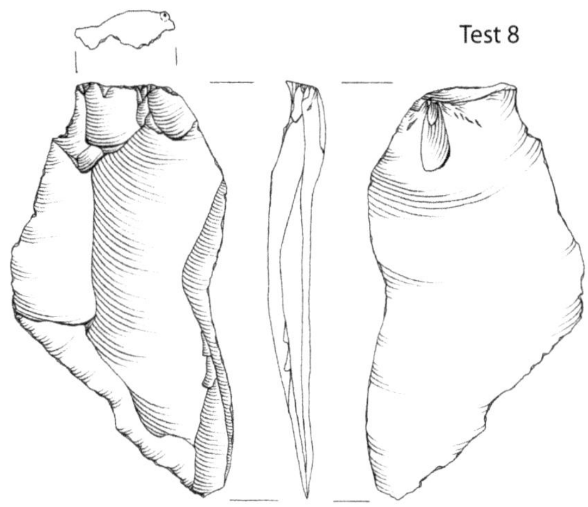

Test 9

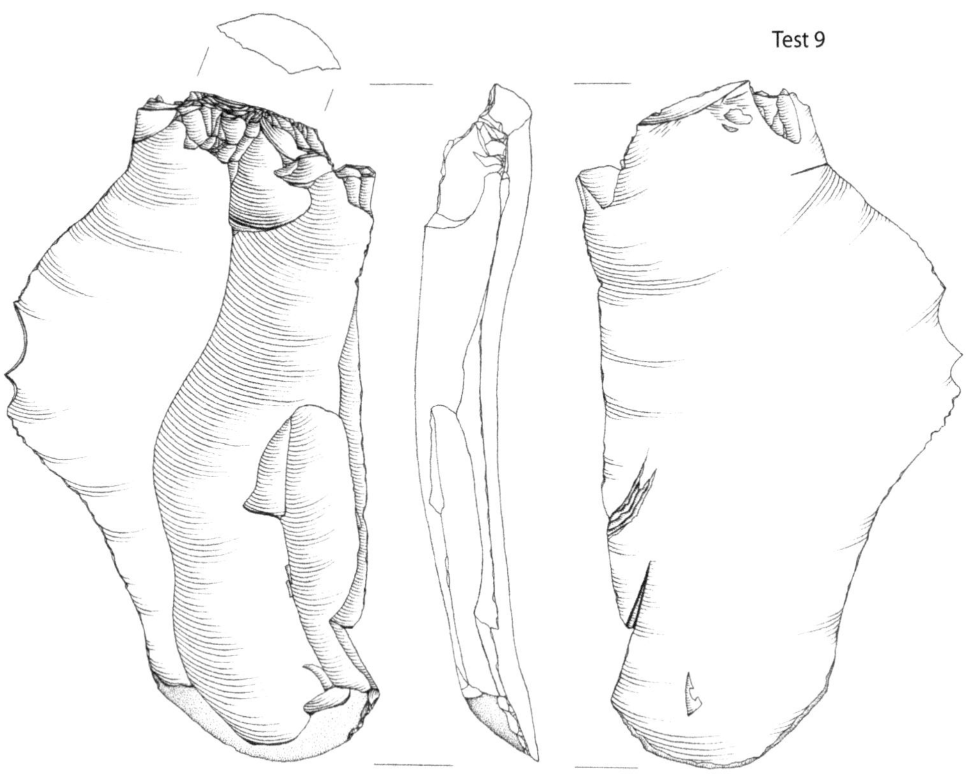

Test 10

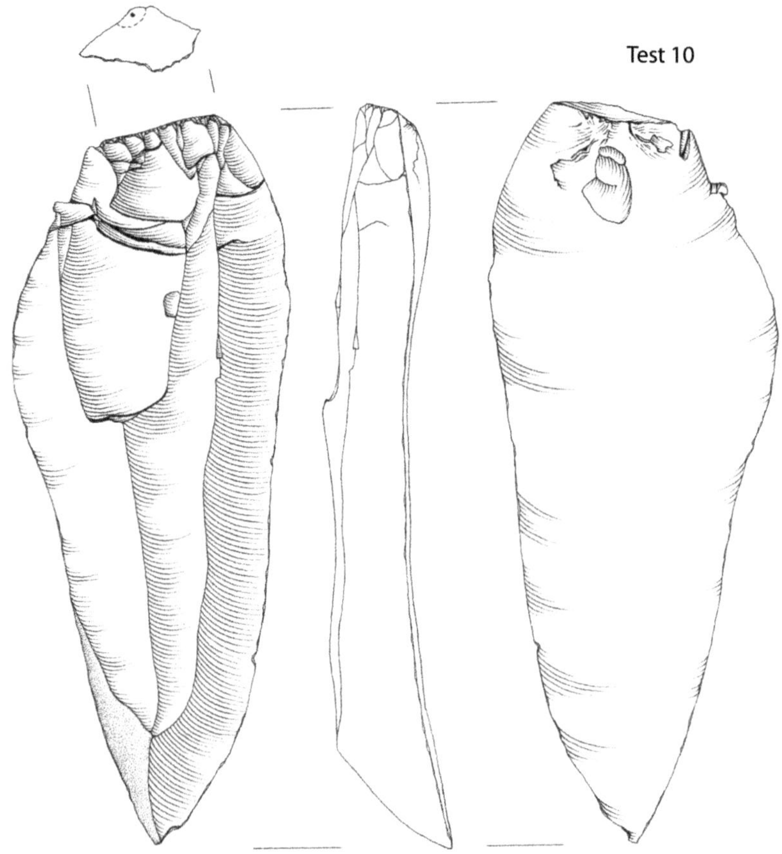

Test 11

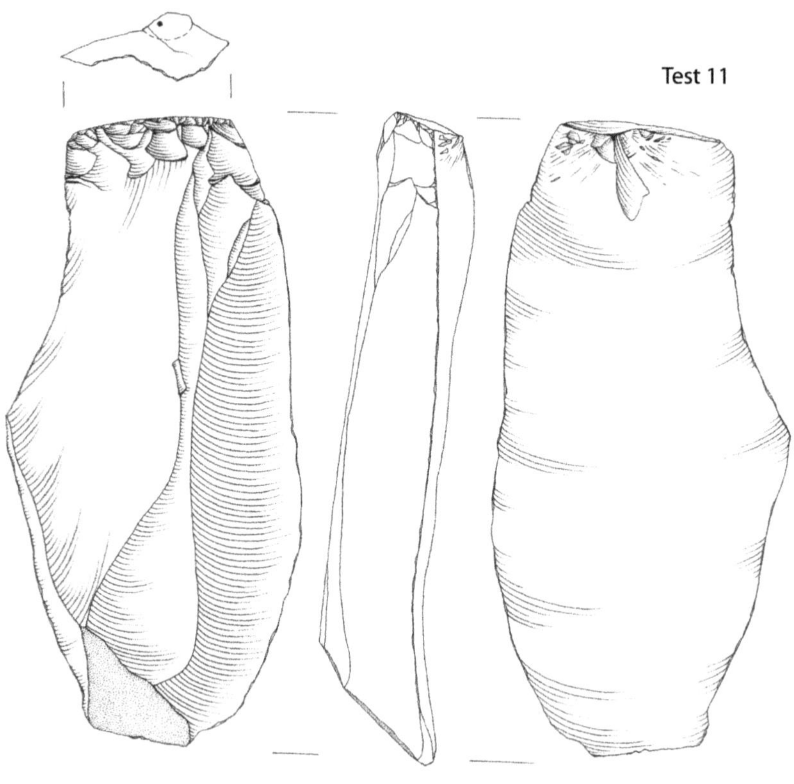

Figure 2.43. Continues on the following pages.

Test 12

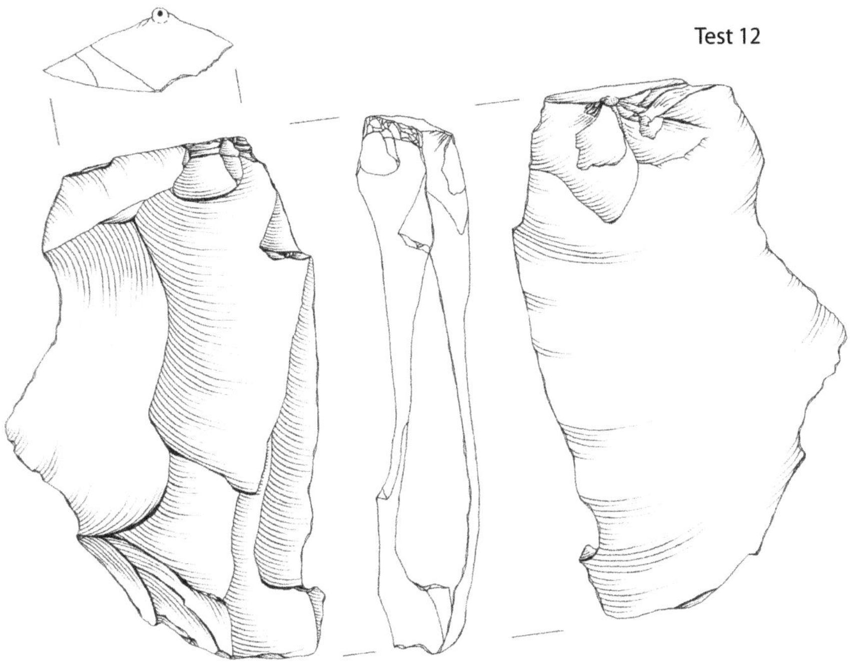

Test 13a

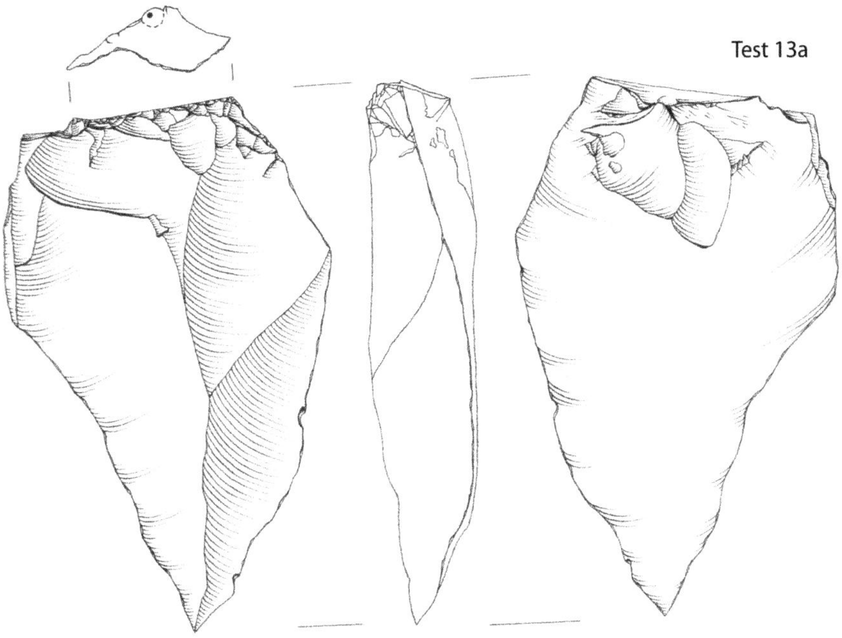

Test 13b

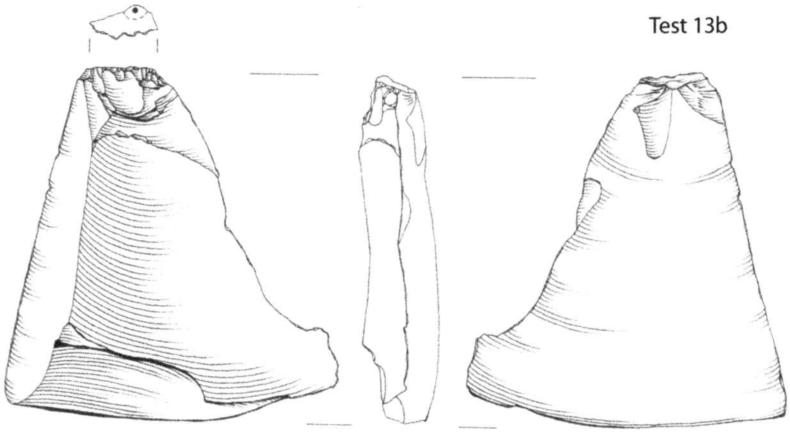

Test 14a

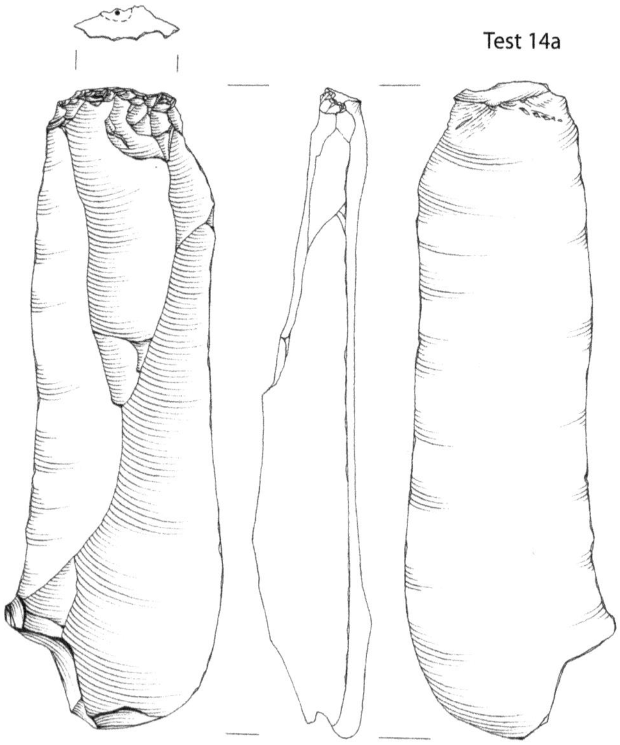

Test 14b

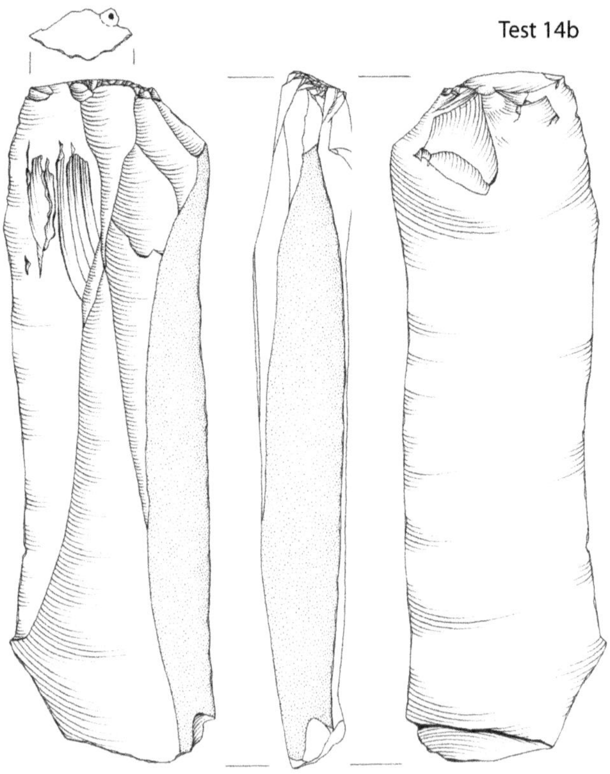

Test 15

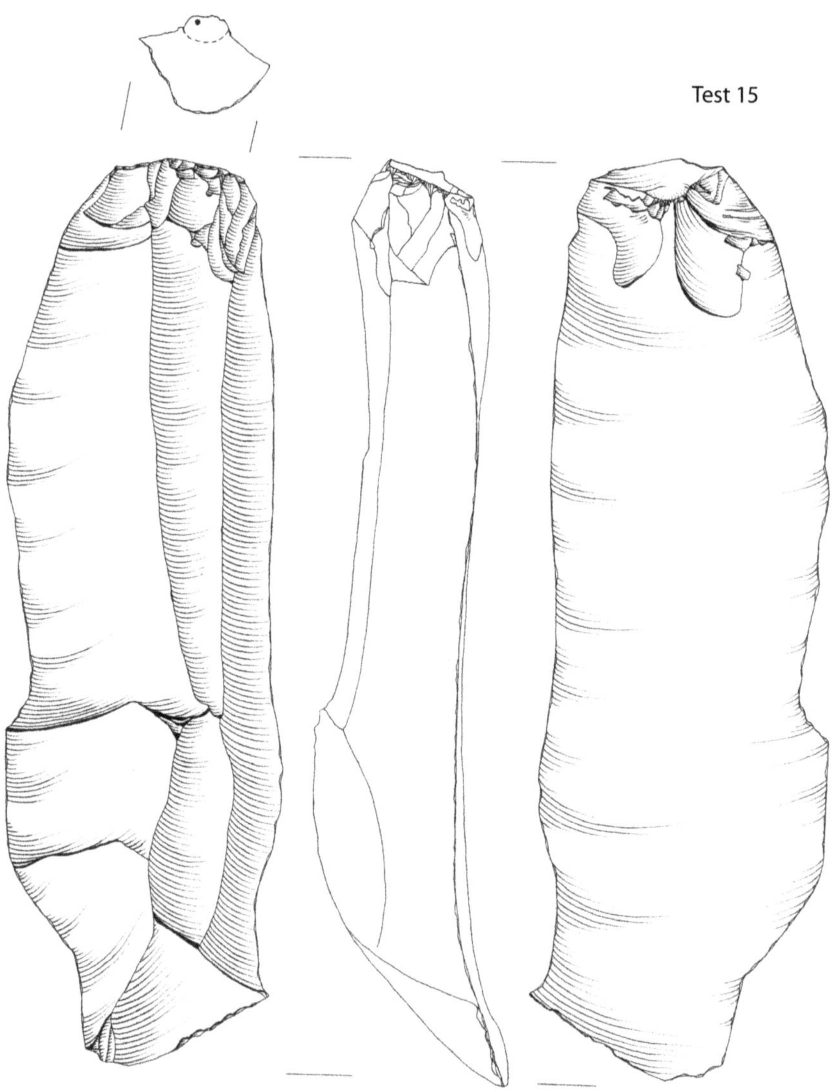

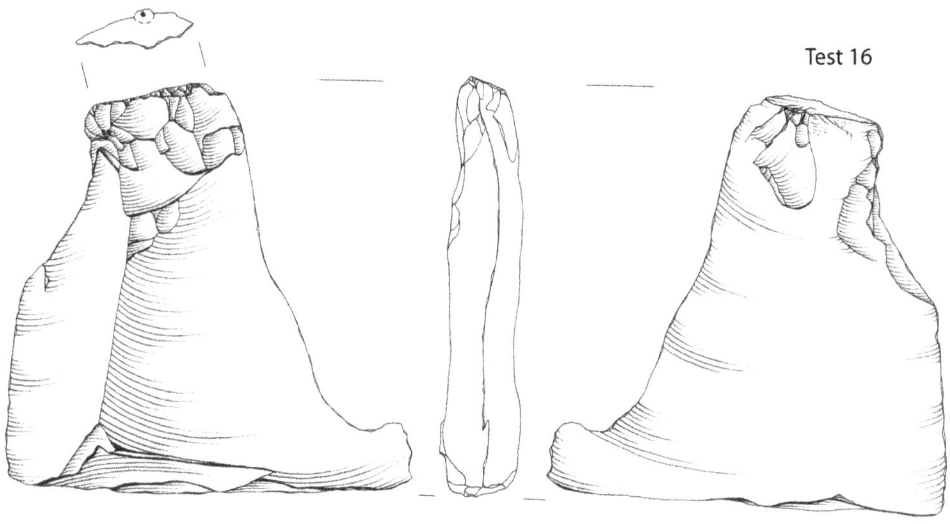

Figure 2.43. The first series of experiments resulted in eighteen flakes differing in the morphology of the platform and the bulb of percussion. It should be emphasized that even though each part of the first series of experiments is presented here with one flake, five or more flakes were produced in each part of the series. The point of this was to ensure that the attributes of the platform and bulb of percussion we achieved could be reproduced and had not arisen by chance. The flake in test 1 has a maximum width of 4.5 centimetres. Drawings: Björn Nilsson.

Manufacture of type I large blade knives with a well-prepared platform

As stated above, there are a number of criteria to satisfy in the manufacture of large blade knives of flint. These concern morphology and properties such as edge curvature and size. This makes specific demands of technique and method. The preform for the large blade knife should have the right length and thickness. The width is not so important since the width of a finished blade knife can be adjusted when it is retouched. But to get the right length and thickness the preform has to be detached from the core in such a way that the force of the blow runs through the flint so that the intended amount of the core is split in a balanced way all through the flaking and gives a preform for a large blade knife with the right proportions and properties. The edge of a large blade knife should be as straight as possible, which means that the core front from which a preform for a blade knife is flaked must be straight. The core should not curve away at the distal part. The edge should also be continuous, strong, and have a useful edge angle over the whole perform. When it comes to the manufacture of type I large blade knives with a well-prepared platform, it is also necessary that the experimental blade knives that are produced have the specific attributes that are characteristic of this knife, in the form of a well-prepared platform and preparation flakes running from the platform edge down over the dorsal side, figure 2.45.

To begin with we used relatively small flint nodules for the manufacture of the cores. We discovered almost immediately, however, that it takes really large flint nodules to cope with the manufacture of large blade knives of type I with a well-prepared platform. A large, wide platform, and also a wide front and a wide distal end are required. This is so that the blade knife will be straight-edged and equally thick and wide at the proximal end as the distal end. We made comparisons with the production of, for example, Scandinavian Mesolithic blade types and noted that these usually taper, becoming thin and narrow, and that they have a heavy curve inwards in the distal part (Sørensen 2006a). This is a form that permits a core

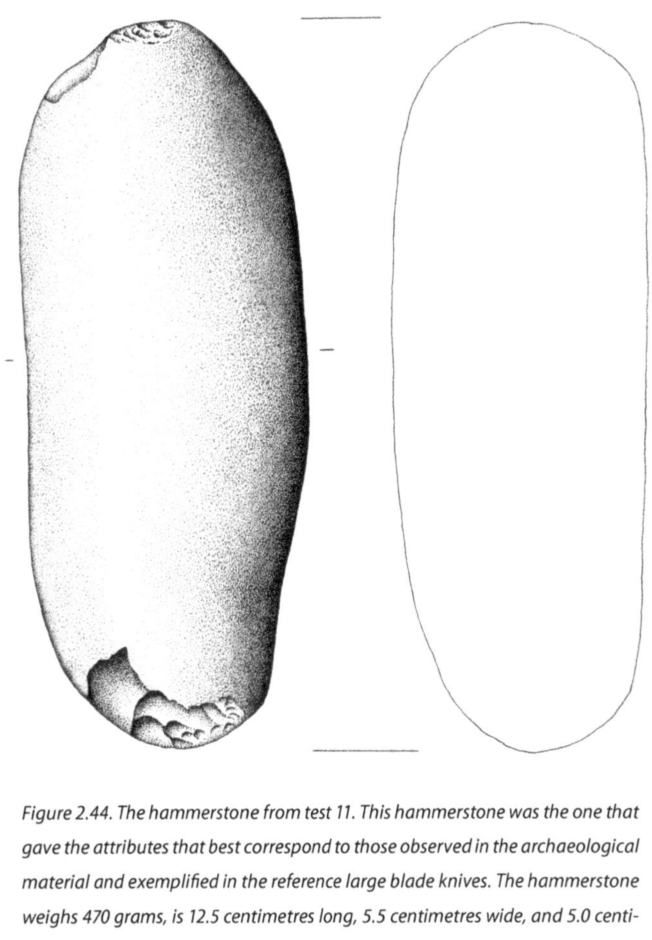

Figure 2.44. The hammerstone from test 11. This hammerstone was the one that gave the attributes that best correspond to those observed in the archaeological material and exemplified in the reference large blade knives. The hammerstone weighs 470 grams, is 12.5 centimetres long, 5.5 centimetres wide, and 5.0 centimetres thick. Drawing: Björn Nilsson.

Figure 2.45. The work with the core is important. Another extract from the experiment diary: "It is not entirely easy to find the right flint nodule to proceed from. It takes a big, flat platform which can 'contain' many potential preforms for large blade knives. It also takes a volume 'under' this platform, that is, a distal end, which can contain as many preforms for large blade knives. Large quantities of flint are consumed in preparing the core, since the larger preparation flakes must resemble blade knives. Platform preparation in the form of trimming of the edge is something we don't need to work with. It 'comes by itself', so to speak. Platform preparation takes place with the hammerstone, and preparation and simple cleaning is done with intense preparation and trimming of the platform edge. This procedure gives the same traces of platform preparation as found in the archaeological material" (Högberg 2000b).

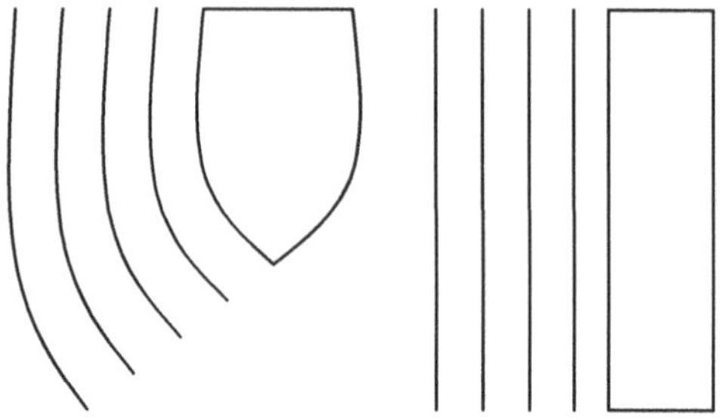

Figure 2.46. Sketch illustrating the differences in flint cores for production of Late Bronze Age large blade knives of flint and other blade production. The production of large blade knives requires a core of large volume. A difference worth noting is that when blades are produced from conical blade cores, the core and hence also the blades gradually become shorter during the production, to the left in the figure. This is because the blades curve away and then take with them a part of the length of the core. Since Late Bronze Age large blade knives do not curve away and in principle are equally as thick at the proximal end as the distal end, to the right in the figure, the core does not become shorter during the production, and hence nor do the preforms for large blade knives.

which is pointed or conical. The production can thus proceed from flint nodules with a small volume. This is not the case with large blade knives of type I with a well-prepared platform. Their size and straight edge requires flint blocks of large volume, see figure 2.46.

The large volume of the cores means that this type of manufacture of large blade knives demands considerable resources. To get some idea of this, a sequence of preforms for blade knives was made from a flint block. The block weighed about 6 kilos when the experiment began. From this block we made six preforms for blade knives. Together these weighed about 800 grams. In this experiment, then, each large blade knife of 130 grams required roughly a kilo of flint to make. Even if this particular experiment cannot be regarded as representative of all production of large blade knives, it still gives an indication of how much raw material this manufacture consumes.

With this experience gained, we worked further with the reduction process. By flaking off whole series of large blade knives we were able to make some important observations:

Figure 2.47. Preparation along the whole edge.

Figure 2.48. The location of the preparation flakes on the block.

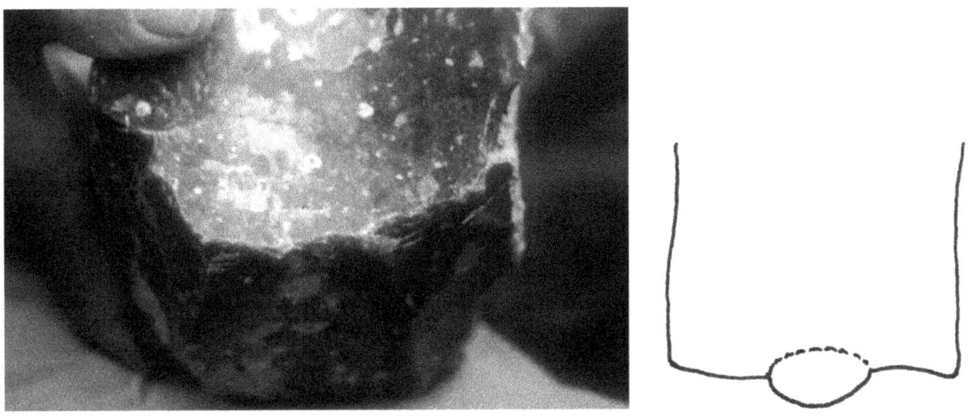

Figure 2.49. Isolating the platform. The platform viewed from above.

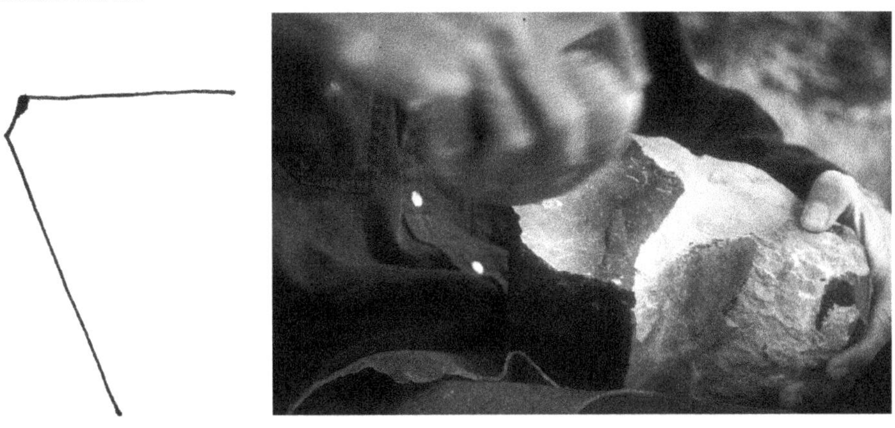

Figure 2.50. The overhang must be taken in so that the edge of the platform is inside the line of the desired preform for a blade knife.

- Platform preparation in the form of trimming must be done along the whole edge towards the front of the core.
- The distal end of the core, opposite the platform, should be fairly flat.
- In the reduction sequence a flat core front should be aimed for.

These observations gave us experience that finally enabled us to produce entire series of large blade knives from the same core. The observations are discussed below separately.

Platform preparation

It is important that the platform is prepared along the whole edge. It is not enough that it is prepared where the preform for a blade knife is to be detached, see figure 2.47. We discovered that the two types of platform preparation flakes defined above, preparation flakes with a hinged termination and preparation flakes with a feathered termination, are crucial for the production, see figure 2.48. These are flaked off to get rid of the overhang on the platform edge, and also to isolate the platform, see figure 2.49. It is important that the overhang is taken in so that the edge of the platform is within the line of the desired blade, see figure 2.50.

If these preparations are not done properly, the removal of a preform for a large blade knife is unsuccessful. For example, the blow can go straight through the platform and crush it without removing a single flake. Or the force of the blow may not be sufficiently concentrated, which results in either cracks in the core without removing any flakes, or the flake gets a hinge or feather termination. It can also result in the blow not following the intended direction through the core, instead extending out over the front of the core. The effect is the removal of a flake that is too big and too wide, figure 2.51.

All of these unsuccessful removals risk destroying the core front entirely, so that the work of preparation done by the flintknapper is wasted. Also, the mistakes do not result in the production of a preform for a large blade knife.

Figure 2.51. Examples of mistakes in the manufacture of large blade knives. A crushed platform, a flake removed without the force running through the whole core, resulting in a hinge, and a flake extending out over the whole front of the core.

The distal end of the core

The base of the core should be reasonably flat. This is so that the preform for a blade knife will not curve, become pointed, or too thin at the tip. This is important for getting the right edge curvature on the finished large blade knife. Sørensen thinks that this is controlled from the platform. However, the core must first be knapped to shape at the distal end so that as flat a surface as possible is achieved.

The reduction sequence

The reduction process that gives best result is to proceed from a flat front. The flakes are removed in a sequence that can most easily be described as: first one side, then the other side, then the middle, see figure 2.52. The flake in the middle is what becomes a preform for a large blade knife. When this three-step reduction sequence is performed, you once again have a virtually flat front to proceed from, and the process can begin again with the same sequence of one side, the other side, and the middle. The blade-like flakes from the sides resemble preforms for large blade knives but have properties such as the wrong edge curvature or a broken edge line, which means that they cannot be modified into a large blade knife flint.

The form taken by the reduction sequences shows that two large blade knives from the same production cannot be refitted. This is because the blade-like flakes from the sides lie between the removals of the actual preforms for the blade knives. To be able to refit large blade knives produced on the same occasion, you have to have the blade-like flakes. This result shows that refitting is not a suitable method of analysis for large blade knives, unless other parts of the reduction process are included in the material.

The way of preparing the core for removing a preform for a large blade knife is important for several reasons. Relieving the core of mass is essential to be able to remove a preform for a large blade knife at all. When large flakes are removed on the sides of the core, one or more ridges are created in the middle of the core front. The ridge or ridges will be on the dorsal side of the intended large blade knife and are thus followed by the force of the blow in the removal of the preform. The ridges also determine the thickness of the blade knife and are thus important in that the relationship between the ridge and the edge of the preform determines the edge angle and the stability of the edge, figure 2.53. This means that the angle and form of the edge are determined when the core is shaped. This demonstrates the importance of a high degree of knowledge in the manufacture.

It is important to stress here that the manufacture of large blade knives is not a matter of pure blade production. The majority of the preforms for large blade knives of flint that are produced should be regarded more as long and narrow flakes than as blades (see also the discussion in Sørensen 2006c). Although large blade knives of flint look like blades, they are not always made of blades. They are just as often made of flakes which are not given the form of blades until they are retouched, figure 2.54.

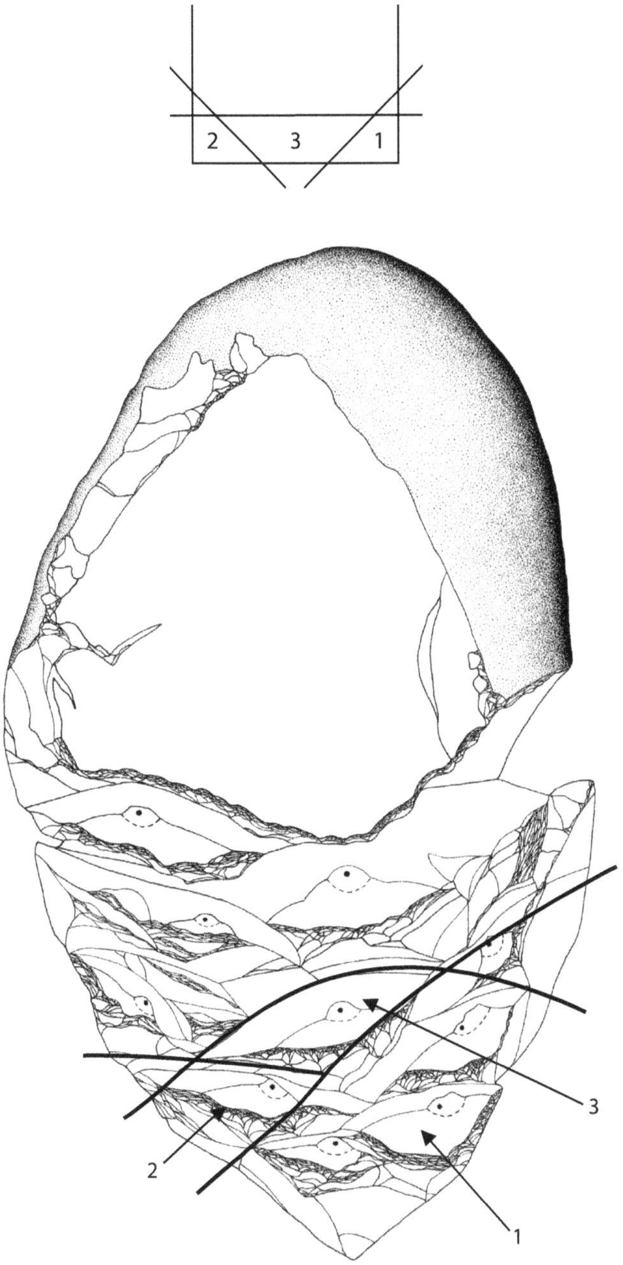

Figure 2.52. The reduction sequence for manufacturing a type I large blade knife with a well-prepared platform. Note that, although the sketch shows a sequence that involves first removing a flake from one side and then a flake from the other side, in order to prepare the detachment of a preform for a large blade knife in the middle, in practice it usually means that more than one flake is removed from each side before the preform can be flaked off. This is clearly visible on the photograph and the drawing of the refitted core. Drawing of the core: Björn Nilsson.

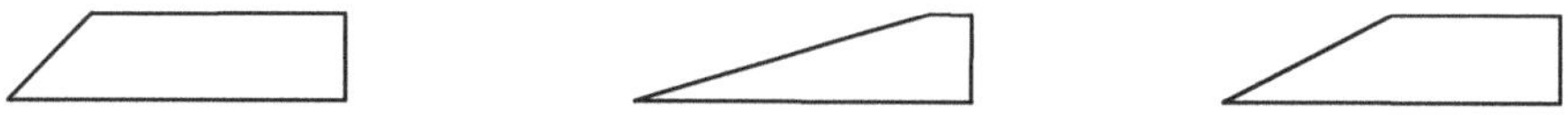

Figure 2.53. Three different sections of a large blade knife and the relationship between the ridge, the thickness of the blade knife, the edge angle, and the stability of the edge. To the left the ridge has ended up too far to the left on the dorsal side, resulting in an edge with too steep an angle; in the middle the ridge has come too far to the right, resulting in too fragile an edge; to the right the ridge has been placed in such a way as to give a stable edge with the correct angle.

Retouching

The first part of the experiment ended with a series of preforms being processed into large blade knives. This was done by retouching. The ventral side of the preforms was used as a platform. First slightly larger flakes were removed and the retouch was then cleaned up by detaching small flakes. The retouch is a backing and not an edge, so there is no need to be so accurate. Presumably the preforms for the large blade knives were retouched more so that they could be shaped to fit a shaft than just to achieve a nice retouch. Often the retouch creates something that can be compared to a ridge on the dorsal side of the blade knife, figure 2.54. This phenomenon has also been observed in the archaeological material.

It is extremely rare for a preform for a large blade knife to have two usable lateral edges. The preforms in figure 2.54 are examples of this. The preform to the left in the figure has a right-hand edge. The left side is covered with cortex. The preform in the middle also has a right-hand edge. The left side is irregular and does not have the right relationship between ridge and edge, which gives an uneven edge. The preform to the right in the figure has a left-hand edge. The right side is irregular and lacks a continuous edge. Whether the retouch is to be on the right or left side of a preform for a large blade knife is thus not something one can choose. The morphology of the preform, that is, whether there is a suitable edge on the right or left side, decides where the retouch should be placed. The morphology of the preform is in turn determined by where on the core the preform was removed. Suitable edges are virtually always formed on the part of the preform facing the middle of the core front, and very rarely on the part of the preform facing the side of the core. Referring back to the discussion in the previous chapter about right- and left-retouched large blade knives of flint, it is obvious that the placing of the retouch on the blade knife was dictated by the manufacture, not by a wish to place the retouch on one or the other side, figure 2.55.

Manufacture of type I large blade knives with a well-prepared platform – a chaîne opératoire

The experiment has shown that the making of type I large blade knives with a well-prepared platform requires a systematic technology that enables the manufacture of standardized products. A blade knife cannot be made in just any way. Although we managed after the initial trials to produce occasional flakes that could be retouched into blade knives, it was not until we reached experiment series number 22 that we began to learn and understand how the systematic and standardized production of large blade knives can be done, figure 2.56 and 2.57. The production begins with the selection of a suitable flint nodule. This should be of high quality without too many cracks or impurities. The flint nodule must be sufficiently large that

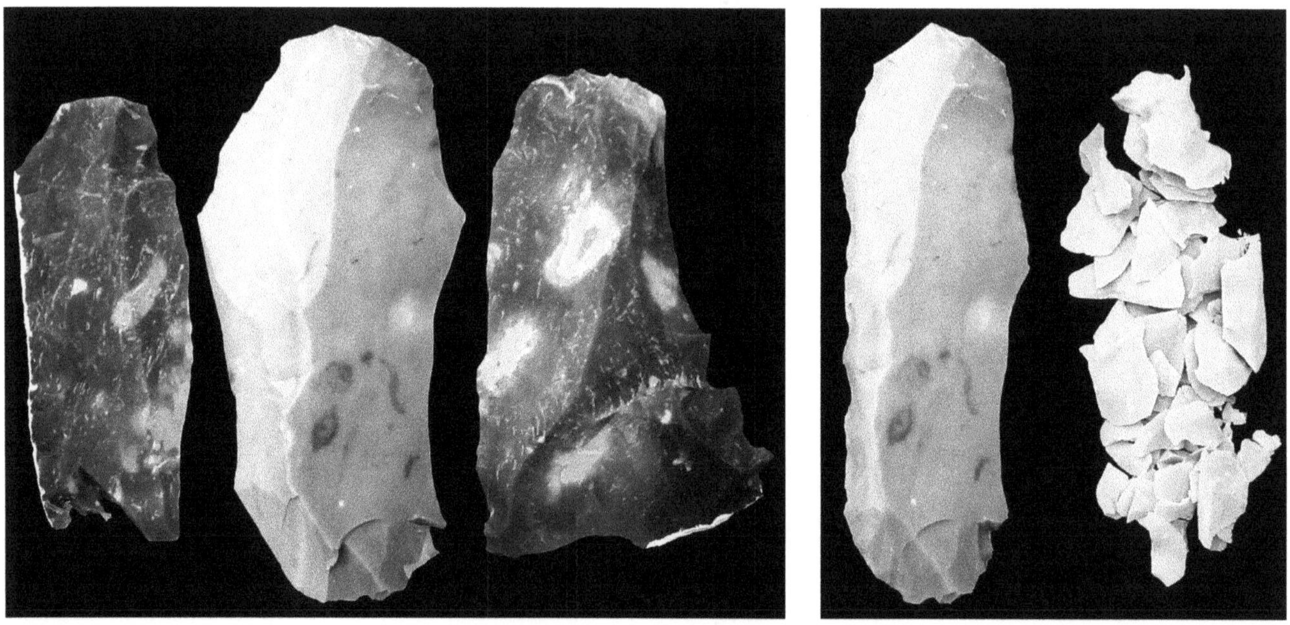

Figure 2.54. To the left three preforms for large blade knives before retouching; the one in the middle is 14 centimetres long; to the right a preform for a large blade knife after retouch together with the flakes from the retouching. Note how the retouch has created a "ridge" on the blade knife.

Figure 2.55. A recently removed preform for a large blade knife and the core from which it was removed. Notice that the side of the preform that has a suitable edge is the one towards the middle of the core.

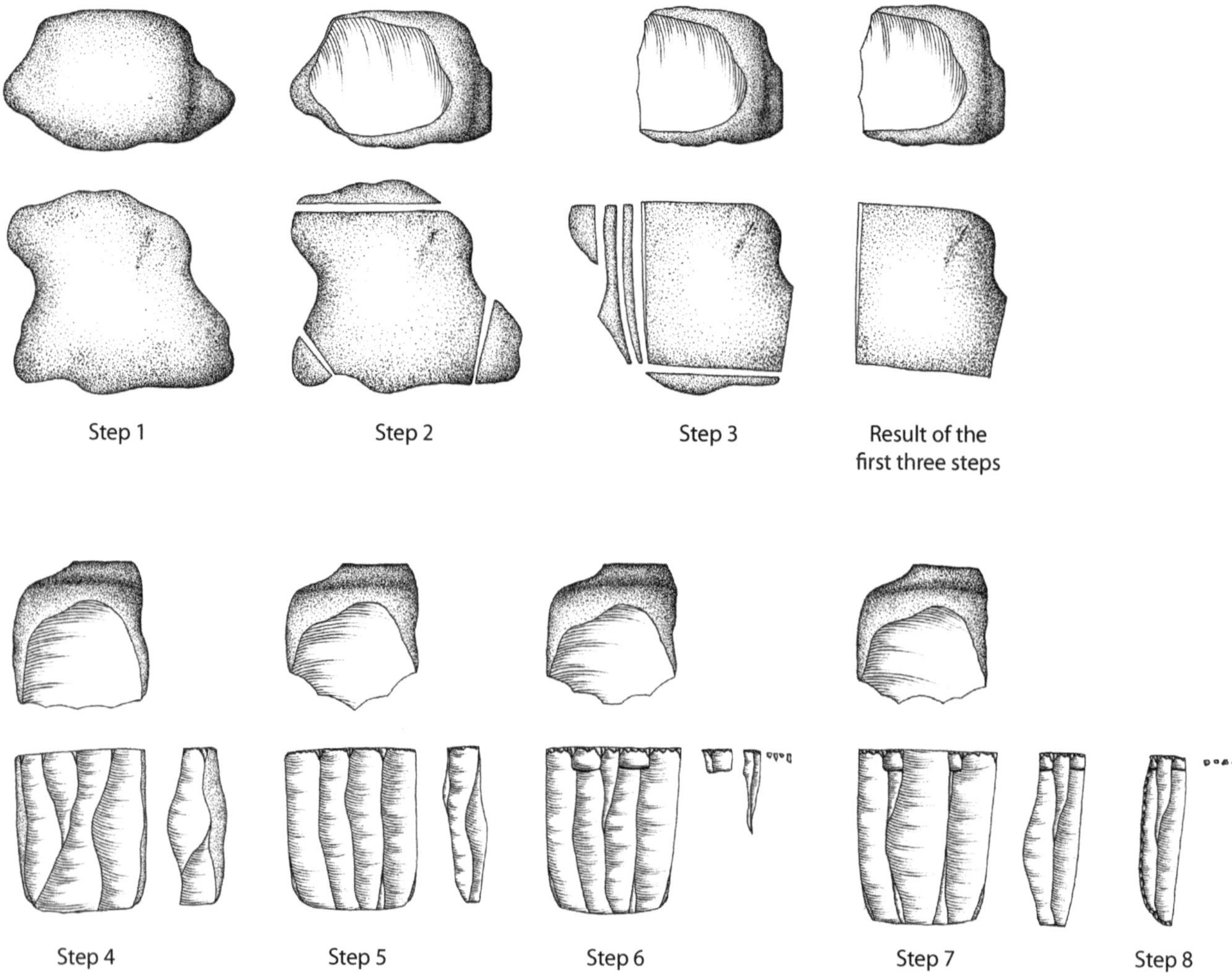

Figure 2.56. A chaîne opératoire for the experimental production of type I large blade knives with a well-prepared platform. Drawing: Björn Nilsson.

a core of large volume can be made from it, step 1 in figure 2.56. Then the first shaping of the block is the removal of the opening flakes, step 2 in figure 2.56. Now impurities and protruding parts of the flint nodule are removed, and clean flint surfaces are exposed by removing large cortex flakes. The platform is created by detaching one large platform flake, and the shaping of the flint nodule gradually proceeds to step 3, figure 2.56, when the front is shaped by removing a number of large flakes from the newly knapped platform. In this stage of the manufacture the base is also created, that is, the distal part of the core. The result of the first three steps in the manufacture is a large core with a square appearance, a flat platform, a reasonably level base, and a broad, flat front, figure 2.56.

The manufacture then proceeds with the shaping and stabilizing of the front through removals from the platform, step 4 in figure 2.56. This gradually gives way to step 5, figure 2.56, when large blade-like flakes are struck off from the sides of the front in order to prepare the removal of a preform for a large blade knife. In step 6 the finishing touches are put to the platform edge, and the proximal part of the front of the core is prepared through the removal of preparation flakes with hinged and feathered termination. When this is completed,

Figure 2.57. Photograph from the series of experiments exemplifying details of the schematic chaîne opératoire in figure 2.56.

Figure 2.58. Two prehistoric large blade knives made of Kristianstad flint, from the collections of Lund University Historical Museum. The blade knife to the left is approximately 15 centimetres long and has find number LUHM 20007. The blade knife to the right has find number LUHM 24567. Photo: Bengt Almgren.

the preform for the large blade knife is detached, step 7 in figure 2.56. If the core is to be used for the removal of more preforms for large blade knives, you now go back to step 4 in the reduction sequence and repeat the work with steps 4 to 7, figure 2.56. If the preform produced in step 7 is to be further processed, the production goes on to step 8. At this step the preform for a large blade knife is shaped by retouching the parts of the preform that need to be retouched.

Manufacturing experiments with Kristianstad flint

Kristianstad flint was used during the Late Bronze Age for the production of large blade knives (Högberg 2005a), and blade knives made of this type of flint are known from archaeological collections (Strömberg 1952), figure 2.58.

Several archaeologists have discussed aspects of Kristianstad flint as a raw material, and some of them find it difficult to handle and not so useful, while others find it excellent to use (Högberg & Olausson 2007). It is clear that the properties of Kristianstad flint make it partly different from other south Scandinavian flint (Högberg 1997).

There is thus some uncertainty as to whether large blade knives of Kristianstad flint can be produced in the same way as with other block flint. The aim of this part of the manufacturing experiments was to investigate this. The point of departure was the experiences of producing blade knives described above. The questions were:

- Is there a difference between Kristianstad flint and the other types of block flint used above when it comes to making large blade knives?
- Are there technological or methodological differences?
- How easy is Kristianstad flint to handle in relation to other south Scandinavian block flints?

Figure 2.59. Manufacture of large blade knives of Kristianstad flint. Extract from the experiment diary: "Round 1. Mikkel knaps a nodule to test the flint. He gets into the production almost immediately. He is surprised that the flint flakes so well. But when the preform for a blade knife was to be detached, it did not work. It failed. He struck through the platform and crushed it. This shows the difficulty of Kristianstad flint. It requires well-prepared platforms and precision in the blow ... Round 3: The nodule is very large. Mikkel has to spend a lot of time knapping the block into shape. When the core is finally prepared and the front sufficiently narrow, Mikkel knaps the first blade knife. The nodule is still a little wide at the front, so Mikkel continues narrowing it. He then removes two preforms for large blade knives in a row, and yet another one after a little preparation of the core. This shows that the production of large blade knives of Kristianstad flint is like the other forms of production with which we have previously worked" (Högberg 2005b).

Figure 2.60. A well-prepared core of Kristianstad flint together with a newly removed preform for a large blade knife. Note the even distribution of greyish-yellow spots in the flint.

The flint used in the experiments, which were performed in 2005, came from the disused limestone quarry at Hanaskog in north-east Scania. The method was to manufacture series of large blade knives based on experiences described above, figure 2.59.

The result of this part of the experiment shows that the raw material Kristianstad flint reacts both like other flint and in its own unique way. It is similar in that it splits in the same way as Senonian or Danian flint. Our assessment of the properties of the flint was that it was surprisingly good. But Kristianstad flint has also special distinctive features that affect the production. The flint requires well-prepared and stable platforms. If the platform is ever so slightly wrongly prepared or placed so that it is a little weak, the blow will go straight through the platform, so that it is crushed or split when the hammerstone hits it. The spots in the flint can cause problems. The many greyish-yellow spots in Kristianstad flint are characteristic and unique for this type of flint. If these spots are evenly distributed in the nodule, it usually causes no problems for the production, figure 2.60. But in many of the flint nodules there are accumulations of spots. When these occur on the core it is difficult to achieve faultless production. Often there is no problem in getting the force of the blow to run through the spots and it is therefore easy to maintain control of the core. But the flakes, and in the case of the manufacture of large blade knives the actual preform for the tool, break at the part of the flint where there is an accumulation of spots.

Manufacture of type I large blade knives without platform preparation

The experimental manufacture of type I large blade knives without platform preparation must satisfy the same criteria as regards morphology and properties as for the manufacture of type I large blade knives with a well-prepared platform. As described above, these criteria concern size, proportions, and edge curvature. These criteria are the same for both variants of type I large blade knives of flint. The most evident difference is the platform preparation. In the series of experiments described above, platform preparation has been stressed as a crucial factor for success. This obviously does not apply to type I large blade knives without platform preparation, since they lack this preparation. Another difference is the occurrence of cortex. Blade knives that have no platform preparation have more cortex on the dorsal side than type I large blade knives with a well-prepared platform have. Yet another difference is the raw material. The majority of type I large blade knives with a well-prepared platform are made of block flint. Type I large blade knives without platform preparation are made of tabular flint.

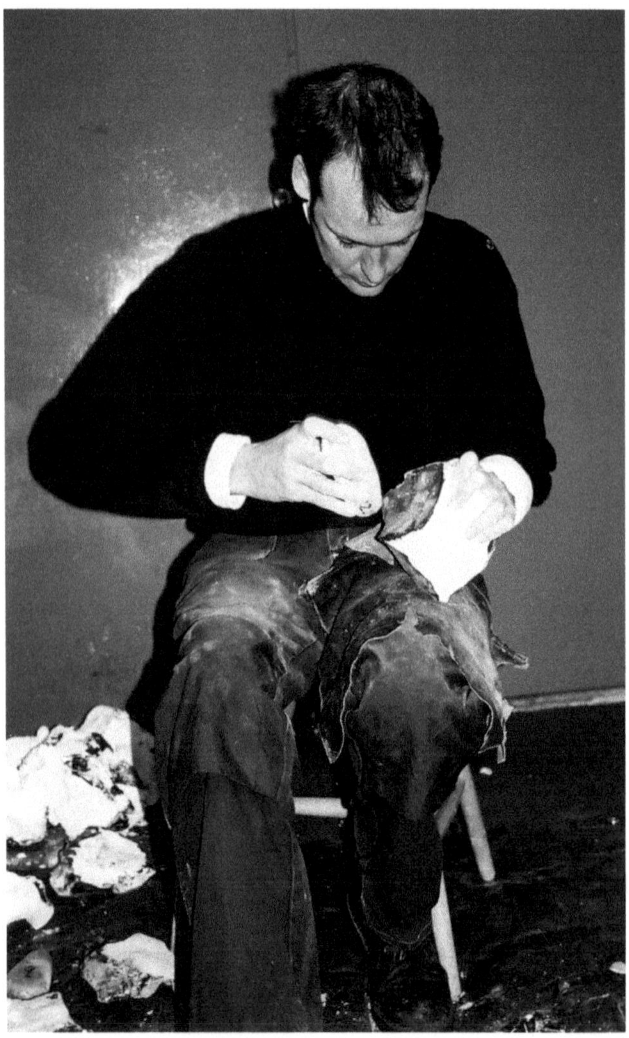

Figure 2.61. Sørensen manufacturing type I large blade knives of tabular flint without platform preparation.

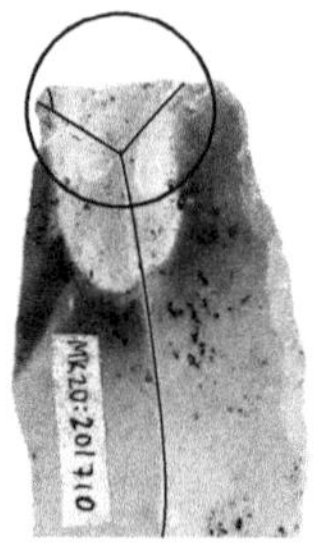

Figure 2.62. Close-up of the proximal part of a type I large blade knife without platform preparation. The part of the knife that is circled is the part that is characteristic of the technology with which this type of blade knife was made, and this is also what helped us to choose the right course in the manufacturing experiments.

This combination of morphological similarities and differences in raw material and technology makes other demands of the manufacture than those presented hitherto. The aim of this part of the experimental study was to investigate this. The experiment was performed at Malmö Heritage in 2003, figure 2.61. The point of departure was the experiences described above. The questions asked were:

- What is specific for the production of type I large blade knives without platform preparation?
- How does the production of the different type I large blade knives differs?
- What are the reasons for the difference?

The flint for the experiment was taken from the disused chalk quarries at Södra Sallerup in Malmö the site of Sweden's only archaeologically excavated prehistoric flint mines (Rudebeck

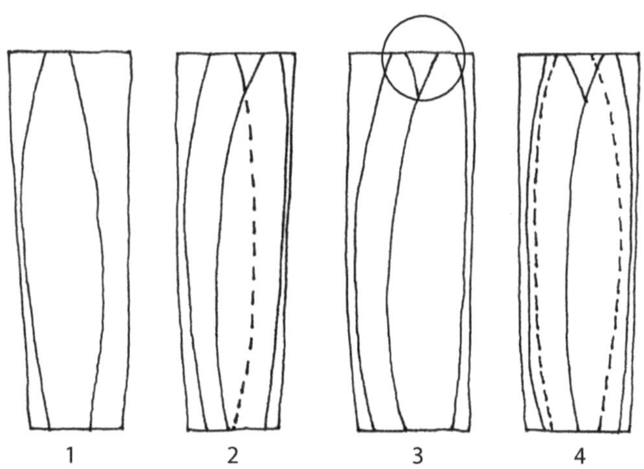

Figure 2.63. Sketch showing of how two flakes are removed in such a way that the characteristic morphology of the proximal part of the blade knife is achieved. (1) The first flake is detached. (2) The second flake is removed so that it runs in over roughly half of the detachment area of the first flake to give the characteristic morphology of the proximal part at the platform is formed, circled in number 3 in the figure. (4) The dashed line shows a preform for a blade knife in the core before the preform is removed.

1986). The series of trials began with us feeling the flint and discussing how our previous experiences could be translated into the manufacture of large blade knives in new conditions. We immediately observed that tabular flint could not be used for making cores of large volume. There was quite simply no room for large volume in a nodule of tabular flint. We also realized that the only way to get a large blade knife of the right length out of a nodule of tabular flint is to place the nodule upright. With these observations as a point of departure, we continued the experiments and tried to make large blade knives.

Sørensen began by striking off a large flake in order to make a platform and a flake to shape the distal end of the block. We both agreed that this must be the right way to start the reduction process. We then continued by trying in different ways to make a core front as we had done in the experiments with block flint. Both Sørensen and I were convinced that this was the next step in the manufacturing process. But no matter how we tried, we were unable to achieve a well-formed straight core front as from which a preform to a large blade knife could be detached. The volume of the flint was insufficient, and the few times we managed to get a reasonably neat core front it was still impossible to remove a perform to a large blade knife since the platform was not prepared. We soon realized that we would have to think again.

One series of experiments was devoted to preparing the core front from the base of the block and thus preparing the block for removal of a preform for a large blade knife. We quickly understood that this was the wrong approach. Even if we managed to detach preforms in this way, these had bidirectional dorsal flaking scars, that is to say, running opposite to the flaking direction of the preform. This is very unusual in prehistoric large blade knives of flint.

After several series of trials, however, we began to get a grasp of the manufacture. It was the morphology of the proximal part of the prehistoric large blade knives that helped us, figure 2.62. The proximal part clearly shows that two negative removals shaped the appearance of the blade knife and that the second of these negatives is struck so that it goes in over more than half of the surface of the first negative, figure 2.63.

Based on this new knowledge, we were able to go further in our experiments. We discovered that what is needed to succeed in making type I large blade knives without platform preparation is a nodule with a surface that can be shaped into a platform and with the opposite side of the platform – that is, the distal end of the block – shaped in such a way as to permit the flakes to run through. This means that the distal end of the block does not need to be shaped in advance through removals. It also means that it is important to place the platform at the right point on the nodule.

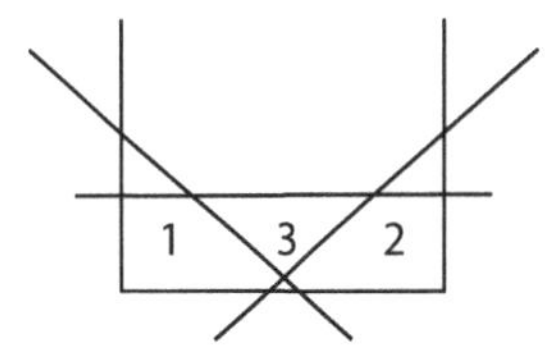

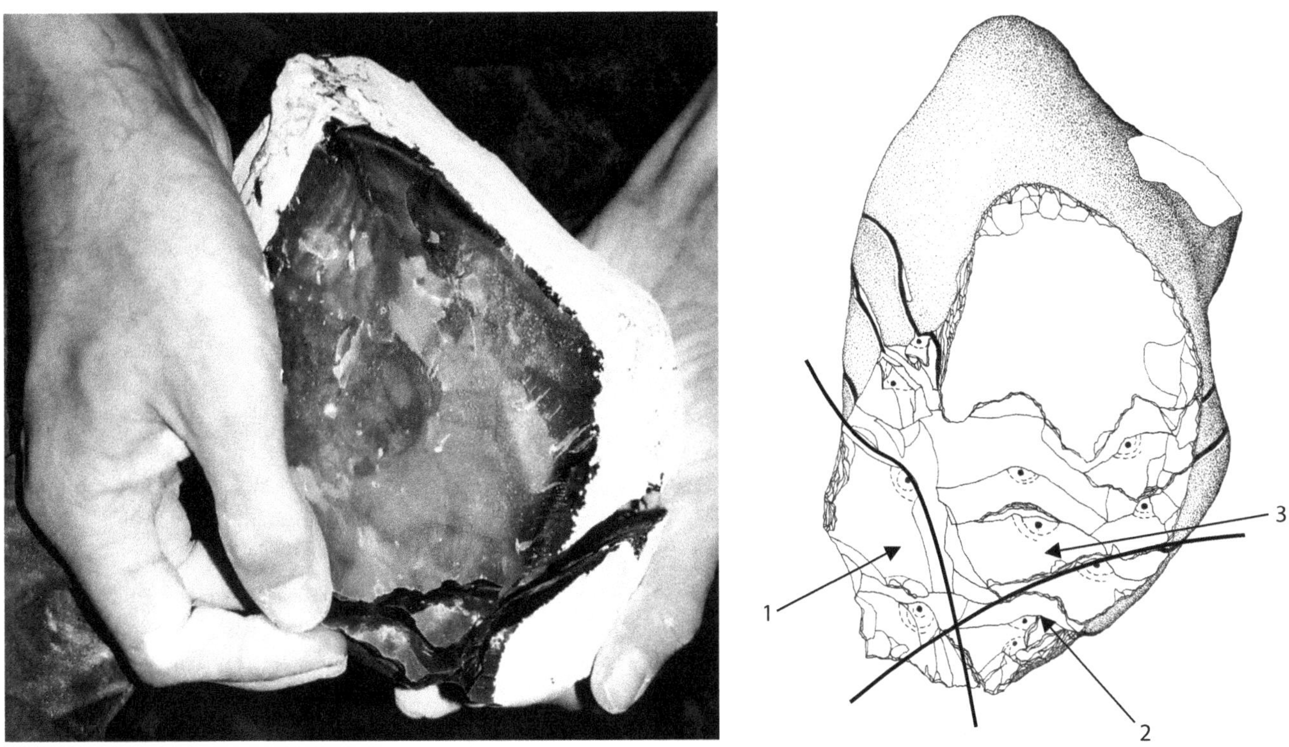

Figure 2.64. The reduction sequence, manufacture of a type I large blade knife without platform preparation. Note that, although the sketch shows a sequence which consists of first detaching a flake from one side and then a flake from the other side, all in order to prepare the removal of a blade knife preform in the middle, in practice it sometimes but not always means that more than one flake is removed from each side before the preform can be removed. This is clearly visible on the photograph and the drawing of the refitted core. Drawing: Björn Nilsson.

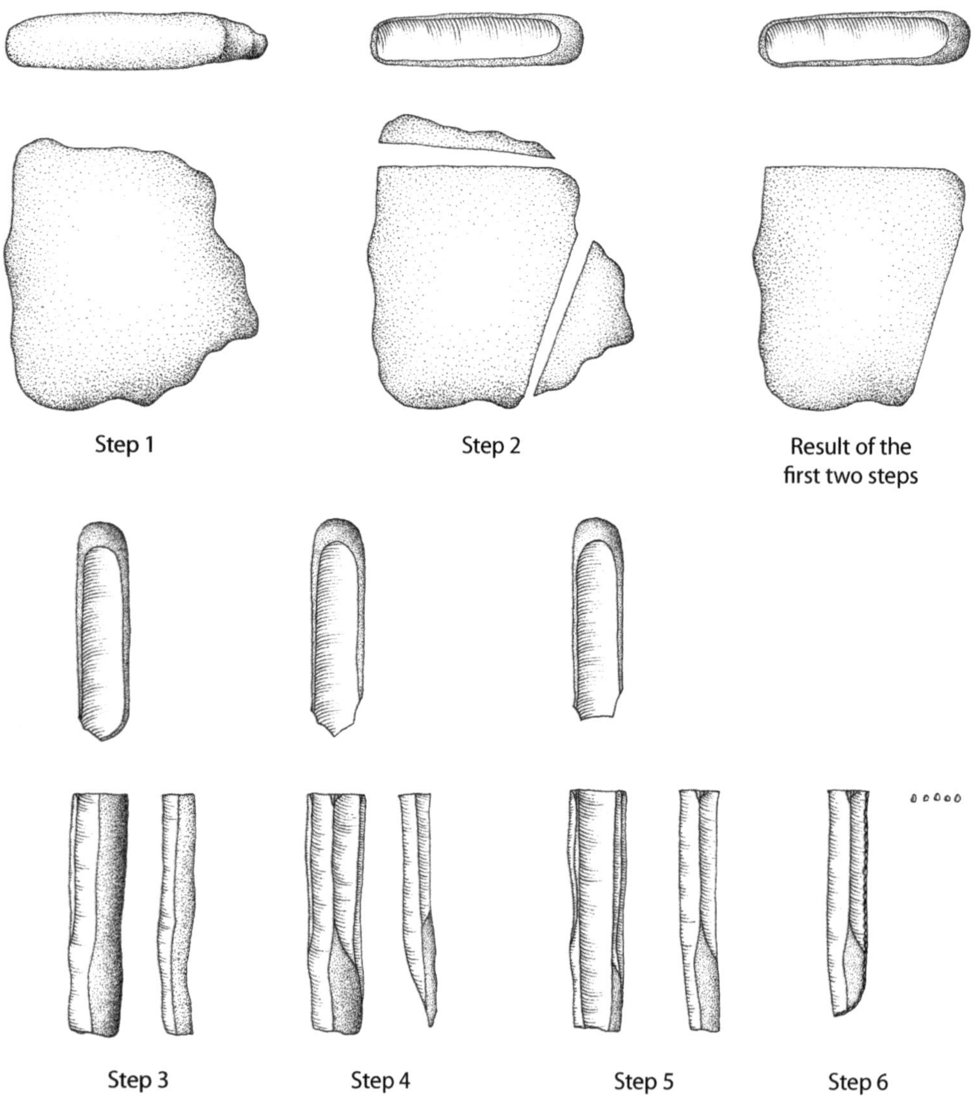

Figure 2.65. A chaîne opératoire for the experimental production of type I large blade knives without platform preparation. Drawing: Björn Nilsson.

The reduction sequence that gives the best results is that, once the platform is shaped, a number of blade-like flakes are removed from this down over the surface that is selected as the front of the block. First one or more flakes are detached along one side of the block, followed by one or more flake along the other side of the block. These run in over roughly half of the surface formed by the earlier removals. Then a flake is removed with a platform in line with the ridge formed between the lateral removals. This flake is the preform for the large blade knife, figure 2.64.

After the preform for a blade knife has been removed it is sometimes possible to continue the reduction and make yet another preform. But just as often the core is already exhausted after one preform has been removed. The reason for this is that the volume of the flint is often insufficient for more removals. The flat flint is not big enough. This means that the manufacture of type I large blade knives without platform preparation is resource-demanding. Several times we got only one preform for a large blade knife per nodule.

The similarity between the reduction process for flat flint and the one defined for block flint is that both consist of a three-step reduction sequence. The differences between the reduction sequences are nevertheless considerable. The platform preparation, the morphology of the block, and the way in which the removals are done all differ.

Manufacture of type I large blade knives without platform preparation – a chaîne opératoire

Like the manufacture of type I large blade knives with a well-prepared platform, the making of type I blade knives without platform preparation requires a systematic technology that enables the manufacture of standardized products. Type I large blade knives without platform preparation are thus a product of a standardized technology just as much as type I blade knives with a well-prepared platform, figure 2.65 and 2.66.

Figure 2.66. Examples of parts of the schematic chaîne opératoire illustrated in figure 2.65. Note that the photographs come from different manufacturing series.

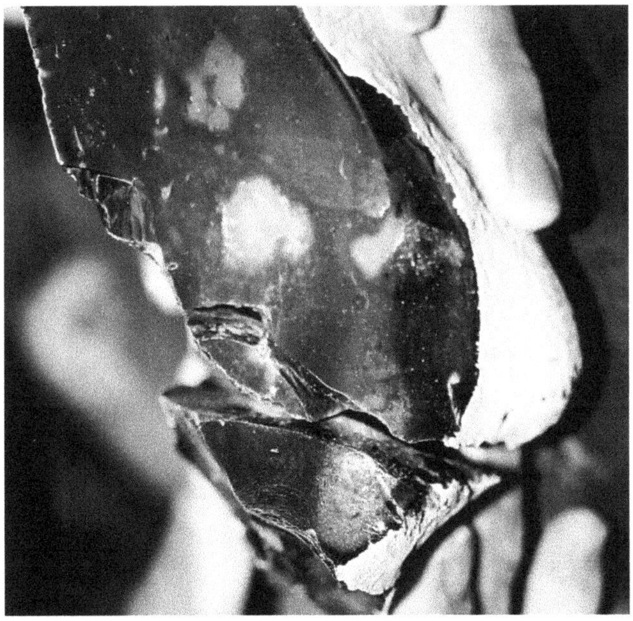

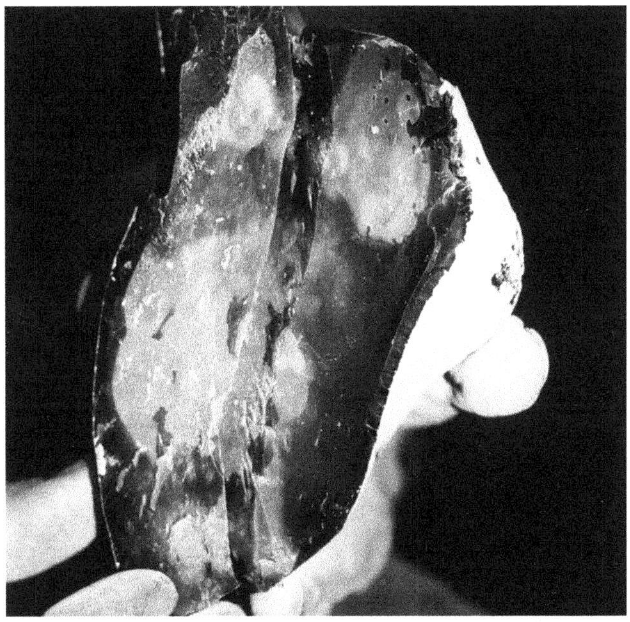

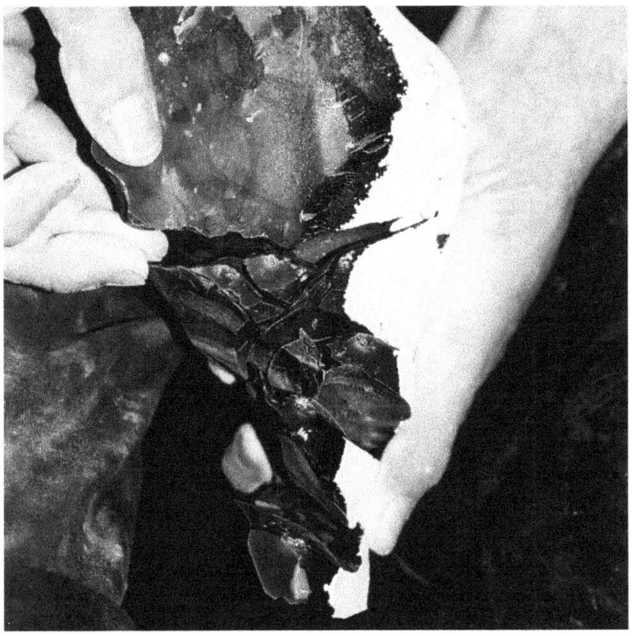

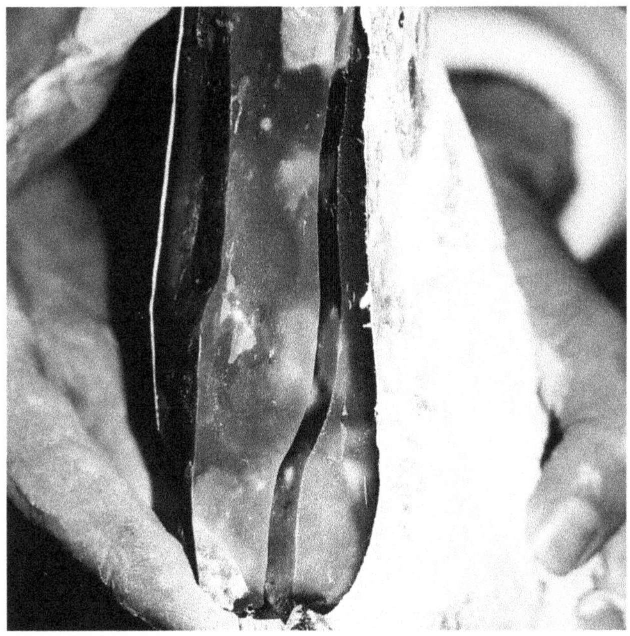

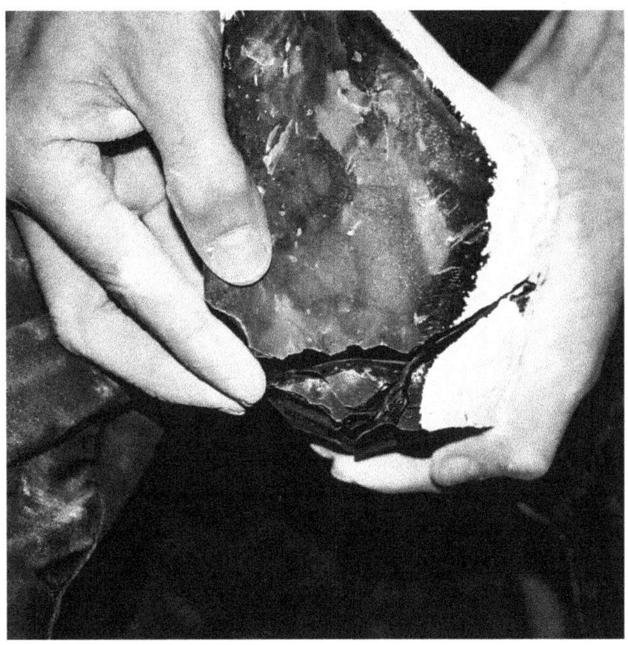

Production begins with the selection of a suitable flint nodule. This should be of high quality without cracks or impurities. The flint nodule must have a morphology that makes it possible to create a platform in a location so that there is enough flint to manufacture performs to large blade knives at the front of the block, and the base at the opposite end of the nodule from the platform must be shaped so that the flakes can come off without problems, step 1 in figure 2.65. The platform is then made by detaching an opening flake, step 2 in figure 2.65. In step 2 one can also knock off irregularities and protruding parts of the block. The result of the first two steps in the manufacture is a core with a flat platform but lacking a shaped front and base, figure 2.65.

In step 3 of the manufacture one or more flake is removed from the platform down along one side of the front of the core, figure 2.65. In step 4 one or more flake is removed from the platform along the other side of the front of the core. These flakes run in over roughly half of the surface from the earlier removals. The preform for the blade knife is then struck off, step 5 in figure 2.65. If the core can be used for continued manufacture, you now go back to step 3 in the reduction sequence. If the preform produced in step 5 is to be processed into a large blade knife, retouch is applied to the parts of the preform that need it, step 6 in figure 2.65.

The first part of the experiment – results

The first part of the experiment has taught us how large blade knives of flint are made. The series of trials shows that they are produced by standardized manufacture. They are removed with direct technique using a hammerstone. The shape of the block and the core is important if the production is to succeed. The reduction follows necessary steps and patterns throughout the manufacture. If the steps are not followed, the blade knife cannot be made. The production consumes large quantities of raw material.

We thus managed to recreate the technological and morphological attributes of the large blade knives of flint in the archaeological material which were detected by the attribute

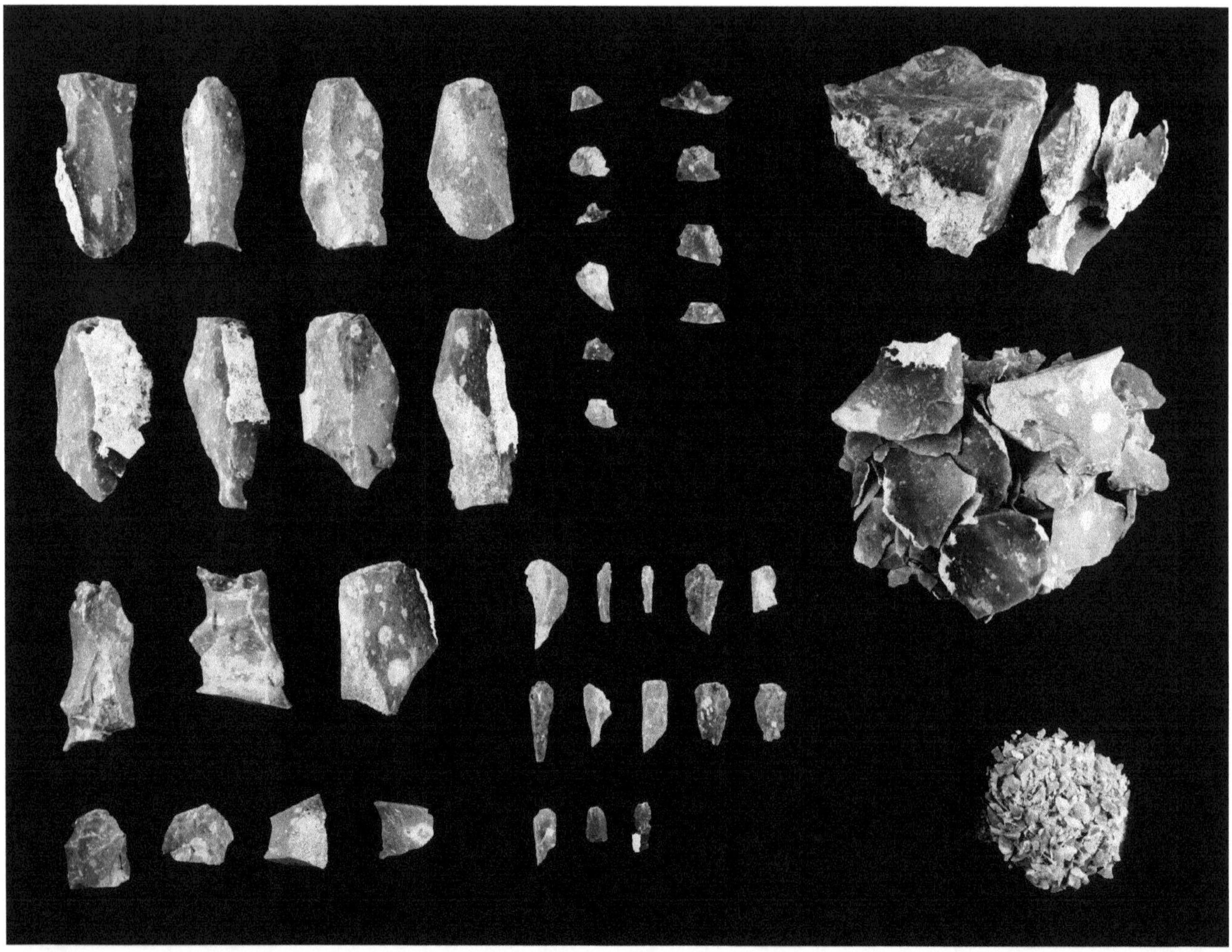

Figure 2.67. All the flakes from manufacture number 24 in the experiment series Lejre HAF 6/00, making type I large blade knives with a well-prepared platform. Top left: four preforms for large blade knives in a row. Under these are seven blade-like flakes. Bottom left: four fragments of blade-like flakes. Top centre: ten preparation flakes with hinged termination and under them thirteen preparation flakes with feathered termination. Top right: five flakes from the primary shaping. Under these 201 non-diagnostic flakes. Bottom right: debris and flint dust. Note that the exhausted core is not shown. Flint-knapper: Mikkel Sørensen. The preform at the top left is 12 centimetres long. Photo: Jenny Thornell, Malmö Museums.

Table 2.44. A description of products and debitage from each stage of production in the chaîne opératoire for manufacture of type I large blade knives with a well-prepared platform. Note that the table describes idealized manufacture with no mistakes or complications. Note also that the exhausted core is not included in the table.

Type I large blade knives with a well-prepared platform

Step	Products and degitage
1 to 2	Opening flakes, cortex-covered or with patina. Irregular size.
2 to 3	Large and small irregular flakes, varying degree of cortex, and original surfaces. Sometimes flakes without cortex. Irregular size.
3 to 4	Irregular blade-like flakes together with flakes of varying shape. Several of the flakes are longer than they are wide. Sometimes with, sometimes without cortex.
4 to 5	Regular blade-like flakes. If there is cortex it is either at the distal part of the flake or along one lateral side, and then the opposite side has no cortex, often with an irregular edge extending along parts of the flake.
5 to 6	Preparation flakes, either small flakes which are often wider than long and end in a hinge termination, or long thin flakes with a feather termination.
6 to 7	A preform for a blade knife in the form of a blade or a flake with a long and continuous edge along one lateral side and with a proportion between edge and ridge that gives a stable edge wtth a good edge angle. The preform is as thick as a blade knife.
7 to 8	A newly manufactured blade knife and flakes from the retouching.

analysis, to define reduction processes, and to establish a chaîne opératoire for block flint and one for tabular flint. We brought this knowledge into the second part of the experiment.

The second part of the experiment

The aim of the second part of the experiment was to create a set of reference material showing what the debitage from the manufacture of large blade knives of flint can look like, and on this basis to define diagnostic flakes from the production. This was done by producing large blade knives based on the results of the first part of the experiment, from raw flint nodules to finished products. Questions that we asked were:

- What does the waste material from the manufacture of large blade knives look like?
- Can any part of the material be characterized as diagnostic?

The focus in the second part of the experiment was thus on the waste material formed when large blade knives are made. The method involved analysing the residue from the entire manufacturing series of blade knives in the experiment. This resulted in the definition of a number of different types of flakes from the manufacture of type I large blade knives with a well-prepared platform, table 2.44 and figure 2.67, and of type I large blade knives without platform preparation, table 2.45 and figure 2.68.

Table 2.45. A description of products and debitage from each stage of production in the chaîne opératoire for manufacture of type I large blade knives without platform preparation. Note that the table describes idealized manufacture with no mistakes or complications. Note also that the exhausted core is not included in the table.

Type I large blade knives without platform preparation

Step	Products and degitage
1 to 2	Platform flakes, cortex-covered or patinated opening flakes. Varying size. Sometimes also large cortex-covered flakes and parts knocked off the nodule.
2 to 3	Large cortex-covered blade-like flakes.
3 to 4	Sometimes regular, sometimes irregular blade-like flakes, often with one lateral side with cortex and the other without cortex and with an irregular edge extending along parts of the flake. Almost always cortex on the distal part.
4 to 5	A preform for a large blade knife in the form of a blade or a flake with a long and continuous edge along one lateral side and with a proportion between edge and ridge that gives a stable edge with a good edge angle. The preform is as thick as a blade knife. Often cortex on the distal part.
5 to 6	A newly manufactured large blade knife and flakes from the retouching.

Figure 2.68. All the flakes from manufacture number 32 in experiment series MK 03, making type I large blade knives without platform preparation. Top left: two preforms for large blade knives in a row. Under them five blade-like flakes. Bottom left: three fragments of blade-like flakes. Top right: eight flakes from primary shaping. Under these nineteen non-diagnostic flakes. Bottom right: the exhausted core. Note that debris and flint dust are not shown. Flint-knapper: Mikkel Sørensen. The preform for a large blade knife at the top left is 13 centimetres long.

Figure 2.69. Cores from the manufacture of large blade knives, top: one from the production of type I large blade knives with a well-prepared platform made of block flint, bottom: one from the manufacture of type I large blade knives without platform preparation, made of tabular flint. Several of the cores left over from the manufacturing experiments show no negative removal after a preform for a blade knife. This is because the cores were not discarded after the removal of a preform, but after no further preparation or manufacture could be done because of unsuccessful flaking attempts. When a preform is detached, the core can usually continue to be used for making more preforms. It is only when manufacture has failed and we have realized that a core was exhausted that we have discarded it. The front of these cores therefore shows negative flaking scars from our failed attempts to make preforms for large blade knives, not from successfully detached preforms.

From our experience of working with the different reduction processes we can state that waste material from the manufacture of blade knives has several characteristic and diagnostic features. The waste material generally shows the same types of technological attributes that can be observed on the large blade knives. This is one of the characteristics of the waste material from the production. More specifically, the waste material can be divided into four groups: cores, large blade-like flakes and preforms for large blade knives, preparation flakes, and non-characteristic flakes. These four groups differ depending on whether the type I large blade knives were made with or without a prepared platform. I discuss them below separately.

The cores

The cores from the manufacture of type I large blade knives with a well-prepared platform are large, have a flat platform with traces of platform preparation, and a front with more than two long negative removals. The front of the core has no residue of the nodule's cortex or other original surface. The core from the manufacture of type I large blade knives without platform preparation is as wide as the tabular flint originally was, has a flat platform lacking preparation, and a front with one or two negative removals. If there are remains of cortex on the front of the core it is on the distal part, figure 2.69.

Large blade-like flakes and preforms for large blade knives

Since both reduction processes require that at least two blade-like flakes are removed from the core before a preform for a large blade knife can be made, these flakes are a characteristic feature of the waste material. In shape and appearance these flakes resemble preforms for large blade knives but they cannot be modified into blade knives. The reason for this is that, even though they look like preforms, they lack one or more of their characteristic properties in the form of the right size, the right edge angle, the right edge curvature, or a continuous edge line. On the dorsal side of the blade-like flakes there can be one or two negative removals. They can also be partly or wholly covered with cortex. There are distinct left and right flakes in the material. This is because the blade-like flakes are removed on the sides of the core. The flakes struck from the right side of the core usually have cortex on the left lateral side and those struck from the left side of the core have cortex on the right lateral side, figure 2.70.

The preforms for large blade knives are thus like the blade-like flakes. What distinguishes them is that the preforms have a shape and appearance allowing them to be retouched into large blade knives, figure 2.71.

The experiment with Kristianstad flint showed that there are distinctive features that can be linked to manufacture with this particular kind of flint. Since it is easy to strike through the platforms when using this flint, if they are not well-prepared and stable, a large proportion of the waste flakes consists of crushed as well as large, well-prepared platforms. This mixture is also found in waste material from production with

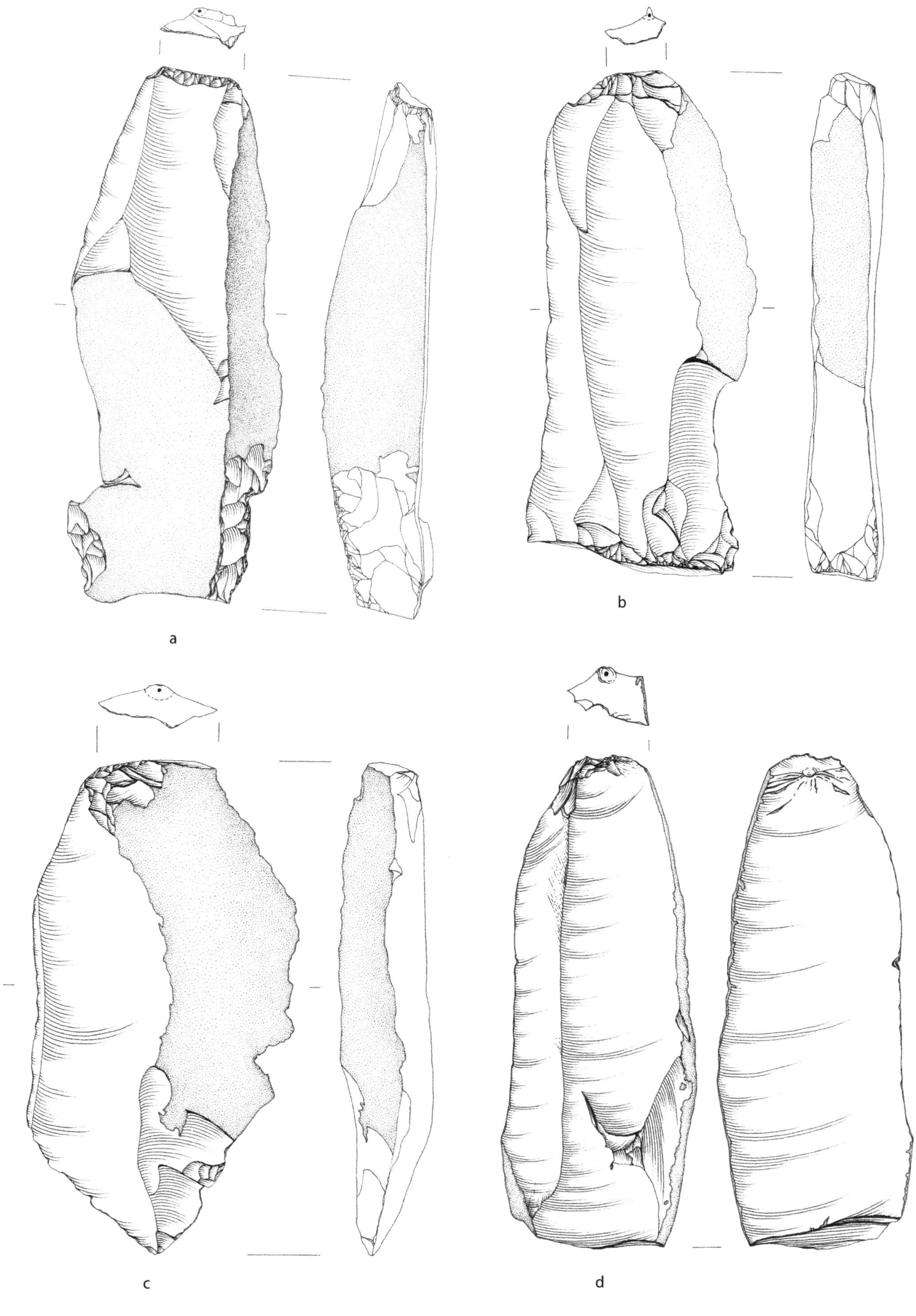

Figure 2.70. Continues on the following page.

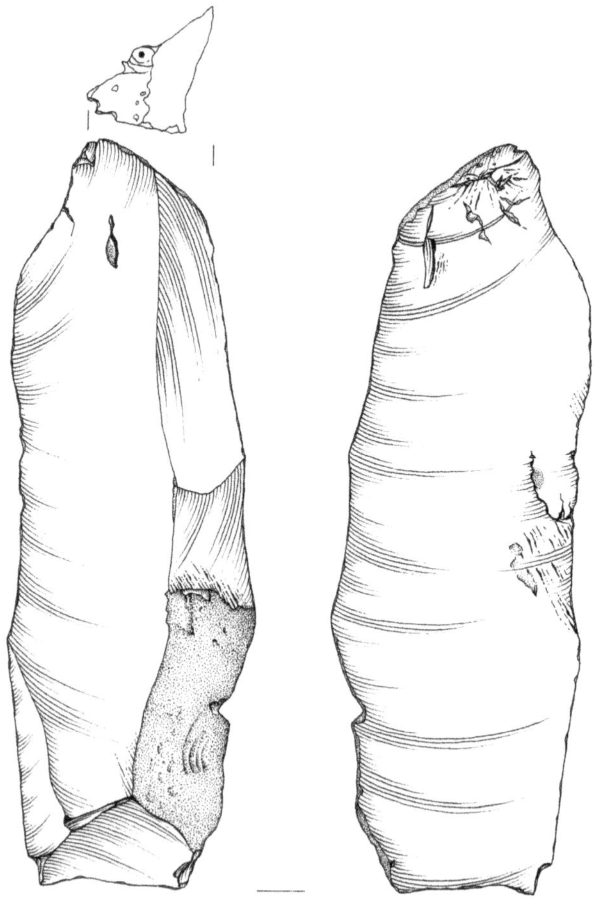

e

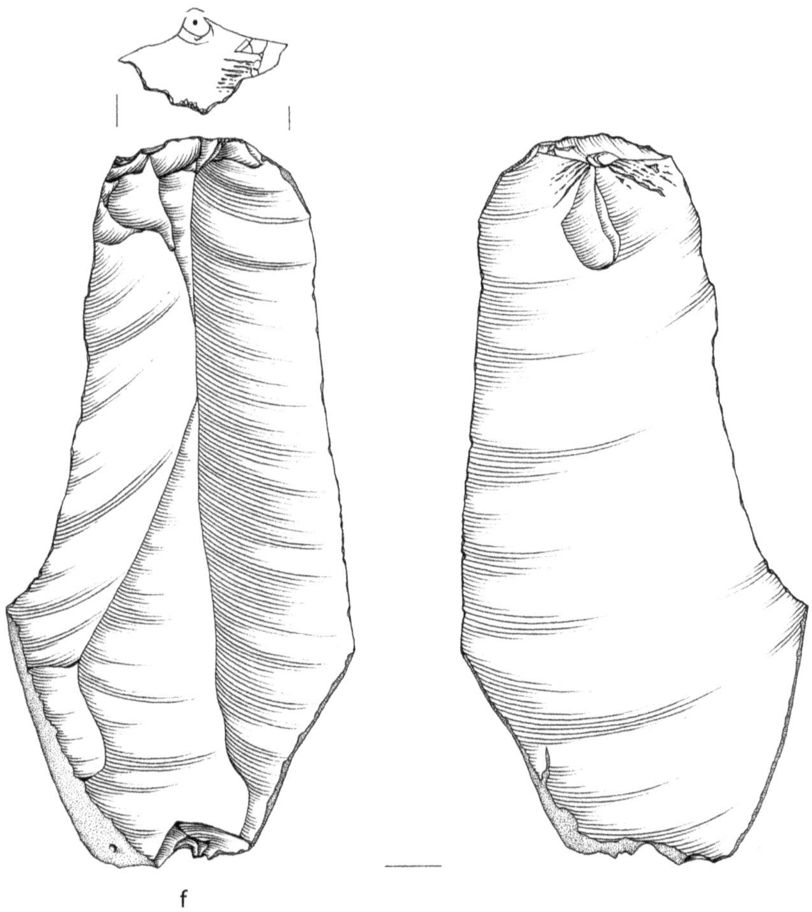

f

Figure 2.70. Blade-like flakes, diagnostic for the manufacture of large blade knives. a–c, three from the making of type I blade knives with a well-prepared platform, d–f, three from the making of type I blade knives without platform preparation. The flake at the top left, a, is 13 centimetres long. Drawings: Björn Nilsson.

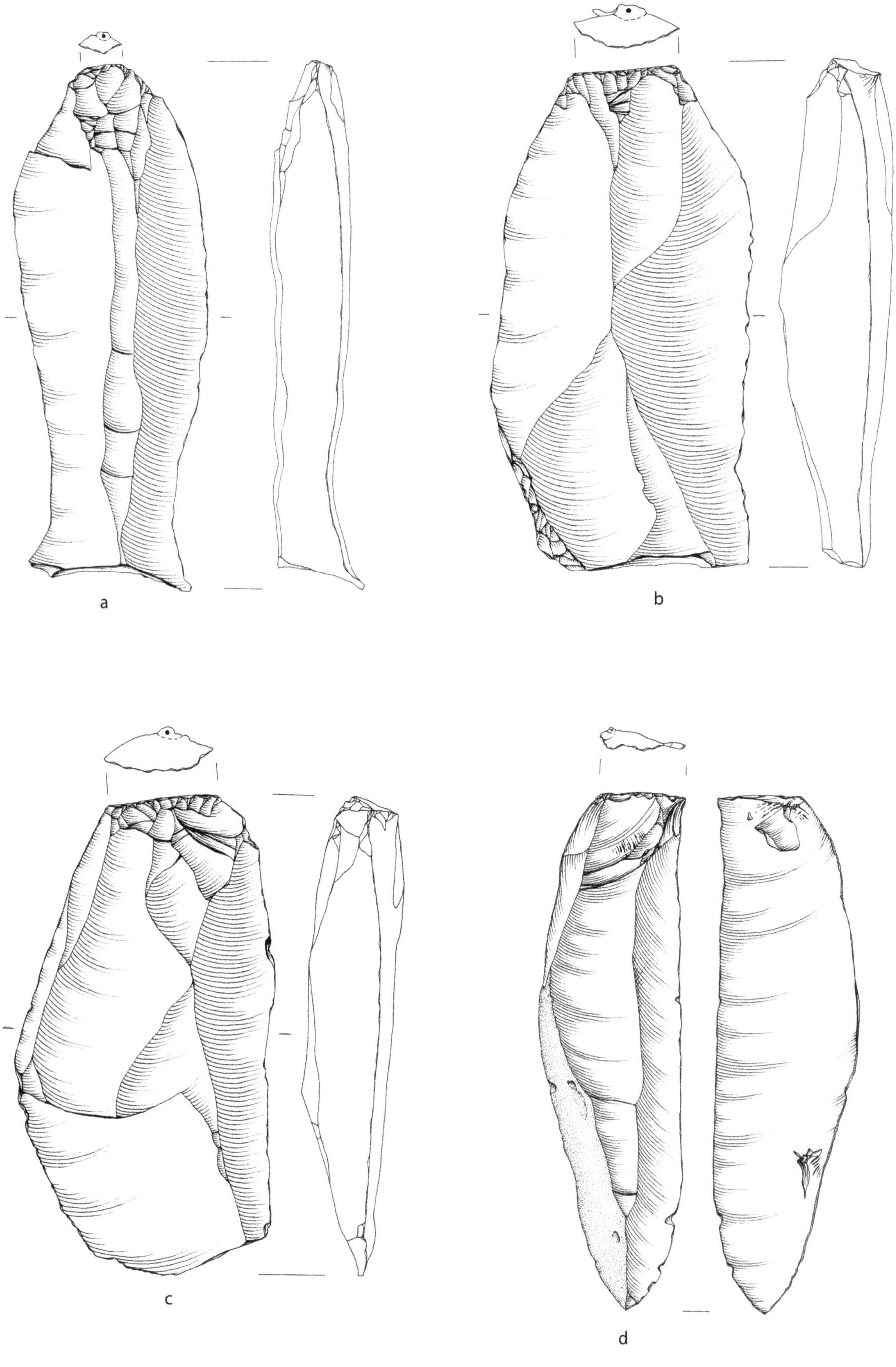

Figure 2.71. Preforms for large blade knives. a–c, three from the making of type I blade knives with a well-prepared platform, d, one from the making of type I blade knives without platform preparation. The preform at the top left is 11.5 centimetres long. Drawings: Björn Nilsson.

Figure 2.72. Manufacture of type I large blade knives with Kristianstad flint, distinctive features in the form of crushed platform, well-prepared and large platform, and broken waste flakes.

other flint types, but Kristianstad flint accounts for a larger proportion than other flint types. Another distinctive feature is that, since the flakes crack when the nodule contains accumulations of spots, a large share of the waste material will also be broken when a nodule with accumulations of spots has been used for production, figure 2.72.

Preparation flakes

Two characteristic preparation flakes have been defined in the manufacture of type I large blade knives with a well-prepared platform. One is a long and narrow flake with a feathered termination, the other a smaller, rather wide flake with a hinge termination, see figure 2.73.

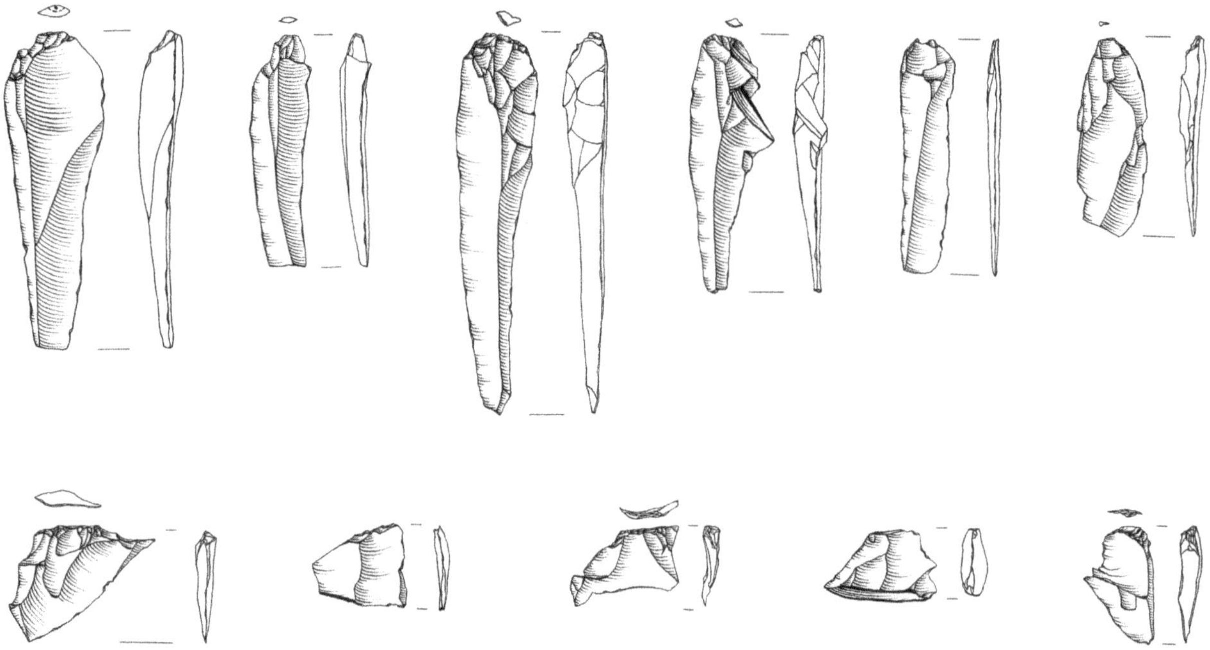

Figure 2.73. Characteristic preparation flakes from the production of preforms for type I large blade knives with a well-prepared platform. Six preparation flakes with a feathered termination and five with a hinge termination. See also figure 2.48 and 2.67. Drawings: Björn Nilsson.

Figure 2.74. Non-diagnostic flakes from three different experiments. These flakes display attributes which cannot be associated with any specific kind of production, being found in debitage from many different types of manufacture. To the left, non-diagnostic flakes from the manufacture of a square-sided Neolithic axe, in the middle non-diagnostic flakes from the manufacture of a Late Mesolithic core axe, and to the right, non-diagnostic flakes from the manufacture of large blade knives. Note that the manufacture of both core axes and square-sided axes also generated diagnostic flakes. These flakes, however, are not pictured here. The big flake to the far left in the picture is about 4 centimetres long. Flint-knappers: Neolithic axe Thorbjørn Petersen, core axe Anders Högberg, and blade knife Mikkel Sørensen. Photo Jenny Thornell, Malmö Museums.

Non-characteristic flakes from the production of large blade knives

A large share of the flakes that come from the manufacture of large blade knives is not characteristic. Throughout the production series, the core is worked in different ways. In this process flakes are repeatedly removed which look like any flakes at all. The attributes they display cannot be associated with any specific production method, but occur as waste material in a great many different contexts when flint objects are made, figures 2.74 and 2.75.

The proportion of diagnostic and non-diagnostic flakes

An important question is how many of the flakes from manufacture are diagnostic and how many are not. In one of the experiments at making type I large blade knives with a well-prepared platform, four preforms for blade knives were produced. Table 2.46 shows that 16% of the flakes counted by number and 43% of the flakes counted by the weight were diagnostic. The rest were flakes which could not be linked to any specific production. This shows that a considerable amount of flakes from this manufacture of blade knives are not diagnostic.

The second part of the experiment – results

The aim of the second part of the experiment was to investigate what the waste material from the production of large blade knives looks like and to ascertain whether any part of it is diagnostic. The results show that a large share of the waste material from the manufacture is not diagnostic. The study also found that the parts of the waste material that are diagnostic are easy to recognize. These are blade-like flakes, preforms for large blade knives, and preparation flakes with a hinged or feathered termination. Another finding is that the cores are diagnostic if the production ceases immediately after the removal of a preform for a blade knife. However, few cores in the experimental material are diagnostic. This is because the production did not cease after the removal of a preform for a large blade knife, but only when the core was completely exhausted after unsuccessful attempts to remove preforms.

Table 2.46. The number of diagnostic flakes, that is, preforms for large blade knives, blade-like flakes, and preparation flakes, and of non-diagnostic flakes from manufacture number 24 in the series of experiments performed at Lejre Research Centre, making type I large blade knives with a well-prepared platform. The flakes from the initial working of the core, what is labelled in the table as opening flakes, are diagnostic in so far as they can generally be identified as coming from the initial working of a core (Inizan, Roche & Tixier 1992:45ff), but they are not specifically diagnostic for the manufacture of large blade knives.

	Number	Weight in grams
Preforms for large blade knives	4	527
Blade-like flakes	7	1 182
Fragments of blade-like flakes	4	98
Preparation flakes with hinged termination	10	32
Preparation flakes with feathered termination	13	49
Opening flakes	5	1 319
Other flakes	201	1 090
Debris and flint dust	not counted	107

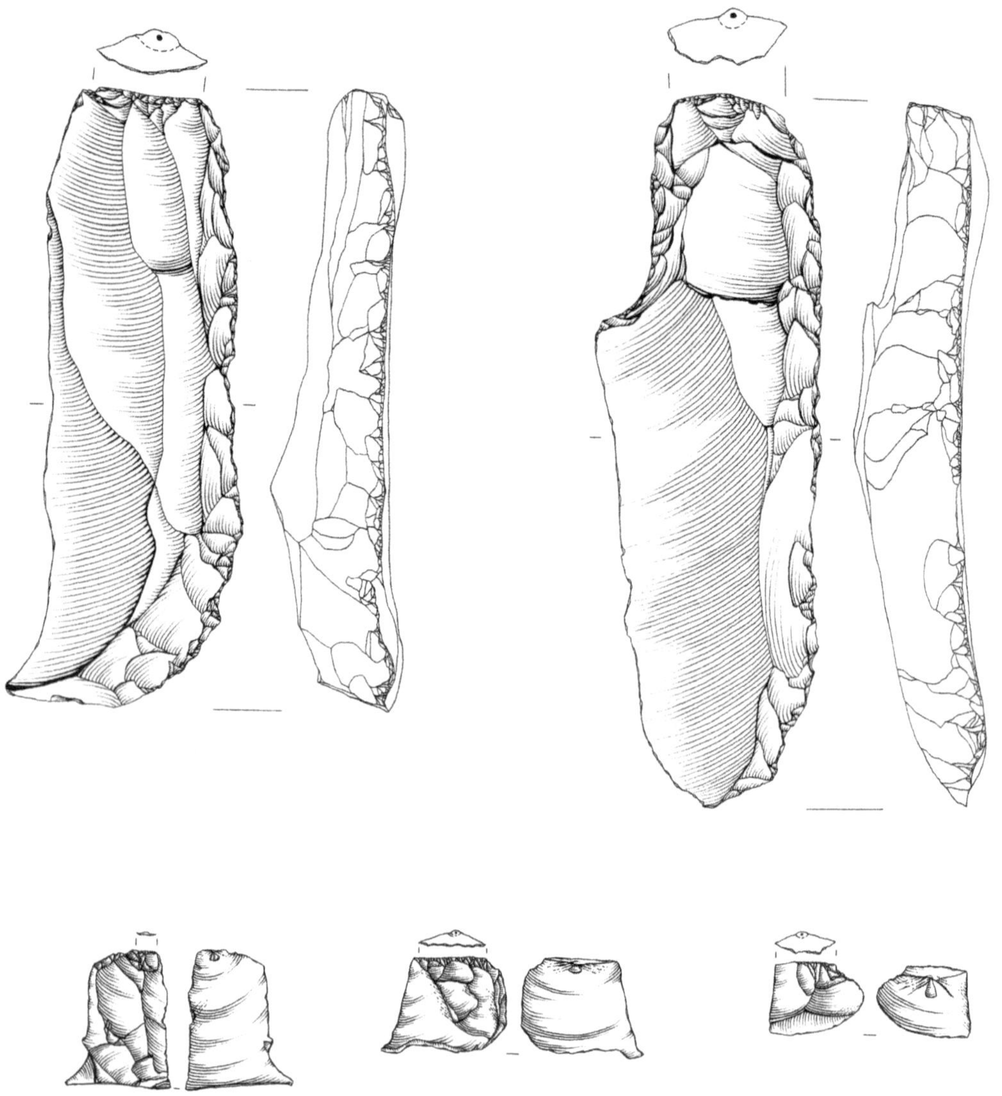

Figure 2.75. Flakes from the retouching of preforms for large blade knives when these are shaped into large blade knives, here together with blade knives from the experimental manufacture. These flakes are not diagnostic in themselves but if they are found together with other flakes from the manufacture of large blade knives they can be diagnosed within this manufacture as coming from the final shaping of preforms for large blade knives. The flake to the far left is 2.5 centimetres long. Drawings: Björn Nilsson.

It is important here to point out that individual flakes can never be regarded as diagnostic. It must be the overall impression of an assemblage of waste material that is decisive. The flakes highlighted here as being diagnostic are however clear in that they do not resemble flakes from any other south Scandinavian prehistoric flint-knapping technology.

The third part of the experiment

The aim of the third part of the experiment was to fragment parts of the waste material in order to determine whether it is possible to recognize diagnostic flakes in fragmented waste material from the production of large blade knives. The blade-like flakes are excellent as preforms for making tools such as scrapers and awls. They can also be used as raw material for cores for flake production. It is therefore reasonable to assume that if large blade knives of flint were produced at settlement sites or activity areas in the Late Bronze Age, the residue from the production may have been used to produce something else. Two important questions for the third part of the experiment were therefore:

- If the blade-like flakes from the production of large blade knives are reworked and broken, what do they look like?
- Is it possible to recognize the blade-like flakes even if they are fragmented?

There are many ways to break or reshape blade-like flakes. In the series of experiments I tried to divide, reduce, and rework them. This was done at Malmö Heritage in 2000 and 2007.

Dividing

The flakes can be systematically divided into smaller parts by being placed against an anvil and hit with a hammerstone, figure 2.76.

Figure 2.76. To divide a blade-like flake into smaller bits one can place it on an anvil and hit it with a hammerstone at the point where one wants it to split. Photo: Jenny Thornell, Malmö Museums.

Figure 2.77. From the left: three flakes broken at the moment of removal in the manufacture of large blade knives. To the right: three parts of a blade-like flake which has been deliberately broken after the production of large blade knives. The medial part of the flake split in several pieces on division, so that the proximal and the medial parts do not fit. Photo: Jenny Thornell, Malmö Museums.

The result of this is pieces of the blade-like flake which have edges that can be used unmodified to cut, scrape, and plane things with. These pieces can be recognized in different ways. Since the pieces consist of a proximal and a distal part of a blade-like flake, and a number of medial parts if the flake is split more than once, these parts have the same characteristic attributes as those found to be diagnostic for blade-like flakes. Since most of these characteristic attributes can be found on the proximal part of the flakes, the part of a broken flake that has a platform and a bulb of percussion is easiest to recognize. The medial and distal parts can be slightly harder to recognize since they show fewer characteristic attributes.

Blade-like flakes can be broken in a lateral fracture, that is, parts of the flake are split lengthwise. The pieces resulting from this fracture usually show no characteristic attributes and are therefore difficult to recognize.

An important thing in this connection is to be able to distinguish whether fragments were deliberately knapped to pieces or broken off by mistake during the production of large blade knives, figures 2.77 and 2.78. This distinction would allow us to separate waste material from production which was treated like waste and remains which were reused as raw material for other production. These flakes are most easily distinguished by looking at the fracture surface formed when the

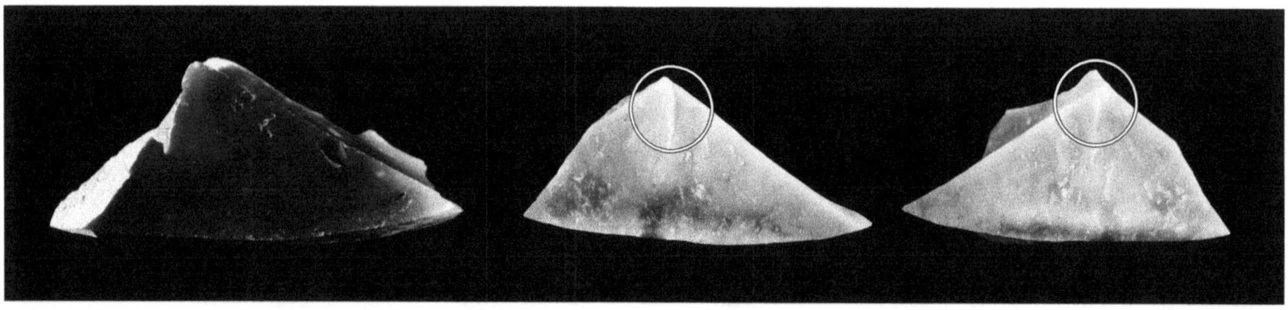

Figure 2.78. To the left one can see a fracture surface on a flake that was broken at the moment of removal during the production of large blade knives. To the right two blade-like flakes which have been deliberately broken after the production of large blade knives. They both have a positive fracture surface. The bulb of percussion is clearly visible on the fracture surface of these flakes. Photo: Jenny Thornell, Malmö Museums.

flakes were divided. If this displays a negative or positive bulb of percussion, or a hertz cone, it means that the flake was split with one blow. Sometimes one can also see fissures radiating concentrically from the point of impact on these flakes, and there is often a ring crack at the point of impact on the part of the original blade-like flake that was used as a platform. Flakes broken off at the moment of removal in the production do not display these attributes.

Reducing

The flakes can also be reduced as cores in order to produce flakes. In the series of experiments I used the ventral side as a platform and the flakes were removed in a series from the distal towards the proximal end of the flake, figure 2.79. This resulted in a flake and a core, in other words, the core is the blade-like flake which has been reduced from a flake to a core.

The core is what is left of the original blade-like flake. If the flake reduction is done from the distal part of the flake towards the proximal end, the core will have the characteristic attributes described above as being diagnostic for blade-like flakes.

The front of the core also takes on a special appearance. Since the reduction is done with the ventral side as the platform and the flakes run down over the dorsal side of the blade-like flake, which always has one or more ridges, the reduction must be done alternately from the right and the left. This gives the front a distinctive look that is characteristic of this reduction method, see figure 2.79.

The flakes that are removed from the core, that is, from the original blade-like flake, are not characteristic. They are of a type that occurs in the manufacture of many different tools.

Reworking

The blade-like flakes can also be used as raw material for making tools such as awls, scrapers, or burins, figure 2.80. If so they are reworked by retouching to give the desired shape of the tool. It should be stressed here that there are few known tools such as awls and scrapers in the Late Bronze Age flint toolkit. Burins are totally unknown (Högberg 2001a). But since it is possible to use blade-like flakes for making these tools, it is important to include them in this series of experimental production.

The result of the working is tools that have the type of characteristic attributes described above as being diagnostic for blade-like flakes from the production of large blade knives. However, if the flakes have been totally reshaped when a tool is made, for example, by being knapped or retouched along several edges, it is no longer recognizable.

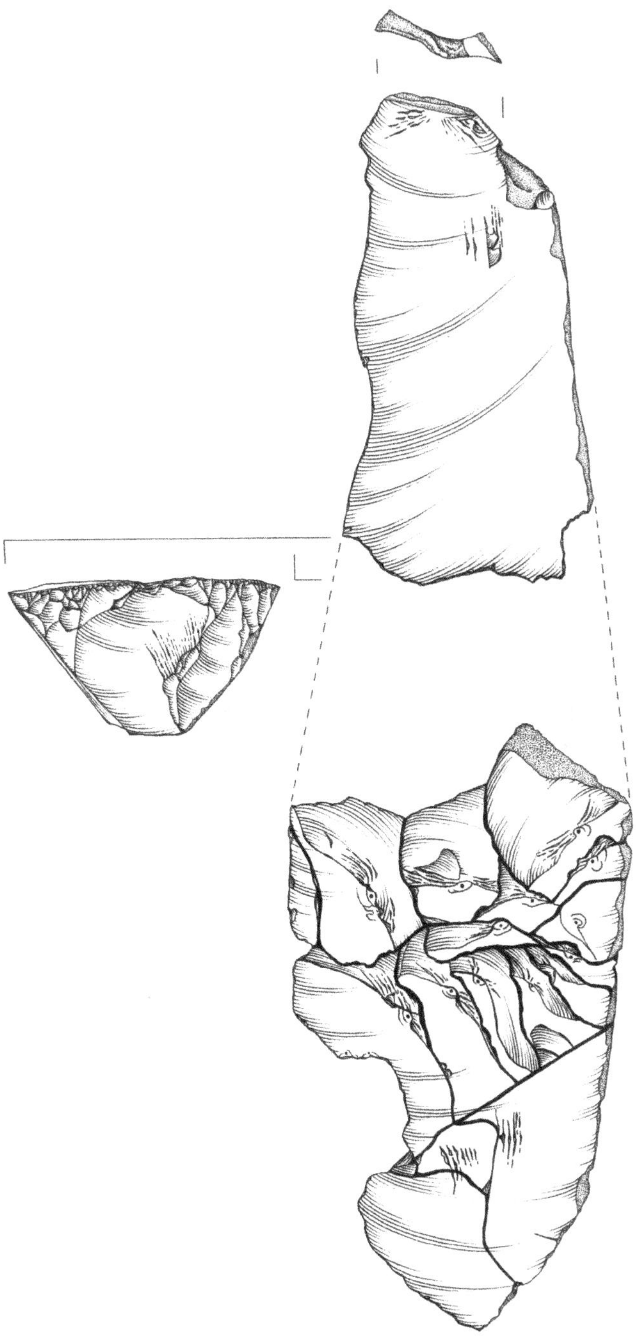

Figure 2.79. Part of a blade-like flake reduced as a core and then fitted together. I have used the flake as a core by starting the removal from the distal part and working my way towards the proximal part of the flake. The picture shows the proximal part of the original blade-like flake, displaying clearly how the core was struck alternately on the right and the left. This way of reducing the core gives the core front a characteristic appearance. The core front has a maximum width of 5 centimetres. Drawing: Björn Nilsson.

The third part of the experiment – results

The aim of this part of the series of experiments was to process blade-like flakes to see what the flints look like and whether they can be recognized despite being fragmented. The result shows that several of the characteristic attributes still survive even though the flints have been reworked, that some of the flints become more difficult to recognize, and some new attributes arise.

If the blade-like flakes are divided into several pieces, each piece shows the same characteristic attributes identified as being diagnostic for the blade-like flakes. Proximal bits are easiest to recognize. Medial and distal bits are harder to recognize. If the blade-like flake is divided so that a lateral break arises, the pieces cannot be recognized. An important observation in this part of the experimental series is that it is possible to

Figure 2.80. An awl, a scraper, and a burin, all made of blade-like flakes from the production of large blade knives. Far right: a scraper which has been retouched along three edges. This scraper was made from a broken blade-like flake which has been heavily retouched. If blade-like flakes have been heavily retouched in this way they are no longer recognizable. The awl is 7 centimetres long. Photo: Jenny Thornell, Malmö Museums.

distinguish flakes divided by reworking from flakes that were broken or split during the actual production. This is done by looking at the fracture surface. Flakes divided by reworking have a positive or negative bulb of percussion on the fracture surface, sometimes fissures radiating from the point of impact, and a ring crack around the point of impact on the part of the original blade-like flake used as a platform. Flakes which were broken during the manufacture of large blade knives do not show these attributes. If a blade-like flake has been used as a core for flake production, this gives a core with a characteristic conical front and attributes identified as being diagnostic for blade-like flakes. The morphology of the core front is an attribute for recognizing reworked flakes from the manufacture of large blade knives. If a blade-like flake is reworked into a tool, the attributes that are identified as diagnostic for blade-like flakes remain, unless there is extensive reshaping. Then the flake cannot be recognized.

Results of the manufacture of large blade knives – experimental analyses

The experiment has shown that large blade knives of flint were made by standardized production. The manufacture requires a knowledge of lithic technology; not just anyone can make large blade knives. The manufacture requires good access to high-quality raw material. We managed to recreate the technological attributes observed in the archaeological material. For this we used a direct technique with an oblong and hard, but elastic, hammerstone with a rounded, not pointed or blunt, impact surface. Of the different types of tool and techniques we used, this hammerstone was the only one that gave the attributes we were aiming for. Type I large blade knives with a well-prepared platform and type I large blade knives without platform preparation have many similarities in the way they are manufactured. When the manufacture is described in a chaîne opératoire for the two groups of blade knives, however, it is clear that they were made differently. This difference is mainly due to the fact that type I large blade knives with a well-prepared platform are usually made from large flint nodules with a large volume, while type I large blade knives without platform preparation are usually made from tabular flint.

The production of large blade knives generates a considerable amount of flint waste. Four types of waste material can be distinguished: cores, large blade-like flakes, preparation flakes, and other material. The blade-like flakes and the preparation flakes are diagnostic for the manufacture. Even if they are fragmented, the majority of the waste material is still diagnostic.

Summing up – morphology and manufacture

The aim of performing extensive attribute analyses and experimental analyses was to investigate what the large blade knives of flint look like, how they can be manufactured, and what the waste material from the manufacture looks like. The results show that blade knives should be regarded as an artefact in which the various types and ways of manufacture show both great similarities and significant differences.

Based on the attribute analyses, four groups of large blade knives have been defined: type I large blade knives with a well-prepared platform, type I large blade knives without platform preparation, type II large blade knives, and blade knives from Rügen. Most common are type I large blade knives with a well-prepared platform. They are usually 11 to 15 centimetres long, weigh 60 to 120 grams, are three to four centimetres wide and one to one and a half centimetres thick. They usually have one or two ridges and two or three negative removals on the dorsal side and lack cortex there. The majority of the blade knives have a straight and well-formed edge and an exterior platform angle of 70–80 degrees. The platform is flat and rhombic with a ring crack around the impact point, usually 1.5 to 3.0 centimetres wide and 0.6 to 1.5 centimetres thick. Platform preparation in the form of trimming along the platform edge and thin preparation flakes running down over the dorsal surface commonly occur. As part of the preparation, several type I large blade knives with a well-prepared platform also have preparation flakes with hinged or feathered termination.

Type I large blade knives without platform preparation are similar to those with a well-prepared platform. What differentiates the two groups is the platform preparation. On the blade knives without platform preparation the platform edge consists instead of a ridge formed by two negative removals running from the platform down over the dorsal surface. Type I large blade knives without platform preparation often have cortex on the dorsal side. Type I large blade knives without platform preparation are not as common as type I large blade knives with a well-prepared platform.

Type II large blade knives show partly the same attributes as type I blade knives, but they are shorter and the retouch at the distal part gives a rounded end to the blade knife, unlike type I blade knives which have a more angular end. Attributes that distinguish type II large blade knives from type I are that few of the type II blade knives have any cortex on the dorsal side, a much larger proportion of type II than type I knives have bidirectional dorsal flake scars, and there is a difference between the number of ridges, as a larger share of type II blade knives have two ridges while the majority of type I blade knives have one ridge. Type II large blade knives have greater density than type I. Type II blade knives make up a small proportion of my material.

Large blade knives from Rügen have technological attributes that resemble type I blade knives but differ in size. They are smaller. The majority are between 8 and 10 centimetres long, weigh between 20 and 60 grams, and are slightly narrower and thinner than other type I large blade knives.

The majority of the blade knives, irrespective of which group they are assigned to, have a marked and well-formed bulb of percussion with a bulbar scar, moustache, and bulbar fissures and a continuous, regular, and direct retouch. Roughly half of the blade knives have the retouch on the left side, the other half on the right side. The majority of the blade knives are made of Senonian flint.

It is interesting to note that the large blade knives in my material which come from Schleswig-Holstein in Germany are grouped with type I large blade knives, not with the blade knives from Rügen. There is thus a difference in the appearance of large blade knives from northern Germany. Also interesting to note is that the three large blade knives of Zele type from Poland in the material have platform angles of 80, 85, and 90 degrees. They thus differ from the majority of the other large blade knives.

The experimental analysis has demonstrated that the large blade knives were made according to standardized production and a systematic chaîne opératoire. The manufacture requires a good knowledge of lithic technology and access to high-quality raw material. Although the manufacture of type I large blade knives with and without platform preparation has a great deal in common, each group of large blade knives follows a different chaîne opératoire. The major difference is in the selection of raw material. Type I large blade knives with a well-prepared platform are usually made of block flint. Type I large blade knives without platform preparation are usually made of tabular flint. This difference in raw material is directly related to the differences in the morphology of the blade knives.

The manufacture of large blade knives of flint is not a pure blade technology. Although large blade knives look like blades, several of the preforms for large blade knives are more like flakes than blades. Large blade knives only acquire a blade-like form when they are retouched. Right- and left-retouched blade knives are an effect of the manufacture. Roughly half of the preforms for large blade knives have a suitable edge on the left side, the other half on the right. The reason for this is that the edge on the preform is almost always on the side facing the middle of the core front during manufacture.

The blade-like flakes and preparation flakes resulting from the manufacture are diagnostic for the production of large blade knives and can be used to search for traces of production and manufacturing sites. Even if the blade-like flakes are fragmented, the majority of them are still diagnostic.

Few earlier studies have focused on issues to do with the manufacture of large blade knives. Ethelberg (1982) interpreted blade knives as part of a reduction process whereby all knapped flint from a settlement site assemblage, in his opinion, was part of the same chaîne opératoire. The results of the attribute analyses and the experiments show that this is incorrect. Large blade knives are standardized products with different chaînes opératoires which do not include the manufacture of anything else.

Björhem and Säfvestad (1993) defined blade knives as standardized. They believed that type I and type II blade knives were made separately, in other words, that type II large blade knives are not made from modified type I knives. The result of my studies confirms this.

Large blade knives of type I and type II have previously been described by Björhem and Säfvestad (1993). The difference which I have defined between type I large blade knives with a well-prepared platform and type I large blade knives without platform preparation should be viewed as a development of Björhem and Säfvestad's type I. This means that, even if the two groups of type I blade knives that I have defined both fit in Björhem and Säfvestad's definition, they are different kinds of blade knives in terms of the way they are manufactured. Blade knives from Rügen, Germany, have not been described before. Nor have the different chaînes opératoires for the making of large blade knives or the waste material from the manufacture been described previously.

Use-wear analysis of large blade knives

After having focused on the morphology and manufacture of large blade knives of flint, it is time to examine aspects of their function and ask what the blade knives were used for. Were they harvesting implements in arable fields and meadows? Were they tools for gathering reeds to thatch houses? Were they used to cut leaves as winter fodder for stabled livestock? Or were they multifunctional tools employed for all these purposes and much besides? The function of large blade knives has been discussed for a long time, and all of these uses have been suggested.

In this chapter I shall discuss aspects of this in a number of sections, each one building on the last. First I survey previous functional analyses with the focus on what has been said about the uses of blade knives and how the question has been analysed. This leads to a specification of the questions that are relevant to this chapter and of the different contact materials to be investigated, table 2.47. This is followed by an in-depth study of previous investigations of a number of specific contact materials and a report on my experiments in the use of the tool.

These different parts serve as a foundation for a focused use-wear analysis of 64 large blade knives of flint from excavations in the Malmö area, south-western Scania, Sweden, in the last twenty years. The number of large blade knives makes this use-wear analysis the hitherto largest study of large blade knives ever published. The chapter ends with a presentation and discussion of the results of these analyses. The aim of the work presented in this chapter is to investigate what the large blade knives were used for.

Table 2.47. Terms and definitions specific for use-wear analysis as employed in this study (from Jensen 1994:12ff).

Terms	Definitions
Contact material	The material which the flint tool is interpreted to have been used to work.
Polish	A modification of the original flint surface as a result of contact with other material. Appears as an altered surface that reflects more light in the microscope than the surrounding area.
Gloss / Sickle gloss	A brilliant, lacquer-like alteration along the edge of a flint implement.
Generic weak polish	A non-diagnostic polish generated by the initial working of virtually all sorts of material, i.e. an early stage in polish development on tools. It is a polish which can be interpreted as coming from use of the implement, but not allowing identification of the contact material.
Micro scarring	Microscopic edge damage.
Striations	Different forms of grooves, scratches, and linear depressions identified in the polish or on the flint surface

Previous functional analyses of large blade knives

As we saw above, Blinkenberg (1898:147), more than a hundred years ago, identified the function of a shafted blade knife as a sickle for reaping grain. Based on a consideration of the durability of the shaft, it was his opinion that the implement could not have been a tool for woodworking. He noted shiny traces of wear on the edge, which he thought were strangely distributed over the knife. He also noted that the shaft was worn shiny on the underside at the part where the blade knife was held in place. Based on these observations, above all the distribution of use wear over the underside and up over the ridges of the blade knife, he declared that it must have been straws being drawn over the surface during reaping that produced the shiny traces of wear (Blinkenberg 1898:148). As previously discussed, this shows that large blade knives, as soon as the first finds were noticed, were interpreted as reaping tools.

This is also clear in Steensberg's (1943) study of prehistoric reaping implements. Based on modern experiments and extensive studies of archaeological material, he discussed a great many aspects of the harvesting tools in early agriculture. Two series of experiments were performed using copies of the find from Stenild. In the first experiment a prehistoric (!) large blade knife was used, shafted in a copy of the Stenild shaft. This large blade knife was described before the start of the experiment as "not very sharp, otherwise intact" (Steensberg 1943:13). The tool was used for 80 minutes to reap a field with modern barley with an admixture of oats and weeds. The work is described as not having been particularly effective. The second experiment was performed with a fresh, sharp flake glued in the shaft. This gave a good-quality result, meaning that the straws were cut off systematically and evenly without being pulled up by the roots. Steensberg noted, however, that the work took a long time with the implement (Steensberg 1943:18).

In a comparison of the two series of experiments, it was noted that the differences between them were large. This was explained by differences in the sharpness of the blades and the fact that different people had done the experiments. Steensberg also noted that the first experiment was performed in a field with many thistles. This had a considerable impact on the efficiency of the work (Steensberg 1943:23f).

Steensberg summed up his observations by classifying the copies of the Stenild find as being among the worst he had used. He thought that it was not a good reaping tool, even though it could be used. These observations about the efficiency have influenced later studies of prehistoric agricultural implements (Welinder 1998:146). Moreover, Steensberg thought that the edge was far too delicate for cutting twigs and foliage. This statement, however, is not based on any experiment using the tool. One hypothesis propounded by Steensberg is that the implement was not used for cutting grain but for weeds. This hypothesis is grounded on the observation that the tool easily cut through thistles and other weeds in the series of experiments.

The study shows that shafted large blade knives could be used as a reaping tool. Yet the study is problematic in its design, since Steensberg in his first series of experiments used a blade knife from the Bronze Age which obviously was not sharp, and in the other series an ordinary flake glued into the shaft. Neither of these series of experiments can be said to be reliable for assessing the efficiency of a newly manufactured large blade knife.

Focused use-wear analysis

There have not been many focused use-wear analyses of large blade knives. In 1979 Mark Wyszomirski published a study in which he did microscope analysis of a number of blade knives from Scania, Sweden, using the low-power approach (LPA) (Wyszomirski 1979). Helle Juel Jensen has published analyses at high magnification (HPA) of an assembled find of six blade knives from Spjald and an analysis of the shafted blade knife from Stenild. Both find spots are in Jutland, Denmark (Jensen 1990). As part of a major project about Bronze Age settlement sites in central and north-western Jutland, Merete Christensen has performed use-wear analysis at high magnification of assemblages from the Bronze Age. The study includes a number of large blade knives of flint (Christensen 1996). A study of thick bifacial sickles from the Netherlands by Anne Lou van Gijn (1988) is also important to mention here, even though these implements are not large blade knives; they are contemporary with blade knives, and they have a gloss that is visible to the naked eye, similar to that observed on large blade knives of flint. A Polish study has described the extent of use-wear on large blade knives (Kruk 1994).

These previous studies are important as a point of departure for the present studies of large blade knives. They also provide a foundation of knowledge for the function-related experiments that I have performed. The following questions are considered:

- What contact materials have been demonstrated in these studies?
- Do the studies show that large blade knives of flint were shafted, and if so how?

A number of blade knives from Scania

In 1979, as mentioned above, Wyszomirski published a brief study where he discussed the typological classification, function, and dating of large blade knives. The functional analysis was done at low magnification (LPA) and the work was inspired by S. A. Semenov's (1964) studies. Wyszomirski presents observations of gloss, microflaking, edge scarring, and striations on 85 large blade knives from the collections of Lund University Historical Museum. These observations are summarily described: "on most of the tools there is gloss on elevated sections, and striations were found on some of them" (Wyszomirski 1979:146). He then describes the direction of the striations on 22 of the blade knives. Photographs illustrate the gloss on seven of the analysed blade knives (Wyszomirski 1979:146ff).

Wyszomirski thinks that the gloss he observed is quite different from that found on Neolithic bifacial sickles, and he defines the gloss as being "of the type that arises from woodworking" (Wyszomirski 1979:146). The arguments for this are that the gloss is visible only on elevated sections of the flint surface and not in the hollows between them. Having identified woodworking, Wyszomirski goes on to analyse the striations and deduces that they come from working with hard material such as fresh oak, bone, antler, or groundstone. He adds, however, that certain blade knives also seem to have been used as sickles, since intensive gloss suggests this.

His conclusion is that large blade knives were presumably not used solely for collecting foliage, but also as tools for working "hard material such as wood, antler, bone, and possibly other things" (Wyszomirski 1979:148). The study briefly mentions the shafting problem; Wyszomirski believes that blade knives were not always shafted as in the example from Stenild, since traces of use, in this case striations, were found all over the surface of some blade knives.

Wyszomirski's study made a considerable impact and is still cited in discussions of the functions of large blade knives. Reference is made above all to his opinion of blade knives as implements for harvesting foliage and as multifunctional tools (see e.g. Björhem & Säfvestad 1993:76f; Jensen 2006:104). This makes it important to examine Wyszomirski's work more closely.

There are a number of source-critical problems with the study. These concern methodological aspects, the status of the material, and the interpretations of polish, striations, and working operations. The statement that the gloss is different from what is observed on bifacial sickles is left unsupported since there are no descriptions or pictures of the gloss on these sickles. The difference as described is that the gloss is visible only on elevated sections of the blade knives. This argument does not hold water given what we now know from analyses by high-magnification microscope of polish arising from work with silicaceous plants. Several authors have discussed, described, and illustrated the polish that develops on the edge of the flint when a tool is used. These authors agree in their description of the way in which polish from the contact materials mentioned by Wyszomirski develops first on the raised parts of the flint and subsequently – to varying extents depending on the length of the operation, the intensity of the work, and variations between different contact materials – through time can also arise in the depressions between elevated sections and finally merges into a polish that covers more or less the entire surface (Keeley 1980; Vaughan 1985; van Gijn 1989, 1994; Jensen 1994, Levi Sala 1996). It may also be questioned whether it is possible at all to assess such differences and assign them to different contact materials through analyses by low-magnification microscope. Blind tests have shown that LPA works well for identifying the occurrence and distribution of use-wear, the direction of movement of the implement, and the type of operation – for example cutting or scraping – for which the implement was used. On the other hand, it has been shown that LPA cannot be used to determine exactly what the contact material was (Odell 2004:146f).

The study was also palpably influenced by the function indicated by the Swedish designation for blade knives, meaning literally "leaf knives". There is nothing in Wyszomirski's ana-

Table 2.48. The seven analysed artefacts from Lund University Historical Museum which were published with photographs of use-wear by Wyszomirski (1979:148) and have now been examined by me. The table states the find number, year of acquisition, designation, and the status of the large blade knives as I perceive it.

LUHM number	Year of acquisition	Designation	Status
2334	1855	Late Neolithic dagger shaft	
3324	1855	Large blade knife	Heavy, thick white patina covering the entire surface of the object.
18153	1914	Late Neolithic axe	
18159	1914	Large blade knife	Slight patina covering the entire surface of the object. Severe post-deposition damage to the edge.
19017	1915	There are 26 different finds under this number. None of them is a large blade knife.	
19234	1916	Neolithic blade knife	
27231	1935	Large blade knife	Heavy gloss on parts of the blade knife. Minor post-deposition damage on the edge. No patina.

lysis of gloss on blade knives to indicate that they were used for cutting leaves. Nor is there any analytical discussion of working operations. What Wyszomirski does is to identify the contact material as wood, not to show that the working operation was foliage harvest. Despite this, the function of the blade knives is specified as having been used to harvest leaves. In my opinion, the results of the study are not reliable given today's knowledge of different kinds of polish and contact materials (Levi Sala 1996).

Another problematic aspect is the status of the implement. Several of the large blade knives that I have studied in museum collections have a flint surface that is so transformed and patinated that it is impossible to do use-wear analyses. This problem has also been noted by other scholars (Jensen 2006:104). It is therefore essential to investigate the status of the seven blade knives that Wyszomirski examined and published pictures of (Wyszomirski 1979:148). I have done this by retrieving each of these artefacts from the museum collection and analysing them as regards year of acquisition, designation, and status.

Table 2.48 shows that, of the seven blade knives that Wyszomirski analysed and illustrated with photographs of use-wear, only three are tools that I regard as large blade knives of flint. One is a Late Neolithic axe, another is a shaft for a Late Neolithic dagger. This is remarkable, since neither a dagger shaft nor an axe can be mistaken for a large blade knife. Evidently some mistake arose in the analysis or the publication of the results. Of the three artefacts that I regard as large blade knives, two have patina and post-deposition damage which, in my opinion, makes use-wear analysis impossible. Only one of them has a status that permits use-wear analysis, number 27231 in table 2.48.

The three large blade knives in table 2.48 were acquired by the museum between 1855 and 1935, in other words, 70 to 150 years ago. Since then they have underwent handling, examination, being shaken around in find boxes, placed together with other flint objects, rubbing against them, being packed and repacked. Several studies have demonstrated how damage can be caused by post-deposition and post-excavation treatment (Levi Sala 1986, 1996:125; Unrath et al. 1986; van Gijn 1989:54). There is nothing to say that the striations observed by Wyszomirski are of a kind that arose before the artefacts ended up in the earth.

After this reconsideration of Wyszomirski's analysis of a number of large blade knives from Scania, I find that I cannot use the results at all. The study is not reliable.

Spjald and Stenild, Denmark

At Spjald in western Jutland in Denmark, a site with remains of houses from the Late Bronze Age was excavated in 1969–1971. In a house foundation, more specifically in a post-hole for a roof-bearing post from house XII, six large blade knives of flint were found together with a flint scraper, the remains of two moulded splint boxes, a wooden vessel, pottery, and lumps of resin (Becker 1972, 1990). Helle Juel Jensen has performed use-wear analyses (HPA) of these blade knives. Her study also includes an analysis of the previously mentioned large blade knife from the shafted find from Stenild (Jensen 1990).

On the edges on the large blade knife from Stenild and those from Spjald there is a band of gloss that arose from working with silicaceous plants such as cereals, reeds, and grass, but also to a minor extent fresh small branches. Jensen's study was intended to determine which of these contact materials the blade knives were used on. No experiments were performed to answer specific questions for the use-wear analysis. The study is based on published works and reference material created by general trials.

The study shows that the polish on the large blade knives is found on the edge, on the distal and medial part of the ridge on the dorsal side, at the distal part of the dorsal and ventral sides, and along the retouched back. The six blade knives from Spjald show a high-reflecting gloss with a domed surface. The ridge on the dorsal side of the blade knives, however, is fluted. The polish forms a broad string along the edge and extends far in over the surface. Striations are relatively rare. The orientation of the polish shows that the direction of working was oblique or at right angles to the longitudinal axis.

The overall impression of the polish on the six large blade knives from Spjald, according to Jensen, is that they were used for cutting reeds. When cutting annual shoots on trees, as for example when harvesting foliage, polish can arise which resembles that from cutting reeds, but the extent of the polish caused by harvesting leaves does not look like what is seen on the six large blade knives from Spjald (Jensen 1990:209ff).

To the naked eye, the extent of the polish on the blade knife from Stenild looks like that on the implement from Spjald. At high magnification, however, one can see differences. The blade knife from Stenild has a smooth, matt, and scratched polish, usually associated with working with cereals or straw. The direction of working is generally the same as for the six blade knives from Spjald, oblique or at right angles to the edge (Jensen 1990:210).

The six large blade knives from Spjald and the one from Stenild display the same type of shafting retouches. Parts of the blade knives are free of gloss and polish. These parts were probably hidden during use and thus not in contact with the material on which the blade knives were employed. On these "clean" surfaces there is polish which can be attributed to shafting, known as friction polish. Since the traces on the edge and the ridges extend further towards the proximal part of the blade knife than the traces on the back, the shaft must have covered more of the back than the edge. Jensen's interpretation is that the way in which the blade knife from Stenild was shafted can also be said to be representative of the shafting of the six large blade knives from Spjald.

Jensen's study of six large blade knives from Spjald and one from Stenild shows that they were used for working with siliceous plants in the form of reeds and cereals. The large blade knives were shafted in the way indicated by the find of the blade knife from Stenild. Becker (1990) has interpreted the blade knives from Spjald as having been used to cut reeds for thatching, an interpretation also cited in later studies (Adamsen 2002; Jensen 2002; Jensen 2006).

Large blade knives from Late Bronze Age settlement sites in central and western Jutland

A special study from the project "Bronze Age Settlement Sites in Viborg County" (Bertelsen et al. 1996) conducted by Christensen (1996) describes use-wear analyses of artefacts from eight different excavation sites in Jutland, Denmark. The study comprises about thirty objects, mostly knives and scrapers. No experiments geared to specific questions were performed for the use-wear analysis, which was based entirely on research by Keeley (1980) and Jensen (1990). The study describes the many kinds of polish observed, and in the presentation of the results the author suggests that large blade knives of flint were used on reeds, cereals, wood, and hides, and that some of the blade knives were shafted (Christensen 1996).

The study is problematic in several ways. The majority of the pictures are published at 100× magnification but the author does not say what magnification was used for the interpretation. The result of the study is therefore difficult to assess critically. Several scholars, which whom I would agree, think that it is impossible to interpret certain types of polish at such low magnification as 100× (see e.g. van Gijn 1989:12). As Jensen has shown, the scale of the magnification depends on what type of contact material is to be analysed. Work with siliceous plants gives a polish that can be recognized at 100× magnification, whereas polish from work with, for example, wood or hides, requires at least 200× magnification (Jensen 1994:19). This means that several of the photographs in the study of polish which is interpreted as having arisen from working with silicaceous plant material are clear and easy to recognize, while polish which has been interpreted as coming from work with different kinds of hides is harder to assess critically. It is quite simply impossible to assess the results presented.

One large blade knife has polish that Christensen interprets as having been caused by contact with wood (Christensen 1996:86, fig. 19). The picture of the polish is one of two reproduced at 200× magnification and therefore capable of assessment. It shows polish which is heavily worn down to a level surface at the edge of the object. This type of polish is not usually interpreted as wood polish (see e.g. Keeley 1980; Vaughan 1985).

As regards the studied shafting traces, these have been observed on two blade knives and a fragment of a blade knife. The polish, however, is of a character that, according to Christensen, cannot be used as a basis for further discussion of the design of the shafting (Christensen 1996:88).

To sum up, it can be said that the study of large blade knives from Late Bronze Age settlement sites in central and western Jutland suggests that they were used on reeds, cereals, wood, and hides. This shows that it is important to keep an eye out for polish from these contact materials when analysing large blade knives of flint. It also shows that there are traces of shafting on certain blade knives. However, the results of the study are difficult to assess critically since the photographs of the different traces of wear are rendered at low magnification. The polish that is interpreted as coming from contact with wood is in my view assessed wrongly.

Large blade knives from Poland

In an analysis of large blade knives, some of them from Rudnik in south-east Poland (Moskwa 1976), Karol Kruk has performed use-wear analyses of a number of blade knives, several of which show close morphological similarities to the large blade knives from Rügen that are part of my study (Kruk 1994). In the presentation of the analysis there are pictures of the direction and extent of the polish, showing close agreement with the Danish analyses by Jensen (1990) and Christensen (1996). This indicates uniformity in the way large blade knives were shafted and used.

Bifacials from the northern Netherlands

From the northern parts of the Netherlands there are finds of a number of crescent-shaped thick bifacial tools classified as sickles. These are contemporary with large blade knives (Woltering 2001:364). Annelou van Gijn has performed functional analyses of some of these implements (van Gijn 1988:197ff). The analyses are based on extensive series of experiments and microwear analyses (HPA) of prehistoric artefacts. The results show that these tools were used for cutting turf sods or comparable work.

The analysed bifacials are contemporary with the large blade knives of flint. This, together with the fact that they have gloss that is visible to the eye and resembles what can often be observed on large blade knives, makes them important to include in this survey. The study shows that it is essential in an analysis of the function of large blade knives to include turf as a hypothetical contact material.

Results of previous studies

The survey of previous studies of function and use-wear analyses of large blade knives has yielded important results for continued investigation. First and foremost it may be observed that even if blade knives were recognized early in research to be reaping tools, few functional analyses have been performed. No analysis with experiments dedicated to answering specific questions has been attempted. Earlier studies have proved inadequate to use as a foundation, either because they are based on isolated rather than general observations, or were performed using methods and materials that are not adequately related to the results and interpretations put forward. The study by Jensen (1990) was performed systematically and is presented in a manner that enables source-critical scrutiny. It is therefore important as a basis for the analyses in this chapter.

The studies show considerable agreement as regards observations of use-wear on large blade knives of flint which is interpreted as coming from cutting silicaceous plants. Cereals and reeds are mentioned as contact materials. Apart from this, wood and hides have been specified as contact materials. This has been done in studies which, in my opinion, do not withstand critical scrutiny, but it shows that we should keep our eyes open for these contact materials in any analysis.

As regards discussions of shafting, the survey shows that it is clear that large blade knives were shafted in the same way as the implement from Stenild, Jutland. Yet there are observations which show that it cannot be taken for granted that all large blade knives of flint were shafted. A Polish study demonstrated that large blade knives there were used in a way that has resulted in a shafting pattern closely resembling what has been observed on Scandinavian blade knives (Kruk 1994).

Anne Lou van Gijn's (1988) study of thick bifacial tools from the Netherlands has been included even though these tools are not large blade knives of flint. She clearly demonstrates that these bifacials were used to cut sods. This shows that turf should be borne in mind as a hypothetical contact material for large blade knives.

Information for analysis: what should I be looking for?

It is obvious that it is chiefly polish arising from work with silicaceous plants that has been observed on large blade knives of flint. It is therefore important here to study what this polish looks like in order to have the necessary information for the analyses of the blade knives that I intend to study. This is done in two parts: as a comparative analysis of published observations, and as a presentation of observations made in connection with a small series of experiments. It proceeds from two questions:

- What is the appearance of polish arising from work with silicaceous plants?
- Is it possible to distinguish polish from work with different silicaceous plants through use-wear analyses by HPA?

It is clear that large blade knives were shafted, and therefore a study of shafting traces is also performed here. It proceeds from the question:

- What is the appearance of shafting traces on artefacts which were shafted in the way that large blade knives of flint seem to have been shafted?

Comparative analyses of published observations – polish from silicaceous plants

The comparative analysis of published observations is mainly based on three works. Patrick C. Vaughan's "Use-wear Analysis of Flaked Stone Tools" (1985) contains studies of an extensive series of experiments and an application of these series in analyses of Magdalenian flints from the cave of Cassegros in south-western France. The study deals with polish from a wide variety of contact materials. Anne Lou van Gijn's study "The Wear and Tear of Flint" (1989) is an analysis of experimental material and of some Neolithic artefacts consisting of Linear Band assemblages from Beek-Molensteeg and a number of Late Neolithic assemblages from the Vlaardingen culture. Like Vaughan, van Gijn bases her studies on a large corpus, with polish from a great many different contact materials. A few years before that study, van Gijn (1988) published her observations of polish arising from cutting sods, see above. That study is also included here. Helle Juel Jensen's work "Flint Tools and Plant Working" (1994), unlike the others cited here, is not a broad study of different contact materials but a focused analysis of polish from plants. Jensen bases her observations on analyses of experimental material and applies this to studies of different cutting and scraping tools from prehistoric south Scandinavia. Jensen's study, together with her article cited above reporting analyses of seven large blade knives (Jensen 1990), provides an important basis for my studies.

Although these authors have been inspired by each other (see e.g. the cross-references in Jensen 1994:32f and van Gijn 1989:40f), they have all performed independent studies of both experimental and prehistoric material. The studies show expertise and a critical attitude to the state of research prevailing at the time of publication. They are among the most comprehensive and all-embracing studies ever done of polish from silicaceous material and hence establish a firm foundation for the work done here.

Shared attributes

Polish from different silicaceous plants displays several shared properties and attributes (Vaughan 1985; van Gijn 1994; Jensen 1990, 1994).

- The polish develops relatively quickly and soon passes the stage of generic polish.

- Within a "reasonable lifetime" for a tool, the polish develops so much that is it visible to the naked eye in the form of what is called sickle-gloss.
- Variations with the category of polish from silicaceous plants are due to the length of time the tool was used, the intensity of the work, and differences between different individuals of the same plant species and variations in the character of different species. This makes it hard to distinguish one plant from another solely on the basis of the characteristics of the use-wear.

Despite all the common features, there are a number of subtle characteristics that have been distinguished as specific for different plants. These characteristics have been demonstrated with both experimental and archaeological material and are primarily related to:

- Variations in the topography of the polish.
- Traces indicating direction, such as striations and depressions.
- The extent of the polish from the edge inwards over the tool.

There is consensus between different researchers in their observations about this, but there are also some differences. Let us examine these by looking at the contact materials reeds, cereals, grass, and sods.

Reeds

Jensen (1994:29ff) defines contact with reeds as producing a bright or very bright polish, soft in structure and with domed topography. At intervals the polish is interrupted by linear grooves, giving these areas a fluted or rippled topography. The polish quickly becomes joined, covering a wide extent of the tool. The initial phase of reed polish is like that of wood, but it rapidly develops beyond this similarity. In addition, there is no fluted or rippled topography on wood polish.

Vaughan (1985:34f) has also compared the polish from reeds and wood, concluding that the tool quickly develops a polish that is very different from wood polish. The edge area and high parts of the microtopography are covered with domes, and although these become linked together the pits between them remain. He is also of the opinion that the domes of reed polish distinguish it from other silicaceous plants which cause a flatter, less undulating surface on the polish. This is also one of several attributes on which Jensen bases her interpretation of the polish on the blade knife from Stenild. The smooth, flat surface, in combination with scratches and striations and a matt appearance, are according to her what shows that the blade knife was used to cut cereals, not reeds (Jensen 1990:212).

Anne Lou van Gijn (1989:40) defines the polish from reeds as a well-defined and extended band of highly reflective polish, soft in texture, with domes but no striations.

Cultivated plants, cereals

Jensen (1994:32) thinks that, when analysing polish from cultivated plants such as cereals, it is important to understand that the polish comes from a variety of components apart from the plant being reaped. Earth and weeds are examples of things that can affect the look of polish. This means that there are great variations between different kinds of polish depending on where, how, and when the tool was used. How much earth there is in and on the straws that are cut, and how much weeds there are round about differs from one occasion to another. This was also something that Steensberg noted in his functional experiments, as we saw above (Steensberg 1943:23f)

Despite this, Jensen believes that there are characteristics which can be described as diagnostic for cereals (Jensen 1994:32). In comparison with reed polish, cereal polish develops slowly. It is not as bright and soft as reed polish, and the topography is flat instead of domed when the polish develops to give continuous surfaces. Cereals together with weeds give a more matt polish, broad striations, and coarse polish with pits (Jensen 1994:136).

Van Gijn (1989:40f) notes that the polish from cereals is matt and soft. On tools which do not have well-developed polish the topography is domed, whereas on tools with heavy and joined polish it is flat. In a text published in 1988 van Gijn says that the polish from reeds is very like that from cereals: "the resulting wear attributes are rather similar to those caused by harvesting of cereals, casting some doubt on claims that we can distinguish the two plant groups in archaeological context" (van Gijn 1988:205). In the study from 1990 she qualifies this statement with the words "it is possible, at least experimentally, to differentiate between polishes from cutting reeds and those from reaping domesticated cereals, in the sense that the former have a 'wet' fluid-like appearance, while the latter have a somewhat rougher and flatter polish" (van Gijn 1989:40).

Vaughan (1985:36) writes that cereals give a polish with bright domes which are united after use into a solid mass with a flat surface, which he describes as floating on the flint surface. The polish reaches far in from the edge to cover much of the surface of the tool. Other than Vaughan (1985), none of the authors explicitly describes the extent of cereal polish on tools. We have to read between the lines to see that it is extensive and reaches far over the surface of the tool. In the analysis of the large blade knife from Stenild, however, Jensen describes the spread as being identical with that of reed polish (Jensen 1990:210).

Grass

Jensen (1994:33) diagnoses grass polish as a high-reflecting, bright to very bright, soft, domed polish. It develops quickly. The polish is gathered in a band along the working edge. No striations or fluting occur in the polish. According to Jensen, the extent of the polish and the fact that it has no striations distinguishes grass polish from reed polish (Jensen 1994:33f). These observations are also confirmed by van Gijn (1989:40).

Sods

The researcher who has best described the polish caused by contact with sods is van Gijn (1989:48). This polish develops very quickly. Gloss that is visible to the eye arises on the edge after a short spell of work. The polish is highly reflective, has a matt, rough surface extending over much of the area of the flint, has a flat topography and long, narrow striations of varying depth.

Results, comparative analyses of published observations – polish from silicaceous plants

To sum up, it can be said that it is rather difficult to grasp the descriptions of the different attributes. Although the authors use set terms which show great agreement, the descriptions are written with many words and nuances in their uses which require close-up reading to interpret. Another thing that makes it difficult is that several of the descriptions are made as negative projections. This means that some kinds of polish are not described solely in terms of the observations of that specific polish but also in relation to polishes from other contact materials (see e.g. Vaughan 1985:34; van Gijn 1989:40; Jensen 1994:31). There is nothing intrinsically strange about this, and indeed it is often necessary for establishing the relationship between observations of different polishes. Yet it still makes it more difficult to interpret the diagnoses by the different authors.

Despite this, I find these authors' descriptions sufficient as a basis for my analyses. I have summed up descriptions in four tables 2.49, 2.50, 2.51, and 2.52. These show which characteristic attributes each author singles out for the different contact materials.

When selected attributes from the different tables are combined, it is observed that there are characteristic attributes that can be used to distinguish polishes from different silicaceous plants, table 2.53.

Of the silicaceous contact materials it is reeds, cereals, and grass that have common attributes at certain stages in the de-

Table 2.49. Characteristic attributes of polishes from contact with reeds. Described on the basis of experiments in which reeds were harvested or cut.

Reeds	Bright or very bright highly reflected light	Smooth texture	Domed topography	Fluted or rippled topography	Pit depressions, even in highly linked areas	Quickly linked and invasive distribution, very developed polish	No striations	Thin and shallow striations
Jensen (1994:31)	x	x	x	x		x		
van Gijn (1988:205)	x	x	x			x		x
van Gijn (1989:40)	x	x	x	x		x	x	
Vaughan (1985:34)	x	x	x	x	x	x		

Table 2.50. Characteristic attributes of polish from contact with cereals. Described on the basis of reaping experiments.

Reaping cereals	Matt reflected light	Bright to extremely bright polish	Smooth texture	Flat topography, when polish has developed into linked surface	Pockmarked aspect until completely linked	Wet, fluid-like appearance
Jensen (1994:32f)	x		x	x		
van Gijn (1989:40)	x	x		x		x
Vaughan (1985:36f)	x	x		x	x	

Table 2.51. Characteristic attributes of polish from contact with grass. Described on the basis of reaping experiments.

Grasses	Bright to very bright polish	Smooth and domed	Polish concentrated along the immediate edge forming a narrow band	No striations, no fluting
Jensen (1994:33)	x	x	x	x
van Gijn (1989:40)			x	x

Table 2.52. Characteristic attributes of polish from contact with sods. Described on the basis of experiments.

Sod-cutting	Extreme brightness	Rough and matt texture	Flat topography	Spread-out distribution	Quickly forming macroscopic gloss	Long and narrow striations with varying depth and a random or parallel directionality
van Gijn (1989:48, 1988:209)	x	x	x	x	x	x

Table 2.53. The different contact materials classified according to four polish attributes. Data come from tables 2.49, 2.50, 2.51, and 2.52.

Reflexivity of the surface		Topography of the surface		Extent of the polish		Texture of the surface	
High	**Low**	**Domed**	**Flat**	**Large**	**At the edge**	**Soft**	**Rough**
Reeds	Cereals	Reeds	Cereals	Reeds	Grass	Reeds	Sods
Grass		Grass	Sods	Cereals		Grass	
Sods				Sods		Cereals	

Table 2.54. Attributes characteristic of each contact material and thus distinguishing the contact materials from each other.

Contact material	Reflexivity of the surface	Topography	Extent of polish on the tool
Reeds	High	Domed	Over large areas
Cereals	Low	Flat	Over large areas
Grass	High	Domed	Band along the edge

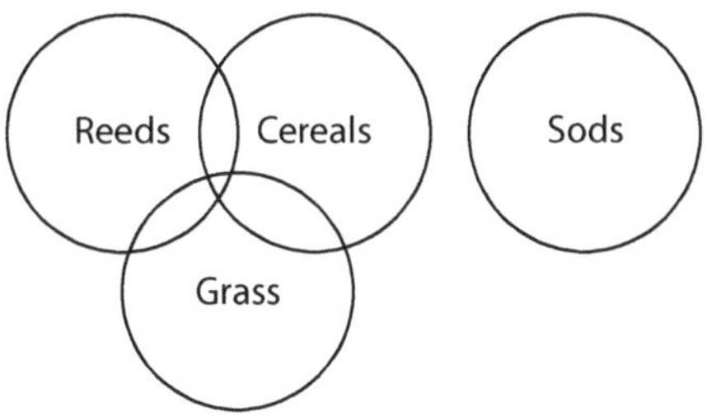

Figure 2.81. Schematic illustration summing up the descriptions in tables 2.49, 2.50, 2.51, 2.52, and 2.53. The figure illustrates how the contact materials can be confused with each other, but also that they can be distinguished.

velopment of polish. But if the different tables are compared it will be seen that there are some characteristic attributes which, if the polish is highly developed, make it possible to distinguish these contact materials, figure 2.81. Table 2.54 defines the attributes characteristic of each contact material which enable them to be distinguished.

It is important not just to describe characteristic attributes of different contact materials but also the similarities and differences which overlap and which can make it easier to distinguish different kinds of polish. In table 2.55 these differences are formulated as similarities and differences where the contact materials are viewed in relation to each other.

Table 2.55. Similarities and differences between different contact materials which can be used to distinguish different kinds of polish.

Contact material	Similarities and differences
Reeds vs. cereals	Unlinked polish caused by both. Both are domed and bright. When linking occurs, cereal polish develops into a flat, slightly matt surface while reed polish remains bright and domed.
Grass vs. reeds and cereals	Grass is distinguished from reeds and cereals through the extent of polish, with grass polish in a band along the edge as opposed to reed and cereal polish which is found on other parts of the tool. Grass is also distinguished from reeds and cereals through the absence of striations and fluting.

To sum up, the survey shows that it is possible both to define characteristic polish from different silicaceous plants and to distinguish these.

Comparative analyses of published observations – shafting of large blade knives

As we saw above, polish and other macro- and microtraces that have been interpreted as coming from the shafting of large blade knives has been observed in several studies (Jensen 1990; Christensen 1996). Two different observations serve as a foundation for identifying shafting: the extent of the polish from the material the large blade knife was used on, and observations of what has been interpreted as traces of the shafting itself.

Observations of the extent of polish from contact material proceed from the hypothesis that if large blade knives were shafted when used, the parts of the knife blade enclosed in the shaft were protected from the contact material and thus did not develop the type of polish that is found on the other unprotected parts of the large blade knife. If the blade knife was used in such a way that the polish from the contact material is well-developed, a clear difference is seen between the areas on the large blade knife which were enclosed in the shaft and those which were not.

Shafting traces on flint tools are considered by many scholars to be difficult to interpret (Unrath et al. 1986). Although they are emphasized as important for understanding the use of tools (Keeley 1982) and observations of shafting traces are regularly reported in analyses (see e.g. Keeley 1980; Vaughan 1985), this is a field of use-wear analyses that has not been studied in detail. Few experiments have been performed with the specific ambition to understand different ways of shafting tools, different types of shaft, and different ways of using them (Odell 2004:152f). In blind tests traces of shafting have often been confused with traces of use (Keeley 1980; Rots 2004).

This situation has changed considerably as a result of studies conducted by Veerle Rots (see e.g. Rots 2004 and works

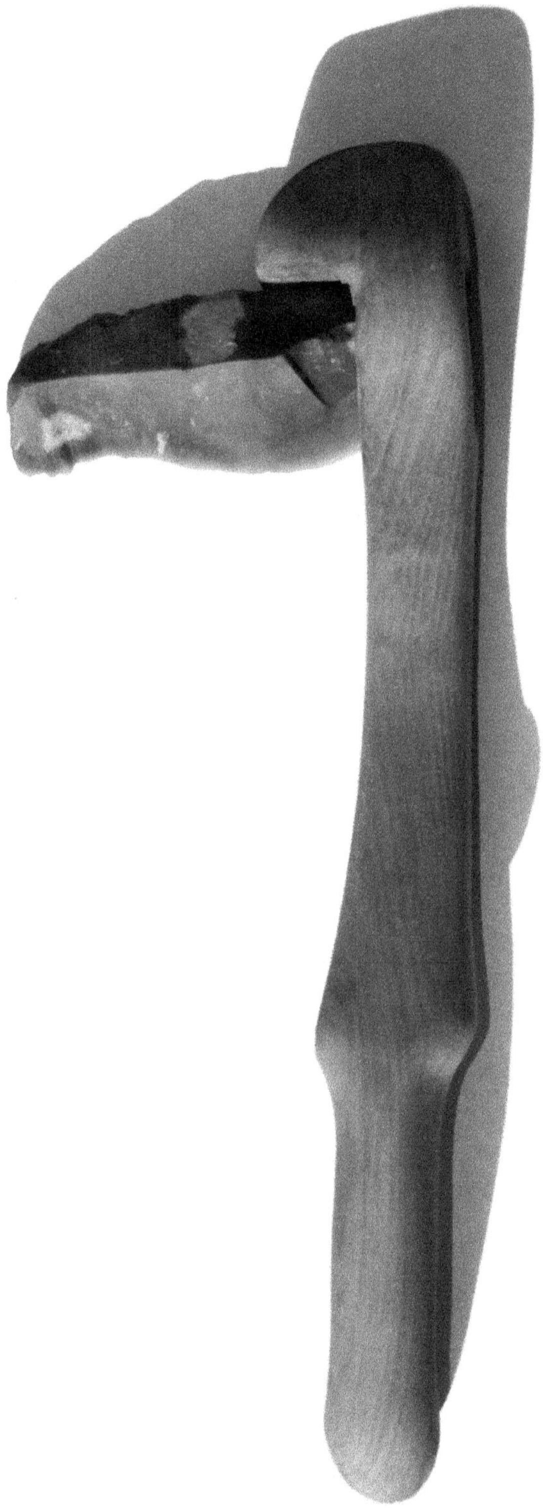

Figure 2.82. The tools used for the experiments at Lejre Research Centre in summer 2003 are reconstructions of the Stenild find. The knife blades were made by Mikkel Sørensen, SILA, National Museum in Copenhagen. The flint that was used was a fine-grained high-quality Senonian flint, black to grey in colour. The shafts were made by Jenny Eliasson, Malmö Museums, modelled on the find from Stenild using Blinkenberg's (1898) detailed description. The shaft is 36 cm long. The handle is round, roughly 2.4 cm thick, and is curved at the bottom. The central section of the shaft is more flat than round and has a width of 2.5 to 3.8 cm while the thickness here is 1.3 cm. The head of the shaft is 3.3 cm thick. It has an opening shaped like an irregular quadrilateral. On the underside the head of the shaft widens to 5.5 cm. A groove is cut in this in line with the hole in the shaft (Blinkenberg 1898:144f), see also figure 1.1. Blinkenberg identifies the wood in the shaft as "probably birch" (1898:144). For my experiment I contacted the National Museum in Copenhagen and received confirmation of the identification of the wood as birch (personal communication, Claus Malmros). When the Stenild find was discovered, the knife blade was in the shaft and the blade was held in place with small triangular wooden wedges, which were the same length as the cross-section of the shaft. The wedges were lost at the time of the excavation. No string or fixing material seems to have been used (Blinkenberg 1898:142, 146). The blade knives in the experiment were therefore shafted solely with the aid of wooden wedges. The photograph shows one of the tools that was used.

cited there). In extensive experimental studies she has analysed traces on flint tools caused by a great many different ways of holding, shafting, and using them. She has distinguished different variants of hand-held tools, tools held by wrapping part of the flint in protective material, and composite tools consisting of shaft and blade (Rots 2004:12ff).

Jensen (1990), Kruk (1994), and Christensen (1996) have shown that the way in which the Stenild find is shafted could also have been used for other large blade knives. Rots has demonstrated that tools which were shafted in these and similar ways display clear shafting traces. These consist of well-developed polish, edge scarring, and bright spots (Rots 2004:18ff). The polish from shafting is visible as a lighter or darker area compared to surrounding areas on the knife blade. The extent and the intensity of the polish depends on how the flint blade is shafted. In cases where the implement is stuck into a hole in the shaft, which is relevant in the case of large blade knives, there is polish on all the shafted parts of the flint. Edge scarring, just like polish, is located on the parts of the tool that were in direct contact with the shaft, and it consists of microflaking along the edges of the blade knife. In my judgement, however, it is difficult to use edge scarring as an indication of shafting, since prehistoric large blade knives of flint usually show such heavy post-deposition and post-excavation damage. Bright spots are small, isolated, and limited areas with intensively developed polish on the surface of the flint. Bright spots are what Jensen described on large blade knives and what she defines as friction polish (Jensen 1990).

A recent study, however, has suggested another type of shafting than that represented by the Stenild find. In an analysis of large blade knives from a settlement site in north-western Scania, Sweden, Bo Knarrström has made observations which indicate that a type II large blade knife was shafted in a way that enclosed the entire retouched back of the blade knife in the shaft (Knarrström 2007:103).

Result, comparative analyses of published observations – shafting of large blade knives

Both earlier use-wear analyses of large blade knives and general studies of shafting issues show that it is possible to find traces of shafting on blade knives. Three attributes are particularly important:

- Several studies have shown that the extent of the polish caused by the contact material can indicate whether the tool was shafted or not. The parts of the tool enclosed in the shaft do not exhibit the polish found on the rest of the tool.
- The polish from the actual shaft is visible as a lighter or darker area on the parts of the tool that were enclosed in the shaft in relation to other parts of the tool.
- Friction polish consists of small, well-defined areas with intensively developed polish on the surface of the flint.

Edge scarring has also been held up as an important attribute to determine shafting. As mentioned above, large blade knives of flint usually have severe post-deposition and post-excavation damage, so that attribute has been ignored in this study.

A comparative experimental analysis

In use-wear analyses, focused experiments are important as a starting point for analyses, chiefly for two reasons. By using modern reconstructions of prehistoric tools in controlled experiments, use-wear can be produced from different contact materials and different ways of shafting tools. These traces are necessary as study material and for reference in use-wear analyses of prehistoric tools (see e.g. Keeley 1980). Focused experiments also give the researcher experience of practical work demonstrating the potential and limitations of the tool for specific uses (see e.g. *EuroREA* 2005).

Although practical experiments with large blade knives of flint have been used in previous research (Steensberg 1943), none of the use-wear analyses of blade knives which have been published are based on focused trials employing reconstructions of blade knives. They have been based on general experiments with other kinds of tools. It was therefore necessary for me to conduct a special series of experiments using large blade knives (Högberg 2003). This was done in summer 2003 at Lejre Research Centre in Denmark (reference number HAF 08/03) and sought to answer four questions:

- What are the possibilities and limitations in the use of large blade knives?
- What traces are left by different uses of blade knives?
- Can the same material cause varied use-wear?
- Can different materials cause similar use-wear?

The ambition was to perform a series of practical tasks. Based on the earlier findings of use-wear analyses of large blade knives presented above, modern blade knives were mainly used to cut and harvest silicaceous plants such as reeds, cereals (barley), meadow grass, and sods. Large blade knives have long been closely associated with the operation of harvesting foliage and therefore this was also included in the experiment. Nettles were likewise cut as part of the trial.

I used a total of eleven different blade knives. Since both Jensen's (1990) studies and my own preparatory studies (Högberg 2004a) had shown that large blade knives were shafted, the knives had to be shafted for the trials. The design of the shaft was modelled on the find from Stenild (Blinkenberg 1898), figure 2.82. The shafting problem was not studied beyond what was necessary to perform the experiments. The idea was to use the blade knives on different materials for a sufficiently long time to allow a characteristic polish to arise. The actual experiments thus consisted of working with large amounts of the materials specified above. The results are presented below with descriptions of working operations and use-wear analyses.

Operations

The large blade knives worked very well as sickles for cutting cereals, grass, reeds, and nettles. With my left hand (I am

Figure 2.84. The blade knife from experiment 3 together with newly reaped barley.

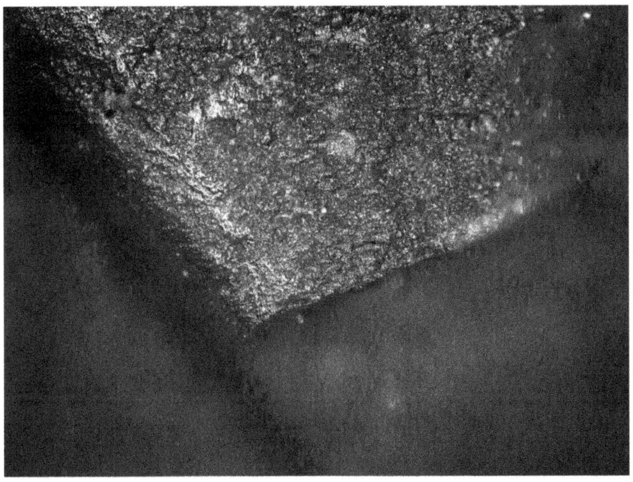

Figure 2.83. Polish on blade knife number 1 from the experiment, bright with low reflectivity, soft, continuous, and level on the surface. Original magnification 200×, EFI 11 pictures.

Figure 2.85. Polish on blade knife number 4 from the experiment. It is well-developed along the edge, highly reflective, bright, soft, and domed. It is distinctly confined to edge of the blade knife where it is visible as a band. Original magnification 100×.

right-handed) I grabbed hold of a bunch of what was to be cut, and with my right hand I swung the tool in a movement simultaneously pulling it to one side and towards me. The tool was easy to use and the blade cut the plants with no problem. In cases where cereals were growing together with a lot of weeds, problem did arise. Some weeds have very thick stems and the blade could not cut through them so easily.

Foliage was hard to cut. It took a lot of force, and the annual shoots I was harvesting were broken off by the force as often as they were cut off. If I tried to cut a shoot that was too thick, I failed. Yet even if the implement felt downright unsuitable for cutting foliage, I had quite soon gathered a bunch of branches with leaves.

Cutting turf was no great problem. The knife blade cut, or rather chopped, easily through the sods. However, it was impossible to cut deeper than the distance the knife blade protruded from the shaft, which made this work more difficult.

To sum up, the large blade knives functioned for all the operations but were best for cutting reeds, cereals, grass, and nettles and worst for weeds and leaves. The blade knives worked well for cutting sods, but the operation was restricted by the length of the blade. As regards weeds, this result is the opposite of what Steensberg (1943) found in his studies, which led him to the opinion that large blade knives were excellent for cutting weeds.

If the knife blade is carefully fixed in the shaft with the aid of wedges, it stays in place with no problem during the work. I quickly learned how best to wedge it fast. In one situation, however, the knife blade easily comes loose from the shaft. The times when I accidentally hit the shaft against something without cutting it off, the knife blade flew away, sometimes several metres. This happened, for example, if I was harvesting reeds and missed when I swung the tool against the bunch of reeds gathered in my hand and instead of hitting them with the blade hit them with the shaft a centimetre under the blade. If I was moreover standing with water to my knees, it could take a lot of searching before I found the knife blade again. An insight that I gained from fitting knife blades into wooden shafts was the importance of the morphology of the large blade knife. One of the blades I used was very thick and I was forced to knap it into shape so that it would fit in the shaft. Another knife blade was very small. It took me a long time to fix it in place in the shaft so that it would not fall out. It is thus obvious that large blade knives were shaped during manufacture so that they would fit a shaft.

To sum up, the shaft and the knife blade work well together. It is no problem to get the blade to sit fast with the aid of wedges.

Result – analyses of polish caused by contact material on the experimental large blade knives

Before the use-wear analyses the newly used large blade knives were cleaned in a bath of KOH, 5% solution, for four hours. Before the bath they had lain in water for twenty-four hours in order to become saturated, and after the bath they were carefully washed with water. This cleaning is in accordance with the recommendations discussed by Jensen (1994:17ff). The results of the analyses of the polish caused by each contact material are presented in table 2.56.

Results of the analysis of shafting traces from contact material on the blade knives from the experiment

Although the series of experiments did not have any specific question about shafting traces, I have nevertheless examined the blade knives to find traces of the shafting. In the discussion above, three different attributes to look for crystallized: the extent of polish from the contact material; polish from the shaft itself; and friction polish.

On the large blade knives used in the experiment, no overall polish developed. It is therefore not possible to see any clear difference on the surface of the blade knives. Along individual edges, however, a difference can be observed between the parts

Figure 2.86. Photo of newly cut reeds. To the left is a larger area to the right of which is a close-up. To the far right a photo of the blade knife from experiment 5 together with newly cut reeds.

of the edge that were used against a contact material and the parts that were enclosed in the shaft.

As regards polish from the actual shaft, this can be observed on the blade knives that were used for the longest time, numbers 5 and 10. On the bulb of percussion there is a slight change in the surface of the flint which is not found anywhere else on the blade knife. This is however impossible to define and demarcate. On other blade knives in the series of experiments I could see no such polish.

On the majority of the blade knives there is friction polish at different points, mainly on the bulb of percussion and on areas beside it. The size of these points varies from 0.1 up to 1.0 millimetre.

As discussed above, edge scarring is an attribute that Rots (2004) held up as a diagnostic trace of shafting. The experimental series confirms this finding. All large blade knives show some form of microflaking along the parts of the unmodified edge enclosed in the shaft. It is impossible to determine whether these arose from the shafting itself, that is, when the knife blade was fixed in the shaft, or during the work with the tool. As discussed above, however, this is an attribute that I chose not to use in this study for reasons of source criticism.

Results of the experiment

The experiment has given experiences of working with a tool like the Stenild find. Within the framework of the trial it has brought an understanding of the potential and limitations of this type of tool. Quite simply, I have increased my experience by spending a couple of days in the company of large blade knives of flint. The series of experiments have left use-wear and polish on the modern blade knives which I can use as references in the work of analysing use-wear on prehistoric blade knives. In a comparison with the results of previously published studies discussed above, it can be observed that the

Figure 2.87. Harvesting foliage along a small road at Lejre. To the left before the work began, to the right afterwards.

Table 2.56. Result of the analyses of the experimentally used large blade knives.

Knife number	Contact material	Operation	Use-wear and polish
1	Sparsely-growing barley with a lot of non-silicaceous weeds, Lejre's experimental field.	Reaping everything growing in the field, except thistles. Cutting at about 30 cm above ground level. Reaped a total area of about and with a 50 square metres.	Heavy microflaking along the whole edge. The polish is well developed in places, less so in others. It is bright with low reflexivity, soft, linked, flat surface, figure 2.83. The polish extends a bit in over the surface from the edge.
2	Only barley in Lejre's experimental field.	Reaping only barley, directly under the ears. Worked for ten minutes, during which time I filled half a wood basket. I quickly realized that there was no point in continuing with this method, that I was reaping barley but not in a way that the HPA analyses I had previously performed showed that the blade knives had been used.	Only slight generic polish and some small microflaking could be observed along the edge.
3	Barley from a modern field.	Worked for 30 minutes. The barley was cut about 30 cm above ground level. The field had no weeds at all, figure 2.84.	The polish is well developed in places and has starting to link together. It is bright with low reflexivity, soft, continuous, and with a level surface. The polish extends a bit in over the surface from the edge.
4	Meadow grass.	Worked for 25 minutes. The grass was cut about 30 cm above ground level.	Well-developed polish along the edges. The polish is highly reflective, bright, soft and domed. It is distinctly limited to the edge, where it is visible as a band, figure 2.85.
5	Straw-like reeds.	Cut reeds about 60 cm above the surface of the water. worked for 45 minutes, figure 2.86.	Mostly weakly developed polish. Some areas of well-developed polish along the edge. These are not linked. The polish is highly reflective, bright, soft and domed, with polish both on and between the domes. The polish extends far in over the edge, the surface of the tool, and up on to its ridges.
6	Foliage cut from annual shoots, various species of tree.	Worked for 25 minutes. Chopped as much as I cut. Cut off about 50 metres of brushwood and annual shoots along a small road. Cut branches up to 15 mm thick. I had to use a lot of force and often cut or chopped the blade through half the branch while the force of the stroke broke off the rest, figure 2.87.	Heavy edge damage could be observed during the work. Large and small edge flaking arose. Well-developed polish along the edge of the ventral side. This is reflecting and domed with polish on the top of the domes, not in the troughs. Several heavy friction striations along the edge. These are located in microflakes from pieces of flint which came loose from the blade knife during use, got stuck in the contact material and there scratched the surface of the tool, figure 2.88. This is not usually observed on tools used for woodworking and can in this case presumably be linked to the highly specific operation that leaf cutting is.
7	Nettles.	Worked for 15 minutes. Reaped an area of 25 square metres. The nettles were cut about 15 cm above ground level, figure 2.89.	The polish is domed, matt, slightly fat on the surface, and weakly reflecting. It is well-developed in places but less so on parts of the edge. The polish extends a bit in over the surface of the tool. As has also been observed by other researchers (e.g. van Gijn 1989:40), this polish differs significantly from polish from silicaceous plants.
8	Turf and soil.	Cut with chopping movements. Worked for 15 minutes. The turf consisted of thick-growing grass with wet soil under. Sometimes, however, I hit stones and thick branches not visible above ground. This caused severe damage to the point and the edge of the knife blade, figure 2.90.	After 5 minutes' work the edge was worn slightly round. Heavy macroflaking along the edge. The polish is well developed. It is rough, level, slightly matt on the surface, and reflecting. Heavy striations are visible in the polish. The polish extends far in over the surface and up on to the ridges.
9	Grass-like, broad-leaf reeds.	Worked for 5 minutes. Cut the reeds about 50 cm above ground level.	Microflaking along the edge. Well-developed polish on parts of the blade knife, other parts less developed polish. The polish is highly reflective, bright, soft, and domed, with polish both on and between the domes, figure 2.91. The polish extends far in over the edge, the surface of the tool, and up on to the ridges.
10	Several different plants, and other materials.	This blade knife was used by interested visitors to Lejre. I left it lying out at every place where I was working and encouraged passers-by to try using it. It was used a great deal both on the material that I was working with and on other things. Examples of other materials are dandelions, flowers, twigs on bushes, shoe laces, a gym shoe, and hair.	Some microflaking. The blade knife has well-developed polish that is difficult to interpret. In places it looks like reed polish, figure 2.92, in places like grass polish. The polish extends far in over the edge.
11	Turf and soil.	Chopping sods and damp earth in partly drained wetland. Worked for 5 minutes.	The edge was quickly worn round. Heavy macroflaking along the edge. Well-developed polish on both the dorsal and the ventral side. The polish is rough, level, slightly matt on the surface, and reflecting. Heavy striations are visible in the polish. The polish extends far in over the edge and up on to the ridges.

similarities as regards polish and use-wear between my experiment, table 2.56, and the published observations are very large. The experiment has neither qualified existing knowledge nor contributed any new findings. It has merely confirmed what was already known.

As we saw above, several of the operations did not lead to any well-developed polish on the edges of the modern large blade knives. There are also large differences between different series of experiments. In experiment number 5, for instance, I worked for 45 minutes cutting reeds. This time was not sufficient for any continuous polish to develop on the blade knife. In experiment 9 when I worked for five minutes cutting grass-like, broad-leafed reeds, well-developed polish arose on that blade knife. The reason for these differences is something that I cannot determine. Also interesting here is the polish on the blade knife that visitors were allowed to use. Although this part of the experiment was not controlled, the result gives an indication that certain contact materials dominate over others as regards the polish that is developed and observed, as other researchers have also pointed out (see e.g. Jensen 2006:104).

The results of the experiment show that different uses give different use-wear, that these traces vary with the material that the large blade knives were used to cut, and that these different contact materials can be distinguished. The results also show that several of the blade knives after use displayed a slight polish that does not allow the contact material to be identified. This is probably due to the short time that I used the tools. The fact that generic and weakly developed polish from similar contact materials is hard to distinguish has been well documented in the past (see e.g. Keeley 1980; van Gijn 1989). As regards shafting traces, the results of previous studies have been confirmed.

Use-wear analyses, prehistoric large blade knives of flint

The survey of previous studies and the experiences gained from the experimental use of large blade knives gives a good foundation for use-wear analyses of prehistoric large blade knives of flint. In this section I present analyses of a total of 64 prehistoric blade knives. These were all found by excavations in the area around Malmö in south-western Scania, Sweden, figure 2.93. The excavations have all been published (Ödman 1992; Hadevik & Gidlöf 2003; Eliasson & Kishonti 2003; Rostoványi 2005; Steineke et al. 2005; Celin 2006; Grehn 2006). The analysis is thus confined to a specific geographical area within the region covered by this book.

Figure 2.89. Newly cut nettles.

Figure 2.88. Heavy friction striation along the edge, located at a spot with micro-flaking. Original magnification 50x.

I have chosen to analyse virtually all the large blade knives of flint from excavations conducted in the last twenty years in this area. There are many advantages of analysing blade knives from recently excavated sites. The find circumstances are controlled. The use-wear analysis can be linked to a context. Since the blade knives were found recently, they have not circulated in collections and exhibitions. They are mostly in the state they were in when they left the excavation site. Post-excavation damage is small, and damage that arose during the excavation is easy to recognize. Also, the blade knives have not developed any patina to speak of. It is well known that prehistorically knapped flint can be quickly patinated when exposed to sunlight, modern air, or archaeological treatment (Luedtke 1992).

The starting point for the analysis is two questions: What contact materials were the large blade knives used on? Were the blade knives shafted? The use-wear analyses involved four stages. These concern aspects of status, edge, polish, and shafting. Under "Status" I describe what the blade knife looks like, whether it is intact or broken, and whether it is possible

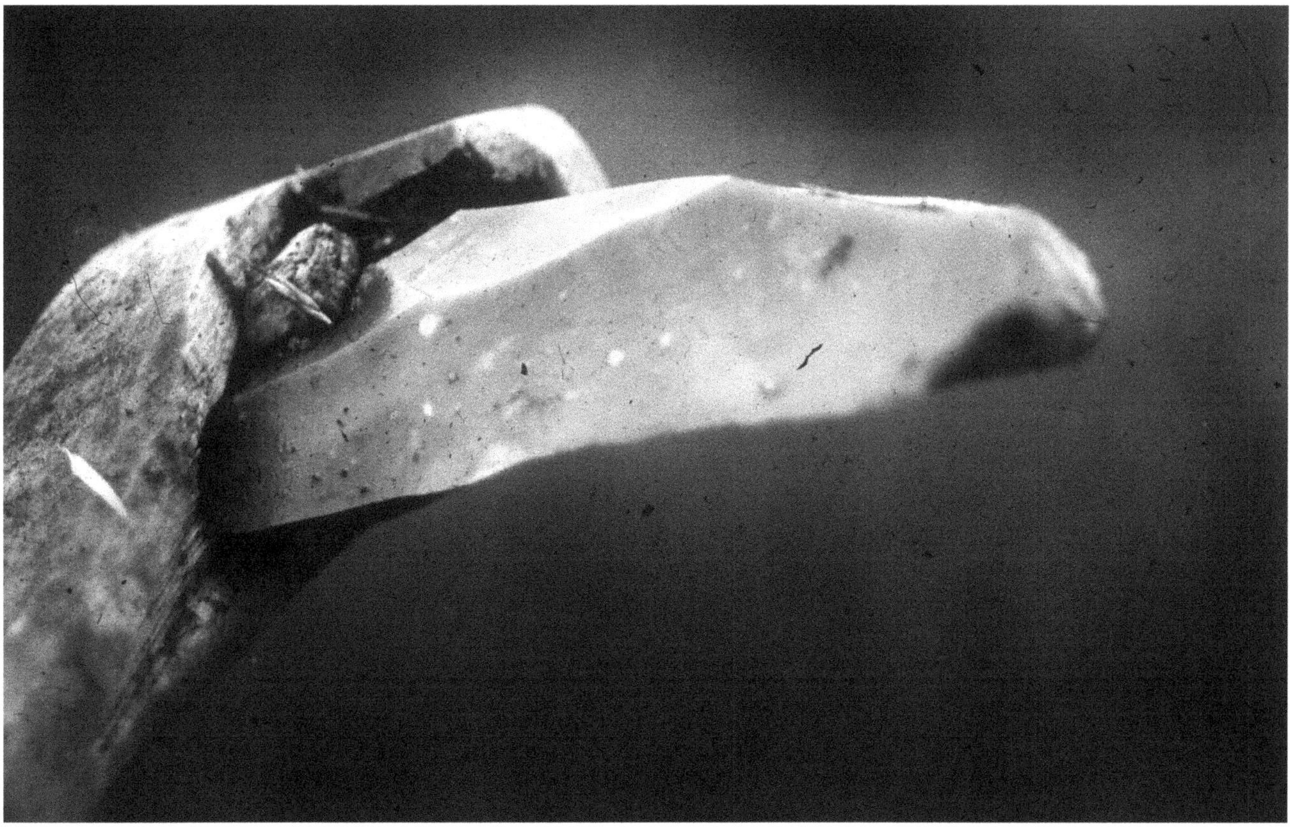

Figure 2.90. Close-up of the blade knife from experiment 8. At the far tip there is clear macrodamage from hitting stones and branches not visible above ground when I was cutting sods.

to see if the flint is patinated. Under the heading "Edge" I describe the status of the edge, whether there is gloss that is visible to the eye, and whether the blade knife has post-deposition or post-excavation damage. This damage is defined in terms of four observations which all refer to damage arising after the blade knife was moved from its original context of deposition, for example, by ploughing, in connection with archaeological excavation, or in storage. The damage consists of fresh fractures which are visible as a difference in the colour of the flaking scars in comparison with areas around the scar, clean surfaces where soil and dirt have not made their way into fissures and microcracks, sharp edges on a knife edge or retouch which is

Figure 2.91. Polish on blade knife number 9 from the experiment, less developed polish. The polish is highly reflective, bright, soft, and domed, with polish both on and between the domes. Original magnification 200×.

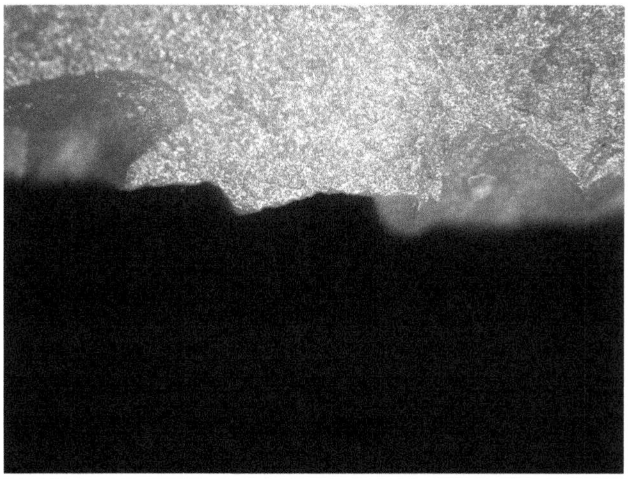

Figure 2.92. Knife blade number 10 from the experiment was used for many different things. This yields a well-developed polish that is difficult to interpret. In places it looks like reed polish, in other places like grass polish. The polish extends far in beyond the edge. Original magnification 200×.

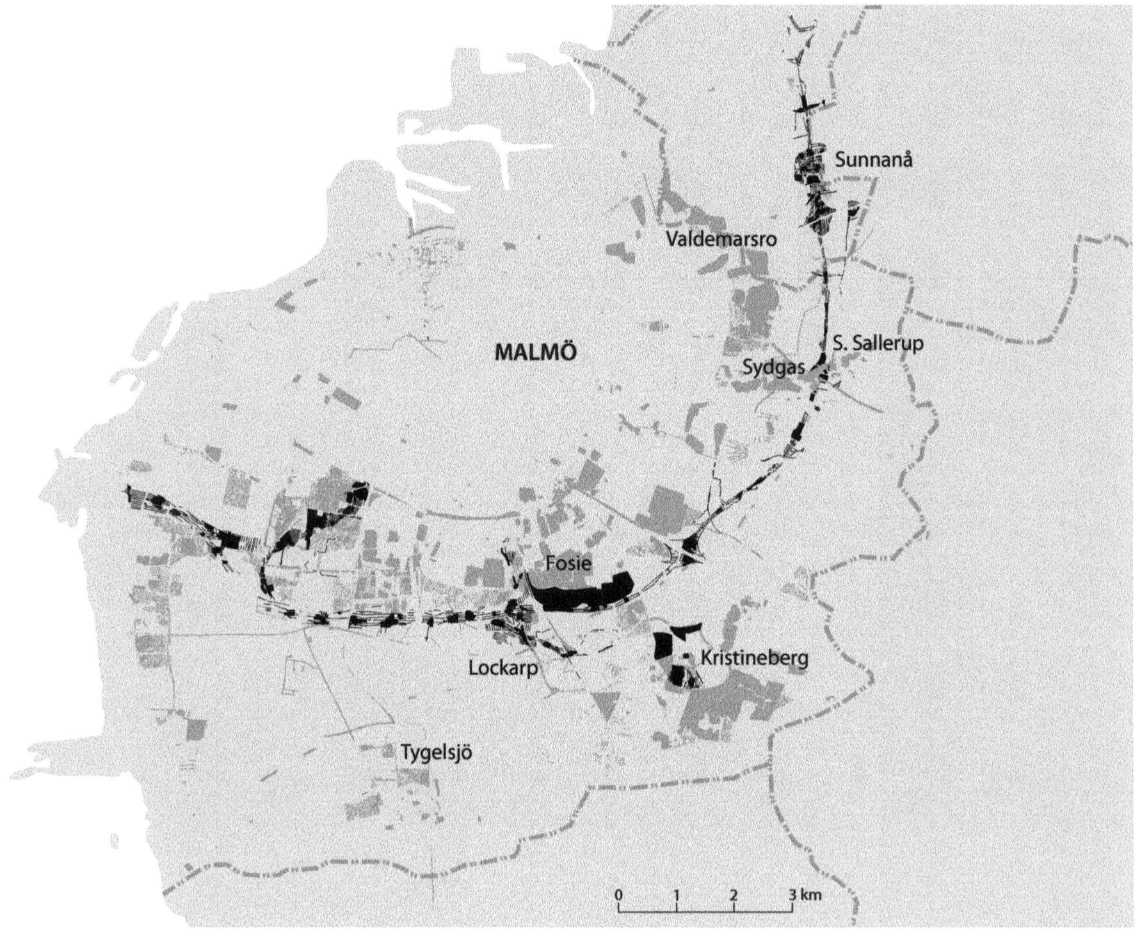

Figure 2.93. Map of Malmö and the surrounding area, marking the place names from where the large blade knives in this analysis come. All these find spots have been excavated in the last twenty years.

otherwise soft and rounded, or flaking scars that lack polish on an edge that otherwise has well-developed polish. Through these different observations it is possible to distinguish post-deposition and post-excavation damage from flaking scars that were on the blade knife from the beginning or arose when it was being used. The latter flaking scars are recognized by being covered with the same type of polish as other parts of the edge. Under the heading "Polish" I describe the analyses of the polish observed under the microscope. The description is based on the criteria stated in the studies discussed above. Under this heading I define the contact material. Shafting is the last step in the analysis. Here I examine the blade knife to see any traces of shafting, also based on the criteria used in the studies discussed above.

Examining large blade knives by HPA is time-consuming. These are big tools and there is a large area on each blade knife to scan under the microscope. I have not systematically registered how much time was spent on each analysis but I estimate that it takes between thirty minutes and an hour to analyse, photograph, and describe the use-wear on a large blade knife.

Analyses of sixty-four large blade knives

The analyses of the large blade knives are presented below with descriptions, table 2.57, and photographs, figure 2.94, in cases where a photograph was taken. These observations are then summed up in a concluding section. All contact materials are interpreted in 200x magnification.

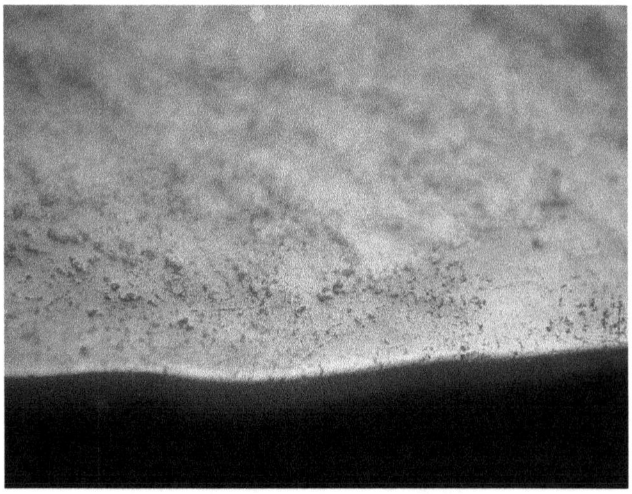

MK 20:206866a. 200×, at the edge. Reeds.

Figure 2.94. Results of use-wear analyses of sixty-four prehistoric large blade knives of flint, HPA-photographs. Original magnification specified (continues on the following pages).

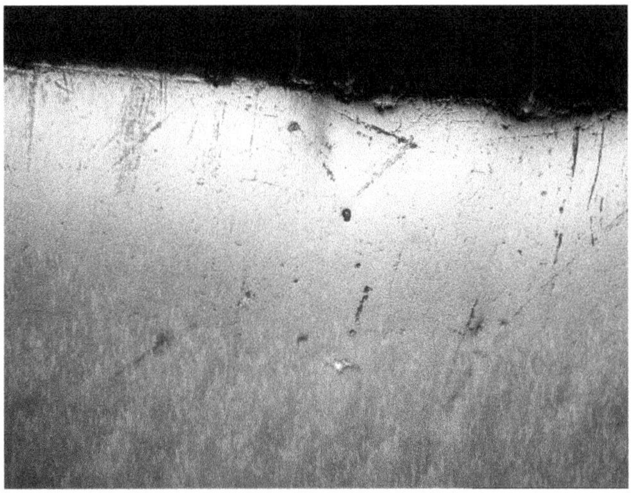

MK 20:206866b. 200×, by the dorsal ridge. EFI, 5 pictures. Reeds.

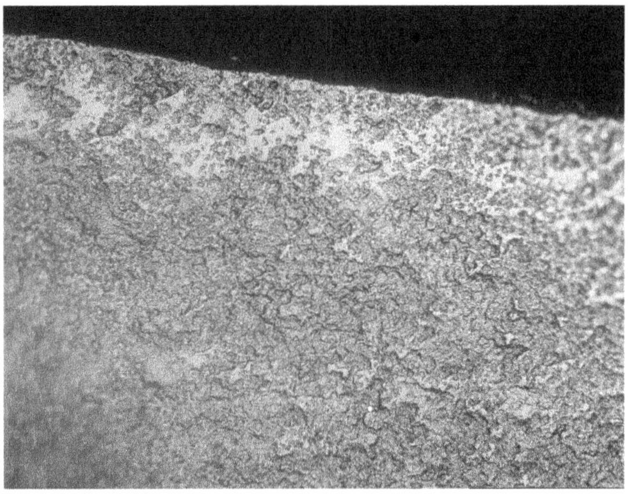

MK 20:206866c. 200×, shafting traces. EFI 3 pictures.

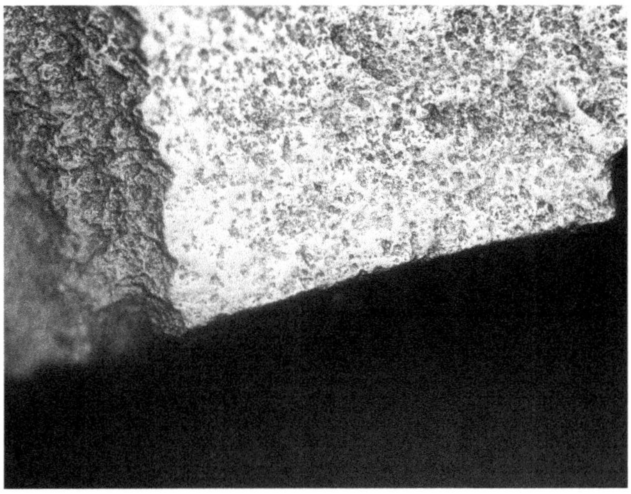

MK 20:281868. 200×, at the edge. EFI 3 pictures. Reeds.

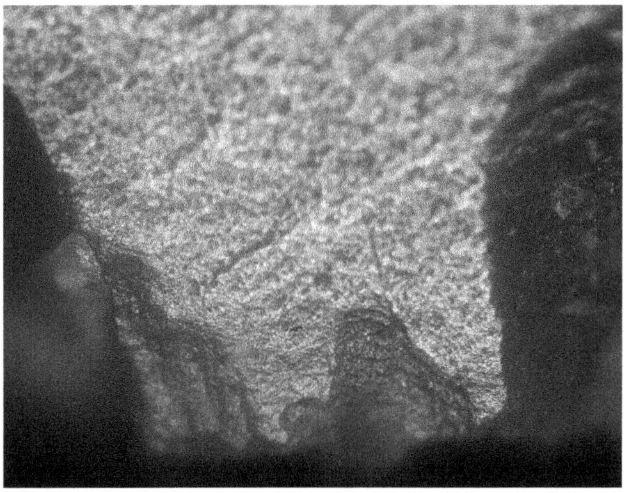

MHM 7373:345. 200×, at the edge. Reeds.

MK 20:201697. 200×, at the edge. EFI 5 pictures. Reeds.

MK 20:200409. 200×, at the edge. EFI 5 pictures. Reeds.

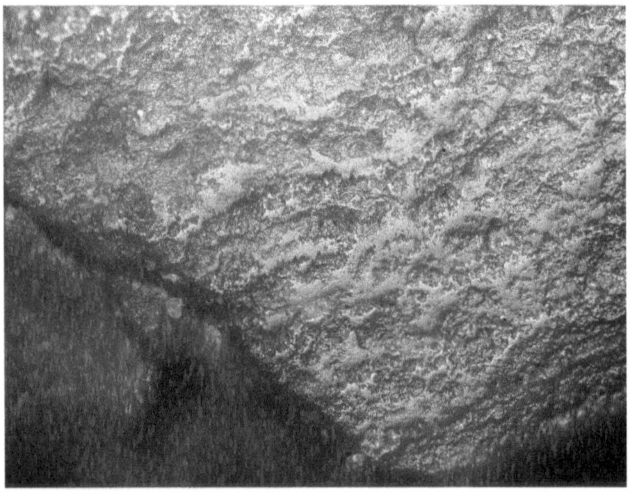

MK 20:206480. 200×, at the edge. EFI 7 pictures. Reeds.

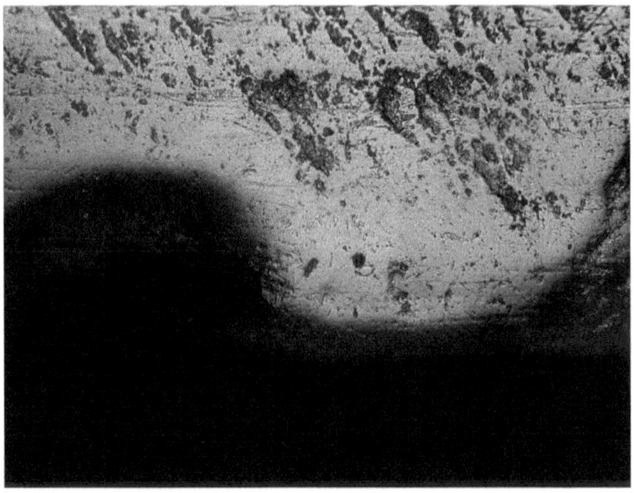

MK 20:206013. 200×, at the edge. EFI 5 pictures. Cereals.

MK 20:201876a. 200×, at the edge. Cereals.

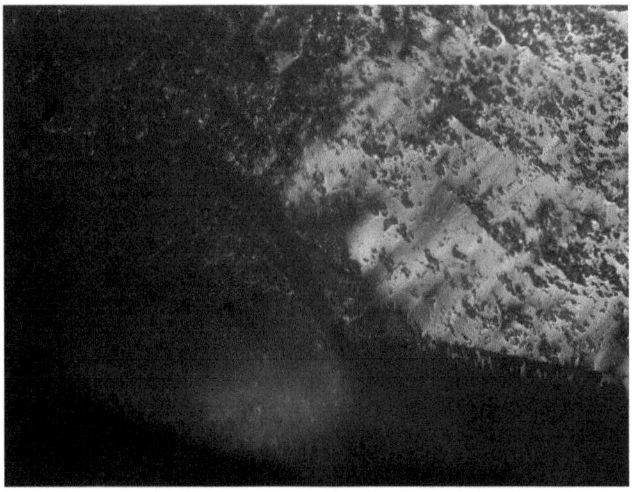

MK 20:201876b. 100×. Transition between edge polish and shafting traces. EFI 5 pictures.

MHM 9128:96a. 200×, at the edge. EFI 3 pictures. Reeds.

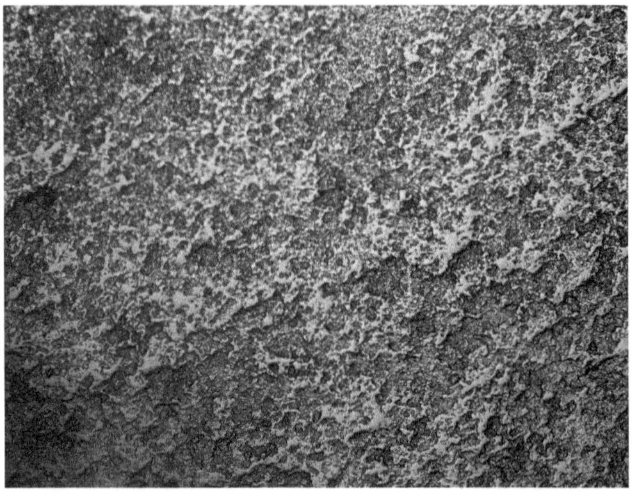

MHM 9128:96b. 200×, 1 cm in from the edge. Reeds.

Figure 2.94. Continues on the following pages.

MHM 9127:95a. 100×, at the edge. Reeds.

MHM 9127:95b. 100×, shafting traces.

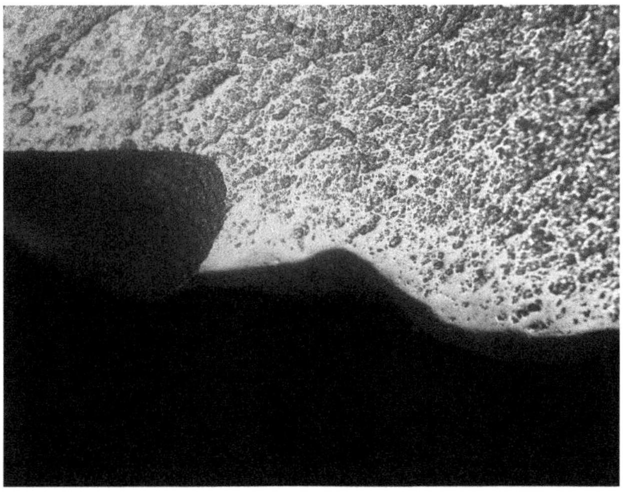

MHM 9128:208. 100×, at the edge. Polish together with micro scarring from post-deposition damage. Reeds.

MHM 9128:94. 100×, at the edge. Cereals.

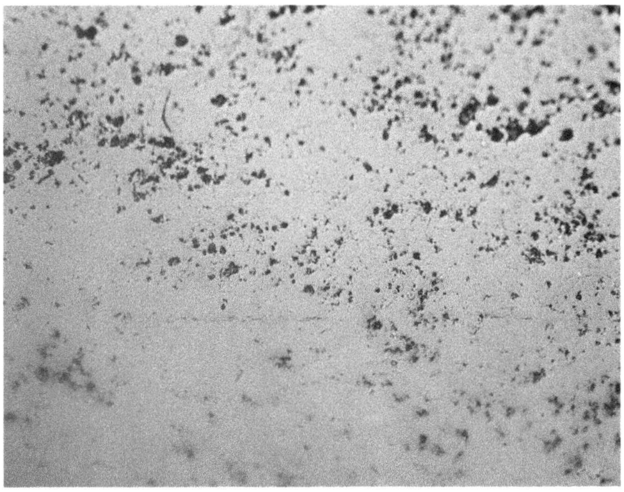

MHM 9128:66. 200×, 5 mm in from the edge. Cereals.

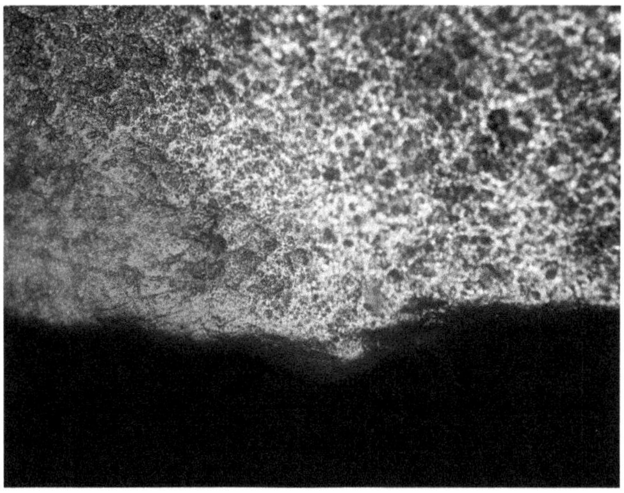

MHM 9064:195. 200×, at the edge. Reeds.

MHM 9151:3232a. 100×, 1 cm in from the edge. Cereals.

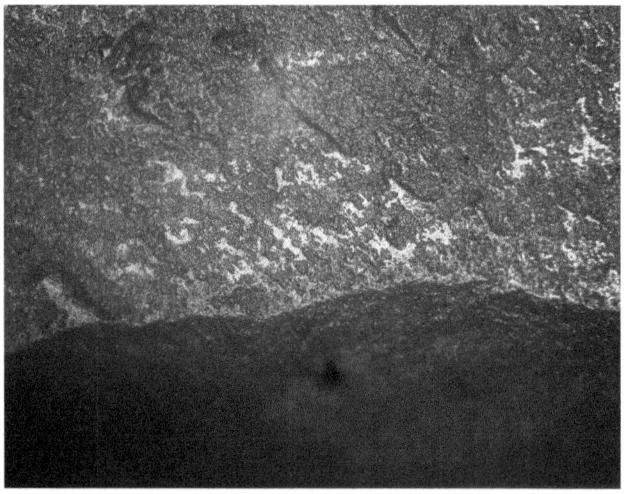

MHM 9151:3232b. 100×, shafting traces.

MHM 9145:757. 200×. Patina.

MHM 9145:797. 100×, at the edge. Reeds.

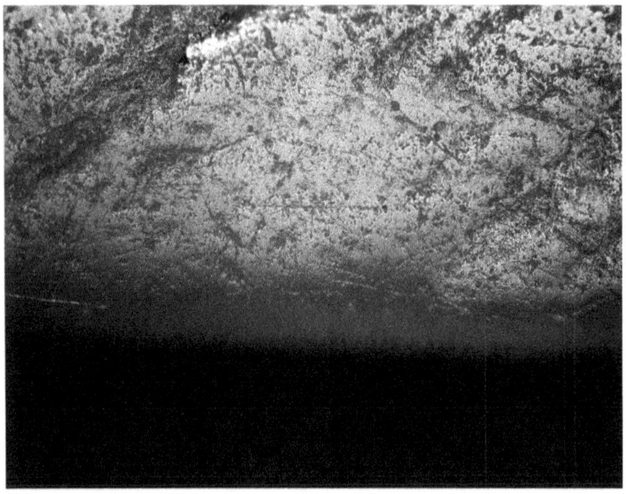

MHM 9145:1215. 200×, at the edge. EFI 6 pictures. Sods.

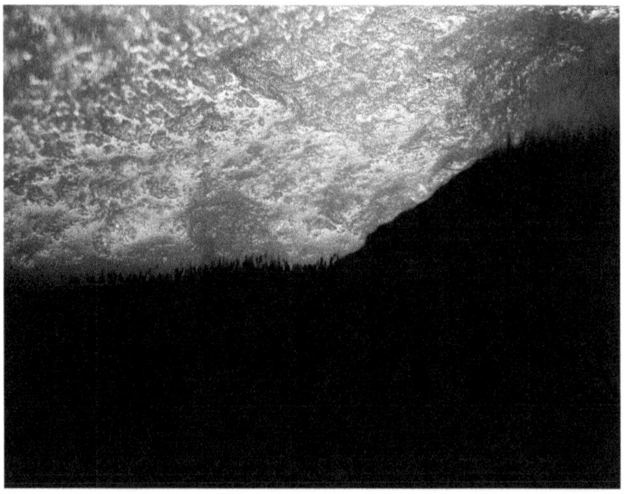

MK 141:126. 200×, at the edge. EFI 4 pictures. Reeds.

Figure 2.94. Continues on the following pages.

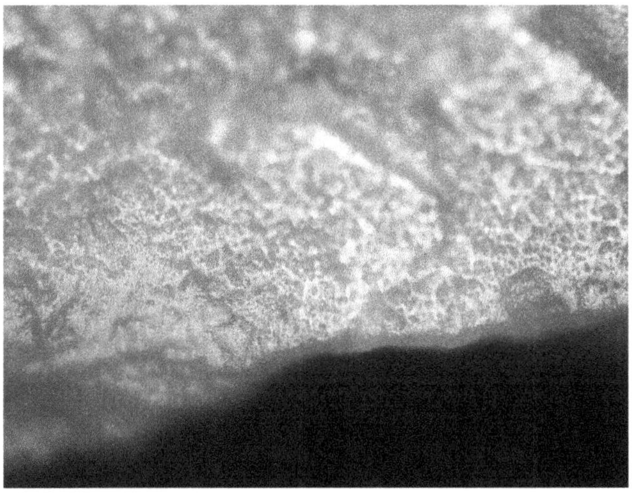

MK 141:72. 200×, at the edge. Reeds.

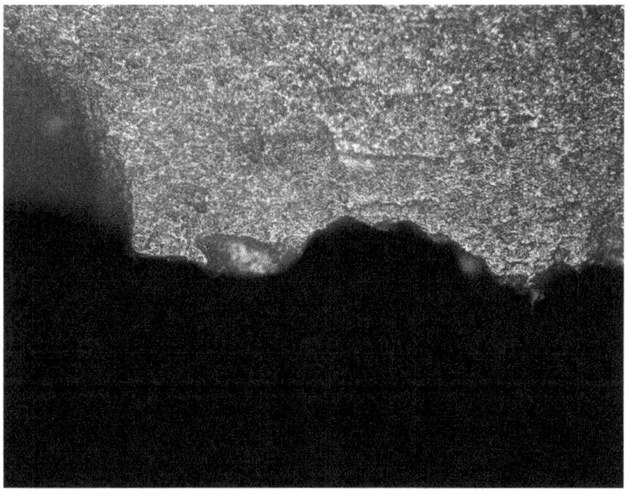

MK 141:486. 200×, at the edge. Cereals or reeds.

MK 141:191a. 200×, at the edge. Cereals.

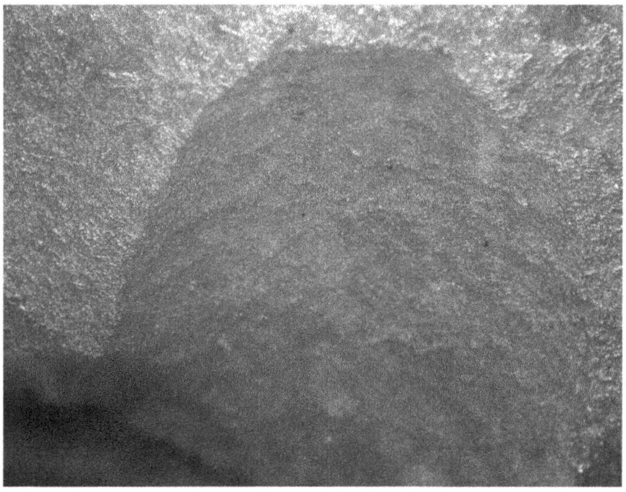

MK 141:191b. 50×, at the edge. Scarring from post-deposition damage.

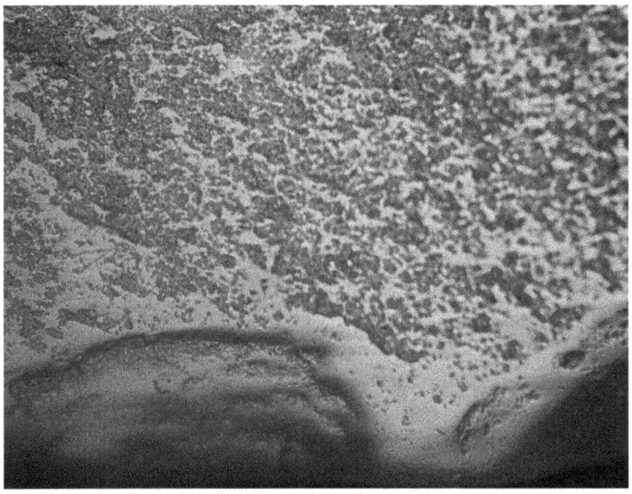

MK 141:499a. 200×, at the edge. Cereals.

MK 141:499b. 100×. Shafting traces.

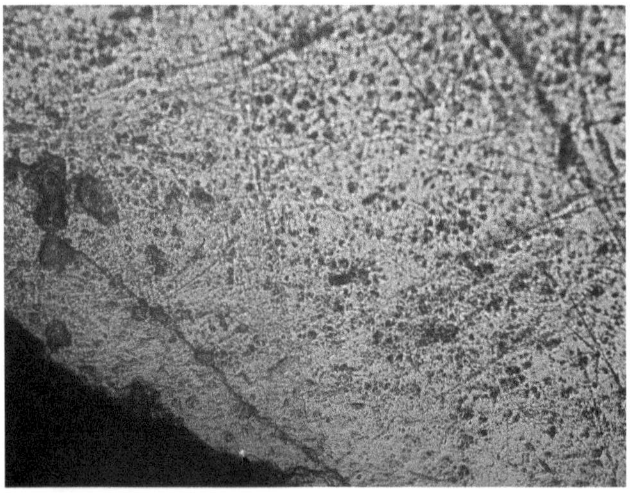

MHM 12587:15. 200×, at the edge. Reeds.

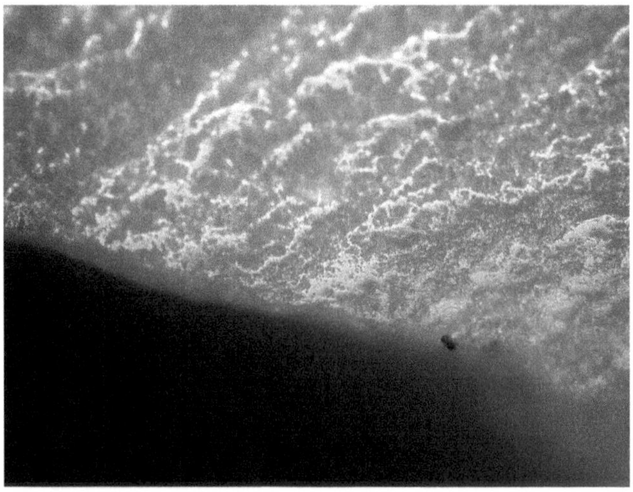

MHM 12881:945. 200×, at the edge. Reeds.

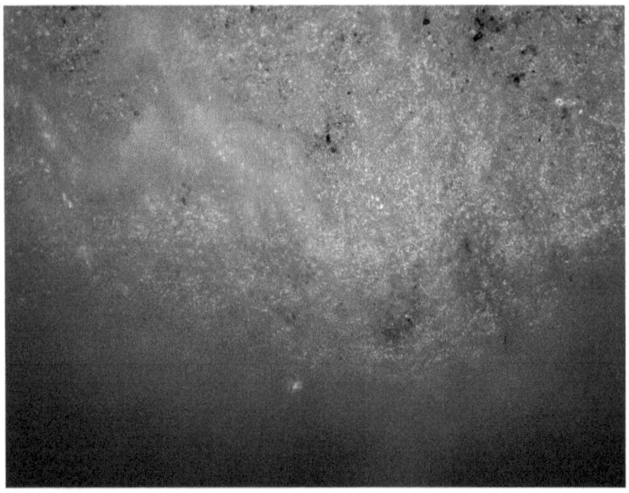

MHM 12755:166. 200×, at the edge. EFI 6 pictures. Hides.

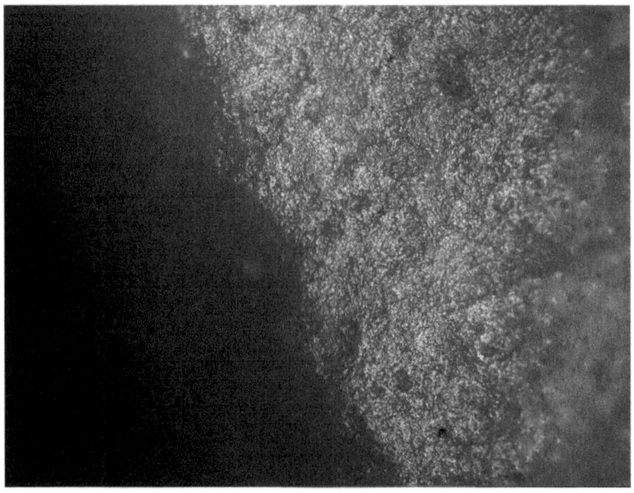

MHM 12755:167. 200×, at the edge on the broken part of the blade knife. The edge is rounded. EFI 6 pictures. Hides.

MHM 12881:966. 200×, at the edge. Cereals.

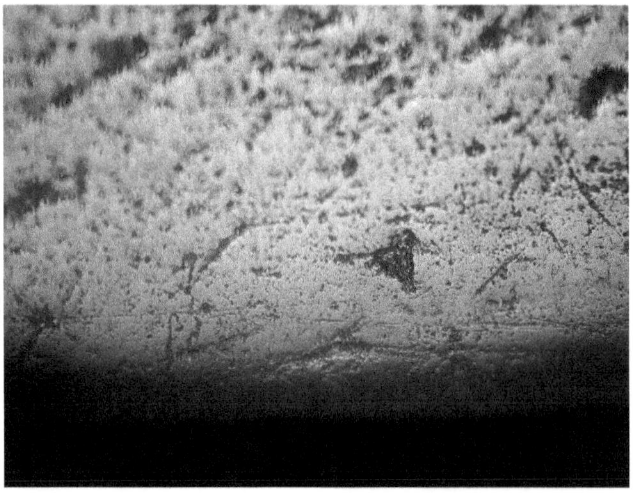

MHM 12873:1887. 200×, at the edge. EFI 2 pictures. Cereals.

Figure 2.94. Continues on the following pages.

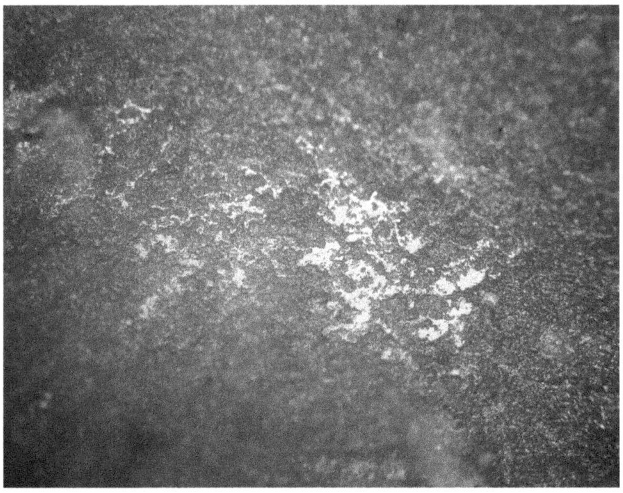

MHM 12755:168. 200×, shafting traces.

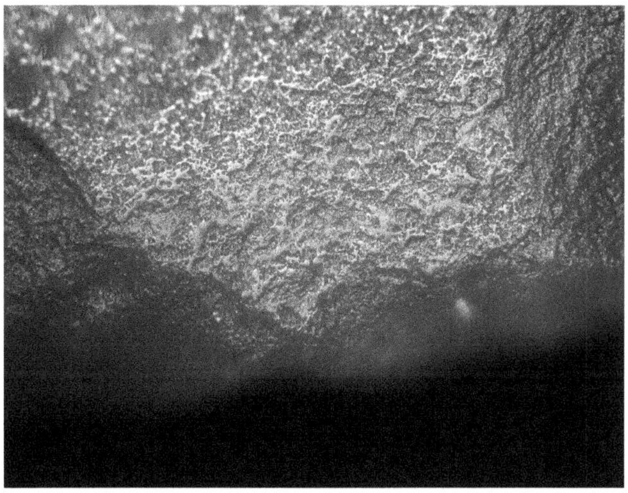

MHM 6650:XVII. 200×, at the edge. EFI 5 pictures. Reeds.

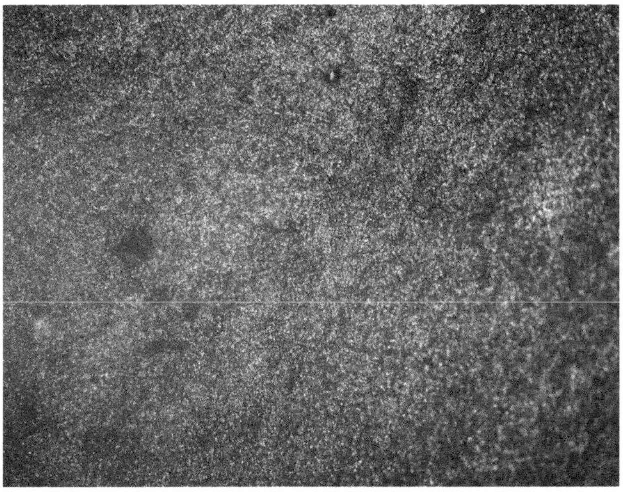

MHM 6650 no. 2. 100×, shafting traces.

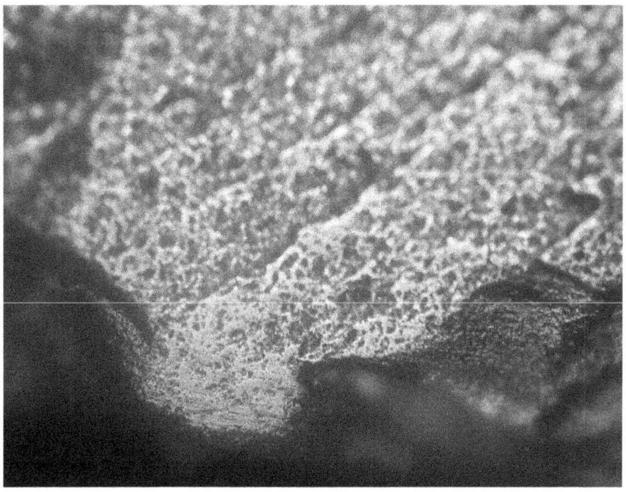

MHM 6650 no. 3. 200×, at the edge. Reeds.

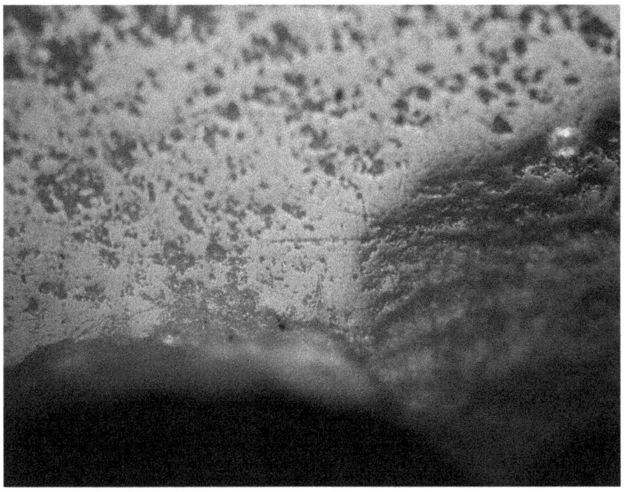

MHM 6541a. 200×, at the edge. Cereals.

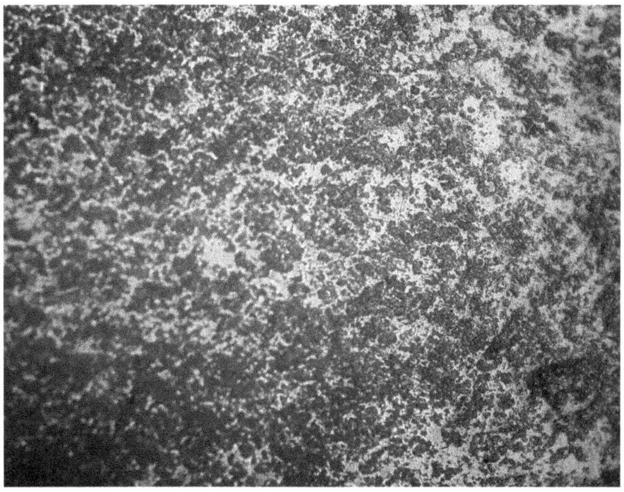

MHM 6541b. 100×, shafting traces.

Table 2.57. Results of the analyses of sixty-four large blade knives of flint from excavations in the area around Malmö in south-western Scania, Sweden. From the left: in the first column the find spot and inventory number. The second column states whether the analysis was done on a whole knife or parts of a knife. Columns three to six describe the occurrence or absence of patina, damage, gloss, and rounding. Column seven defines the development of the polish. Column eight shows the contact material as interpreted from the polish, see also figure 2.94. Column nine states whether or not there are shafting traces on the blade knife, and any other comments are given in column ten. Explanations: MK and MHM are prefixes to the inventory number of the finds in the Malmö Museums collections, – = no analysis performed.

Find spot and inventory number	Status	Visible patina	Post-deposition damage	Gloss visible to the of naked eye	Rounding	Development polish	Contact material	Traces of shafting	Comments
Kristineberg MK 20:201876	Whole	No	Yes	Yes	Yes	Heavy	Cereals	Yes	
Kristineberg MK 20:206013	Whole	No	No	Yes	Yes	Heavy	Cereals	Yes	
Kristineberg MK 20:203043	Whole	No	No	No	No	Slight	Cereals or reeds	No	
Kristineberg MK 20:206480	Whole	No	Yes	Yes	Yes	Heavy	Reeds	Yes	
Kristineberg MK 20:201710	Whole	No	No	No	No	Slight	Cereals or reeds	No	
Kristineberg MK 20:205722	Whole	No	Yes	No	No	Slight	Cereals or reeds	–	
Kristineberg MK 20:203051	Whole	Yes	Yes	No	No	–	–	–	Patina renders analysis impossible
Kristineberg MK 20:205724	Whole	No	Yes	No	No	–	–	–	Patina renders analysis impossible
Kristineberg MK 20:200409	Whole	No	No	No	No	Heavy	Reeds	Yes	
Kristineberg MK 20:206769	Whole	Yes	Yes	No	No	–	–	–	Patina renders analysis impossible
Kristineberg MK 20:201697	Distal part	No	Yes	Yes	No	Heavy	Reeds	–	
Kristineberg MHM 7373:345	Distal part	No	Yes	Yes	No	Heavy	Reeds	–	All-over patina
Kristineberg MK 20:201886	Proximal part	No	No	No	No	No polish	–	Yes	
Kristineberg MK 20:281868	Distal part	No	Yes	Yes	No	Heavy	Reeds	–	
Kristineberg MK 20:203045	Medial part	No	No	No	No	Slight	Cereals or reeds	–	
Kristineberg MK 20:206866	Proximal part	No	No	Yes	No	Heavy	Reeds	Yes	
Kristineberg MK 20:203097	Distal part	No	Yes	Yes	No	Heavy	Reeds	–	
Tygelsjö MK 141:499	Whole	No	No	Yes	Yes	Heavy	Cereals	Yes	
Tygelsjö MK 141:191	Whole	No	Yes	Yes	Yes	Heavy	Cereals	Yes	
Tygelsjö MK 141:486	Whole	No	No	No	No	Slight	Cereals or reeds	Yes	
Tygelsjö MK 141:122	Whole	No	Yes	No	No	Slight	Cereals or reeds	Yes	
Tygelsjö MK 141:72	Proximal part	No	Yes	Yes	No	Heavy	Reeds	–	
Tygelsjö MK 141:126	Proximal part	No	No	Yes	Yes	Heavy	Reeds	Yes	
Tygelsjö MK 141:496	Proximal part	No	No	Yes	Yes	Heavy	Reeds	Yes	
Tygelsjö MK 141:384	Distal part	No	Yes	No	No	Slight	Cereals or reeds	–	
Tygelsjö MK 141:154	Proximal part	No	No	No	No	Generic	–	Yes	
Tygelsjö MK 141:140	Proximal part	No	No	No	No	Generic	–	No	
Lockarp CT 7 MHM 12755:168	Whole	No	Yes	No	No	Generic	–	Yes	
Lockarp CT 7 MHM 12873:1887	Whole	No	Yes	Yes	Yes	Heavy	Cereals	Yes	
Lockarp CT 7 MHM 12881:966	Distal part	No	No	Yes	Yes	Heavy	Cereals	–	

Find spot and inventory number	Status	Visible patina	Post-deposition damage	Gloss visible to the of naked eye	Rounding	Development polish	Contact material	Traces of shafting	Comments
Lockarp CT 7 MHM 12755:167	Distal part	No	No	No	Yes	Heavy	Hides	–	Reused fragment
Lockarp CT 7 MHM 12755:166	Distal part	No	Yes	No	Yes	Heavy	Hides	–	Reused fragment
Lockarp CT 7 MHM 12881:945	Distal part	No	Yes	Yes	Yes	Heavy	Reeds	–	
Lockarp CT 7 MHM 12881:940	Medial part	No	No	No	No	Heavy	Cereals	–	
Lockarp CT 7 MHM 12881:795	Proximal part	No	No	No	No	–	–	Yes	No edge preserved, only shafting part
Lockarp CT 7 MHM 12755:108	Proximal part	No	Yes	No	Yes	Heavy	Hides	No	Reused fragment
Lockarp CT 7 MHM 12881:946	Proximal part	No	No	No	No	–	–	Yes	No edge preserved, only shafting part
Lockarp CT 7 MHM 12881:771	Proximal part	No	No	No	No	Slight	Cereals or reeds	Yes	
CT YT-spåret MHM 12587:13	Distal part	Yes	No	No	No	–	–	–	Patina renders analysis impossible
CT YT-spåret MHM 12587:15	Whole	No	Yes	Yes	No	Heavy	Reeds	Yes	
S.Sallerup 15I MHM 9145:1215	Whole	No	Yes	Yes	Yes	Heavy	Sods	Yes	
S.Sallerup 15I MHM 9145:471	Whole	No	Yes	No	No	Slight	Cereals or reeds	Yes	
S.Sallerup 15I MHM 9145:416	Distal part	No	No	No	No	Slight	Cereals or reeds	–	
S.Sallerup 15I MHM 9145:756	Medial part	No	Yes	No	No	–	–	–	Patina renders analysis impossible
S.Sallerup 15I MHM 9145:797	Distal part	No	No	Yes	Yes	Heavy	Reeds	–	
S.Sallerup 15I MHM 9145:495	Distal part	No	Yes	No	No	Generic	–	–	
S.Sallerup 15I MHM 9145:757	Distal part	No	No	No	No	–	–	–	Patina renders analysis impossible
Sunnanå 19A MHM 9151:3232	Whole	No	Yes	Yes	No	Heavy	Cereals	Yes	
Sunnanå VGV 1:9 MHM 9064:195	Whole	Yes	Yes	Yes	No	Heavy	Reeds	Yes	
Fosie 11C MHM 9128:66	Distal part	No	Yes	Yes	Yes	Heavy	Cereals	–	
Fosie 11B MHM 9128:94	Whole	No	Yes	Yes	Yes	Heavy	Cereals	Yes	
Fosie 11B MHM 9128:208	Medial and distal part	No	Yes	Yes	Yes	Heavy	Reeds	–	
Fosie 11B MHM 9127:95	Whole	No	No	Yes	Yes	Heavy	Reeds	Yes	
Fosie 11B MHM 9128:96	Whole	No	No	Yes	Yes	Heavy	Reeds	Yes	
Lockarp 7B MHM 8544:1704	Whole	Yes	No	Yes	Yes	Heavy	Cereals or reeds	Yes	
Lockarp 7B MHM 8544:1706	Whole	No	Yes	Yes	Yes	Heavy	Reeds	Yes	
Lockarp 7B MHM 8544:1707	Whole	No	Yes	No	No	Generic	–	No	
Valdemarsro MHM 6541	Whole	No	Yes	Yes	No	Heavy	Cereals	Yes	
Sydgas MHM 6650, no. 1	Distal part	No	No	No	No	Heavy	Reeds	–	
Sydgas MHM 6650, no. 2	Proximal part	No	No	No	No	Generic	–	Yes	
Sydgas MHM 6650, no. 3	Distal part	No	Yes	Yes	No	Heavy	Reeds	–	
Sydgas MHM 6650:IX	Medial part	No	Yes	No	No	Generic	–	–	
Sydgas MHM 6650	Proximal part	No	Yes	No	No	Slight	Cereals or reeds	Yes	
Sydgas MHM 6650:XVII	Whole	No	Yes	Yes	No	Heavy	Reeds	Yes	

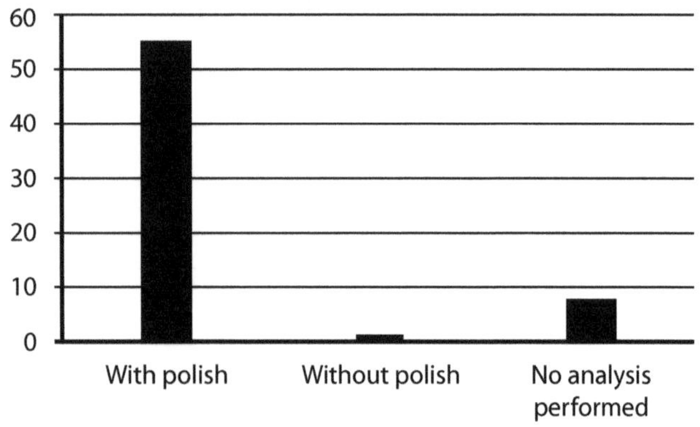

Figure 2.95. The number of large blade knives with polish, without polish, and not capable of being analysed. The chart is based on analyses of a total of 64 large blade knives of flint.

Results

A total of sixty-four large blade knives have been subjected to use-wear analysis in this part of the study. The blade knives come from excavations conducted in the last twenty years in the area around Malmö in south-western Scania, Sweden. The analyses sought answers to two questions: What contact materials were the blade knives used on? Were they shafted?

Table 2.57 shows that, of the 64 large blade knives included in the analysis, polish from contact material has been observed on 55. One blade knife has no polish and eight blade knives could not be analysed, figure 2.95. The latter either have a patina that renders analysis impossible or they lack a preserved edge. All but one of the blade knives that could be analysed thus have polish showing that they were used.

The result shows that the vast majority of the large blade knives were used on cereals and reeds, with 44 of the 55 large blade knives displaying polish from contact with one of these plants. These contact materials dominate, with 21 blade knives showing reed polish and 11 cereal polish. Twelve of the blade knives have polish that is interpreted as "cereals or reeds", meaning that I have not been able to determine which of these materials the large blade knife was used on. The polish on these 44 blade knives is well-developed. It forms a broad band along the edge and reaches far in over the surface of the tool. Often the polish has merged to cover the whole surface of the flint. It extends up over the ridges of the blade knife and is seen along the retouched back. It is usually more developed on the dorsal than the ventral side. Seven blade knives have generic polish, revealing that they were used, but since the polish is not sufficiently developed it is impossible to determine the contact material. It could be that there are among these blade knives some that were used on a contact material that is not otherwise represented in this study. One blade knife has polish that I interpreted as coming from contact with sods, figure 2.96.

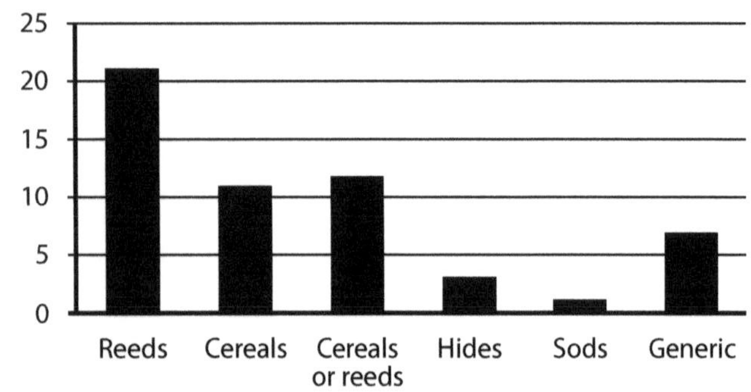

Figure 2.96. Contact materials identified on the 55 large blade knives of flint in the analysis which show polish.

Three fragments of large blade knives, all from the same excavation site (Lockarp CT 7), have use-wear which I interpret as coming from contact with hides. On all of these implements this polish is found both on the edges of the blade knives and on the edge of the fracture surface. Striations show that the direction of working was at right angles to both these edges. The interpretation of these three artefacts is that they were used as scrapers for working hides. These three fragments are the only examples of large blade knives included in the analysis which were used as tools after they were broken.

As discussed above, several authors have suggested that large blade knives were used for harvesting foliage. No blade knives in this study have polish indicating any contact with wood. There is thus nothing in the analysis to support the assumption that large blade knives of flint were used to cut leaves, despite the Swedish name for the tool, *lövkniv*, literally "leaf-knife".

A total of 31 of the 64 blade knives have gloss visible to the naked eye. Of these, 19 also have visible rounding on the edge. The three blade knives that were used on hides have heavy rounding but lack gloss. Of the 21 blade knives that were used on reeds, 19 have gloss visible to the naked eye. Of the eleven blade knives used on cereals, ten have gloss visible to the naked eye. Of the twelve blade knives where the contact material is interpreted as cereals or reeds, only one has gloss visible to the naked eye (Lockarp 7B MHM 8544:1704). The reason I have not been able to specify the contact material on that blade knife is that it has a faint brown patina which makes analysis difficult.

These results show a clear association between well-developed polish and the possibility of determining the precise contact material. The large blade knives which were used in such a way that gloss visible to the naked eye developed, in other words, those with heavy polish, are the ones on which I have been able to identify the contact material as either cereals or reeds. On the blade knives that have polish but no gloss visible to the naked eye, that is to say, where polish has developed but is not heavy, I have not been able to specify whether the contact material was cereals or reeds.

There are traces of shafting on 32 of the large blade knives. The shafting traces consist of a matt surface with low reflectivity, differences in the extent of polish from the contact material and the shaft, and friction polish. It is not always the case that all of these shafting traces are found on one and the same blade knife, figure 2.97.

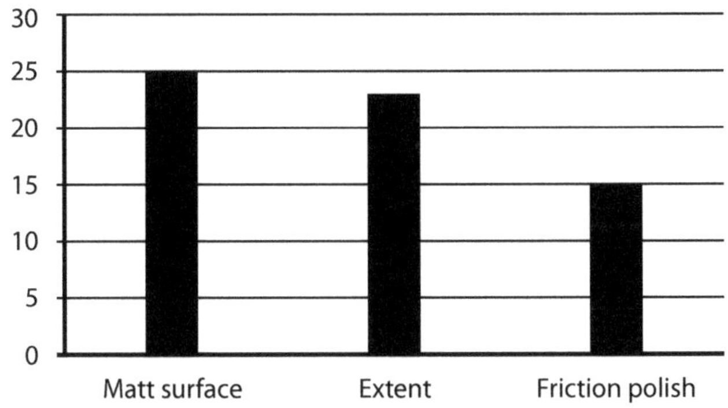

Figure 2.97. The number of observations of three shafting traces – matt surface, differences in extent, and friction polish – on the analysed large blade knives. Note that the chart states the number of observations, not the number of blade knives with occurrences of these attributes.

Five blade knives lack shafting traces. Of these, one was used as a hide scraper, two have a slight polish from contact with cereals or reeds, and two display generic polish. All of the blade knives with developed polish from contact materials and capable of being analysed for traces of shafting do indeed display shafting traces. The results show that the majority of the large blade knives of flint were shafted. There is nothing in the study to indicate that large blade knives were used unshafted. Twenty-seven blade knives have not been analysed for shafting traces, either because they have patina which renders analysis impossible, or the proximal part is not preserved or has suffered post-deposition or post-excavation damage rendering analysis impossible.

The analyses show that all of the blade knives with shafting traces were shafted with the proximal part enclosed in the shaft. The extent of the shafting polish and the difference between polish from the shaft and polish from the contact material on the surface, along the edge, and the back of the blade knife show that the knife was fixed in the shaft in such a way that parts of the flint surface were protected by the shaft, while the rest of the knife blade was not protected. On blade knives which have shafting retouch and which also have shafting polish, the extent of the polish corresponds to the position of the shafting retouch.

These findings agree well with both Rots' (2004) studies of traces of shafting, and the studies of shafting traces on large blade knives by Jensen (1990), Kruk (1994), and Christensen (1996). The results show that the thirty-two blade knives with shafting polish included in the study were shafted in a way that is similar or identical to the Stenild find. The type II large blade knives in the analysis have the same type of shafting traces as type I large blade knives and thus were not shafted in the way suggested by Knarrström (2007:103).

On twenty-three of the blade knives, great differences have been observed between areas with polish from the contact material and areas with shafting polish. It is important to stress that these dividing lines between polish from the contact material and from the shafting are visible but not distinct. In other words, there is no blade knife where the boundary between the two kinds of polish is exact. Within an area of 5–10 millimetres there is a diffuse transition between the polish from the contact material and that from the shafting, see figure 2.94: MK 20:201876b. The large blade knives were thus shafted in such a way that both the shaft and the contact material have left traces on the surface of the flint at the transition between the parts of the knife blade that were enclosed in the shaft and the parts that were not protected by it. In the series of experiments it happened often that small straws got stuck under the shaft when the implement was being used. This could explain the diffuse transition between the different kinds of polish observed on the prehistoric blade knives. Another explanation could be that the blade knives were not fixed in the shaft but moved back and forth slightly in it.

Several authors have observed that large blade knives often have serious macrodamage on the edges and have ascribed this damage to the use of the blade knives (Björhem & Säfvestad 1993; Jensen 2006). A total of thirty-six blade knives, or more than half of the blade knives in this study, display various types of post-deposition or post-excavation damage, despite the fact that many of them came direct from the excavation to my microscope. It is obvious that large blade knives are a type of artefact that, for some reason, easily suffers damage. This finding makes me doubtful about interpretations of function based on ocular examination of the edges of blade knives. I am not claiming that systematic use-wear analyses by low-magnification microscope (see e.g. Odell 2004:146f) cannot be done. What I mean is that assessments of the function of large blade knives are not credible if they are based on observations of macrodamage where so-called use-retouch is automatically linked to use.

Comparison of the sites

Thus far into the analysis of the results of use-wear analyses, the large blade knives from all the sites have been treated together. But it may be wondered whether there are differences between the sites. The three places with finds of most blade knives have been selected in order to study this: Kristineberg (17 blade knives), Tygelsjö (10 blade knives) and Lockarp CT 7 (13 blade knives). The question is investigated using comparisons of contact material and shafting traces. The results are summed up in table 2.58.

The table shows that there are both similarities and differences. From all the sites there are large blade knives that were used on cereals or reeds, and all the sites have blade knives that were shafted. There are some differences as regards the proportion of blade knives with the different kinds of polish. From Kristineberg and Tygelsjö there are more with reed polish than cereal polish, while from Lockarp CT 7 there are more with cereal polish than reed polish. These differences are small, however, and based on a limited number of observations. A striking difference is that all three fragments of large blade knives which display hide polish come from Lockarp CT 7. There are no such blade knives from the other areas.

Table 2.58. Contact material and shafting polish from the sites of Kristineberg, Tygelsjö, and Lockarp CT 7, stated as the number per occurrence.

Site			Contact material			Shafting polish		
	Cereals	Reeds	Cereals or reeds	Hides	No analysis	Yes	No	No analysis
Kristineberg	2	7	4	x	4	6	2	9
Tygelsjö	2	3	3	x	2	7	1	2
Lockarp CT 7	3	2	1	3	4	6	1	6

The results show that the overall impression of all the large blade knives in the analysis also applies to the sites individually, with minor variations. The study has also found that, apart from this general picture, there are places where the use of fragments of large blade knives looks completely different.

Comparisons of different groups of large blade knives

An interesting question to consider is whether there is a difference as regards contact material and shafting traces between the different types of large blade knives that I have defined earlier. Table 2.59 shows that there is no difference.

Summing up – functional analyses of large blade knives

This chapter has been devoted to functional analyses of large blade knives of flint. The study proceeded from a survey of previous research on the function of large blade knives. Through a detailed examination of how different kinds of polish from silicaceous plants have been described, combined with experiments in use, a foundation was created for analysing sixty-four prehistoric large blade knives from excavations in the area around Malmö in south-western Scania, Sweden. These use-wear analyses were intended to ascertain what the large blade knives of flint were used for and whether they were shafted.

The results show that a large blade knife should be regarded as an artefact that, in the Late Bronze Age, was part of a composite tool consisting of a knife blade of flint and a shaft of wood. The knife blade was fixed in the shaft in a way identical or similar to the shafting of the Stenild find. The tool was mainly used for cutting cereals or reeds. The experiments show that the tool works very well for these uses. The results also demonstrate that occasional blade knives were used for other work and that fragments of blade knives which got broken were used for scraping hides.

Table 2.59. Polish and shafting traces on type II large blade knives and type I large blade knives without platform preparation. The table is extracted from table 2.57, showing the blade knives that could be analysed by HPA.

Type II large blade knives	Contact material	Shafting traces
20:200409	Reeds	Yes
8544:1707	Generic weak polish	No
12881:945	Reeds	–
12881:966	Cereals	–
12587:15	Reeds	Yes
Type I large blade knives without platform preparation	**Contact material**	**Shafting traces**
9127:95	Reeds	Yes
9145:1215	Sods	Yes
9151:3232	Cereals	Yes
12755:168	Generic weak polish	Yes

These results both agree and disagree with earlier functional analyses. Several researchers have shown that large blade knives were used for cereals and reeds and that they were shafted (Jensen 1990; Christensen 1996), as this study has confirmed. It has also been suggested that blade knives were used for flax or nettles (Knarrström 2007), wood (Christensen 1996), and leaves (Wyszomirski 1979; Björhem & Säfvestad 1993), and that large blade knives were not shafted (Wyszomirski 1979; Christensen 1996), or that type II large blade knives were shafted in a different way from type I large blade knives (Knarrström 2007). None of this has been confirmed by the results of this study. Several scholars have claimed that the large blade knives are multifunctional tools (Wyszomirski 1979; Björhem & Säfvestad 1993; Christensen 1996; Jensen 2006). There is nothing in my study to support this. Christensen (1996) shows in her study that large blade knives were used for working hides. Three blade knives in my study were used on hides. However, all of these were broken during use and thus were not employed as intact shafted knife blades, and should thus not be regarded as large blade knives of flint in the same way as the other large blade knives analysed here. Earlier researchers have also claimed that large blade knives were ineffective when used as reaping tools (Steensberg 1943). The present study shows that this is not correct.

The result of this functional analysis of large blade knives has yielded knowledge of the contact materials the blade knives were used on. But the study has not investigated what operations the tools were used for. To clarify this difference let us return to the questions from the introduction to this chapter.

- What were large blade knives used for?
- Were they harvesting implements for arable fields or meadows?
- Were they tools for gathering reeds to thatch houses?
- Were they tools for gathering leaves as winter fodder for stabled animals?
- Or were they multifunctional tools used for all this and much more?

The first question has been answered; we known what contact materials most of the large blade knives in the analysis were used on. But we know very little about the second question. We can exclude meadows since there are no signs that blade knives were used for grass-like plants. And we know that many of the blade knives were used for reaping cereals. But we do not know whether they were used as reaping tools in arable fields. Purely hypothetically, they could have been used for straw crafts. Nor have we answered question number three. We know that several of the blade knives were used for cutting reeds but we do know in what operation this took place. As for questions four and five, the analyses have ruled out these functions.

This means that we now know what contact materials the large blade knives were used for. What remains to be studied is the operations, more specifically, what work the blade knives were used for. To investigate this requires other methods than use-wear analyses. This will be done in a later chapter.

Dating, contexts, and geographical distribution

Up to now I have discussed how large blade knives have been treated in archaeological research and practice, what they look like, how they were made, and what they were used for. I shall now turn to issues to do with dating, find contexts, and the geographical distribution of blade knives.

As we saw above, previous studies have placed large blade knives in the last part of the Late Bronze Age, that is, phase B (see e.g. Müller 1919; Thrane 1980; Ethelberg 1982; Björhem & Säfvestad 1993; Rostholm 1999; Ring et al. 2001; Högberg 2004a; Knarrström 2007; Olesen & Eriksen 2007). And in a study from the early 1980s, Sørensen (1984–85) pointed out that blade knives do not occur in the Early Bronze Age or the Pre-Roman Iron Age. The forward and backward chronological limits of the occurrence of large blade knives are thus clear. But what we also found in previous chapters is that there is still some doubt as to whether blade knives should be dated only to the latter part of the Late Bronze Age or if they existed throughout the Late Bronze Age (Björhem & Säfvestad 1993). This doubt is reflected, for example, in the exhibition "Danmarks Oldtid" which opened in May 2008 at the National Museum in Copenhagen, where large blade knives of flint are dated to 1100–500 BC. There is thus reason here to further study aspects of the dating of blade knives: it is obvious that they do not occur together with material from the Early Bronze Age and Pre-Roman Iron Age, but what is not clear is how they should be dated within the Late Bronze Age.

First of all, it may be noted that all of the large blade knives that I have seen which were found in contexts dated to the Late Bronze Age have occurred together with phase B pottery (see e.g. Stamm Forssblad 2001; Rostoványi & Hyden 2002; Eliasson & Kishonti 2003; Grehn 2006). Large-scale excavations of settlement sites with finds of pottery that can be dated to Phase A of the Late Bronze Age have been published (see e.g. Fendin 1999; Edring 2000; Ring et al. 2001; Ingvardson 2005). I have not found a single example from these excavations where large blade knives of flint have been found together with phase A pottery. There is thus a clear distinction between features dated with the aid of ceramic typology to Phase A and those dated in the same way to phase B. Phase A features lack large blade knives of flint. Phase B features do have them. A good example is the Svågertorp 8 B–C excavation site in Malmö. Here several hundred features have been excavated and dated to Phase A of the Late Bronze Age. Not one of these features yielded large blade knives. In the excavation site there were three features with pottery dated to phase B. In one of these features a large blade knife was found (Rostoványi & Hyden 2002:170f).

There is one example, however, of a published find of a large blade knife discovered together with pottery dated to Phase A, figure 2.98. Of the 54 large blade knives and fragments of large blade knives found in dated contexts by the excavations at Fosie IV, 53 occurred together with phase B pottery and one, a fragment, was found together with Phase A pottery (Björhem & Säfvestad 1993:64f). The latter find was used – together with the results of previously published studies where doubts have been expressed about dating (e.g. Becker 1990) – as an argument suggesting that the terminus post quem of large blade knives is uncertain (Björhem & Säfvestad 1993:77); this

Figure 2.98. The proximal part of a knife or flake, previously interpreted as a large blade knife of flint and used as a foundation for a discussion of the uncertain backward chronological limit of large blade knives (Björhem & Säfvestad 1993:64f; Björhem & Magnusson Staaf 2006:173). An attribute analysis demonstrates that the find does not display the characteristic technological attributes that large blade knives of flint have. The attributes that the proximal part of the flake on the photograph have in common with large blade knives of flint are a flat platform, a platform angle of 75 degrees, and a ridge and two negative removals on the dorsal side. The attributes that the proximal part of the flake on the photograph does not have in common with large blade knives of flint are a wing-shaped platform, a lip, no visible impact point, a diffuse bulb of percussion, no moustache, no fissures, and no ring crack, table 2.60. I have performed use-wear analysis of the flake. It has generic polish. The flake was found in feature 331A, Fosie IV (Björhem & Säfvestad 1993:64f) and the excavation has inventory number MHM 6185. The find is stored at Malmö Museums.

Table 2.60. Comparison of the attributes on the knife defined by Björhem and Säfvestad (1993:64f) as a large blade knife of flint and dated to Phase A of the Late Bronze Age and the results of the attribute analyses of large blade knives of flint above. The column to the far left lists the attributes of large blade knives of flint. Other columns state the occurrence of each attribute on large blade knives based on the results of the analyses in the previous chapter.

Attribute	Common on large blade knives	Unusual on large blade knives	Never present on large blade knives
Flat platform	X		
Exterior platform angle, 75 degrees	X		
One dorsal ridge	X		
Two dorsal flake scars	X		
Wing-shaped platform		X	
Lip			X
No impact point on the platform		X	
Diffuse bulb of percussion		X	
No moustache		X	
No bulbar fissures		X	
No bulbar scar		X	

has also been repeated in later studies (Björhem & Magnusson Staaf 2006:173). It has therefore been considered important to analyse this fragment of a large blade knife from Fosie IV here to determine what it looks like, table 2.60.

A detailed attribute analysis shows that this find which has been dated to Phase A does not display the technological attributes which the analyses above found to be characteristic of large blade knives of flint. My interpretation is therefore that the find is not a large blade knife, figure 2.98.

Although the terms phase A and B were initially used as designations in a chronology based on ceramic typology, these terms have come to be employed as designations for periods whose radiometric datings agree with the ceramic phases (Björhem & Magnusson Staaf 2006:173). In calendar years Phase A was initially defined by Björhem and Säfvestad (1993:55) as 1420–825 BC and phase B as 840–390 BC. Recent research shows that phase B directly follows phase A, with both overlaps and the occurrence of certain transitional forms. This shows that it is not possible to draw a dividing line between the phases; the change in the pottery began some time after 900 BC and was completed around 800 BC (Björhem & Magnusson Staaf 2006:174). Radiometric datings on material from features with large blade knives confirm these datings.

What has not been discussed, however, is the forward chronological limit of phase B. As table 2.61 shows, there is only one dating which at 2 sigma extends into the fifth century BC, while several have the sixth century as the ultimate value at 2 sigma. My interpretation of these calibrated radiocarbon values at 2 sigma, together with the results of the pottery dating (Björhem & Magnusson Staaf 2006:174), is that large blade knives of flint were introduced some time just after 900 BC and fell out of used just before 500 BC.

As the values for 2 sigma in 2.61 show, the calibration curve for the time-span defined for phase B is far too jagged and wide to allow precise datings within this interval of 400 years. This makes it impossible for me to date features containing large blade knives with any exactitude. This is a shortcoming. Since the ambition of my work is to study change and stability, it would of course be significant to have a more precise dating foundation. A resolution that would allow, say, a dating to period V or period VI would be desirable. Detailed combination studies of radiocarbon analyses, pottery, large blade knives, and other flint material would probably have given a more sophisticated basis for dating, but it has not been possible to do this within the scope of this investigation. In the detailed studies of pottery that have been published, the flint material has not been examined in relation to the pottery (see e.g. Rostoványi & Hyden 2002), thus making it impossible to use those. It looks as if there is no way to discuss the dating of large blade knives other than that they belong in the 400-year Phase B of the Late Bronze Age. There is work here for future research to do.

In a study from the start of the 1980s Sørensen published a hypothesis that the large blade knives can be typologically divided into those with shaft retouch but no other retouch, and those with retouch along the whole back. This typology, according to Sørensen, could also be chronologically relevant since the former type could be dated to period V and the latter to period VI (Sørensen 1984–85:60). As previously mentioned, Björhem and Säfvestad tested this chronology in their analyses of the material from the excavations at Fosie IV. They found no support for Sørensen's hypothesis (Björhem & Säfvestad 1993:77). In the analyses that I have performed, I have likewise not been able to corroborate Sørensen's chronology. In a new survey of the flint material from the excavations on which Sørensen based his hypothetical chronology, I have found that the patterns presented in Sørensen's article from 1984–85 are not as distinct as the article claims.

After this discussion of dating I shall proceed with an investigation of find contexts and the geographical distribution of large blade knives. In previously published studies of blade knives the find contexts are well described. In excavations they are usually found in contexts interpreted as settlement site material. These are various types of pits for clay extraction,

Table 2.61. Radiometric datings of material from features with finds of large blade knives of flint. All values are calibrated with OxCal v. 3.5–3.10.

Excavation site	Feature number	^{14}C 1 sigma	^{14}C 2 sigma	Lab. no., material	Pottery	Reference
Lockarp 8:4 (MHM 12898)	A24901	800–550	810–520	Ua-20471, oak charcoal	Per. V	Stamm Forssblad 2001
Svågertorp 8C (MHM 8541)	A42640	770–420	780–400	Ua-14719, macrofossil (barley)	Phase B	Rostoványi & Hyden 2002
Lockarp 7B	A1562	840–560	910–530	Ua-13067, oak charcoal (high eigenvalue)	Phase B	Eliasson & Kishonti 2003
Lockarp 7B	A5592	900–790	930–590	Ua-12648, sallow charcoal	LBA	Eliasson & Kishonti 2003
Lockarp 7B	A125	980–830	1030–800	Ua-13065, oak charcoal (high eigenvalue)	Phase B	Eliasson & Kishonti 2003
Kristineberg syd (MK 20)	A1751	920–815	1005–800	LuS 7452 macrofossil (unidentified)	Phase B	Rostoványi 2005
Kristineberg syd (MK 20)	A1751	900–800	930–790	LuS 7453 macrofossil (unidentified)	Phase B	Rostoványi 2005
Kristineberg syd (MK 20)	A1751	910–805	975–795	LuS 7454 macrofossil (unidentified)	Phase B	Rostoványi 2005
Kristineberg syd (MK 20)	A1751	840–795	915–780	LuS 7455 macrofossil (unidentified)	Phase B	Rostoványi 2005
Kristineberg syd (MK 20)	A1751	975–830	1015–805	LuS 7456 macrofossil (unidentified)	Phase B	Rostoványi 2005
Kristineberg syd (MK 20)	A1751	910–810	975–795	LuS 7457 macrofossil (unidentified)	Phase B	Rostoványi 2005
Kristineberg syd (MK 20)	A1751	890–795	915–785	LuS 7458 macrofossil (unidentified)	Phase B	Rostoványi 2005

storage, and work which have been filled with waste material, and the large blade knives are found together with potsherds, fire-cracked stone, and knapped flint (Müller 1919; Moskwa 1976; Ethelberg 1982; Sørensen 1984–85; Björhem & Säfvestad 1993; Auler 1999; Rostholm 1999; Grehn 2006; Olesen & Eriksen 2007), figure 2.99.

Large blade knives of flint also turn up as stray finds in ploughed fields, in excavations of settlement sites from the Late Bronze Age, or in land that was formerly wetland (Adamsen 2002). Few finds of large blade knives have been documented from other contexts; the best-known is the aforementioned find of blade knives in a post-hole at Spjald in Jutland (Becker 1990). Large blade knives have also been discovered as assemblages and at manufacturing sites (Ebbesen 1981). Almost all of these have been retrieved without archaeological investigation. I will return to analyses of assemblages and manufacturing sites later on. To my knowledge there is only one find of a large blade knife of flint in a mortuary context. This was discovered by the excavation of an urn field cemetery in northern Germany and will likewise be considered later. Otherwise I have not found blade knives in any other context from the Late Bronze Age. The result of this brief survey of the contexts in which large blade knives occur shows that hitherto they have mostly been found together with what is interpreted as settlement site remains.

Another question is where they are found. This has been touched on in previous chapters, but will be further specified with an account of finds from the countries and provinces included in my study.

From Scania there are many finds of blade knives, both in museum collections and from excavations (Björhem & Säfvestad 1993; Högberg 2004a). In my material a total of 254 are from Scania. The finds mainly come from the plains and coastal areas.

Figure 2.99. A clay pit at Fosie IV (Björhem & Säfvestad 1993:118, fig. 166).

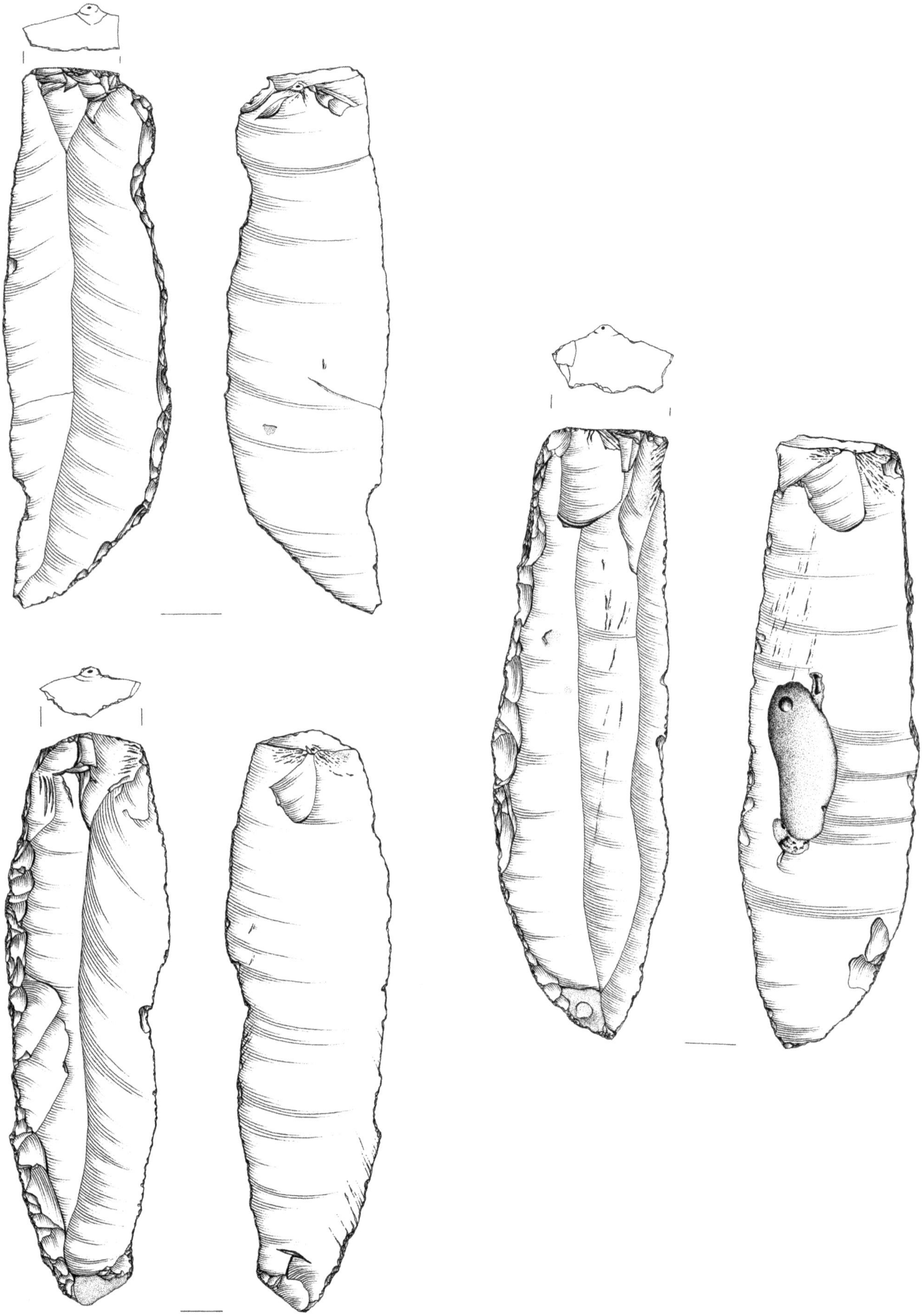

Figure 2.100. Some large blade knives from the collections at the County Museum in Halmstad. Accession numbers: HM 413, HM 11303, HM 8515. The blade knife to the upper left is 13 centimetres long. Drawings: Björn Nilsson.

Table 2.62. Sites in Halland with archaeological material dated to the Late Bronze Age from which I have examined the flint assemblages to see whether they include large blade knives. None of them did. The abbreviation HM stands for Halmstad Museum and refers to the museum's catalogue number.

Site name	Reference
Sannarp 1:1	Streiffert & Strömberg 1998; Streiffert 2001
Stafsinge	Nicklasson 2001
Fornlämning 82, Tvååker Parish	Lundqvist 1992
Veinge Skogaby	HM 25665
Slottsmöllan	HM 24758

From Halland there are several published assemblages of large blade knives (Lindälv 1964), and in my material I have registered 43 large blade knives from the province, figure 2.100. All of these come from museum collections. I do not know of any finds of large blade knives from excavations in Halland. In view of the amount of blade knives in museum collections, there should also be some finds from excavations. I have therefore examined find material from excavations in Halland in search of large blade knives or fragments of blade knives, table 2.62. These excavations have yielded flint material from different periods, quite a lot of which can be dated to the Late Bronze Age, but I have not found any large blade knives or fragments of blade knives among this material.

From Bohuslän there is one large blade knife registered in my material. It lacks the proximal part. It was found during excavations at Herrestad, just west of Uddevalla (Lindman & Ortman 1997:30). An excavation of the remains of a sunken-floor hut discovered it together with coarse-slipped pottery. The pottery, the occurrence of a blade knife, and a radiometric analysis date the context to the Late Bronze Age (Lindman & Ortman 1997:32). I have searched through parts of the collections at Bohuslän Museum without finding any other large blade knives.

From Blekinge there are six large blade knives in my material, figure 2.101. I know of no excavation find of blade knives from the province.

From Småland my material includes two large blade knives. Besides these I know of ten or so finds of blade knives from the area around Kalmar. I have not studied these. Large blade knives are unusual finds at excavations in the province; the large-scale excavations as part of the Möre Project at the end of the 1990s unearthed just a few (Gurstad-Nilsson 1999:208; Ring et al. 2001).

From Västergötland there are two finds of large blade knives in my material. I have not found any examples of blade knives

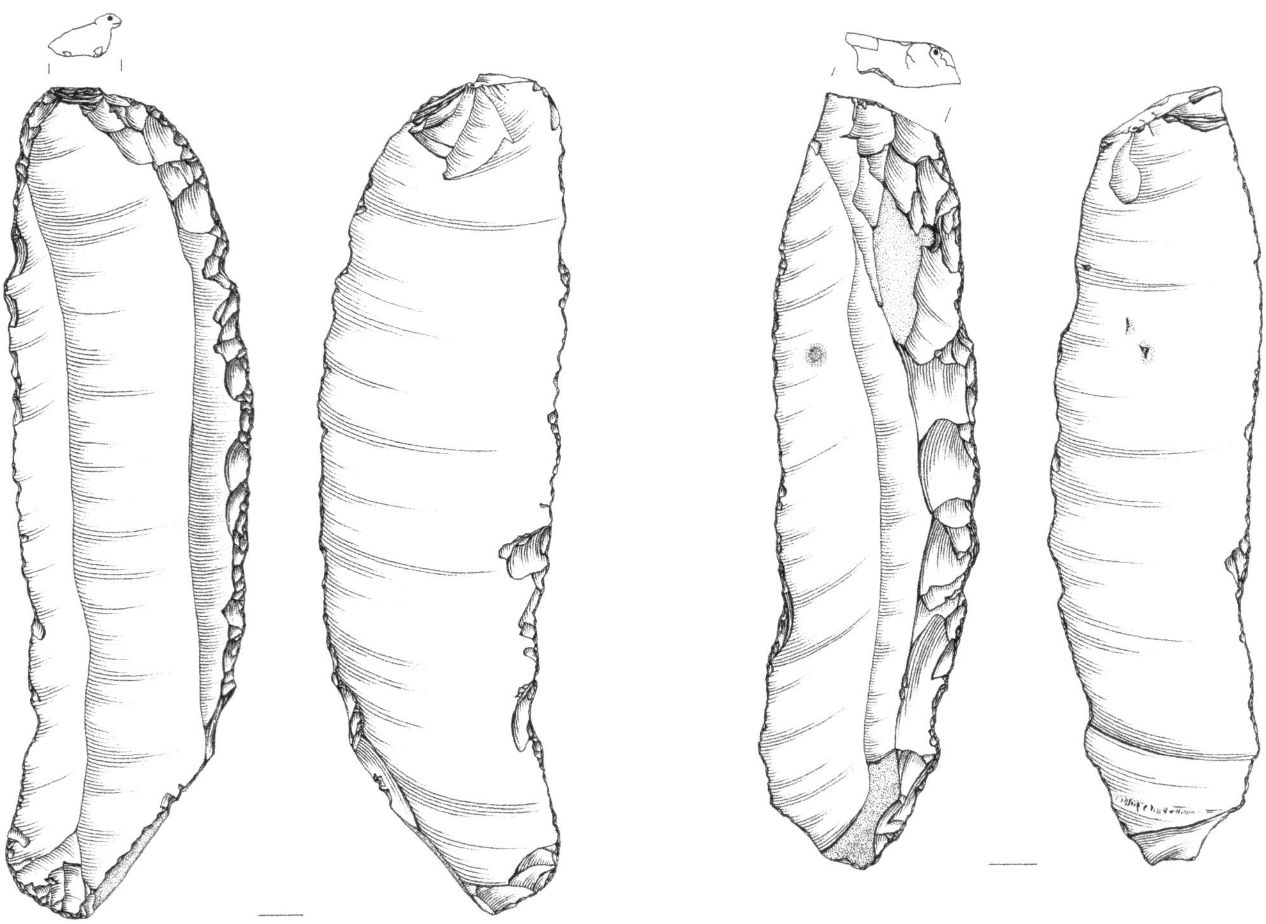

Figure 2.101. Some large blade knives from the collections at Blekinge County Museum. Accession numbers from the left: BL 13479 and BL 5422. The blade knife to the left is 14.5 centimetres long. Drawings: Björn Nilsson.

from excavations. A report on an excavation conducted near Gothenburg mentions the find of a knife that it says could be a blade knife. Judging by the photograph in the report, however, it is not a large blade knife (Nordqvist 2005:28).

From Östergötland there is one large blade knife in my material. In the publications of the extensive excavations at Pryssgården there is no mention of any blade knives (Borna-Ahlkvist, Lindgren-Hertz, & Stålbom 1998; Borna-Ahlkvist 2002). Nor are any reported from the excavations at Vistad (Larsson & Hulthén 2004). I have not investigated the material from either of these excavations.

From the large-scale excavations at Hallunda in Södermanland there are no large blade knives published (Jaanusson 1981:18ff). I have gone through the find material and not been able to discover any blade knife.

An excavation in the parish of Viksta in Uppland retrieved a burnt proximal part of a large blade knife. This is a stray find discovered close to a burnt mound, the upper parts of which have been dated to the Late Bronze Age (Forsman & Victor 2007:213). The archaeological material from the large-scale excavations at Apalle in Uppland have been summarily presented in a report (Ullén 2003). According to this there are no

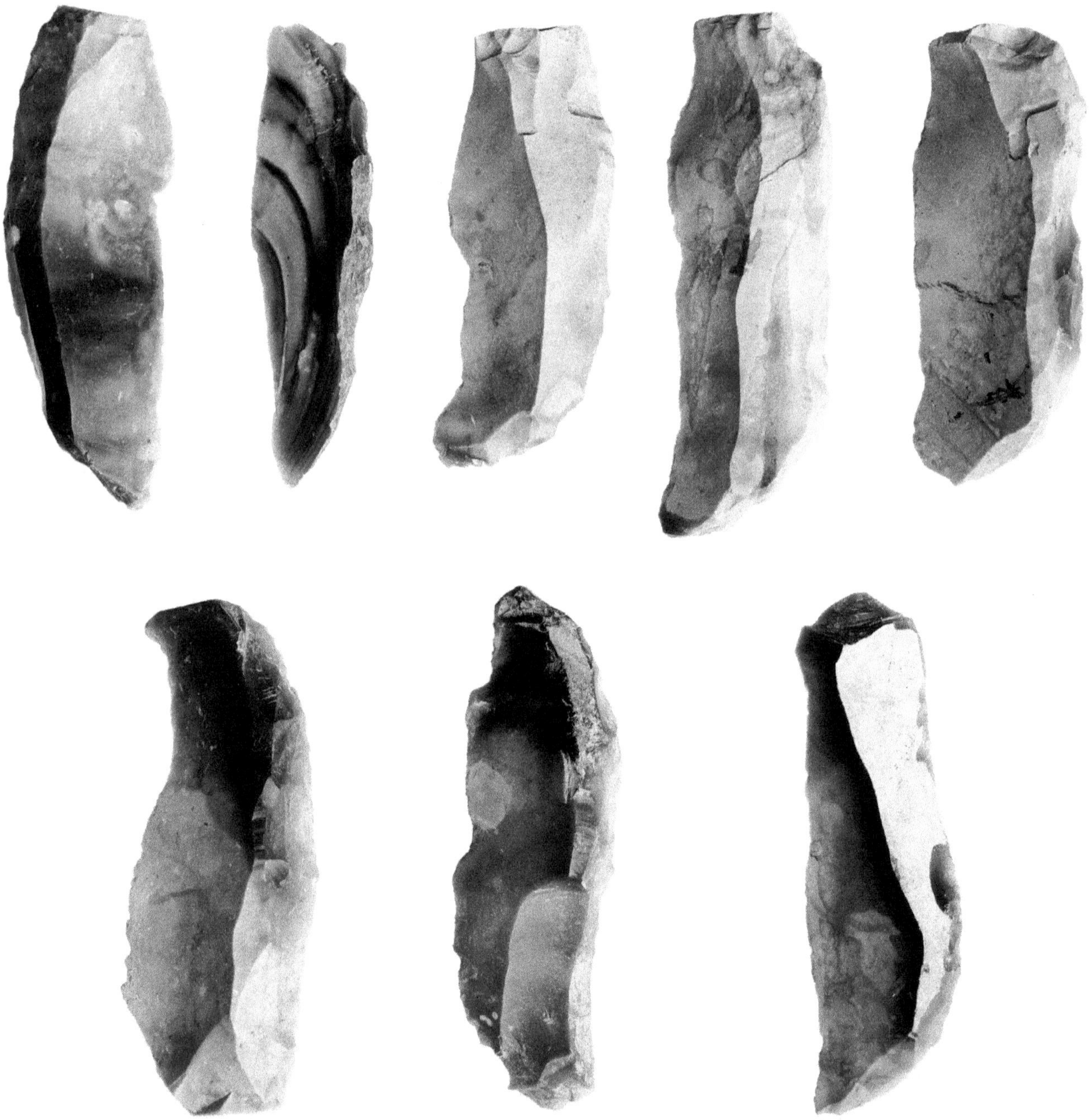

Figure 2.102. Large blade knives of flint found by an excavation at Flædemose at Stevns in Denmark. An excavation in 1969 uncovered several features containing finds that included flint and pottery. A bellows nozzle and fragments of a clay mould show that bronze casting was done on the site. The finds from Flædemose have been dated to the Late Bronze Age period VI (Thrane 1980:32f). The photograph was first published in Thrane (1980:28). Photo from the National Museum in Denmark, accession number B16358.

finds of blade knives, but when going through the flints from the excavation I identified a burnt distal part of a large blade knife. The object is defined as a scraper (Lindholm 2003:179, find number F 55562). The large blade knife was discovered in the remains of a house dated to the excavation's phase 5, which means one of the latest settlement layers from the Bronze Age, *c.* 900 to 600 BC (Inga Ullén, e-mail, 27 March 2007).

From Denmark, large blade knives are known from find spots from all parts of the country. They are common among the finds from excavations of Late Bronze Age settlement sites (Müller 1919; Thrane 1980; Ethelberg 1982; Sørensen 1984–85; Becker 1990; Bertelsen et al. 1996; Rostholm 1999; Nielsen 2005; Olesen & Eriksen 2007), figure 2.102. I have registered 127 large blade knives from Danish find spots. It is important to point out here a difference in my material between Swedish and Danish finds. When it comes to finds from Sweden I have studied many collections, assemblages, and publications in search of blade knives. As regards Denmark I have contented myself with studying some selected collections and publications. This means that the number of large blade knives from Sweden in my material is close to the total number of blade knives found, whereas the number of blade knives from Denmark in my material is only a sample of all those found. For example, the large-scale settlement site material from the excavations at the start of the 1900s, stored in the National Museum in Copenhagen (Müller 1919; Broholm 1949), were not available for me to analyse, because the museum was busy producing a new permanent exhibition, a project that took several years.

In Norway I have found preforms for large blade knives of flint from a find spot at Tjøling in Vestfold. Apart from this I have found large blade knives of flint at the Museum of Science in Trondheim in Sør-Trøndelag, where two or three are displayed in one of the showcases, although labelled Neolithic finds. The blade knives from the museum in Trondheim are not included in my analyses. The large blade knife of flint does not seem to be a known artefact in Norway. I have not found any published references to blade knives from the Late Bronze Age from Norway (Hagen 1967; Johansen 1993; Ballin 1996). In view of the contacts that existed between, for example, the area around Oslo Fjord and southern Scandinavia in the Late Bronze Age (see e.g. Hagen 1967; Goldhahn 2007), and the finds from the museums in Oslo and Trondheim, it is likely that large blade knives of flint exist in Norwegian find material from the Late Bronze Age without having been recognized as such.

From Germany I have studied large blade knives from two areas: Schleswig-Holstein and Mecklenburg-Vorpommern. In my material there are thirteen blade knives from the collections at Stiftung Schleswig-Holsteinische Landesmuseen, Schloss Gottorf in Schleswig, and 42 from the collections at Landesamt für Bodendenkmalpflege in Mecklenburg-Vorpommern in Stralsund and Schwerin, figure 2.103. The blade knives from the collections in Schleswig have not previously been published. Among the blade knives from the Landesamt in Mecklenburg-Vorpommern there are a number deriving from a published excavation of extensive settlement site remains from the Late Bronze Age at Marlow on Rügen (Auler 1997, 1999). Apart from these, I have also studied blade knives from an unpublished excavation at Plüggenthin on Rügen and from collections of stray finds from the sites of Reidervitz and Ruschvitz on Rügen. I shall return to this later.

As mentioned earlier, large blade knives occur from the flint mines at Wierzbica in central Poland (Lech & Lech 1984, 1997). Three blade knives in my study come from there. Large blade knives resembling those from Rügen in northern Germany have been found on settlement sites in central and western Poland. In eastern and southern Poland as well as in western Belarus and Ukraine there are no finds of large blade knives. Here instead one finds thick bifacially knapped knives (Migal 2004:223ff).

I have not found any large blade knives from Sweden north of Bohuslän, Västergötland, and Uppland or from the Norwegian interior. Nor are there any to my knowledge from the German interior, the North Sea coast of Germany, or westwards along that coast towards the Netherlands. Along the North Sea coast there are instead finds of thick bifacial sickles (Woltering 2001). I have searched for finds of large blade knives of flint in publications from England, the Netherlands, Spain, France, the Czech and Slovak Republics, but without success. It is thus obvious that blade knives are mainly found in southern Scandinavia, northern Germany, and central and eastern Poland. It remains to be studied whether there are finds of large blade knives from Finland, Russia, the Baltic countries, Belarus, and Ukraine, all of which border on my investigation area, and which have also been discussed as areas with reciprocal exchange of influences (see e.g. Hjärtner Holdar 1998; Migal 2004; Bolin 2005).

Summing up – dating, contexts, and geographical distribution

The results of the analyses of the dating of large blade knives of flint show very clearly that they occur in Late Bronze Age contexts together with phase B pottery. Large blade knives do not occur in the early part of the Late Bronze Age, in what Björhem and Säfvestad (1993) call Phase A. Nor do they occur in later periods, from the Pre-Roman Iron Age onwards. These results agree well with findings previously published by Björhem and Säfvestad (1993) and by Sørensen (1984–85). The results also eliminate the previous uncertainty as to whether the terminus ante quem for large blade knives should be set back further in time (Becker 1990). Recent studies which have dated large blade knives to the Late Neolithic or Early Bronze Age (Sjöberg 1998) may be regarded as incorrect.

The survey of contexts and the geographical distribution of large blade knives shows that they are usually found by excavations of settlement site material. Some known assemblages of blade knives also exist, as well as a few production sites. I know of only one find from a mortuary context. Large blade knives are found all over southern Scandinavia. Most blade knives from excavations and in museum collections come from Scania and Halland in Sweden, in Denmark, and along the Baltic coast of northern Germany. In Sweden they also occur

Figure 2.103. Large blade knives and preforms for large blade knives in the stores of the Landesamt für Bodendenkmalpflege in Mecklenburg-Vorpommern in Schwerin, from excavations at Marlow on Rügen. The blade knife at the top left is 8 centimetres long.

in smaller numbers along the west coast, in the east up along the coast to the Kalmar area, and at a few places in central Sweden. From Norway there is a small number of finds of large blade knives. Poland has finds of blade knives from excavations of settlement sites and from an area of flint mines.

Assemblages of large blade knives of flint

As discussed earlier, there is uniformity as regards interpretations of assembled finds of large blade knives. Several authors view them as hoards which can be associated with the established Stone Age tradition of depositing objects in wetlands or on dry ground as votive offerings (Wyszomirski 1979:149; Ebbesen 1981:16ff; Nielsen 2005:216ff). Others who have not studied blade knives but deposits of mainly bronze objects think that a change in deposition patterns took place at the transition from the Early to the Late Bronze Age, while simultaneously stressing the tradition and continuity from the Stone Age as regards offering practices and deposition sites (Larsson 1986:169; Karsten 2005:436). What is clear is that earlier studies of deposits of bronze objects have not included large blade knives of flint in the analyses (see e.g. Sprockhoff 1956; Baudou 1960; Levy 1982; Larsson 1986; Sørensen 1989a). From recent years, however, there are examples of studies where assemblages of blade knives have been discussed together with deposits of bronze objects (see e.g. Karsten 2005; Nielsen 2005). Assemblages of large blade knives have been found in the same kind of places as bronze hoards (Jensen 2006:108f).

Several studies have emphasized that the large blade knives in the assemblages show evidence of having been used (Blinkenberg 1898:152; Wyszomirski 1979:149; Ebbesen 1981:17; Becker 1990:201). Based on this observation, the assemblages have been interpreted as representing completed activities, not as tools for future use. This argument has perhaps been expressed most distinctly by Ebbesen. He has discussed the interpretation of the assemblages of blade knives and believes that they can hypothetically be regarded either as stockpiles or as "votive deposits to higher powers" (Ebbesen 1981:17). Based on the opinion that the finds do not represent great value, that they were not serially produced, and that the objects in each find have been used, Ebbesen thinks that the assemblages of large blade knives should be interpreted as hoards and votive deposits (Ebbesen 1981:17). This interpretation has been put forward in both earlier (Blinkenberg 1898) and later (Nielsen 2005) works.

It is thus obvious that there is consensus about several aspects of the assemblages of large blade knives of flint. They contain used blade knives. They were deposited in wetland and on dry ground. They do not contain serially produced objects. Several authors also think that the finds should be viewed in the perspective of a tradition going back several thousand years, the deposition of objects as offerings. It may be wondered whether this is true. To examine this problem I have broken it down into a number of specific questions:

- What do the assemblages of large blade knives look like?
- What do they consist of?
- Are all assemblages of blade knives similar?
- How were the objects in the finds made?
- Are there different kinds of deposits?
- Are the blade knives in the assemblages used?
- In what contexts have they been found?
- Is it right to place assemblages of large blade knives in the same interpretative framework as Stone Age hoards?

The method for working with these questions is a comparative analysis whereby I study the find context, what each find consists of, what type of flint the blade knives were made of, all this combined with technological analyses and functional analyses. The find contexts are investigated with the aid of published data and archival material. I look at information about when, how, and where the assemblage was found. The content of each assemblage is defined on the basis of the analyses in a previous chapter of large blade knives, preforms for large blade knives, and blade-like flakes. Flint type is defined as in other parts of the study. The technological analyses are chiefly of a general kind. Based on the results of the attribute analyses and the series of experiments presented above, blade knives have been defined as showing technological similarities and differences. They are similar as regards size, curvature and the morphology of the negative removals, different if there is variation in these attributes within one assemblage. Apart from this I analyse three of the finds in detail. Since the majority of the objects have a heavy patina and are often affected by post-deposition treatment, only a few of them have been analysed for use-wear (HPA). Functional analyses have above all been performed to establish whether or not the blade knives have gloss that is visible to the naked eye.

An important variable for identifying the assemblages is patina (see e.g. Karsten 1994:20). It is first and foremost a matter of a simple identification. Finds from the same place can be patinated in similar ways, and finds from different places in different ways. If the patina on the objects in the find is similar, it is thus possible that it is an assemblage from the same place. If the patina is different, the find may be a random collection of blade knives gathered from different places. But the look of the patina on the objects is also affected by other variables than the place where the objects lay in the ground (Högberg & Olausson 2007:67ff). Different types of flint are patinated in different ways, and the status of the object when it was deposited affects the appearance of the patina. An old dried-up and well-used large blade knife is patinated in one way, a newly manufactured, fresh, unused blade knife is patinated in a different way (Luedtke 1992:99). The look of the patina can thus also be used to determine whether the objects in an assemblage were similar or different as regards the type of flint and the status when they were deposited. Below I present first the analyses and then a discussion and summary of the results.

It should be pointed out that I have chosen here to study only assemblages of large blade knives. All the single finds of large blade knives of flint discovered by peat cutting and in ploughed fields (Blinkenberg 1898; Adamsen 2002) are not considered in this study.

Analysis of assemblages of large blade knives

A total of twenty-one assemblages are included in the analysis, eleven from Sweden, nine from Denmark, and one from Norway, figure 2.104. I located these finds by searching the literature and archives and by going through museum collections. Eighteen of the finds are summarily analysed, three in detail. Four of the finds in the analysis have not previously been published. A further six are mentioned in find lists but no analyses of them have been published. Of the twenty-one assemblages I myself have studied eighteen. For the other three I use reference data, table 2.63. I know of no other assemblages than these twenty-one.

Four previously published finds are not included in the study. One is from Asnæs Åsevang in Zealand, Denmark (Ebbesen 1981:9ff; find no. NMA 25.135). This find consists of scrapers and flakes, not large blade knives. The other one is a find from Businge on the island of Møn. In the publication this is described as consisting of six large blade knives (Ebbesen 1981:13, find number NMA 12.733), but this is not correct. It consists of four blades from the Stone Age and two Late Bronze Age large blade knives. It is thus a random collection, not an assembled find, figure 2.105. The third is a find from Rolfstorp Parish in Halland, which is mentioned in three texts (Lindälv 1964:62, 1967:89; Wyszomirski 1979:149). All of these texts repeat the same information about a find that is supposed to consist of four large blade knives and a flint block. No descriptions, photographs, or drawings have been published to illustrate this find. Since I have not seen the find and it was first presented in a context where Neolithic blades are mentioned as being the same type of find as Late Bronze Age large blade knives of flint (Lindälv 1964:62), I have chosen not to include this find in my analysis. The fourth previously published find omitted from this study refers to information about a large blade knife, a preform for a large blade knife, and two large flakes which are said to have been found together in wetland during excavations at Löddeköpinge in Scania. The find is mentioned by Wyszomirski (1979:149), citing a personal communication from Tom Ohlsson, the archaeologist in charge of the excavations. With reference to Wyszomirski, the find is also mentioned by Becker (1990:201), Karsten (2005:433), and Jensen (2006:107). This find is otherwise unpublished. In a manuscript in the archive at Lund University

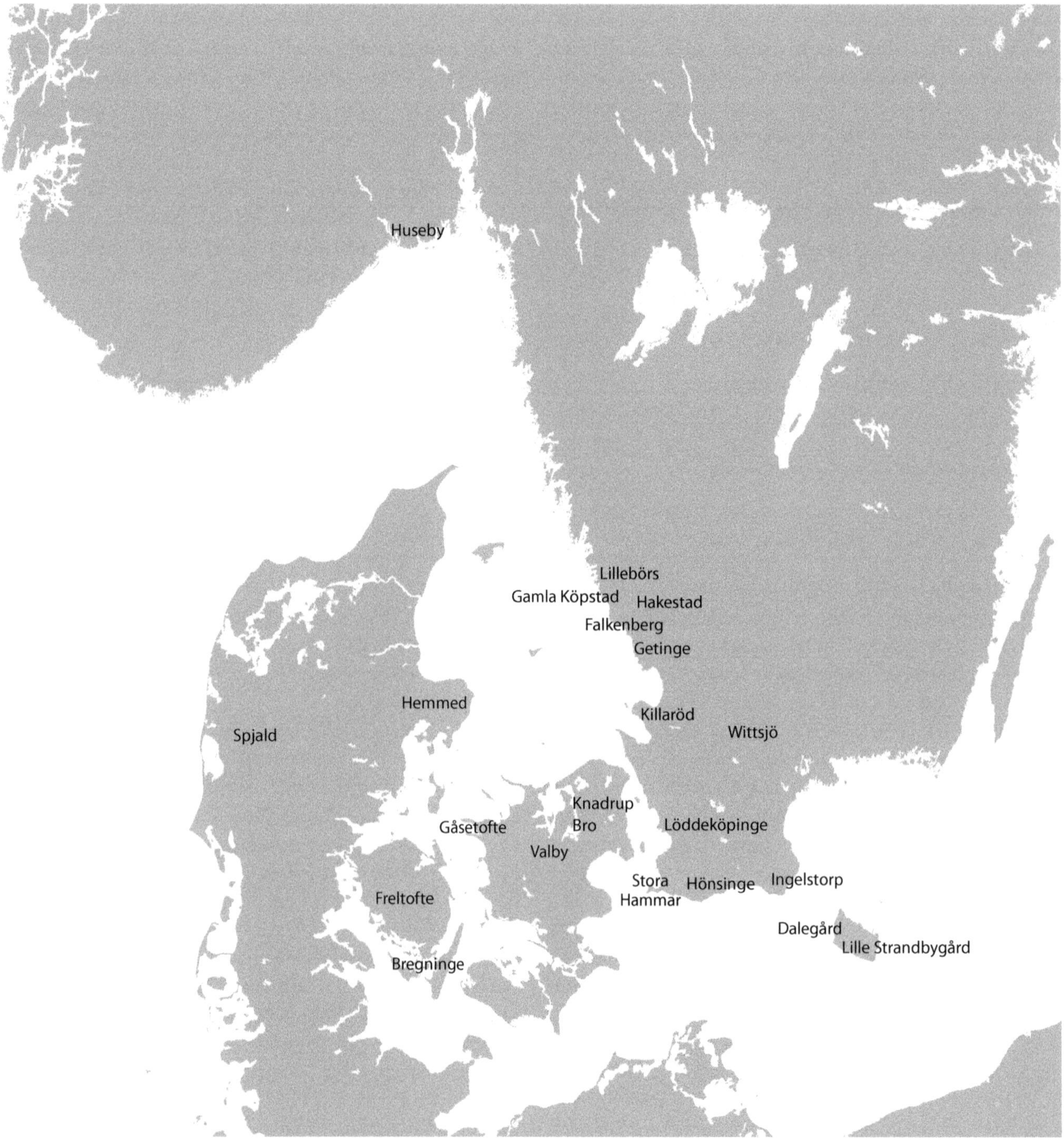

Figure 2.104. The find spots of the assemblages included in the study.

Historical Museum, written by Tom Ohlsson and presenting the results of the excavation, there is no mention of the find (Ohlsson, unpublished manuscript, no date). I have searched through the finds from the excavation in the museum's stores but could not locate this one. I did however come across a bag of stray finds. This contained one large blade knife and a number of flakes. Since there is no documentation or any finds to corroborate the details of the find published by Wyszomirski (1979:149), my interpretation is that the find does not exist. Presumably Wyszomirski confused the data about the bag of stray finds with information about a hoard.

The results of the analyses of each assemblage are presented below.

Ingelstorp Mossar, Ingelstorp Parish, southern Scania

According to the Simrishamn Museum accession catalogue, this find consists of two large blade knives and four blades. These are said to have been found together in a plough furrow in a field some time around 1900. On my visit to the museum I found one of the large blade knives. I have not seen any of the other objects in the find, which has not previously been published.

Table 2.63. The 21 assemblages included in this study. The number of objects in each find is sometimes stated in the table with two figures. The first refers to the number of objects contained in the find when it was discovered. A figure in parentheses indicates how many of the objects are preserved today in the museum collection. The number of finds states the number of large blade knives, preforms for large blade knives, and blade-like flakes in each find. Spjald also contains other finds.

Country	Find spot	Find number	Number of known objects in the find (still in museum collections)	Reference	Finds I have studied
Sweden	Ingelstorp Mossar, Ingelstorp Parish	Simrishamn Museum inv: 3388	6 (1)		Yes
	Wittsjö, Wittsjö Parish	LUHM 24568 & 24569	2		Yes
	Hönsinge, Grönby Parish	LUHM 19641a–c	3	Wyszomirski 1979	Yes
	Killaröd, Grevie Parish	LUHM 17114, 1–2	2	Wyszomirski 1979	Yes
	Löddeköpinge Församlings-hemmet, Löddeköpinge Parish	Provisional inventory number 80250 (LUHM)	6	Wyszomirski 1979	Yes
	Stora Hammar, Stora Hammar Parish	SHM 8651a	40	Wyszomirski 1979	Yes
	Gamla Köpstad, Träslöv Parish	GAM 20.388–405	50–60 (17)	Göteborgs museums årstryck 1921; Lindälv 1964, 1967; Wyszomirski 1979; Sjöberg 1998.	Yes
	Lillebörs, Hunnestad Parish	GAM 5.312–5.314	4 (3)	Göteborgs museums årstryck 1915:76; Lindälv 1964, 1967; Wyszomirski 1979	Yes
	Falkenberg, Klovstenliden	Privately owned, no find number	10 (0)	Göteborgs museums årstryck 1915	No
	Getinge, Getinge Parish		3 (0)	Montelius 1872	No
	Hakestad, Köinge Parish	SHM 15768	3		Yes
Norway	Huseby, Tjølling	Oslo 14675	23 (5)		Yes, two of five objects
Denmark	Hemmed, Djursland, Randers County	Inv. no. R3700	11	Becker 1990	Yes
	Spjald, Ringkøbing County	NMj 436/71	6	Becker 1990	No
	Gåsetofte, Raklev Parish	NMZ 282–3	2	Ebbesen 1981	Yes
	Freltofte, Odense	NMA 14119–14128	10	Blinkenberg 1898; Ebbesen 1981; Becker 1990	Yes
	Dalegård, Nyker, Bornholm	NM A.873	20 (11)	Ebbesen 1981; Nielsen 2005	Yes
	Lille Strandbygård, Nylars, Bornholm	NMA 20879–908	10	Ebbesen 1981; Nielsen 2005	Yes
	Valby, Holbæk County	NMB 14569 & 14570	2		Yes
	Bregninge, Bregninge Parish, Tåsinge	Svendborg County Museum 14.866–873	8	Ebbesen 1981; Jensen 2006	Yes, six of eight blade knives
	Knardrup Bro, Ganløse Parish	NMj 2370/78	10	Ebbesen 1981; Jensen 2006	Yes, nine of ten blade knives

Wittsjö, Wittsjö Parish, northern Scania

The find consists of two large blade knives, one of which I have studied. Details of the find are scant. It is said that the large blade knives were found together, otherwise there is no information. The accession number shows that it was found during the early part of the twentieth century or earlier. The blade knife I studied has a greyish-brown patina and gloss visible to the naked eye.

Hönsinge, Grönby Parish, southern Scania

The find consists of three large blade knives, figure 2.106. They were discovered during farming work in 1917. The flint has no visible patina. The large blade knives are made from different kinds of Scandinavian Senonian flint. They are technologically different. None of the blade knives has gloss visible to the naked eye. An attempt has been made to analyse the use-wear but they all have post-deposition patina covering the entire surface, which makes HPA impossible.

Killaröd, Grevie Parish, north-western Scania

The find is one of three in this study which have been analysed in detail. It consists of two large blade knives. In the archival data it is said that they were "found during levelling of a couple of burial mounds" (LUHM 17114a–b). There is no information about the relationship of the find to the graves or their dating, nor about where they were found. However, the accession number shows that it must have been in the first half of the twentieth century or earlier, figure 2.107.

Technological attributes

The blade knives are technologically different. They are both type I large blade knives with a well-prepared platform, but the length and the size of the platform and the negative removals differ on the two blade knives.

Figure 2.105. When I investigated the find from Businge on Møn which had previously been published as a hoard of large blade knives (Ebbesen 1981:13), I discovered that the find consists of four blades from the Stone Age and two Late Bronze Age large blade knives. The drawing shows the find. Numbers 1 and 3 are large blade knives. Numbers 2, 4, 5, and 6 are blades from the Stone Age. The photograph shows the clear difference in the morphology of the platform and the bulb of percussion on one of the large blade knives, to the left, and one of the blades from the Stone Age, to the right. Blade knife number 1 is 12.5 centimetres long. The drawing comes from Ebbesen (1981:13, fig. 4).

Figure 2.106. Three large blade knives of flint from Hönsinge, Grönby Parish in Scania. An attempt has been made to do use-wear analysis on these large blade knives, but without success. All of them have a post-deposition patina covering the entire surface. The blade knife to the left is 16.5 centimetres long. The blade knives have inventory number LUHM 19.641 a–c.

Figure 2.107. The two large blade knives of flint from Killaröd, Grevie Parish in north-western Scania. The blade knives have inventory numbers LUHM 17114:1 (left) and 17114:2 (right). The blade knife to the left is 15 centimetres long.

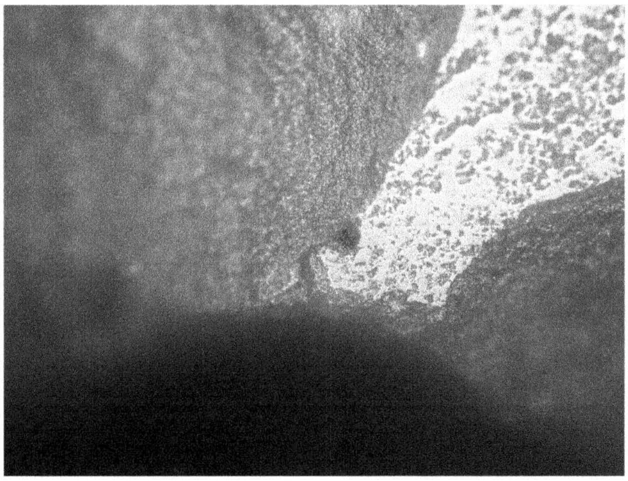

Figure 2.108. Use-wear analyses of one of the large blade knives from the assemblage from Grevie Parish. The polish is well-developed and linked. It has low reflectivity, flat topography, large extent, and a soft surface. I interpret this polish as coming from contact with cereals. On the blade knife there is a clear difference between the shafting polish and the polish from the contact material, which shows that the blade knife was shafted. Original magnification 200×. The tool has inventory number LUHM 17114:1.

Functional analyses

Use-wear analyses show that both blade knives were used. One of them has polish deriving from use on cereals, the other has generic polish, table 2.64 and figure 2.108.

Type of flint

The blade knives are made of different kinds of Scandinavian Senonian flint.

Refit

Neither of the blade knives can be refitted.

Löddeköpinge Församlingshemmet, Löddeköpinge Parish, western Scania

At Löddeköpinge in south-western Scania there have been several excavations since the end of the 1960s. They have mostly attracted attention for the remains from the Late Iron Age and Early Middle Ages (see e.g. Svanberg & Söderberg 2000). But in the archaeological material there are also finds from the Stone Age and Bronze Age.

In an excavation area called "Löddeköpinge Församlingshemmet" an assemblage of six large blade knives was discovered in 1973, figure 2.109. It was in the filling of a small pit and the blade knives were closely packed together (Cinthio, personal communication). No excavation report has been published, but in the archive of Lund University Historical Museum there is, as stated above, an unpublished manuscript by Tom Ohlsson. The find is mentioned briefly here: "In the bottom sand there was in addition a number of smaller features including a hoard of leaf-knives from the Late Bronze Age" (Ohlsson, unpublished manuscript, no date, page 42). No analyses of the find have been published. The find is one of the two uncovered by an archaeological excavation. It is the second of the three that I have analysed in detail.

Wyszomirski mentions the find, citing a personal communication from Tom Ohlsson and writes that it was discovered in a waterlogged field (1979:149). Wyszomirski's data have in turn been cited by Becker (1990:201), Karsten (2005:433), and Jensen (2006:107). The unpublished excavation report, however, says that the find was discovered near the remains of a Viking Age or early medieval sunken-floor hut within an area that has been interpreted as the northern part of the ridge on which the Viking Age and early medieval remains were located. The report also states that the northern part of this area

Table 2.64. Result of use-wear analyses of the two large blade knives in the assemblage from Grevie Parish. The terms in the right-hand column are the same as those used above in the chapter on functional analyses of blade knives. The two large blade knives are illustrated in figure 2.107.

	Status	Edge	Polish
LUHM 17114:1	Whole blade knife. No visible patina.	Post-deposition macrodamage. Gloss visible to the naked eye.	Well-developed polish, linked. Low reflectivity, flat topography, large extent, and soft surface. Interpreted contact material cereals. Clear difference between shafting polish and polish from the contact material, figure 2.108.
LUHM 17114:2	Whole blade knife. No visible patina.	Post-deposition macro-damage. No visible gloss.	Generic polish.

Figure 2.109. The six large blade knives of flint from Löddeköpinge Församlingshemmet. For the analyses in this study the blade knives have been numbered 1–6, from left to right. The find has a provisional inventory number 80250 (LUHM). Knife number 1 is 13.5 centimetres long.

slopes down towards a waterlogged depression. Yet the plans of the excavation stored in the archive of Lund University Historical Museum clearly indicate that the assemblage of large blade knives was found in the southern part of the excavation site. The find thus does not come from a waterlogged area but from dry ground beside a waterlogged area.

Technological attributes

The blade knives are technologically both similar and different. They are all type I large blade knives with a well-prepared platform, but their length and the size of the platform and negative removals are different for some of them, table 2.65.

Table 2.65. The length of the six large blade knives of flint from Löddeköpinge Församlingshemmet shows that the assemblage contains objects which are both similar and different.

Length	13.5	14.0	14.5	17.0
Number	2	2	1	1

Functional analyses

Use-wear analyses show that all of the blade knives were used, table 2.66. One of them has a clear difference between shafting polish and the polish from the contact material, showing that it was shafted.

As is evident from table 2.66, all the blade knives were used. Two have polish showing that they served to reap cereals. One has polish showing that it was used on cereals or reeds. Three have generic polish which does not allow interpretation of the contact material.

Type of flint

The six blade knives are similar in colour. They are made of Scandinavian Senonian flint.

Table 2.66. Result of use-wear analyses on the large blade knives in the assemblage from Löddeköpinge Församlingshemmet. The terms in the right-hand column are the same as those used above in the chapter on functional analyses of blade knives. The large blade knives are illustrated in figure 2.109.

	Status	Edge	Polish
Knife 1	Whole blade knife. No visible patina.	Whole edge. No visible gloss.	Weakly developed, not linked. Low reflectivity, domed topography, large extent, and soft surface. Since the polish is not well-developed I cannot determine whether the contact material was cereals or reeds.
Knife 2	Whole blade knife. No visible patina.	Whole edge. No visible gloss.	Generic polish, figure 2.110.
Knife 3	Whole blade knife. No visible patina.	Whole edge. Gloss visible to the naked eye. Rounding.	Well-developed polish, linked. Low reflectivity, flat topography, large extent, and soft surface. Contact material interpreted as cereals. Clear difference between shafting polish and polish from the contact material, figure 2.111.
Knife 4	Whole blade knife. No visible patina.	Whole edge. Gloss visible to the naked eye. Rounding.	Well-developed polish, linked. Low reflectivity, flat topography, large extent, and soft surface. Contact material interpreted as cereals. Clear difference between shafting polish and polish from the contact material.
Knife 5	Whole blade knife. No visible patina.	Whole edge. No visible gloss.	Generic polish.
Knife 6	Whole blade knife. No visible patina.	Whole edge. No visible gloss.	Generic polish.

Refit

The blade knives in the find cannot be refitted.

Stora Hammar, Stora Hammar Parish, south-western Scania

The find is the third in this study which has been analysed in detail. It was discovered at the end of the nineteenth century. The only information about the circumstances of the discovery is that the objects were "found lying close together during soil-turning and reclamation". However inadequate this statement may be, it is important in that it shows that the flint artefacts belong together in one find context. The objects were donated to the Museum of National Antiquities in Stockholm by an inspector named Fritz Flygare. They consist of 28 large blade knives, eight preforms for large blade knives, and four large blade-like flakes, figures 2.112 and 2.113. The blade knives are between 11 and 16.5 cm long and in total the 40 artefacts in the find weigh 4.2 kg. This makes it the largest of the assemblages in this study. The flint is slightly patinated. None of the objects have a transformed or otherwise badly affected surface.

The find was not professionally excavated. At the time when it was discovered it could happen that objects found simultaneously were separated and only selected parts of the find were handed in to a museum, see for example the details in the analysis below of the find from Gamla Köpstad, Träslöv Parish. However, there is nothing in the information about the find circumstances to suggest that the find from Stora Hammar has been split up. It may therefore be regarded as one find, and the forty artefacts now in the Museum of National Antiquities are the ones that were left on the site some time in the Late Bronze Age.

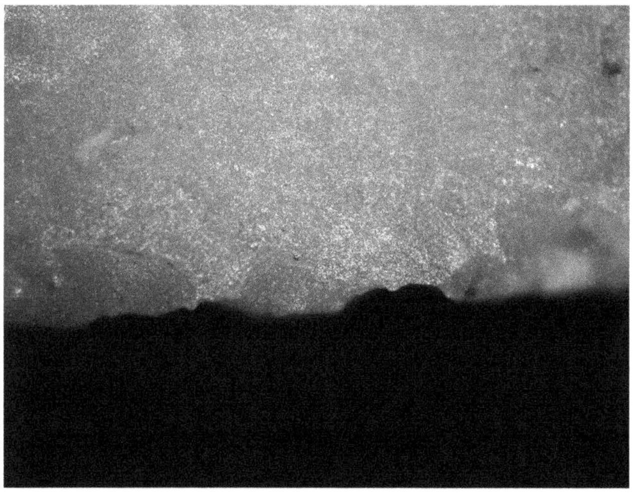

Figure 2.110. Generic polish on the edge of large blade knife number 2 from Löddeköpinge Församlingshemmet. The polish shows that the blade knife was used, but the polish is not developed enough to allow identification of the contact material the blade knife was used on. Original magnification 100×.

Figure 2.111. Polish on the edge of large blade knife number 3 from Löddeköpinge Församlingshemmet. The polish shows that the blade knife was used on cereals. Original magnification 200×.

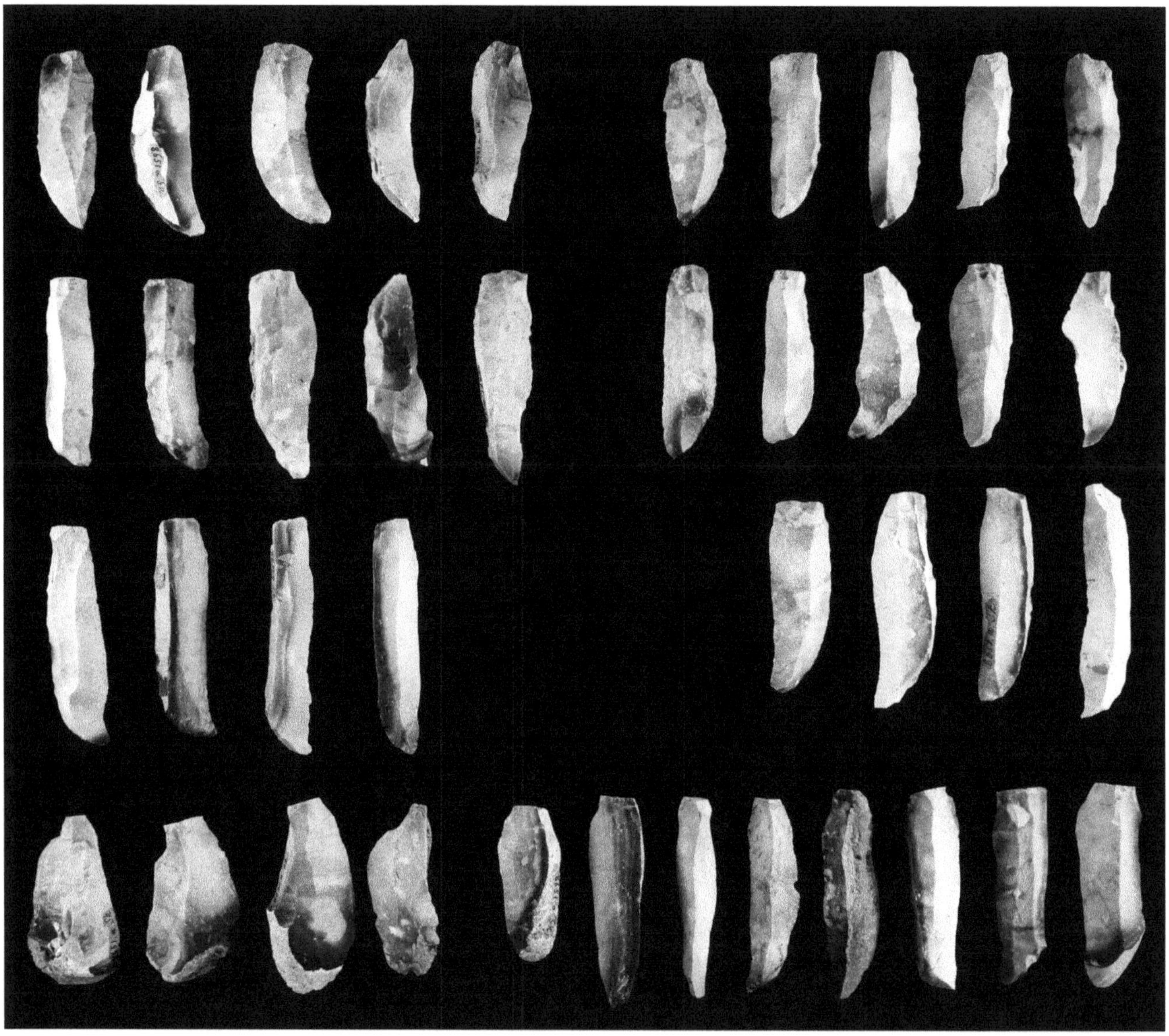

Figure 2.112. Twenty-eight large blade knives (the three upper rows), eight preforms for large blade knives (bottom right), and four large blade-like flakes (bottom left), all found together at the end of the nineteenth century at Stora Hammar in south-western Scania, Sweden. The inventory number of the find is SHM 8651a. Apart from this number, the majority of the artefacts in the find also have an individual number written in pencil beside the inventory number of each object. I use this individual numbering in the text when referring to particular objects. The blade knife at the top left is 13 centimetres long. Photo: Jenny Thornell, Malmö Museums.

Questions

Since the find from Stora Hammar is the only accessible and united find of so many large blade knives, preforms for large blade knives, and blade-like flakes that I know of, it deserves considerable attention. Apart from the general questions posed for the study of assemblages of blade knives, the find from Stora Hammar has also been analysed to answer some questions specific to this find:

- How were the flakes and the preforms made?
- Were they made at the same place with the same technique and method or were they made at different places and with different techniques and methods?
- Have they been used?

Technological attributes

The twenty-eight large blade knives have been analysed to ascertain the following technological attributes: the morphology and size of the platform, the appearance of the platform preparation, the platform angle, the morphology of the bulb of percussion, the number of negative removals on the dorsal side, the curvature of the edge, and the morphology and position of the retouch. The results are summed up in table 2.67.

The results of the technological analyses show that there is great similarity between the large blade knives as regards six of the seven analysed technological attributes. When it comes to the morphology of the platform preparation, there is variation. The analysis of the technological attributes shows that all the large blade knives were manufactured using a technology typical of type I large blade knives with a well-prepared platform.

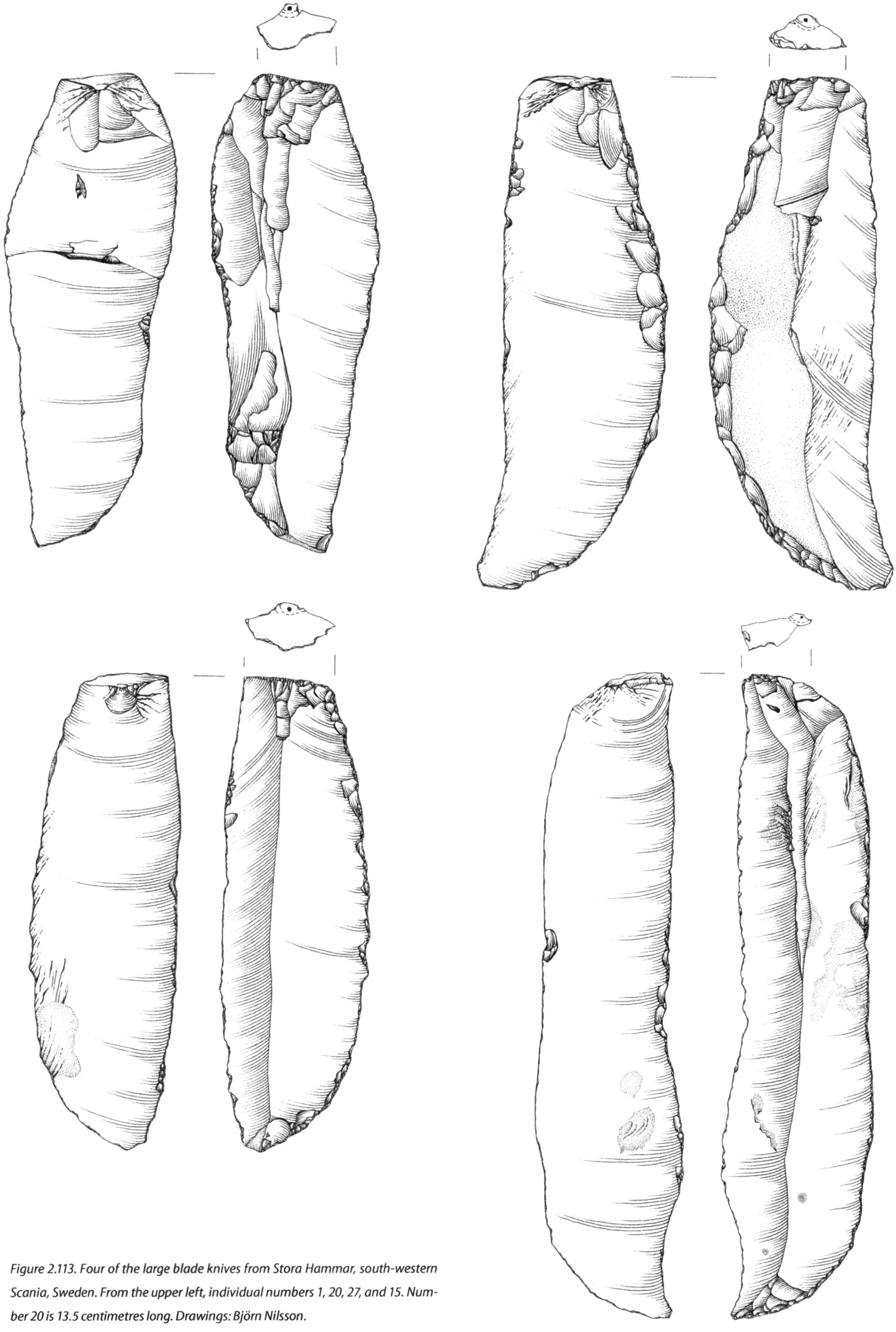

Figure 2.113. Four of the large blade knives from Stora Hammar, south-western Scania, Sweden. From the upper left, individual numbers 1, 20, 27, and 15. Number 20 is 13.5 centimetres long. Drawings: Björn Nilsson.

Table 2.67. Results of the analyses of technological attributes of the 28 large blade knives in the find from Stora Hammar.

The morphology and size of the platform	All 26 blade knives with preserved proximal part have a rhombic platform. One of these has cortex on the platform. Others have a flat platform. All but one have a ring crack around the point of impact on the platform. Figure 2.114 clearly shows that the size is similar.
The morphology of the platform preparation	The platform preparation differs in appearance between the different blade knives, with everything from totally lacking preparation to being intensively prepared, figure 2.115. Ten have preparation flakes with a hinged termination and twelve have preparation flakes with a feathered termination. There is thus no clear uniformity in the look of the platform preparation.
Platform angle	Of the 26 blade knives with preserved platform, 23 have an angle of 75 degrees and 3 have an angle of 80 degrees, figure 2.116. There is thus a great similarity in the material as regards the platform angle.
The appearance of the bulb of percussion	Of the 26 blade knives with a preserved proximal part, all have a well-developed bulb of percussion with a "moustache" and bulbar fissures. All but one have a marked bulb of percussion. Half of the blade knives have a bulbar scar.
The number of negative removals on the dorsal side	All blade knives but one have one, two, or three long negative removals on the dorsal side and one or two ridges that separate these negatives. None of them have a bidirectional dorsal flake scar.
The edge outline	Of the 28 blade knives, 19 have a virtually straight edge, 7 have an extremely slight curve, and 2 have a slightly curved edge, figure 2.117. There is thus great uniformity in the curvature of the edge.
The morphology and position of the retouch	All blade knives have a direct and continuous retouch. This means that they were all retouched with the ventral side as the platform and that the retouch was applied systematically and simultaneously along the retouched edge.

All blade-like flakes are of the kind that derive from the reduction process that goes with the making of large blade knives.

Two large blade knives differ technologically from the others. One has a hinge fracture plateau on the dorsal side and is retouched so that the proximal part has disappeared. One is broken off at the platform. This took place at the time of manufacture. The other 26 large blade knives are highly uniform as regards the technological attributes.

A typical large blade knife of flint in the find from Stora Hammar has a flat rhombic platform with a ring crack around the point of impact. The blade knife was knapped with a platform angle of 75 degrees and has a marked and well-developed bulb of percussion with moustache and fissures on the bulb at the platform edge. On the dorsal side it has one to three negative removals and one or two ridges. The edge is straight or virtually straight and the tool was shaped from a preform into a large blade knife with a systematic retouch applied with the ventral side as the platform. This description agrees with the experimental reference example for type I large blade knives with a well-prepared platform as defined in the chapter about my series of experiments.

All eight preforms for large blade knives have a flat rhombic platform with a ring crack around the point of impact. The platform size is similar. They have platform preparation which varies between those where it does not go down over the shoulders to those with preparation extending some way down on the dorsal side. One preform has both hinge and preparation flakes. Seven were knapped with a platform angle of 75 degrees and one with 80 degrees. All have a well-developed and marked bulb of percussion with moustache and fissures on the

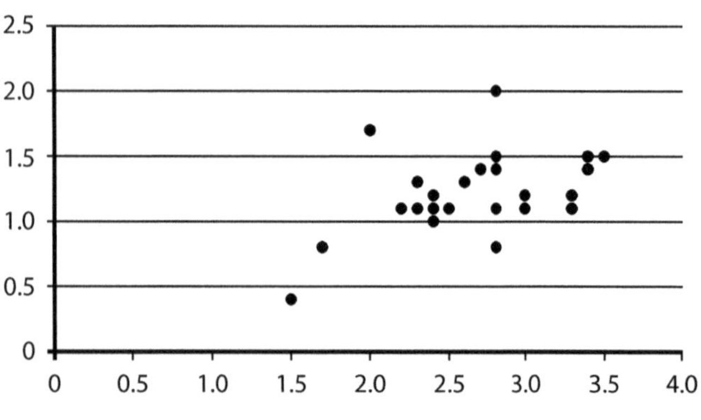

Figure 2.114. The relationship between the length and width of the platform of large blade knives. The chart clearly shows that the majority of the platforms are similar. The values on the axes are stated in centimetres.

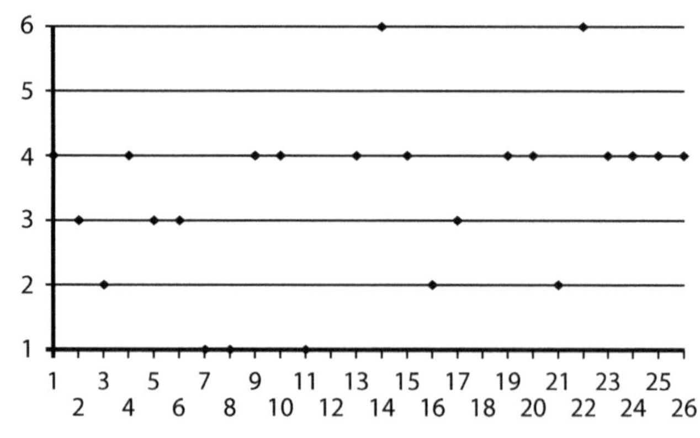

Figure 2.115. The distribution of differences in the morphology of platform preparation on large blade knives according to the attribute analyses above, see figure 2.27. The chart shows that there is variation in this attribute. The vertical axis shows the morphology of platform preparation according to figure 2.27, the horizontal axis states the number of each large blade knife. Note that two of the 28 large blade knives lack a platform and are therefore not included in the chart.

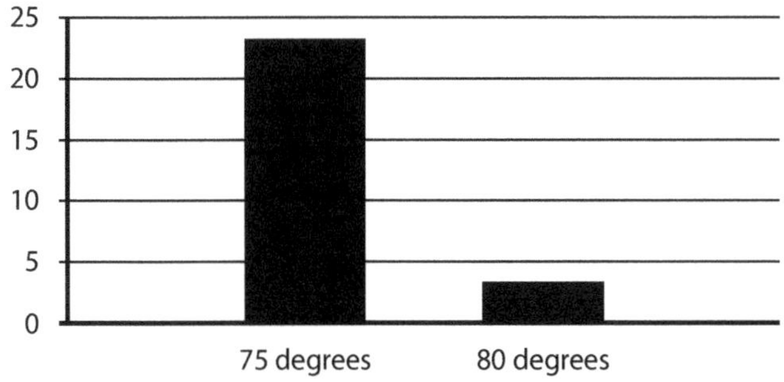

Figure 2.116. The platform angle of large blade knives stated in number and degrees. It is obvious that the overwhelming majority of the blade knives, 23 of 26, have a platform angle of 75 degrees. Note that two of the 28 large blade knives lack a platform and are therefore not included in the chart.

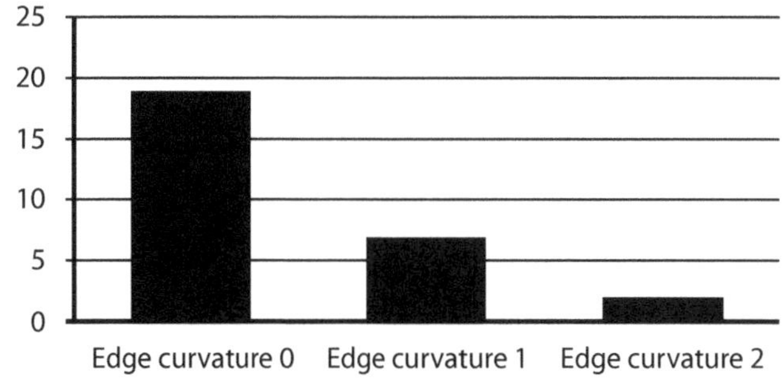

Figure 2.117. The edge curvature of the 28 large blade knives, stated in number per variable in accordance with the attribute analyses above, see figure 2.24.

bulb at the platform edge. Half have a bulbar scar. Of the eight preforms for large blade knives, seven have an almost perfectly straight edge. One has a slightly curved edge.

The four blade-like flakes have a flat rhombic platform with a ring crack around the point of impact. The platform size is similar. Two have platform preparation that does not go down over the shoulders, and two lack preparation. Two flakes were knapped with a platform angle of 80 degrees and two with 75 degrees. All have a well-developed and marked bulb of percussion with moustache and fissures on the bulb of percussion at the platform edge. Three have a bulbar scar.

My interpretation is that the large blade knives, the preforms for large blade knives, and the blade-like flakes in the find from Stora Hammar were made using the same technology and method. The few deviations from the general picture of the technological attributes are within the normal range of variation for the manufacture of large blade knives.

Functional analyses

All the artefacts in the find have been examined in HPA. The aim was to determine whether any of the cutting edges have polish, whether this polish can be interpreted as coming from the use of the artefacts as tools, and whether there is any polish that can come from the shafting of the large blade knives.

When several flint objects are packed together without protection or are handled in some other way that lets them rub and scrape against each other, the surfaces of the objects become worn. This wear causes polish that is visible by high-magnification microscope. Irene Levi Sala (1986, 1996), in some critical studies of the origin of different forms of polish on flint material, has performed a large series of controlled laboratory experiments in order to cause different kinds of polish. Several of these trials involve flint surfaces rubbing against each other. The studies clearly demonstrate the variation in the appearance of these kinds of polish. In another similar but more situationally determined study, Unrath et al. (1986) show how the sur-

Figure 2.118. Large blade knife number 15 from Stora Hammar, see figure 2.113. The photograph was taken on the ventral side, about one centimetre in from the edge. I interpret the polish as having arisen when the blade knife rubbed against other objects in the find. Original magnification 200×, EFI 7 images.

Figure 2.119. Large blade knife number 27 from Stora Hammar, see figure 2.113. The photograph was taken on the ventral side about 0.5 centimetres in from the edge. The polish is interpreted as deriving from antiquarian treatment. Original magnification 50x.

faces of flint objects are altered by abrasion against each other when packed together and carried around in a bag.

From the results of these studies we know that the polish that arises when flint is rubbed against flint is highly reflective at 200× magnification. The polish is visible as shiny points or continuous shiny clusters or worn-down surfaces. The polish is not regularly distributed along or beside edges but occurs at several different places on the flint (Levi Sala 1986:103ff and plate 25, 1996:125; Unrath et al. 1986, plate 51). This type of polish could be observed on all of the forty artefacts in the find from Stora Hammar, figure 2.118.

It is impossible for me to ascertain whether this "flint rubbing against flint" polish arose when the objects were handled or carried together before they were deposited in the Late Bronze Age, or if the polish derives from the way the find was handled after being excavated at the end of the nineteenth century.

Many of the blade knives also have another kind of polish with heavy striations. Several authors have discussed the origin of "post-excavation" polish caused in antiquarian contexts (see e.g. van Gijn 1989:54). When flint objects lie loose, for example, in drawers, sliding back and forth and colliding every time the drawer is opened and shut, a special type of longitudinal polish with striations develops (Levi Sala 1986, plate 25). The majority of the objects in the find from Stora Hammar have polish with striations which I interpret as coming from this kind of antiquarian treatment, figure 2.119.

The only artefact in the find showing polish that I interpret as coming from the use of the blade knife as a tool during the Late Bronze Age is blade knife number 20. Its edge has gloss that is visible to the naked eye, and at high magnification one can see a highly reflective polish that I interpret as typical of large blade knives, figure 2.120. On the blade knife one can see a distinct difference between the polish from the contact material and the polish from shafting. This proves that blade knife number 20 was shafted.

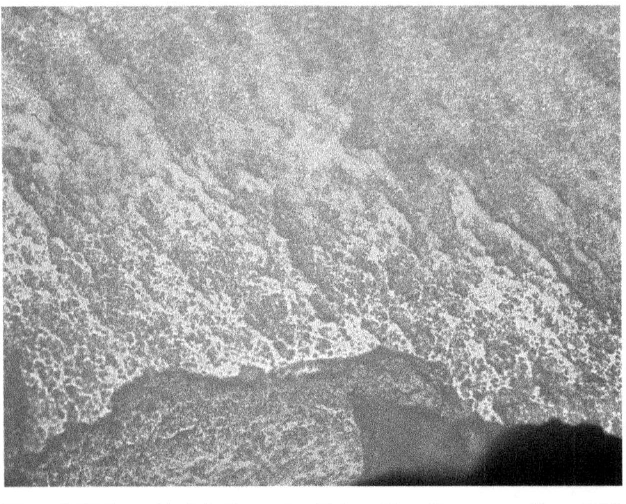

Figure 2.120. Large blade knife number 20 from Stora Hammar, see figure 2.113. The photograph was taken on the ventral side. The polish is well-developed and linked, with low reflectivity, flat topography, and large extent. It has a soft surface. The contact material is interpreted as cereals. This blade knife is the only one of the forty artefacts in the find displaying a polish that I interpret as coming from the use of the blade knife as a tool. Original magnification 200×.

On the other 39 artefacts in the find I have not found any polish that I interpret as coming from use as a tool. Nor have I found any traces of shafting on these large blade knives, preforms for blade knives, or blade-like flakes.

The result of the functional analyses thus show that one of the large blade knives, number 20, has a polish that I interpret as coming from the use of the implements to reap cereals. Other artefacts in the find have no polish that I interpret as coming from use as a tool. All of them have extensive damage caused by post-deposition treatment, in the form of microflaking along the edges, and as irregular polish at scattered points on edges and other surfaces. My interpretation is that one of the large blade knives in the find was used, number 20, while the rest were unused.

Type of flint

The raw material in the find has been analysed to establish the type of flint and the appearance of the cortex in order to determine how many flint nodules the find represents and where the manufacture may have taken place.

All the objects are made of Scandinavian Senonian flint of high quality. Since the blade knives are of type I with a well-prepared platform, it is likely that they were made from block flint. The cortex on all the blade knives, preforms for blade knives, and blade-like flakes apart from number 20 is polished shiny, wind-polished, or wave-washed. The cortex on number 20 is soft and chalky; it is not polished or wave-washed. This shows that the raw material for the manufacture of all the objects in the find except blade knife number 20 is not flint that was quarried or picked direct from a primary deposit. It is probable that the nodules were gathered from a place where large nodules of sufficiently good quality were accessible without needing to be quarried from the chalk. Such places are the beaches below flint-bearing chalk cliffs or below beach ridges containing a great deal of till, stone blocks, and flint nodules (Högberg & Olausson 2007).

The place closest to the find spot where there are large flint nodules of high quality on the beaches today is Stevns Klint in Zealand, Denmark. Another place where there is plenty of flint on the beach, although not of the same high quality as at Stevns Klint, is the beach below Glumslöv in western Scania. The nearest known manufacturing place for large blade knives is Södra Sallerup in south-western Scania (see next chapter). It is unlikely, however, that the objects in the find from Stora Hammar were made there. There are two reasons for this. The flint at Södra Sallerup was quarried from the chalk and the flint from there mainly consists of flat flint. Neither of these factors agrees with the find from Stora Hammar.

Although all the objects are made of Scandinavian Senonian flint, there is some variation. Based on the appearance of the flint, the colour and pattern, it is possible to determine to some extent which blade knives and flakes could have been made from the same nodule. An examination of this shows

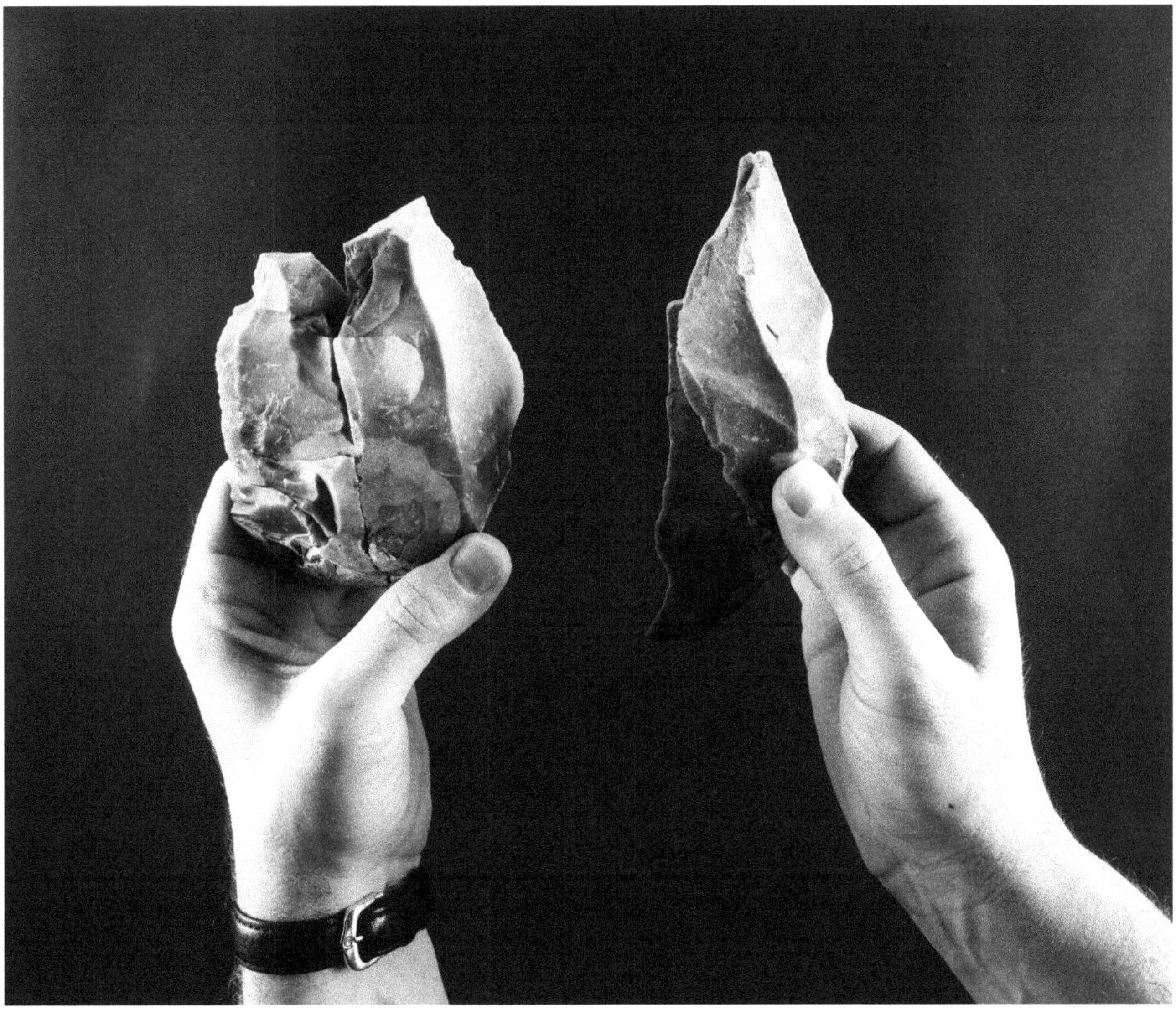

Figure 2.121. Refit of two blade-like flakes, left, and a large blade knife with a blade-like flake, right. Photo: Jenny Thornell, Malmö Museums.

that at least six different nodules were used. Blade knife number 20 is not counted in this.

The results of the analysis of flint types in the find demonstrate that all of the objects are made of Scandinavian Senonian flint. Of the forty artefacts, thirty-nine were made of flint nodules gathered on the beaches just below flint-bearing chalk cliffs or on beach ridges. The closest known place today is Stevns Klint in Zealand, Denmark. Large blade knife number 20 is made of flint from a primary deposit. The find represents at least six different flint nodules.

Refit

Since the find mainly consists of finished large blade knives, it is not likely that much of it can be refitted. As stated earlier, the reduction process for the manufacture of blade knives takes place in such a way that the preforms are not removed in a sequence. Between each preform for a blade knife a number of blade-like flakes are removed to prepare the front of the core. These flakes are as big as a preform for a blade knife. It is thus unlikely that one large blade knife can be refitted with another even if they come from the same core. But since there are both preforms and blade-like flakes in the material from Stora Hammar, there is some point in seeing whether any of the objects fit together. The aim of this is to link different parts of the find more explicitly to each other and to demonstrate whether the individual artefacts come from the same reduction process.

The result shows that two plus two of the objects can be refitted: two blade-like flakes fit together, and one blade-like flake fits a blade knife, figure 2.121. Otherwise I have not been able to refit anything.

Results – detailed analysis of the assemblage from Stora Hammar

The analysis proceeded from a number of questions about the content of the find and the morphology, manufacture, and use of the artefacts. The result of the analysis shows that thirty-nine of the large blade knives, preforms for blade knives, and blade-like flakes are very similar as regards morphology and the technology with which they were made. They were all

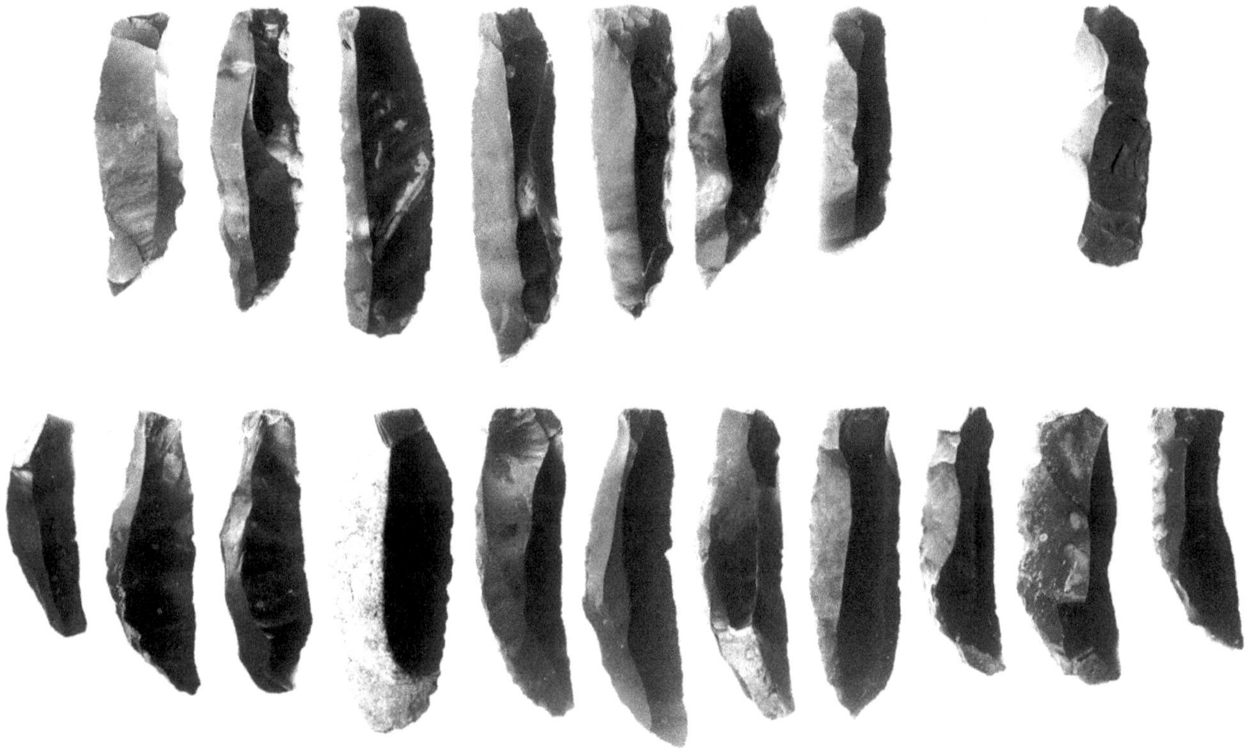

Figure 2.122. Part of the find from Gamla Köpstad, Träslöv Parish in Halland. The photograph was taken in 1920 when the objects were handed in to the museum. Originally the find is said to have consisted of more than fifty objects. The fourth object from the left in the bottom row is the blade-like flake. To the right of this is the large blade knife that is missing today. To the top right is the core that is said to belong to the find. The blade knife to the far left is 12 centimetres long. The photograph was originally published in Göteborgs museums årstryck (1921). Photo: Gothenburg City Museum, negative D. 1617A.

manufactured on a few occasions from the same kind of flint. They are unused.

Blade knife number 20 differs from other objects in two ways. Firstly, it is the only one of the blade knives that displays traces of having been used – for reaping cereals. Secondly, it differs in the appearance of the cortex. All the objects in the find which have some cortex on the dorsal side, except blade knife number 20, have a yellow or yellowish-brown cortex that is polished shiny, wind-polished, or wave-washed. The cortex on number 20 is white and is not polished. These two differences in blade knife number 20 lead me to interpret it as not having been made at the same time as the other objects in the find.

The analysis clearly shows that the majority of the objects in the assemblage from Stora Hammar belong together in a sense that is far deeper than the fact that they were found together. Thirty-nine of the forty objects have a shared biography from production to deposition. One of the large blade knives in the find has a completely different biography from the other objects in the find.

Gamla Köpstad, Träslöv Parish, Halland

The find consists of sixteen large blade knives and one blade-like flake. They were found in 1915, just under ground level beside a large stone block. The objects were packed closely together (Lindälv 1964:59). According to the person who discovered it, the find originally consisted of fifty or more artefacts (*Göteborgs museums årstryck* 1921:90f; Lindälv 1964:59; Sjöberg 1998:62ff). Twenty of these were handed in to Gothenburg City Museum, and today the sixteen blade knives are exhibited, the blade-like flakes are in storage, and three of objects have vanished, figure 2.122.

The large blade knives and the blade-like flake are very similar as regards technology and raw material. They are all made of Scandinavian Senonian flint. My interpretation is that several of the blade knives were made from the same block, figure 2.123. I have tried to fit them together but without success.

Figure 2.123. The sixteen large blade knives from Gamla Köpstad, Träslöv Parish in Halland under analysis. Notice the similarity between the blade knives as regards patina, technology, and flint type.

None of the sixteen blade knives or the blade-like flake have any visible gloss. They all have a heavy reddish-brown patina covering the whole surface. They show extensive post-deposition damage.

A further two artefacts have been ascribed to the find from Gamla Köpstad: a bifacial dagger and a core (Sjöberg 1998:62ff). In view of the dating, it is something of a contradiction that a dagger is found together with an assemblage of large blade knives, as other scholars have pointed out (Jensen 2006:111). This is because daggers are dated to the period from the Late Neolithic to the first phases of the Early Bronze Age (Lomborg 1973), whereas large blade knives are dated to the last part of the Late Bronze Age. This requires a closer investigation.

The dagger is mentioned in archival records at Gothenburg City Museum, both in an inventory journal from 1920 and in the accession catalogue (inventory numbers 20388–20407 & 23266). Here it is stated that the dagger had disappeared when the find was handed in to the museum. In *Göteborgs museums årstryck* from 1921 the assemblage was presented as part of the museum's growth, with the statement that "yet another dagger of flint was found on the site" (*Göteborgs museums årstryck* 1921:91). From this documentation we also learn that the find, after it was discovered in 1915, had been sold by the finder to a dealer, Karl Bengtsson of Träslöv. He had in turn given the find to Elof Lindälv, who turned it in to the museum. The find thus passed through several hands before it ended up in the museum collection. This is evident from Lindälv's own account of the find:

> Down in the subsoil ... was a collection of flint objects so tightly packed together that Albert Svensson [the finder] took it all to be a single stone. After being brought up it turned out that "the stone" consisted of some 50 large, knife-like flint blades. In my survey of ancient monuments in 1920 the flint assemblage had been transferred to the dealer Karl Bengtsson. He donated 19 of the objects to Gothenburg Archaeological Museum. Later yet another flint implement, a dagger, was handed over to the same museum (Lindälv 1964:59).

This brief account has contractions and unclear statements about the find. It changed owner several times, and details of the find were reported by a series of different people. A dagger is said to be missing in 1920. Later a dagger was handed in to the museum and is today ascribed to the find. But there is nothing to say that the latter dagger is the one that disappeared.

I have therefore also studied the dagger and the core that are said to have been found together with the blade knives. They are made of a different type of flint from the large blade knives and they both have a greyish-brown patina that is nothing like the patina on the large blade knives. This indicates that the finds do not come from the same deposit.

My interpretation is that the find when discovered consisted of fifty or more large blade knives and blade-like flakes, that a dagger was later handed in to the museum, and that the museum's collection now treats the find as consisting of large blade knives, a blade-like flake, a dagger, and a core. Yet the dagger was not found together with the blade knives, and presumably neither was the core.

Lillebörs, Hunnestad Parish, Halland

The find consists of three large blade knives (Lindälv 1964:61, 1967:88), figure 2.124. It was discovered in 1880 beside a cairn during digging in a field. The relation of the find to the cairn is not clear from the information, nor is the dating of the cairn. According to data in the Gothenburg City Museum accession catalogue, yet another blade knife was found at the time, but this had vanished by 1914 when the find was handed in to the museum (*Göteborgs museums årstryck* 1915:75ff). The three large blade knives are all made of Falster flint. They are thus similar as regards the type of flint, but the technology differs. They have gloss that is visible to the naked eye.

With reference to *Göteborgs museums årstryck* (1915) and Lindälv (1964), Wyszomirski (1979:149) writes that the objects were found in a bog. There is nothing in the sources to which Wyszomirski refers that confirms this.

Falkenberg, Klovstenliden, Halland

The find consists of ten large blade knives. I myself have not seen the find, which is not in any museum collection. The find is mentioned summarily in *Göteborgs museums årstryck* (1915:76). This says that it consists of ten blade knives and that it was found in 1912, and these details have later been repeated in other publications (Lindälv 1964:60, 1967:87; Wyszomirski 1979:149). In the newspaper *Falkenbergs-Posten* for Saturday 18 May 1912 one can find more information about the find. The ten blade knives are said to have been discovered under a stone slab during work in a quarry. They were "so tightly and

Figure 2.124. The three large blade knives from Lillebörs, Hunnestad Parish in Halland. The blade knife in the middle is 20 cm long. It is the longest of all the large blade knives I have seen. The photograph was originally published in Göteborgs museums årstryck (1915:76). Photo: Gothenburg City Museum, negative D. 1611.

well put together that to the finder they looked like a single, somewhat splintered, piece" (Carlstedt 1912). In Gothenburg City Museum's archive there are copies of correspondence from 1912 between the private collector Carl-Anders Carlstedt, who owned the find, and museum curator Georg F. L. Sarauw. The correspondence shows that the blade knives had a grey patina and were similar in appearance.

Getinge, Getinge Parish, Halland

In 1872 Oscar Montelius published a comprehensive survey of archaeological finds from Halland (Montelius 1872). In it he mentions a find of three large blade knives: "Also found in Getinge Parish, at one and the same place: 3 large, thick flint blades (knives) with knapped back. L. 122 to 110 mm.; W. 41 to 42" (Montelius 1872:159, find no. 201). The find is also mentioned by Wyszomirski (1979:149). I have not been able to locate these blade knives.

Hakestad, Köinge Parish, Halland

According to information about the growth of the collections of the Museum of National Antiquities in Stockholm during 1916, three large blade knives of flint were acquired that year. They had been found together during digging alongside a large erratic. The blade knives are technologically different. One of them is a type II large blade knife. They have patina. One shows gloss visible to the naked eye.

Huseby farm, Tjølling, Vestfold, southern Norway

The find was unearthed around 1900 in a bog at the farm of Huseby at Tjølling in Vestfold. The find originally consisted of 23 objects. Five of them are now in the collections of the Historical Museum in Oslo. These five objects are all preforms for large blade knives, figure 2.125. They have a brown patina and are made of the same type of Scandinavian Senonian flint. Technologically they are similar. There is no visible gloss on the edges. The find has not been described before.

Figure 2.125. The two preforms for large blade knives that I have analysed from the find at Huseby farm, Tjølling in Vestfold, Norway. A further three preforms for large blade knives from the same find are in the exhibition at the Historical Museum in Oslo, see figure 2.3. These three preforms were not accessible to me when I visited the museum. The preform to the left is 14 centimetres long. According to archival information, the find originally consisted of 23 objects.

Hemmed, Djursland, Randers County, eastern Jutland

This find has not been described before. Becker mentions it in passing in his study of the finds from Spjald (Becker 1990:201). It was found in the early 1890s and consists of one large blade knife and ten preforms for blade knives, figure 2.126. The preforms are very similar in technology and raw material. They are preforms for type I large blade knives without platform preparation. The large blade knife in the find is of type I with a well-prepared platform. It thus differs from the preforms. The objects in the find have no visible patina. No gloss visible to the naked eye can be observed on the edges.

Spjald, Ringkøbing County, northern Jutland

The find from Spjald is the second of the assemblages of large blade knives of flint in this study found by an archaeological excavation. It consists of six large blade knives. As stated above, these were discovered during excavations in the 1950s of an area with remains of houses from the Late Bronze Age. In a post-hole for a roof-bearing post in one house, the blade knives were discovered together with a flint scraper, the remains of two moulded splint boxes, a wooden vessel, pottery, and lumps of resin (Becker 1972, 1990). The blade knives were used to cut reeds (Jensen 1990:209ff). They differ in technology. At least three different types of flint were used to make them (Becker 1990:202).

Through stratigraphical observations Becker noted that the different objects in the find had been deposited in a pit that had been dug around a roof-bearing post in the house. The objects were thus not deposited before the post was raised or after it was removed; they were placed in the pit when the post was standing upright and the house was in use. Becker was also able to show that the objects were deposited in the pit on at least three, and presumably five, different occasions (Becker 1990:196f). This means that the find should not be regarded as a single find deposited at one and the same time, but as an accumulated find.

Gåsetofte, Raklev Parish, western Zealand

The find consists of two large blade knives. They were found together under a stone in the latter half of the nineteenth century (Ebbesen 1981:7f). They both have an identical grey patina and are made of a Senonian flint that cannot be specifically identified. They are technologically different. There is no gloss visible to the naked eye on the edges.

Freltofte, Odense, Funen

This was the first assemblage of large blade knives of flint to be published. The author was Blinkenberg (1898:152ff), and in the same article he presented the find of the shafted knife blade

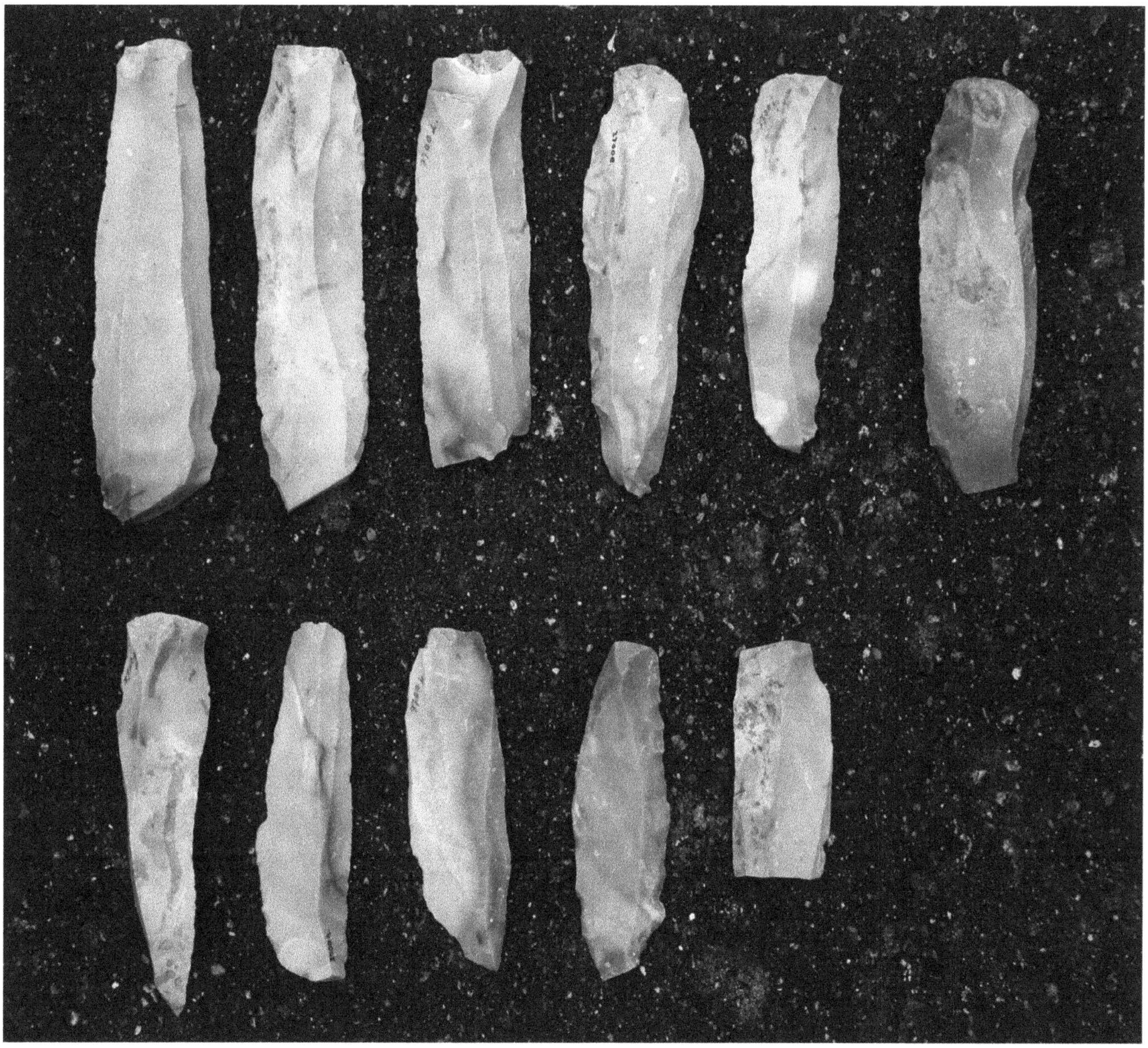

Figure 2.126. The assemblage of ten preforms for large blade knives and one large blade knife from Hemmed in Djursland. The blade knife can be seen to the far right in the top row. It is 13.5 centimetres long.

from Stenild (see also Ebbesen 1981:19; Becker 1990:201). The assemblage was discovered in the second half of the nineteenth century. Ebbesen writes that only one of the objects is a blade knife, the rest are unretouched flakes (Ebbesen 1981:20). This is incorrect. The find consists of ten large blade knives. All of them have gloss visible to the naked eye on the edge and extending in over the surface of the flint. The blade knives all have a heavy brown patina. Eight of them are made of Bryozoan flint, the Funen variety. One is made of Falster flint and one of a Senonian flint that cannot be precisely identified. This similarity between the blade knives was described by Blinkenberg: "the eight resemble each other so much with respect to the colour and structure of the flint that they must be assumed to be made from the same block" (Blinkenberg 1898:152).

The length of the blade knives and the appearance of the negative removals, however, show that, even though the technology and the flint type are very similar for the eight examples made from Bryozoan flint, Funen variety, there is also variation indicating that they were not all made from the same block, table 2.68.

The find is interesting as regards the type of flint. Bryozoan flint of the Funen variety is distinctive in that it can be linked today to just one place where it is found in a primary deposit. That place is Klintholm in south-east Funen (Thomsen 2000:34). The find from Freltofte shows that, even if we do not know of any production site at Klintholm, it is probable that there was one during the Late Bronze Age.

Table 2.68. The length of the eight large blade knives made of Bryozoan flint, Funen variety, in the find from Freltofte, Odense, Funen.

Length	12.0	12.5	13.0	13.5	14.0	14.5	15.0	15.5
Number	1	0	3	2	0	1	0	1

Figure 2.127. The find from Dalegård, Nyker, Bornholm. The blade knife at the top left is 12 centimetres long. The photograph was first published in Nielsen 2005:130. Photo: National Museum in Copenhagen.

Dalegård, Nyker, Bornholm

The assemblage was found around 1870 at Dalegård in Nyker. Originally it consisted of 20 objects. These were discovered close together a little under the ground surface beside a large stone (Nielsen 2005:130). Today the find consists of eight large blade knives and three blade-like flakes, figure 2.127. The blade knives are patinated in a similar way in a yellowish-brown colour. Technologically there are some differences. The large blade knives and the blade-like flakes are made of Scandinavian Senonian flint of three different kinds. No gloss is visible to the naked eye.

Lille Strandbygård, Nylars, Bornholm

The assemblage was found in 1902 in a meadow at Lille Strandbygård in Nylars. The objects were close together in the ground beside a boulder (Nielsen 2005:130). The find consists of seven large blade knives, a preform for a large blade knife, and two blade-like flakes, figure 2.128. The patina on the objects is an identical yellowish-brown. All ten artefacts are made of Scandinavian Senonian flint and several of them seem to have been knapped from the same piece. The blade knives and the flakes are technologically very similar. I have tried to refit them but without success. No gloss is visible to the naked eye.

Valby, Holbæk County, Zealand

The find consists of two large blade knives which, according to information in the Danish National Museum accession catalogue, were discovered together during farming work in the first half of the twentieth century. The find has not been described before. The large blade knives are patinated in the same way, but since each one is made of a different type of flint, they each have a different colour, figure 2.129. They are technologically different. The blade knives have gloss that is visible to the naked eye.

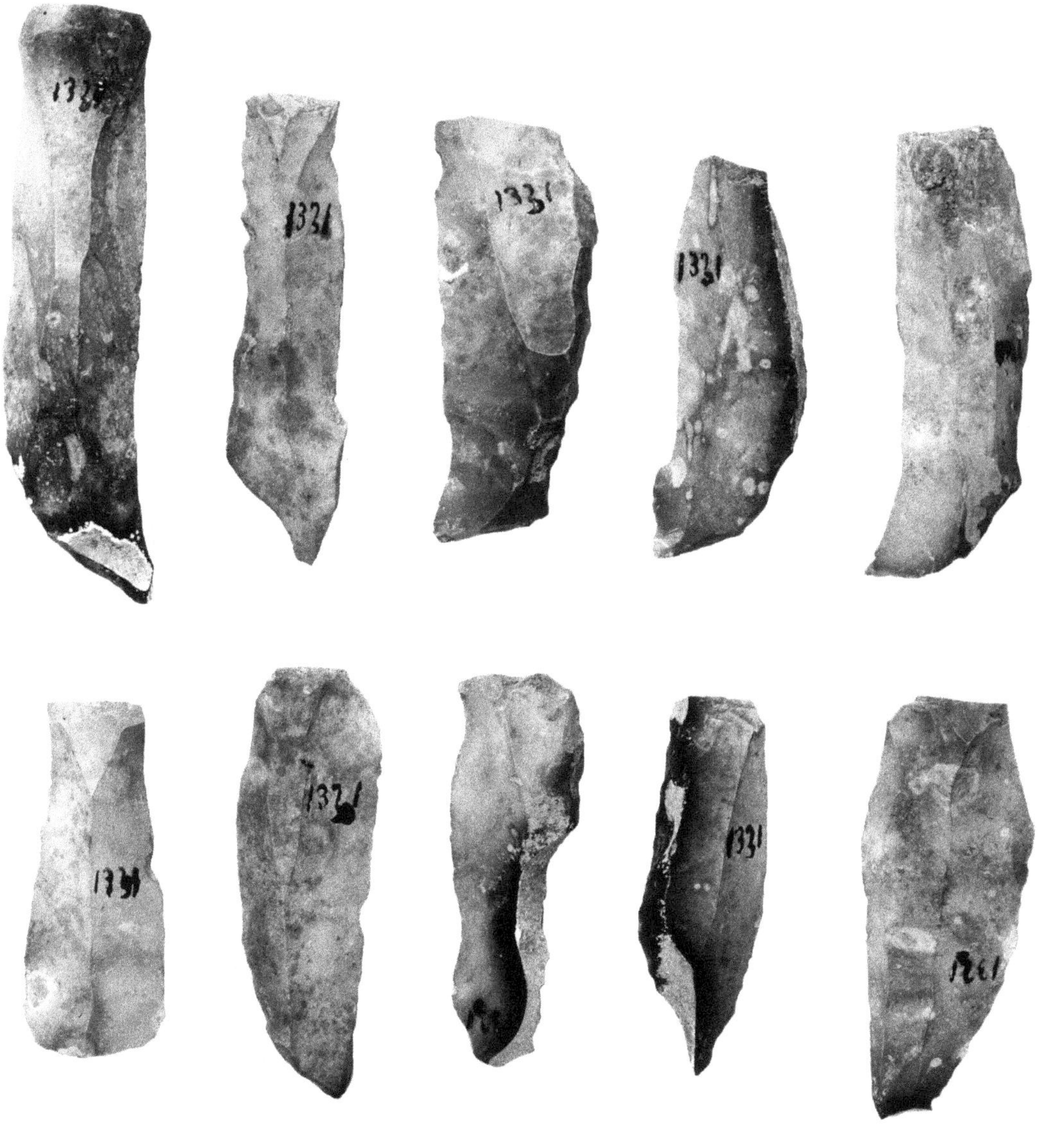

Figure 2.128. The find from Lille Strandbygård, Nylars, Bornholm. The blade knife at the top left is 15.5 centimetres long. The photograph was first published in Nielsen 2005:130. Photo: National Museum in Copenhagen.

Bregninge, Bregninge Parish, Tåsinge

The find consists of eight large blade knives (Ebbesen 1981:14; Jensen 2006:107). I have studied six of them. The blade knives were found in 1935 during construction work, about 70 cm under ground level. They are different in technology and raw material. Five are made of different kinds of Senonian flint which cannot be further specified. Three of them have cortex preserved which shows that the flint nodules used for the manufacture did not come from a primary deposit but were picked out of the till, either on the coast or in a river. One blade knife is made of Bryozoan flint, Funen variety. Three have gloss that is visible to the naked eye.

Knardrup Bro, Ganløse Parish, Zealand

The find consists of ten large blade knives. I have studied nine of them. They were found in 1952 during peat cutting. They lay close together in the peat (Ebbesen 1981:14). They are different in technology and raw material. Six are made of different

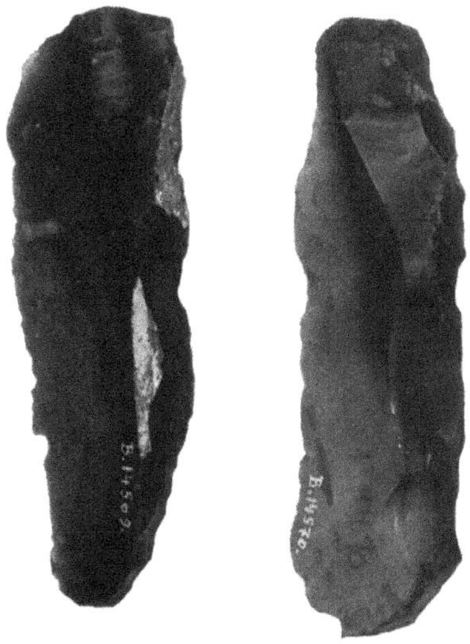

Figure 2.129. Two large blade knives from Valby, Holbæk County, Zealand, found together. The one to the left is 13 centimetres long.

Figure 2.130. The ten large blade knives from Knardrup Bro, Ganløse Parish, Zealand. I have not examined the blade knife to the top left but I have studied the others. The blade knife at the top right is 12.5 centimetres long. The photograph was originally published in Ebbesen (1981:12, fig. 3). Photo: National Museum in Copenhagen,

kinds of Senonian flint which cannot be further identified. Two are made of Brown Bryozoan flint and one of Grey Band Danian flint. Two of the blade knives have cortex preserved showing that the flint nodules used for the manufacture did not come from a primary deposit but were picked from the till on the coast or in a river, figure 2.130. Several of the blade knives have gloss that is visible to the naked eye.

Results – assemblages of large blade knives

The analysis of the assemblages of large blade knives of flint proceeded from a number of questions about the appearance, content, and character of the finds. The results show that there are similarities and dissimilarities both within and between the assemblages. These similarities and dissimilarities concern the find circumstances, find spots, composition, flint types, technology, and function.

Only two of the finds were excavated by archaeologists. Others have been discovered during different kinds of ground work in the late nineteenth and early twentieth centuries, table 2.69. This result is in line with studies of finds from the Neolithic (Karsten 1994:37). Finds both in wetland and on dry ground have been mentioned by several authors (e.g. Ebbesen 1981:16; Becker 1990:201), but the analysis shows that only two of the finds can be securely linked to wetland. Others have

Table 2.69. Periods when different assemblages were found.

Period	Number
1850–1900	9
1900–1950	9
1950–	3

Table 2.70. Character of the find spots of assemblages. The table is based on archival and published data.

Find spot	Number
Dry ground	8
Dry ground beside a stone	6
In a post-hole	1
Wetland	2
Unknown	4

been found on dry ground, often beside a large stone. One of them was found in the hole for a roof-bearing post in a house, table 2.70. Two finds come from beside graves, one cairn and one mound. The information about these is vague, however, as regards the relationship of the finds to the graves and the dating of the graves.

The two finds retrieved by archaeological excavations differ considerably from each other. Although both contain six large blade knives, one of them – from Löddeköpinge Församlingshemmet – consists of blade knives found close together and also interpreted as having been deposited together, while the other – from Spjald – consists of large blade knives which did not lie close together and which are interpreted as an accumulated find, deposited in the same context but at different times. This clearly reveals variation in the manner of deposition.

The find from Spjald is unique in that it contains not only large blade knives but also pottery and organic objects. Several authors state that large blade knives have been found stuck into a pit or beside a big stone (e.g. Nielsen 2005:130). These finds, however, should be regarded as having been deposited together in a specially dug pit, with the points downward and wrapped or tied or deposited in a bag. This shows that other large blade knives than those from Spjald were deposited along with some kind of organic material. It also shows that these blade knives were not deposited shafted; there is simply no room for a shaft in the pit. So even if use-wear analyses show that the large blade knives were parts of shafted tools, the study suggests that it was only knife blades that were deposited in the assemblages, not complete tools in the form of a knife blade in a shaft.

The quantity of objects in the assemblages varies, from just two large blade knives found together to examples containing over fifty objects. If one looks more closely at the composition of the objects in the finds, it is clear that some consist solely of large blade knives while others consists of blade knives, preforms for large blade knives, or blade-like flakes. This difference between the finds is worth a closer examination.

Table 2.71 shows that when the assemblages are categorized in this way, the number of objects in each find is related to the content of the find. Finds with only large blade knives contain at most ten objects. Finds where large blade knives are combined with preforms for blade knives or blade-like flakes contain ten or more objects.

But is there any real difference between the finds merely because one category includes preforms and blade-like flakes? From the Neolithic we have examples of assembled finds with

Table 2.71. The assemblages categorized according to whether the finds contain only large blade knives or these are combined with preforms for blade knives and blade-like flakes. The find from Spjald has been placed in the category of finds with only blade knives. This is because the find does not contain any preforms or blade-like flakes. The Spjald find, however, contains other objects than large blade knives.

Finds with only large blade knives		Finds with large blade knives and preforms for blade knives or blade-like flakes	
Find spot	Number of objects	Find spot	Number of objects
Ingelstorp Mossar	6	Stora Hammar	40
Wittsjö	2	Gamla Köpstad	> 50
Grönby	3	Huseby, Tjølling	23
Grevie	2	Hemmed, Djursland	11
Löddeköpinge	6	Dalegård	20
Lillebörs	4	Lille Strandbygård	10
Falkenberg	10		
Getinge	3		
Hakestad	3		
Spjald	6		
Gåsetofte	2		
Freltofte	10		
Valby Holbæk	2		
Bregninge, Tåsinge	8		
Knardrup	10		

Table 2.72. Similarities and differences between the different categories of assemblages which show clear patterns as regards the difference between assemblages of only large blade knives and assemblages where large blade knives are combined with preforms for blade knives or blade-like flakes. I have not seen the finds from Ingelstorp Mossar, Falkenberg, and Getinge and have not been able to obtain information from published references. These three finds are therefore omitted from the table.

	Place	Raw material		Technology		Function	
		Same	Different	Same	Different	Used	Unused
Finds with only large blade knives	Wittsjö		X		X	X	
	Grönby		X		X		X
	Grevie		X		X	X	
	Löddeköpinge	X			X	X	
	Lilleböre	X			X	X	
	Hakestad	X			X	X	
	Spjald		X		X	X	
	Gåsetofte		X		X		X
	Freltofte	X			X	X	
	Valby Holbæk		X		X	X	
	Bregninge, Tåsinge		X		X	X	
	Knardrup		X		X	X	
Finds where large blades knives are combined with preforms for blade knives or blade-like flakes	Stora Hammar	X		X			X
	Gamla Köpstad	X		X			X
	Huseby, Tjølling	X		X			X
	Hemmed, Djursland	X		X			X
	Dalegård		X		X		X
	Lille Strandbygård	X		X			X

combinations of preforms and finished tools. These preforms have not been interpreted as such but as replacements for finished tools (Björhem & Säfvestad 1989:108). Should the combined finds in this study be perceived in this way? If so, is there any difference between my two categories of assemblages other than the number of objects in the finds?

A closer study of the finds reveals that the differences between assemblages with only large blade knives and those where large blade knives are combined with preforms for blade knives or blade-like flakes are greater than just the number of objects in each category, table 2.72.

The assemblages with only large blade knives consist chiefly of examples that show no overall similarity. They differ in raw material and technology. The large blade knives often have gloss that is visible to the naked eye, and use-wear analyses show that they were shafted and used. These finds consists of large blade knives that were in circulation before they were deposited. The individual blade knives in each assemblage have separate biographies. They were made at different places, they were distributed in different contexts, and they were used. They represent separate life stories that did not meet until they were assembled and deposited together.

The assemblages which contain a combination of large blade knives, preforms for blade knives, and blade-like flakes consist chiefly of objects which are related in more ways than merely having been deposited together. The raw material and technology are the same. The blade knives have no gloss visible to the naked eye. Use-wear analyses show that the majority of the objects lack polish. They have not been used. They were never shafted. They were made at the same places, were assembled at those places, and from there they were distributed together. Several of the finds contain objects manufactured at the same time. The majority of the objects in these finds have shared biographies. They went direct from manufacture to deposition.

The result of the analysis shows that the difference between assemblages with only large blade knives and those with a combination of large blade knives, preforms for blade knives, and blade-like flakes is not just a question of the number or type of objects. It is also about completely different artefact biographies. To use the terminology from the introduction to this chapter, the finds with only large blade knives chiefly represent completed activities while the combined assemblages chiefly represent tools for future use (Jensen 2006:109), figure 2.131.

The finds are also different if data on find circumstances are viewed in relation to the biographies of the finds. This gives a

Table 2.73. Three different types of deposits defined in terms of context in time and space before and at deposition.

Biographies		Spjald	Löddeköpinge	Stora Hammar
Before deposition	**Time**	Different	Different	Same
	Space	Different	Different	Same
At deposition	**Time**	Different	Same	Same
	Space	Same	Same	Same

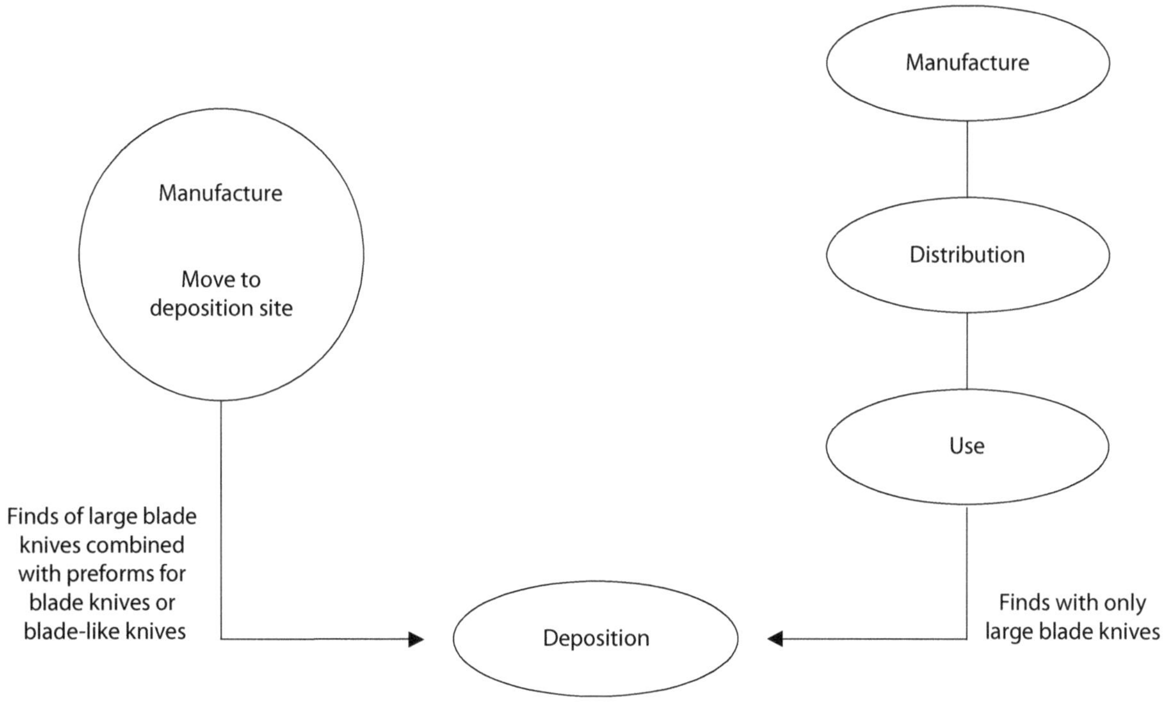

Figure 2.131. An illustration of the different biographies of objects in the assemblages. Finds with only large blade knives are dominated by examples which were manufactured, distributed, and used in different contexts, to the right in the figure, whereas finds with a combination of large blade knives, preforms for blade knives, and blade-like flakes are dominated by objects manufactured and moved to the deposition site of the hoard together, to the left in the figure. The biography illustrated to the left goes direct from manufacture to deposition, whereas the biography illustrated to the right goes through manufacture, distribution, and use before deposition.

result showing three different types of deposits. In table 2.73 this is schematically exemplified with the aid of the finds from Spjald, Löddeköpinge, and Stora Hammar.

The large blade knives in the finds from Spjald and Löddeköpinge had different biographies before they were deposited, that is to say, they were wholly or partly manufactured, distributed, and used in different times and at different places. At deposition they were brought together but in different ways. The place of deposition is the same but the objects in the find from Spjald were deposited at different times while those in the Löddeköpinge find were deposited at the same time. The objects in the find from Stora Hammar mostly have the same biographies as regards time and place, both before and at deposition.

Earlier studies of assemblages of large blade knives of flint have regarded these finds as a uniform group. They have been viewed as hoards of used blade knives which do not represent a high value, or as products not manufactured in series (see e.g. Ebbesen 1981). The result of this analysis demonstrates that this way of looking at assemblages of large blade knives should be revised. The similarities between the finds are confined to the simple observation that all the assemblages contain large blade knives of flint and that the objects were found together. Apart from this there are few similarities. As a group the assemblages of blade knives are varied and complex collections of different objects, forms of deposition, and artefact biographies.

A recurrent theme in earlier studies is the view of assemblages as a tradition from the Neolithic and Early Bronze Age. The word tradition in these studies has not been problematized; the link to older eras is based on the fact that many objects were deposited in wetland or dry ground. To be able at all to take up a stance on the word tradition in this sense, we must first look at what is said to be the origin of the tradition, the votive finds of the Stone Age. This is not entirely easy, since these finds are in no way uniform, besides which they display variation over time (Karsten 1994; Koch 1998). I am therefore forced here to use a generalization of data on Stone Age votive finds in a simple comparison of find spots.

A rough compilation of facts from Per Karsten's studies shows that more than half of the Neolithic finds come from wetland settings and around a third of them from dry ground (Karsten 1994:49–92). As for finds at large stones, Karsten says that two to fifteen per cent of the finds come from such sites (Karsten 1994:141, fig 79).

Table 2.74 shows that assemblages of large blade knives are mostly found on dry ground. Only twelve per cent come from

Table 2.74. Percentage of assemblages of large blade knives found in wetland, dry ground, and at big stones. The table includes 17 of the 21 assemblages. Four finds from unknown locations are omitted. The data in the table come from table 2.70. The percentages are calculated so that the 17 finds together make up 100 per cent, which corresponds to all finds from wetland and dry ground. The percentage for finds beside stones is calculated as the 6 finds from these contexts divided by all 17 finds from known locations.

Find spot	Wetland	Dry ground	At stones
Percentage	12%	53%	35%

wetland. Thirty-five per cent of the finds were discovered beside large stones. A comparison of data on find spots for Neolithic votive finds and assemblages of large blade knives shows that there are large differences between them.

This result agrees well with other studies of Bronze Age hoards. In a study mainly concerning bronze objects in Scania, Karsten has shown that the tradition from the Stone Age, as regards content, intensity of deposition, and votive setting, was broken some time at the transition between the Early and the Late Bronze Age, one factor being that depositions on dry ground increased during the Late Bronze Age compared to earlier periods (Karsten 2005:436, table 8). In a study of the extensive archaeological material from a votive fen at Hindbygården in Malmö in south-western Scania, similar results have been presented (Berggren 2007:301).

It is interesting to note in this context that previous studies have interpreted the decline in deposits during the Late Bronze Age on the basis of assemblages of bronzes (Larsson 1986:169ff; Jensen 2002:518). For example, it has been emphasized that the number of tools in the assemblages fell drastically in the Late Bronze Age (Larsson 1986:169). The assemblages of large blade knives show that it should not be perceived in this way. Depositions of tools did not decline. Instead they changed, from being solely tools of bronze to tools of bronze and flint. There has also been a tendency to interpret the eastern parts of Sweden as being richer in deposits from this time than the rest of southern Scandinavia (Jensen 2002:519). This has been expressed as "a change in the centre of gravity from Scania to the western, eastern and northern regions" (Larsson 1986:169). The eleven assemblages of blade knives from south-western Sweden change this picture.

The result of the analysis shows that assemblages of large blade knives of flint cannot be unquestioningly linked to a millennia-old Neolithic tradition of deposits in the ground, as previous studies have suggested, but should instead be viewed rather in relation to deposits of other objects during the Late Bronze Age.

Special manufacturing sites for large blade knives of flint

As previously discussed, there have been few studies dealing with production sites for large blade knives of flint. Comprehensive investigations of settlement site material have not been able to demonstrate the manufacture of large blade knives (Högberg 2001a; Knarrström 2000, 2001). This has also been pointed out by Björhem and Säfvestad (1993). Despite analyses of a huge amount of finds, they could not detect any traces of production sites. Based on this result, they formulated the hypothesis: "A distinct possibility is that the knives were produced centrally by special leaf-knife knappers at special workshop sites" (Björhem & Säfvestad 1993:77). The question is whether these places exist and if so where one can find them.

To answer these questions I have surveyed several sites and the archaeological material from them. The result of that work is presented below. First I review previously published studies. Then I discuss the premises for the analysis and the work of locating manufacturing sites. This is followed by a presentation of my analyses. The result shows that six places can be held up as special manufacturing sites for large blade knives. These are Wierzbica "Zele" in Poland, Tommestrup in Denmark, Södra Sallerup and Hanaskog, both in Sweden, and Reidervitz and Ruschvitz in Germany, see figure 2.132. A further four places are mentioned where no manufacturing site for blade knives has yet been found, but where it is presumed that one or more existed during the Late Bronze Age. The chapter ends with a discussion summing up the findings. It should be added here that this chapter concerns special manufacturing sites for blade knives. Later on I analyse places where large blade knives of flint were made in other contexts.

Previously known manufacturing sites

Few places are known as special manufacturing sites for large blade knives. The only one that has been excavated and published is Wierzbica "Zele" in the south-east part of central Poland. Flint mines from the Late Bronze Age have been excavated here, and remains have been found of large-scale production of blade knives resembling Scandinavian type I large blade knives, known as "backed blades of Zele type" (Lech & Lech 1984:195). The site is the only known example of an excavated site for raw material extraction and the manufacture of large blade knife-like tools from the Late Bronze Age. Interesting questions are: What does this material look like? What parts of the production are represented?

Within a limited area around Tommestrup at Stevns Klint in eastern Denmark, as we saw above, the private collector Søren Gregersen, during several years of surveys in the 1950s, found large amounts of knapped flint. He published some reflections on the finds in 1960 in a brief note in the journal *Skalk*, stating there that the material consists of some 3,500 tools and more than 15,000 flakes. He defined roughly 1,700 of the tools as large blade knives (Gregersen 1960:28f). The archival material at the Sydsjælland Museum in Vordingborg shows that Gregersen had reported the find to the National Museum in Copenhagen in 1958, and we learn from correspondence from January 1960 between him and Professor C. J. Becker at the National Museum that the find at the time was known to archaeologists (Sydsjælland Museum, inv. no. SMV 7647). Several studies have also referred to the find as a hypothetical production site (Thrane 1980:34f; Björhem & Säfvestad 1993:77). But although the place has been known for 50 years and several references have been made to Gregersen's note from 1960, the find material has hitherto not been subjected to any archaeological analysis. This leads to a number of questions: Was Tommestrup a production site for large blade knives of flint? Are the more than 1,700 large blade knives that Gregersen found really all blade knives? Or were there also remains of production from the manufacture of large blade knives among these finds?

The well-known flint mining area at Södra Sallerup (the archaeological site is also called Ängdala) in south-west Sweden (Olausson, Rudebeck & Säfvestad 1980; Rudebeck 1986,

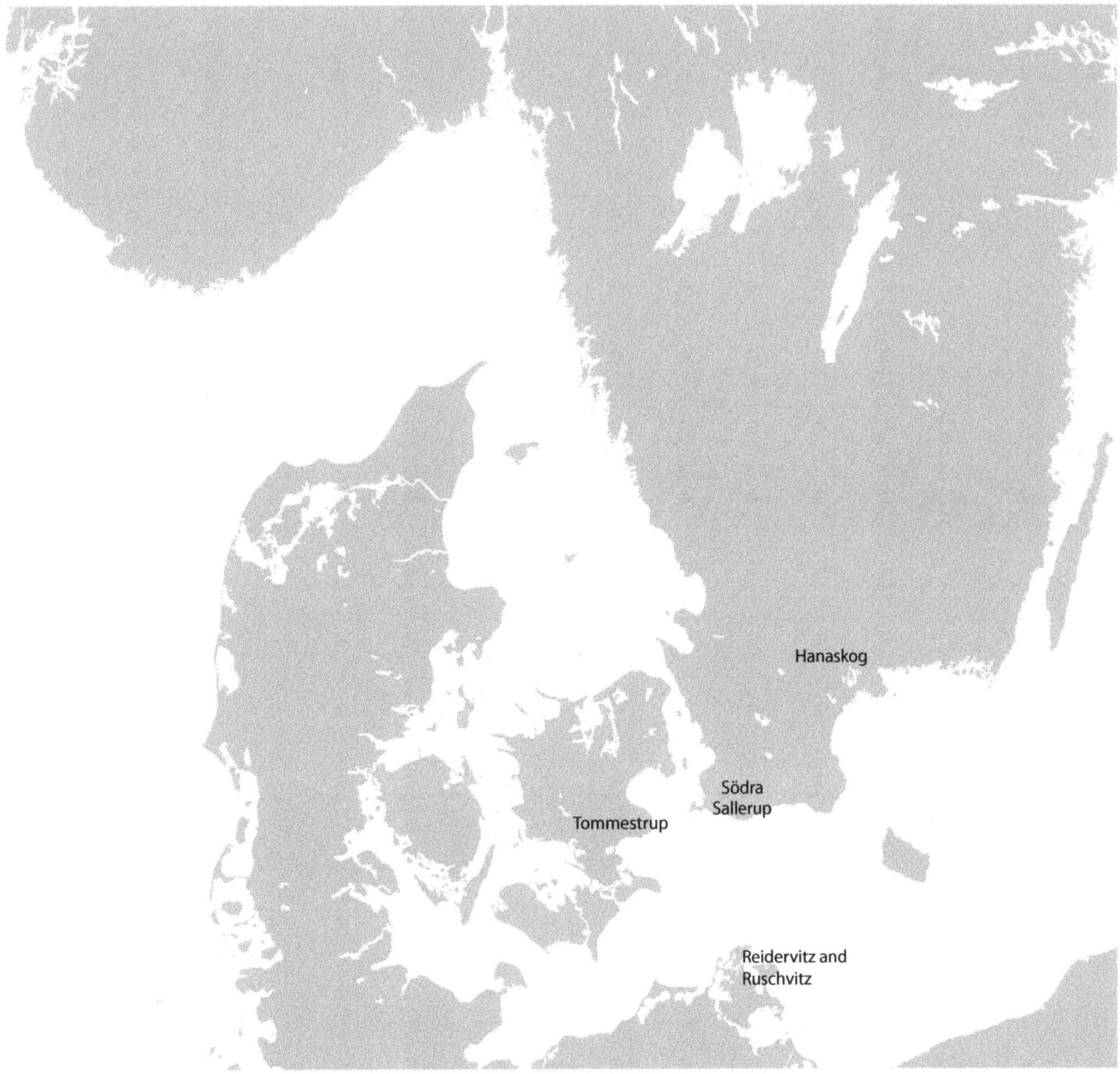

Figure 2.132. The location of five of the six special manufacturing sites for large blade knives at the major sources of raw material defined on the basis of the analyses in this chapter. The site Wierzbica "Zele" in Poland is not included (for location see Lech & Lech 1984).

1987, 1994, 1998; Högberg et al. 2001) has previously been investigated by Björhem & Säfvestad (1993) as a potential production place for blade knives. The investigation was based on the assumption that the existence of cores would indicate a production site:

> It must be stressed that in Ängdala evidence for the production of large blade knives is missing, despite the fact that the site is located within the Neolithic flint mining area and less than 30 metres from the nearest mining shaft (Björhem & Säfvestad 1993:77).

However, at excavations in the area in 1998 a few blade-like flakes which look like those from the experimental production presented above were discovered (Nilsson & Onsten-Molander 2004). These flakes, together with the availability on the site of high-quality Scandinavian Senonian flint in chalk layers, and together with the abundant evidence of archaeological remains dated to the Late Bronze Age, make these finds interesting as probable evidence of a production site. However, this has to be investigated in more detail, and the question is whether these blade-like flakes represent the production of large blade knives in Södra Sallerup.

Tommestrup, Wierzbica, and Södra Sallerup are all located at rich natural sources of high-quality flint. In southern Scandinavia there are several such places where flint is available (Högberg & Olausson 2007), and prehistoric flint mines have also been investigated at some of these (Becker 1959; 1980a, 1980b, 1980c). We may ask whether there are other places than the three mentioned above which should also be examined for traces of the manufacture of large blade knives.

What should I look for in order to find manufacturing sites?

When the Wierzbica "Zele" flint mine area was defined as a production place for "backed blades of Zele type" it was done from a combination of observations: mining activities dating to the Bronze Age, the presence of workshops, and finds of blades. But what can one do if the archaeological material is not as complex? What should I look for?

First of all, one observation may be made. Manufacturing sites for lithic tools are not usually discovered through finds of the tools that were made there. What is produced at a manufacturing site was usually taken away from there (Inizan, Roche & Tixier 1992; Andrefsky 2001). For this reason the specialized production of flaked stone tools must be addressed using manufacturing debitage and the partially finished or broken items discarded during manufacture (see e.g. Hansen & Madsen 1983; Madsen 1992; Högberg 1999; Whittaker & Kaldahl 2001; Pastrana & Hirth 2003). It is thus not primarily large blade knives that I should look for to find manufacturing sites, but the waste material that accumulated during the manufacture of blade knives.

When Björhem and Säfvestad (1993) searched for production sites for blade knives, as we saw above, it was based on a hypothesis that the occurrence of blade cores in the archaeological record from sites dated to the Late Bronze Age could be an indication that blade knives were produced there. The experimental studies presented above, however, show that if the ambition is to locate manufacturing sites for large blade knives, then blade cores are not the most suitable artefact to look for. This is because the blade cores were presumably reduced so much that they cannot be recognized as cores from the production of blade knives, if by this it is meant that the cores must have negative removals with morphological characteristics corresponding to those on a large blade knife.

When searching for special manufacturing sites for large blade knives of flint I have therefore proceeded from the results presented above from the experimental studies of manufacturing blade knives, by focusing on other remains of production than blade cores. In the experimental studies these remains are defined as diagnostic flakes, that is, blade-like flakes, preforms for large blade knives, preparation flakes produced in the manufacture of blade knives, and preforms for blade knives which broke when the retouching was unsuccessful.

Figure 2.133. Stevns Klint in Zealand, Denmark. At the top you can see trees and bushes growing in topsoil and layers of till, and under that limestone with layers of Grey Band Danian Flint, and at the bottom the chalk with bands of nodules of Scandinavian Senonian flint. The beach is full of flint which has eroded from the cliff, along with the till.

Figure 2.134. Four of the more than seventy boxes of finds from Gregersen's surveys at Tommestrup. The finds are kept today at Sydsjælland Museum in Vordingborg, inventory number SMV 7647. In the archives of the Danish National Museum the site is numbered SB39 Store Heddinge Landsogn Parish, Stevns Herred, Præstø County.

Wierzbica "Zele", in the south-east part of central Poland

The Wierzbica "Zele" flint mining area is located in central Poland, on the north-east fringe of the Holy Cross Mountains. In the chalk layers close to ground level, "Chocolate flint" was mined in open shafts up to more than seven metres deep. Chocolate flint is a high-quality black, grey, or chocolate-brown Upper Jurassic flint (Lech & Lech 1984, 1997).

In the 1980s a large mining area with workshops was excavated here. The area around Wierzbica "Zele" has been used in different periods, from the end of the Palaeolithic onwards. Several of the flint mines from Wierzbica "Zele" are dated to the Bronze Age and Early Iron Age. The results of several radiocarbon analyses show that much of the mining and the activities at the workshops can be dated to the Middle and Late Bronze Age. A few datings to the Early Iron Age indicate that the area was used into Hallstatt D and the Early La Tène period (Lech & Lech 1984, 1997; Lech & Leligdowicz 2003). The site has been well published and there is no reason to repeat what has already been written. I have chosen to look at the find material from Wierzbica "Zele" with the aim of determining what it consists of and what was made on the site.

My analyses of the debitage shows that the remains from the manufacture of large blade knives are extensive. It consists of large amounts of different types of flakes, cores, preforms for large blade knives, retouched fragments of preforms for blade knives, and large blade knives of "Zele type" (Lech & Lech 1984:195; Lech & Lech 1997:109f). The entire chaîne opératoire is represented – from the initial primary knapping of nodules just quarried from the mine, to the concluding retouching of preforms for blade knives. No other archaeological material than that from the mining and workshops has been found on the site (Lech & Lech 1984:198). This shows that the area was used solely for flint extraction and tool production.

Tommestrup at Stevns, Zealand in Denmark

At Stevns in eastern Zealand in Denmark, cliffs of limestone and chalk extend for several kilometres along the coast. In the chalk there are bands of nodules of high-quality Scandinavian Senonian flint, and in the limestone there are continuous layers of Grey Band Danian Flint (Högberg & Olausson 2007). Below the cliff the beaches are full of flint eroded from the cliffs. This beach flint is of good quality, figure 2.133.

Up on the cliff at the little village of Tommestrup, as we saw above, a very large number of knapped flints and tools have been found. The finds are stored today at Sydsjælland Museum in Vordingborg. On three large shelves there are more than seventy boxes full of flint from Gregersen's surveys. The material is huge, the largest amount that I have studied, figure 2.134.

The vast majority of the finds are different types of flakes, knapped bits, worked flint pieces, and other remains of production. There are examples of the production of Neolithic axes, bifacial sickles, daggers, scrapers, and various types of

Figure 2.135. Blade-like flakes, whole and fragmentary, and broken retouched preforms for large blade knives from Tommestrup, Stevns.

cores. The area was thus used for the production of flint tools at different times.

The status of the objects in the collection is similar. They all show heavy post-deposition damage. Plough retouches and rust marks are common, and the objects have a heavy patina. The post-deposition damage is so severe that it is often difficult to determine whether the objects were retouched or not. Although the material is surface finds and thus selected, the collection contains more than just large objects. Gregersen collected both small debitage and large primary flakes.

The flint that was used for the manufacture is almost exclusively Scandinavian Senonian flint of the type that is easily accessible in the chalk layers of the cliff and along the beach. A small proportion of the tools were made of Brown Bryozoan flint, which is also found on the beaches and in the till on the site. The Grey Band Danian flint in the layers of limestone at Stevns Klint was not used in the manufacture.

The collection contains huge quantities of residue from the manufacture of large blade knives. The majority of the more than 1,700 large blade knives that Gregersen (1960:28f) claimed to have found are not blade knives. They are blade-like flakes, whole and fragmentary, preparation flakes, and retouched preforms for large blade knives which broke while being retouched. There are only about forty large blade knives in the material, figures 2.135 and 2.136. It is appropriate to clarify the definition here. In other contexts in this study I

Figure 2.136. Twenty-three large blade knives from Tommestrup, Stevns.

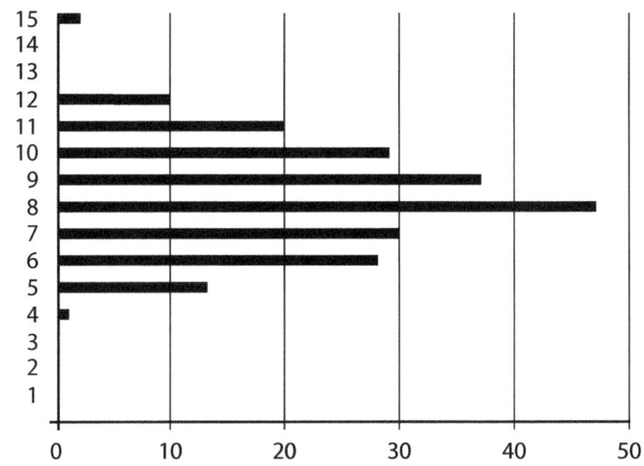

Figure 2.137. Length, in centimetres, of 137 blade-like flakes and 78 retouched preforms for large blade knives in the collection from Tommestrup, Stevns.

have defined fragments of large blade knives as blade knives. I have done so because a fragment of a large blade knife was once part of a whole blade knife. The material from this production site has not been defined in that way. The many fragments of retouched preforms for large blade knives from the site come from occasions when the retouching failed and the preform was broken. These were thus never whole large blade knives and are therefore not defined as such.

The collection is so large that I have chosen to study it at different levels of detail. For roughly half of the material I have analysed each object separately. That work led to the sorting of 30 large blade knives, 673 whole and fragmented blade-like flakes, and 265 retouched preforms for blade knives. These were examined and counted, giving a total of 968 objects. From

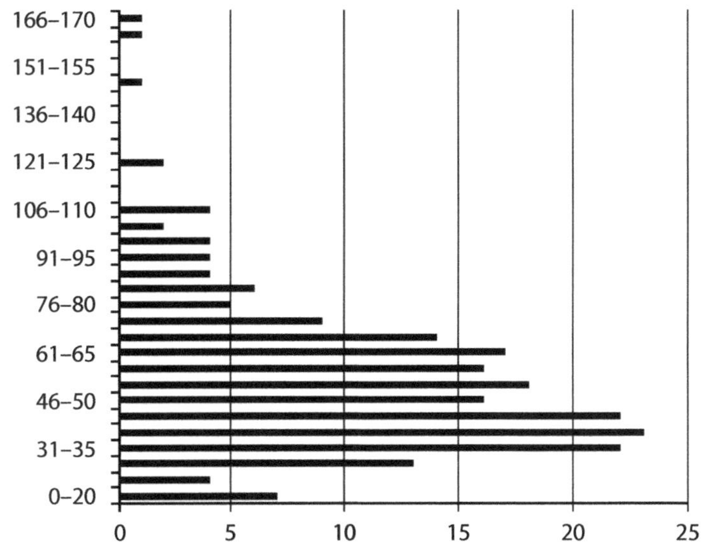

Figure 2.138. Weight of 137 blade-like flakes and 78 retouched preforms for large blade knives in the collection from Tommestrup, Stevns. The weight is stated in grams at intervals of five, with the exception of the first interval which is 0–20 grams.

these 968 objects I also registered the length and weight of 137 blade-like flakes and 78 retouched preforms for blade knives, figures 2.137 and 2.138. This was to give some idea of the size of the individual objects. Of the large blade knives, 17 were registered in detail.

The other half of the collection was examined summarily to see whether there was anything different from what I had seen before. There was not; the other half of the collection is identical to the first.

The large blade knives, the blade-like flakes, and the retouched preforms for large blade knives display uniformity in form and technology. Platforms, bulbs of percussion, curvature, platform preparation, thickness, and width are similar. Length and weight agree generally with the length and weight of large blade knives from southern Scandinavia. Many of the flakes and the preforms are to varying extents covered with cortex on the dorsal side. None of the objects has any gloss visible to the naked eye. I rate the material as having suffered too much post-deposition damage to allow functional analyses by HPA.

Blade-like flakes are those removed during manufacture to shape the front of the core in preparation for the removal of a preform for a large blade knife. There are whole flakes and fragments in the form of proximal, medial, and distal parts. The fragmented flakes were broken during the manufacturing process. The retouched preforms for large blade knives are examples that broke during retouching, that is, when the preform was to be shaped into a blade knife. Either they were broken or excessively large pieces were struck off by the retouching. I regard these as manufacturing failures. There are also some preparation flakes in the material. The large blade knives made at Tommestrup are type I large blade knives with a well-prepared platform. There are no traces of any other type of production from the Late Bronze Age in the material and none which can be associated with what is usually called settlement site remains (see e.g. Sørensen 1980; Ethelberg 1982; Björhem & Säfvestad 1993).

The result of the analysis shows that Tommestrup is a manufacturing site where a large number of large blade knives of flint were produced. The manufacture took the form of standardized mass production using the local supply of high-quality Scandinavian Senonian flint in the cliff and on the shore. The manufacture consisted of making preforms for blade knives and modifying these into large blade knives. If the find material is set in a chaîne opératoire, it is clear that it is steps 1–8 that are represented in the manufacture carried on at the site, as these were defined in figure 2.56.

Södra Sallerup, south-western Scania in Sweden

In the area around Södra Sallerup in south-western Sweden, systematic mining of flint started in the Early Neolithic. This flint is of Maastrichtian age and it occurs embedded in enormous slabs of chalk which were scooped up by glacial ice and re-deposited here. The prehistoric flint-mining area has been known since the start of the twentieth century (Holst 1907; Schnittger 1910), and since the 1950s onwards there have been

Figure 2.139. Map of Södra Sallerup showing all the excavations in the area since 1977. Drawn by Elisabeth Rudebeck.

extensive excavations here (Althin 1951; Olausson, Rudebeck & Säfvestad 1980; Rudebeck 1986, 1987, 1994, 1998; Nielsen & Rudebeck 1991; Nilsson & Onsten-Molander 2004; Winkler 2004; Heimer 2006; Kishonti 2006). These have yielded artefacts dated from the Mesolithic onwards. The most striking archaeological remains in the area are mines and open pits where flint was extracted; the number of these is difficult to estimate but it is certainly over 1,000 (Rudebeck 1994:10). Archaeologists have also excavated several workshops, dumps, house remains, occupation layers, furnaces, residue from bronze casting, and much more. The remains are complex and the area has repeatedly been used during prehistory and later times for a wide range of activities, figure 2.139.

The majority of the flint mines and pits which have been radiocarbon-dated have proved to be Neolithic, while the flint assemblages from the Late Bronze Age have mainly been found in the thick occupation cultural layers that have been excavated in the area. At present it is impossible to determine whether flint was extracted from mines during the Late Bronze Age or if the flint-knapper of that period used the Neolithic dumps on the site as the source of their raw material. There are several European examples of flint mines which were opened and used during the Late Bronze Age (see e.g. Lech & Lech 1997). In

Figure 2.140. Excavation of a large occupation layer at Södra Sallerup. This layer was discovered in the prolongation of the same depression where an occupation layer was excavated in 1978 (Rudebeck 1986). Photograph first published in Kishonti (2006, plate 2). Photo: Patric Ifversen, Malmö Heritage.

Figure 2.141. Blade-like flakes and fragments of retouched preforms for large blade knives from Södra Sallerup. I have interpreted these as residue from the manufacture of type I large blade knives without platform preparation. The flakes and the retouched preforms for blade knives are part of the material found by the excavations in 1978 (Rudebeck 1986, inventory number MHM 6120) and 1983–84 (inventory number MHM 6650:XVII).

view of the enormous number of flint mines that existed in the area around Södra Sallerup in relation to the small proportion that have been excavated (Rudebeck 1994:10), it is not unreasonable to assume that mining was pursued on the site in the Late Bronze Age. However, this remains to be investigated.

The archaeological flint material produced by the excavations is huge. About 65 cubic metres of knapped flint are packed in 1,064 find boxes in the stores of Malmö Museums. As mentioned above, when analysing the flints from an excavation from 1998 I found some blade-like flakes (Nilsson & Onsten-Molander 2004:73). The question is whether there are more of this type of flake in the archaeological material from the site. To explore this I have studied flint from excavations performed in 1978 (Rudebeck 1986) and 1983–84 (Malmö Heritage archive; place-name: Sydgas, inventory number: MHM 6650: XVII, archive number: S09:029). These excavations have investigated large, complex occupation layers and uncovered a rich array of archaeological material dated to the Late Bronze Age, figure 2.140. The finds from the excavations have been interpreted as representing a variety of activities such as dwelling, cooking, craft, and ritual (Nilsson & Onsten-Molander 2004; Kishonti 2006).

The analysis of the finds shows that there is a large amount of flint from the production of large blade knives. This consists

Figure 2.142. Ten whole and fragmentary large blade knives from Södra Sallerup. They come from the 1978 excavation (inventory number MHM 6650:XVII). Notice that the five blade knives with preserved platform can all be identified as type I large blade knives without platform preparation, in other words, the same type of blade knives that were also manufactured on the site.

Table 2.75. Results of functional analyses of ten of the large blade knives from Södra Sallerup from the 1978 excavation (Rudebeck 1986). Five of the blade knives were used on reeds, four on cereals, and one has a post-deposition patina that renders analysis impossible.

Find number	Status	Edge	Polish
MHM 6120, 27/10:I	Distal part of blade knife	Intact edge. Gloss visible to the naked eye	Well-developed polish, linked and fluted. High reflectivity, domed topography, large extent, and soft surface. Interpreted contact material reeds.
MHM 6120, 26/1:I	Whole blade knife	Intact edge. Gloss visible to the naked eye	Well-developed polish, linked and fluted. High reflectivity, domed topography, large extent, and soft surface. Interpreted contact material reeds.
MHM 6120, no data	Proximal part of blade knife	Intact edge. Gloss visible to the naked eye	Well-developed polish, linked and fluted. High reflectivity, domed topography, large extent, and soft surface. Interpreted contact material reeds.
MHM 6120, 25/14:I	Medial part of blade knife	Severe post-deposition damage. Gloss visible to the naked eye	Well-developed polish, linked and fluted. High reflectivity, domed topography, large extent, and soft surface. Interpreted contact material reeds.
MHM 6120, 26/10:I	Whole blade knife	Intact edge. Gloss visible to the naked eye	Well-developed polish, linked and fluted. High reflectivity, domed topography, large extent, and soft surface. Interpreted contact material reeds, figure 2.143.
MHM 6120, 29/7:I	Distal part of blade knife	Intact edge. Gloss visible to the naked eye	Well-developed polish, linked. Low reflectivity, flat topography, large extent, and soft surface. Interpreted contact material cereals.
MHM 6120, 26/5:II	Distal part of blade knife	Post-deposition damage. Gloss visible to the naked eye	Well-developed polish, linked. Low reflectivity, flat topography, large extent, and soft surface. Interpreted contact material cereals.
MHM 6120, 31/10:I	Whole blade knife	Severe post-deposition damage. Gloss visible to the naked eye	Well-developed polish, linked. Low reflectivity, flat topography, large extent, and soft surface. Interpreted contact material cereals.
MHM 6120, from section	Medial part of blade knife	Intact edge. Gloss visible to the naked eye	Well-developed polish, linked. Low reflectivity, flat topography, large extent, and soft surface. Interpreted contact material cereals, figure 2.144.
MHM 6120, 26/13:I	Whole blade knife	Intact edge. No visible gloss	All-over post-deposition patina renders analysis impossible.

of blade-like flakes, whole or fragmented, and fragments of retouched preforms for large blade knives which broke while being retouched, figure 2.141. The blade knives made here were of type I without platform preparation. The flint used for the manufacture is the Scandinavian Senonian flint that is in plentiful supply in the chalk on the site. The type of large blade knives manufactured here corresponds well to the dominant occurrence of tabular flint in the chalk at Södra Sallerup.

In the material there are also many large blade knives, figure 2.142. Five of them have a preserved platform and can thus be identified as type I large blade knives without platform preparation, in other words, the same type of blade knives that were made on the site. All the large blade knives from the site have undergone functional analysis. The aim was to determine whether they were newly produced on the site in the Late Bronze Age or were used blade knives deposited at the same place where they were produced.

The result of the functional analyses shows that the blade knives were used for working with reeds and cereals, table 2.75, figures 2.143 and 2.144. This agrees with the results of the functional analyses of large blade knives presented above.

The result of the analysis shows that large blade knives were manufactured in the area of the flint mines at Södra Sallerup. The blade knives were of type I without platform preparation. The Scandinavian Senonian flint that occurs in the chalk layers on the site was used as raw material. The analysis demonstrates

Figure 2.143. HPA photo, polish interpreted as coming from contact with reeds on a large blade knife from Södra Sallerup (MHM 6120, 26/10:I). Original magnification 200×.

Figure 2.144. HPA photo, polish interpreted as coming from contact with cereals on a large blade knife from Södra Sallerup (MHM 6120, from section). Original magnification 200×.

that blade knives of this type which were used for working cereals or reeds were also deposited on the site. This is therefore a manufacturing site beside a plentiful supply of raw material which shows traces not only of manufacture but also of the use of blade knives. It is also a place with remains of other contemporary activities than the making of large blade knives.

Hanaskog, north-eastern Scania in Sweden

From north-eastern Scania in Sweden there are examples of finds of large blade knives made of Common Kristianstad flint (Strömberg 1952). Sufficiently large and high-quality nodules of Common Kristianstad flint are found in the area north of the town of Kristianstad (Högberg & Olausson 2007:78ff). It is therefore likely that one or more production sites can be found somewhere around here. In the spring of 2004 I did a survey of the area around the modern limestone quarry at Hanaskog, just north of Kristianstad (Högberg 2005a:1ff). Within a limited area beside the quarry there were large quantities of blade-like flakes of the type defined as characteristic residual products from the manufacture of type I large blade knives with well-prepared platforms, figures 2.145 and 2.146. The flakes represent large-scale production. On a later visit to the site, about a hundred flakes in total were observed on the ground (Högberg 2005a). Only a few of these have been collected and there has been no excavation on the site. No traces of any other production have been found on the site.

Figure 2.145. Blade-like flakes of Kristianstad flint found at the Hanaskog limestone quarry in north-east Scania. The flake to the far right is approximately 11 cm long. These flakes are the only ones I collected. There was a large quantity of flakes on the site but I let them lie there for any future excavations. Photograph first published in Högberg (2005a).

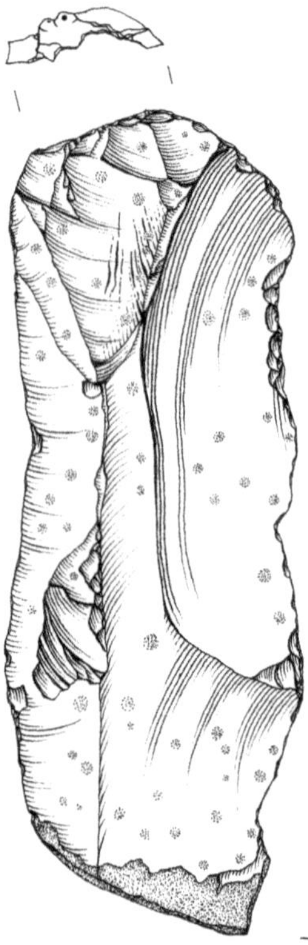

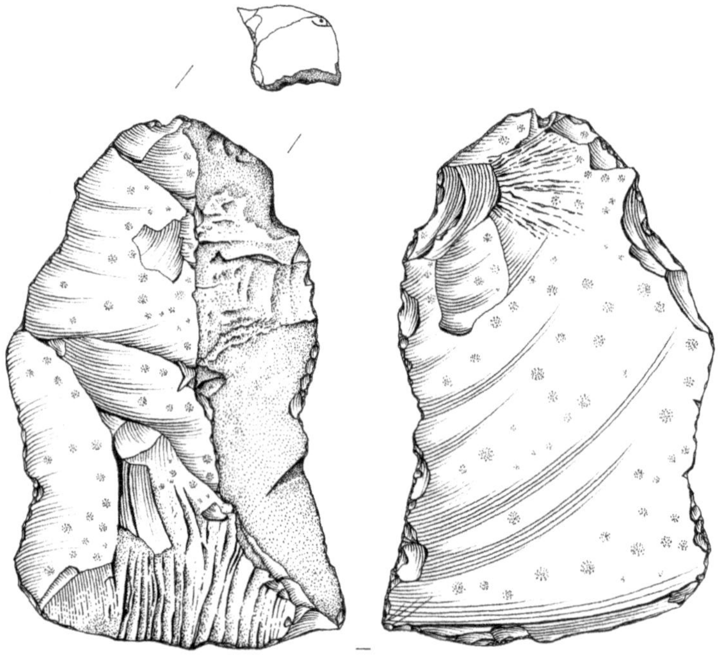

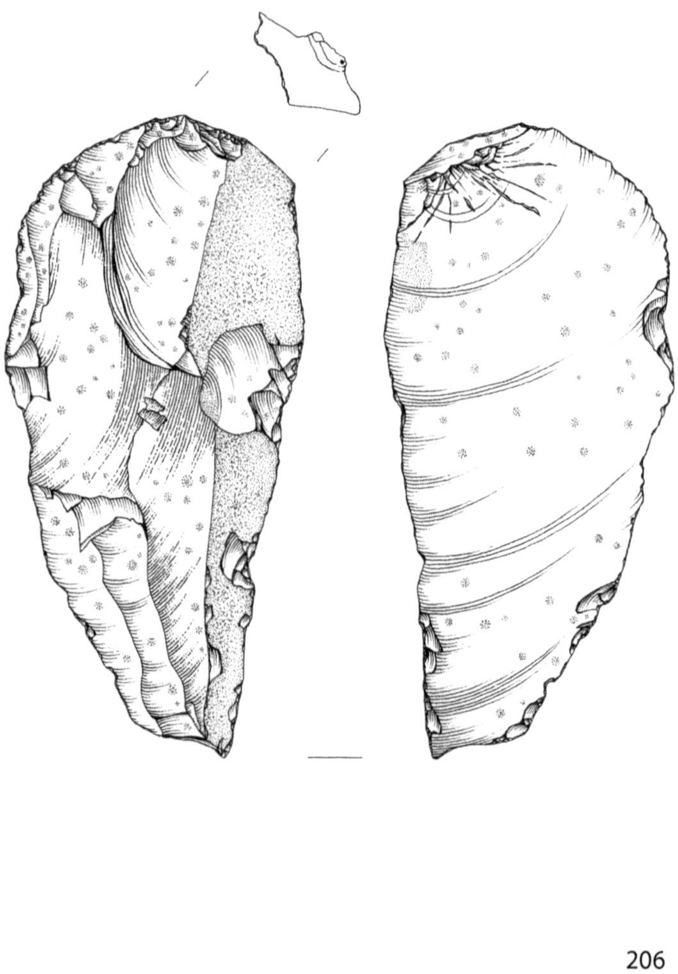

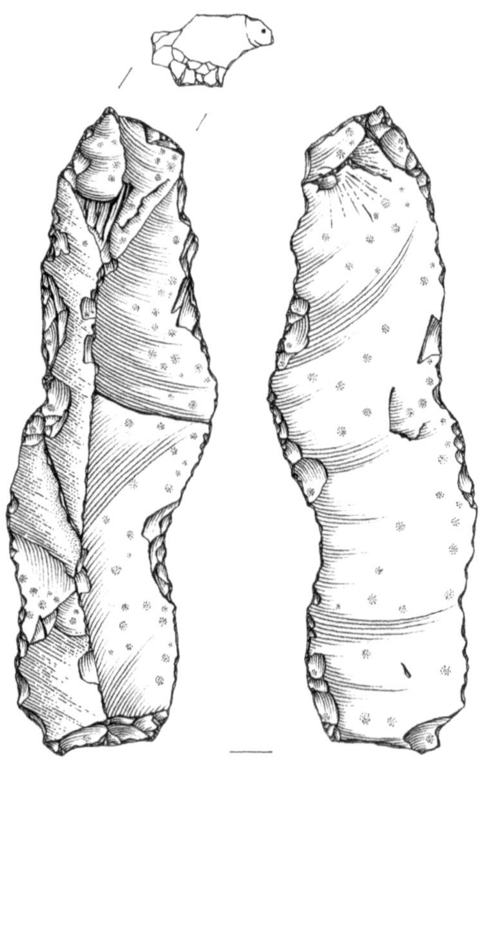

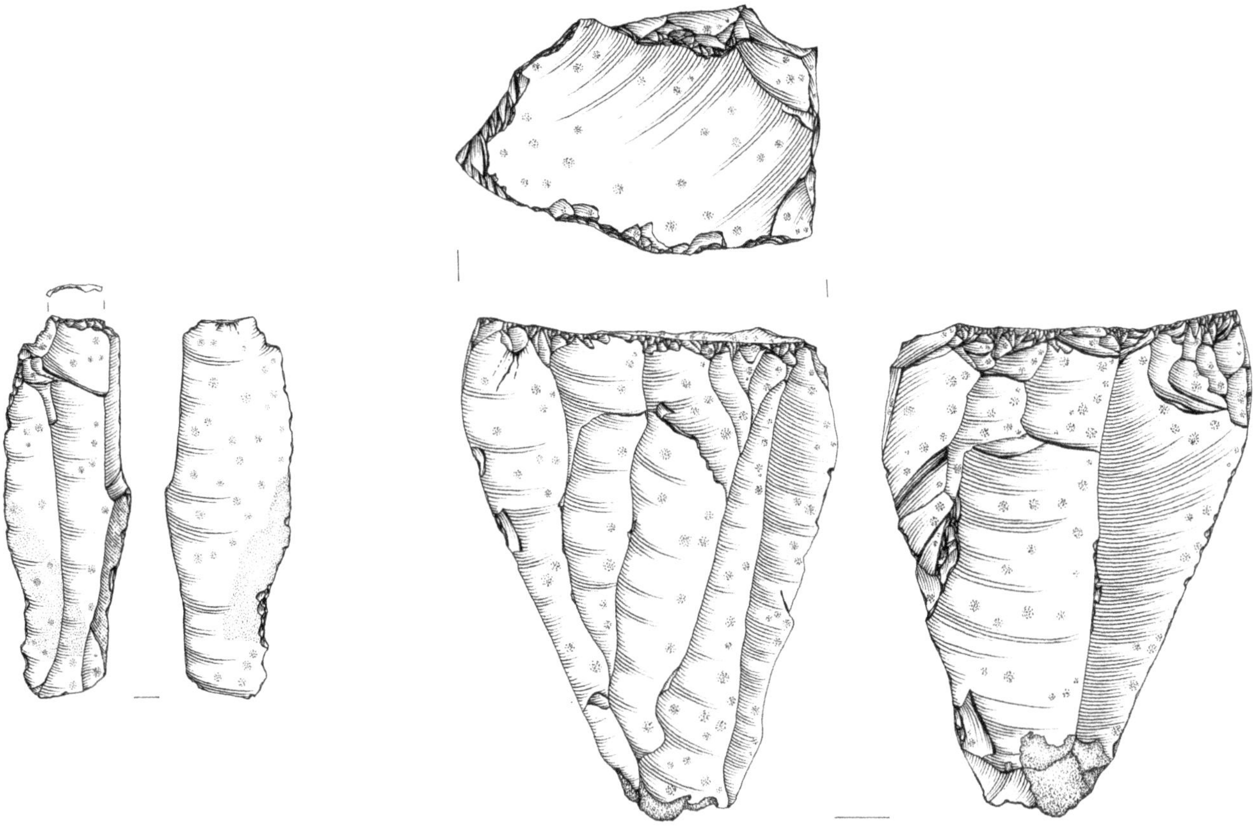

Figure 2.146. Some selected blade-like flakes, a preparation flake with feathered termination, and an exhausted core from Hanaskog. The dots in the drawings mark that the raw material is Common Kristianstad flint. For the size, see figure 2.145. Drawings: Björn Nilsson.

Manufacture of type I large blade knives with a well-prepared platform requires access to high-quality raw material in the form of big, homogeneous flint nodules without cracks, fossils, or other impurities. Nodules of Kristianstad flint meeting these standards can be found today in the disused modern limestone quarry on the site (Högberg & Olausson 2007:84). According to geological records, this flint occurs at a depth of a few metres in the limestone bedrock (Magnusson 1963).

The flakes reveal the scale of production on the site. It used flint of the same type and quality as that found today in the Hanaskog limestone quarry. It is thus obvious that this Common Kristianstad flint was somehow extracted during the Late Bronze Age. It is not known whether this involved mining, as happened for example at Wierzbica in Poland, or if the extraction was done in some other way.

Reidervitz and Ruschvitz, on Rügen in northern Germany

On the German island of Rügen in the Baltic Sea, Königsstuhl rises more than 160 metres above the sea. It is the highest point on the white chalk cliffs that extend several kilometres along the north-east coast of the island. In the chalk there are bands of what is known as Rügen flint, and the shore below the cliffs is littered with flint (Högberg & Olausson 2007:100ff). Yet despite the profuse occurrence of high-quality flint in the chalk cliffs, it was common for prehistoric people to use till flint for tool production on Rügen. There is nothing strange about this, since the till deposits in the area contain plenty of large nodules of high-quality south Scandinavian flint (Auler 1999:18; Rassman 2001:10ff). A few hundred metres inland from the chalk cliffs, and a couple of kilometres from each other, there are two sites where these till deposits were used for extensive prehistoric tool production. The places are called Reidervitz and Ruschvitz, see figure 2.147.

The archaeologist Michael Schirren has drawn my attention to the fact that the stores of the Landesamt für Bodendenkmalpflege in Mecklenburg-Vorpommern in Stralsund have ten full boxes of knapped flint from these sites. In addition, there are said to be as many find boxes with the same type of finds in the collections of the Kulturhistorisches Museum in Stralsund (personal communication Michael Schirren), figure 2.148.

This extensive collection of finds was gathered by amateur archaeologists in periods from the mid-1980s until 2000. The sites have been inspected by professional archaeologists but have not been excavated. The sites have not been published, but the finds were recurrently presented during the 1980s in the chapters on new acquisitions (Kurze Fundberichte) in the annuals of Bodendenkmalpflege in Mecklenburg (see e.g. the years 1983:349ff, 1985:230ff, & 1987:269), figure 2.149.

The finds represent two large production sites which were both used repeatedly at different times. In the material there are examples of the production of Neolithic axes, bifacial sick-

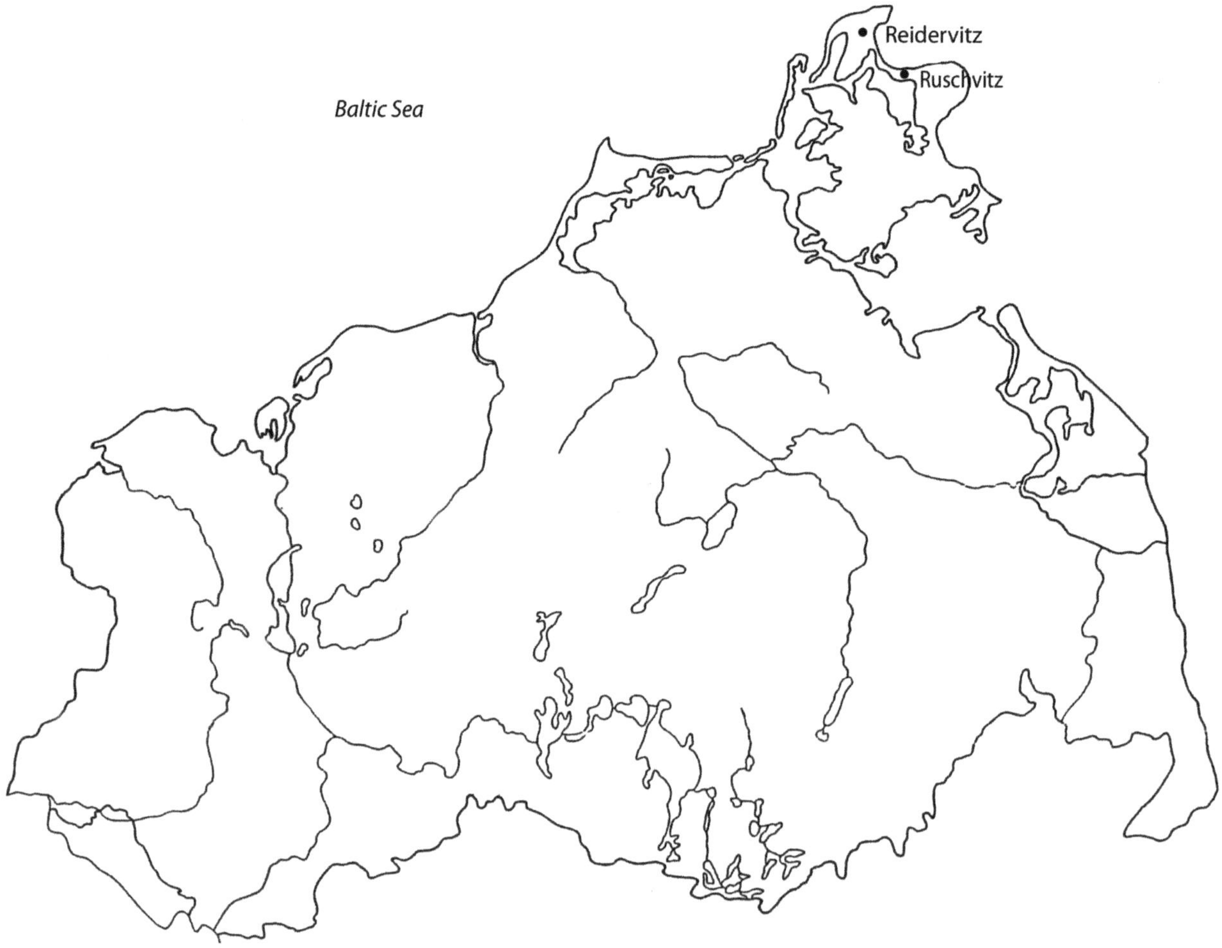

Figure 2.147. Map of the Baltic coast of northern Germany marking Reidervitz and Ruschvitz on the island of Rügen.

les and daggers, scrapers, and various types of cores. There are also traces of extensive manufacture of large blade knives from the Late Bronze Age. The flint that was used for this manufacture is varying types of South Scandinavian Senonian flint. The majority of this is till flint. There are also flakes of Rügen flint but they are rare.

The status of the find material from the sites is very similar. The artefacts have severe post-deposition damage. Plough retouches and rust marks are common, and the majority of the objects show patina and polish from wind and water. Since the material consists of surface finds collected by amateurs, it is highly selected. Only large objects perceived as artefacts have been collected.

I have analysed the parts of the find material that are in the stores of the Landesamt für Bodendenkmalpflege in Mecklenburg-Vorpommern in Stralsund. They contain large amounts of blade-like flakes from the manufacture of blade knives, table 2.76. These flakes display uniformity in form and technology. Platforms, bulbs of percussion, curvature, platform preparation, length, thickness, and width are similar. Many are covered to varying extents with cortex on the dorsal side. In figures 2.150, 2.151, 2.152, and 2.153 the weight and length of the flakes have been used to illustrate the similarities in the as-

Table 2.76. The number of blade-like flakes and fragments of retouched preforms for blade knives from Reidervitz, total 237, and Ruschvitz, total 162, on Rügen in Germany. Landesamt für Bodendenkmalpflege Mecklenburg-Vorpommern, Rügen, Reidervitz: Fpl. Zu 7 and Ruschvitz Fpl. 7003.

Inventory number, ALM	Blade-like flakes	Retouched preforms for large blade knives	Total
Reidervitz 2002/10,1	102	8	110
Reidervitz 2002/30	16	–	16
Reidervitz 2002/39,1	101	2	103
Reidervitz 2002/121	8	–	8
Ruschvitz 2002/1000,1	149	13	162
Total	**376**	**23**	**399**

Figure 2.148. Two of the find boxes with blade-like flakes from Reidervitz, inventory numbers ALM 2002/10,1 and ALM 2002/39,1.

semblages both within and between the two sites. When these figures are compared with the result of the attribute analyses presented above, it is obvious that the large blade knives made at Reidervitz and Ruschvitz are the same kind of blade knives as those found at excavations of settlement sites from the Late Bronze Age on Rügen (Auler 1997, 1999).

The flakes are blade-like flakes detached to shape the front of the core in preparation for the removal of a preform for a large blade knife, figure 2.154. There are both intact and fragmentary flakes in the material. Those which are fragmented I interpret as having been broken during the manufacturing process. There are only a few preparation flakes in the mate-

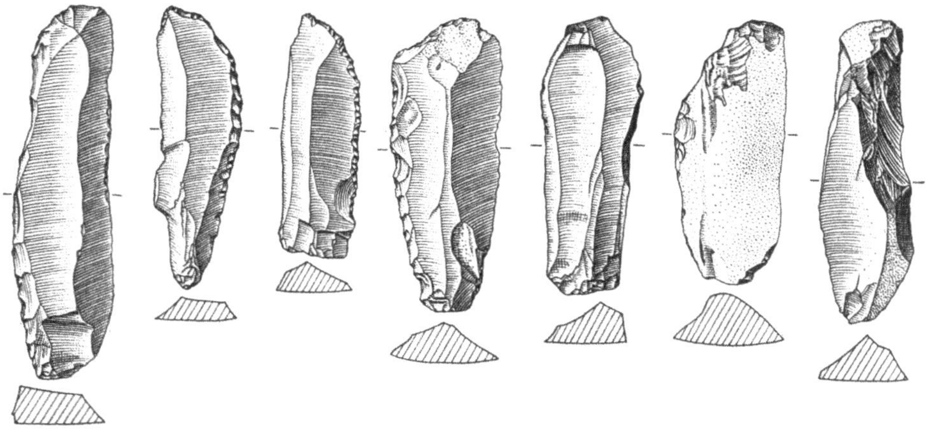

Figure 2.149. Four large blade knives and three blade-like flakes from Ruschvitz on Rügen, first published in "Kurze Fundberichte", Bodendenkmalpflege in Mecklenburg, 1987, page 269.

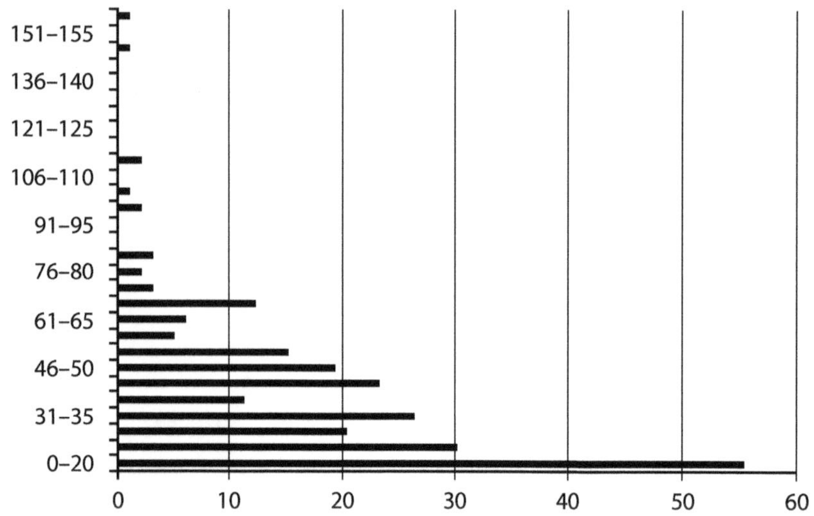

Figure 2.150. Weight of the blade-like flakes and retouched preforms for large blade knives analysed from Reidervitz, Rügen, total 237. The weight is stated in grams in intervals of five, with the exception of the first interval which is stated as 0–20 grams.

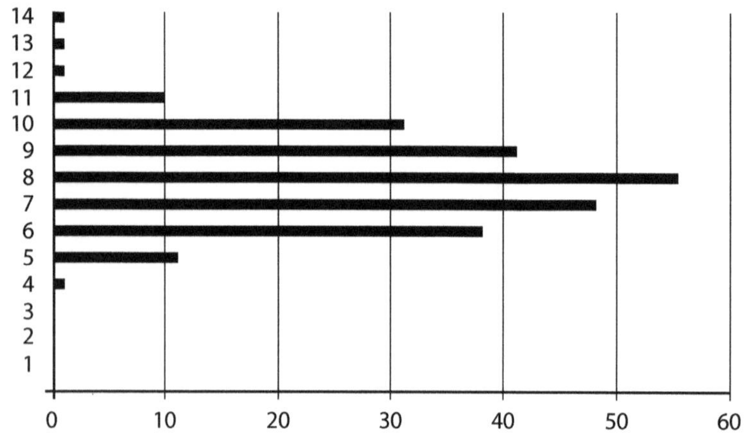

Figure 2.151. Length, stated in cm, of the blade-like flakes and retouched preforms for large blade knives analysed from Reidervitz, Rügen, total 237.

rial. There are also finds of fragments of retouched preforms for blade knives, table 2.76. All of these are preforms for large blade knives which broke during retouching, when being shaped from preforms into large blade knives. Either they were broken off or excessively large pieces were struck off when they were retouched. I regard these as unsuccessful blade knives. It is worth noting that there are no unmodified preforms for blade knives in the material, nor any primary flakes. None of the artefacts I studied has any gloss visible to the naked eye. I rate the material as having suffered too much post-deposition damage to allow functional analyses by HPA.

The result of the analysis shows that Reidervitz and Ruschvitz are sites where extensive manufacture of large blade knives of flint took place. The blade knives are of the kind that I have previously called large blade knives from Rügen. They come from standardized mass production using the local supply of high-quality till flint. Manufacture consisted of making preforms for large blade knives and modifying these preforms into blade knives. No primary shaping, i.e. step one, was done on these sites. If the sites are set in a schematic chaîne opératoire for the manufacture of blade knives, it is obvious that the work done on the site represents steps 2/3–8 as defined in figure 2.56.

Other places worth closer study

As discussed above in the chapter about assemblages of large blade knives, there seem to have been special production sites for blade knives at places where Bryozoan flint, Funen variety, and Falster flint were available in sufficiently large quantities for production.

As stated above, Bryozoan flint, Funen variety, can be linked to just one place where it is found today in a primary deposit. That place is Klintholm in south-east Funen in Denmark (Thomsen 2000:34). Although we do not know of any production site at Klintholm, it is likely that there was one during the Late Bronze Age.

Places where Falster flint was used as raw material are a little more difficult to pinpoint. On the island of Falster in southern Denmark there are large supplies of high-quality nod-

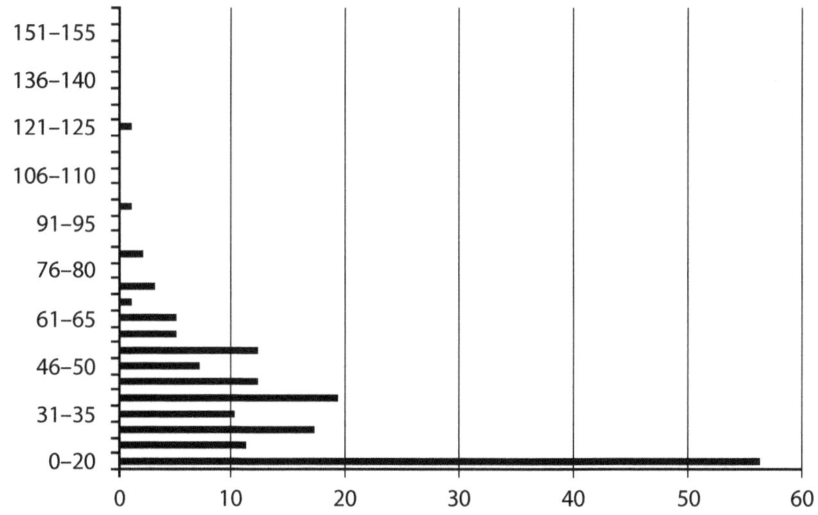

Figure 2.152. Weight of the blade-like flakes and retouched preforms for large blade knives analysed from Ruschvitz, Rügen, total 162. The weight is stated in grams in intervals of five, with the exception of the first interval which is stated as 0–20 grams.

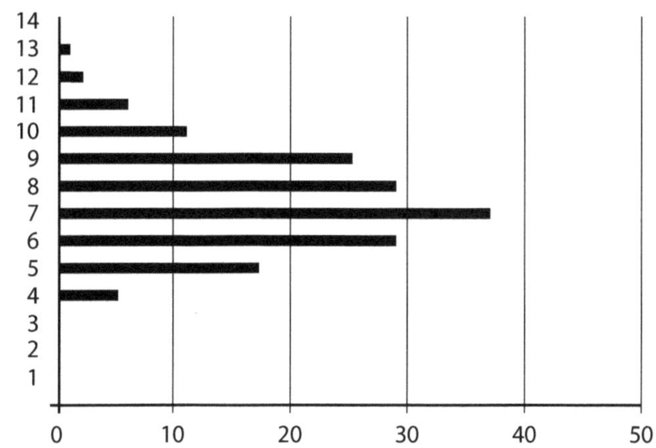

Figure 2.153. Length, stated in cm, of the blade-like flakes and retouched preforms for large blade knives analysed from Ruschvitz, Rügen, total 162.

ules of Falster flint in the till and in the subsoil (Högberg & Olausson 2007:36ff). But Falster flint is also found along the shores on the island of Fehmarn off the Baltic coast of northwestern Germany. Just as in the case of Klintholm on Funen, there are no known production sites for large blade knives on Falster or Fehmarn, but presumably they did exist in the Late Bronze Age.

Finally, Helle Juel Jensen mentions two other places of interest, at Asnæs on Zealand and Sangstrup in Jutland, both in Denmark. Here large blade knives and flakes have been collected by amateur archaeologists (Jensen 2006:111). These places, however, have not been analysed or published.

Result and discussion – special manufacturing sites for large blade knives

In the analyses presented in this chapter I have worked with questions as to whether there were special manufacturing sites for large blade knives and if so where. The analyses show that these places did exist. Six have been located. They are: Wierzbica "Zele" in Poland; Tommestrup in Denmark; Södra Sallerup and Hanaskog, both in Sweden; and Reidervitz and Ruschvitz in Germany. All of these places are at large supplies of raw material, and huge quantities of large blade knives of flint were made at them during the Late Bronze Age. As is evident from

Table 2.77. Types of large blade knives made at the six manufacturing sites.

Place	Type of large blade knives manufactured there
Wierzbica "Zele", Poland	"Backed blades of Zele type"
Tommestrup, Denmark	Type I large blade knives with a well-prepared platform
Södra Sallerup, Sweden	Type I large blade knives without platform preparation
Hanaskog, Sweden	Type I large blade knives with a well-prepared platform
Reidervitz, Germany	Rügen blade knives
Ruschvitz, Germany	Rügen blade knives

Figure 2.154a. Blade-like flakes from Reidervitz, eight from inventory number ALM 2002/10,1.

table 2.77, all except type II of the groups of large blade knives defined above were made at one or more of these places. I have thus been unable to locate any specific manufacturing site for type II large blade knives.

The places display both similarities and differences. Tommestrup, Reidervitz, and Ruschvitz are all surface-surveyed production sites from which thousands of artefacts from the Neolithic and the Bronze Age have been collected. The find material from the sites includes extensive remains of the manufacture of large blade knives in the form of blade-like flakes and retouched preforms for large blade knives, both whole and fragmentary.

Wierzbica "Zele" and Södra Sallerup are both manufacturing sites located at flint mines. In Wierzbica "Zele" the flint was extracted from mines during the Late Bronze Age and this flint was used on the spot to make "backed blades of Zele type". In Södra Sallerup the flint from the site was likewise used for making large blade knives. We do not know whether the flint there was extracted from mines in the Late Bronze Age or if dumps from earlier mining activities were used as raw material.

The special thing about Södra Sallerup is that it is the only one of the manufacturing sites where used large blade knives have been found together with remains of the production of blade knives. It is also the only manufacturing site where extensive settlement site remains from the Late Bronze Age have been found. It is easy to cling to the idea that places must be described as either/or: places where things were produced were those with access to the right raw material and with finds of large quantities of production waste in the form of,

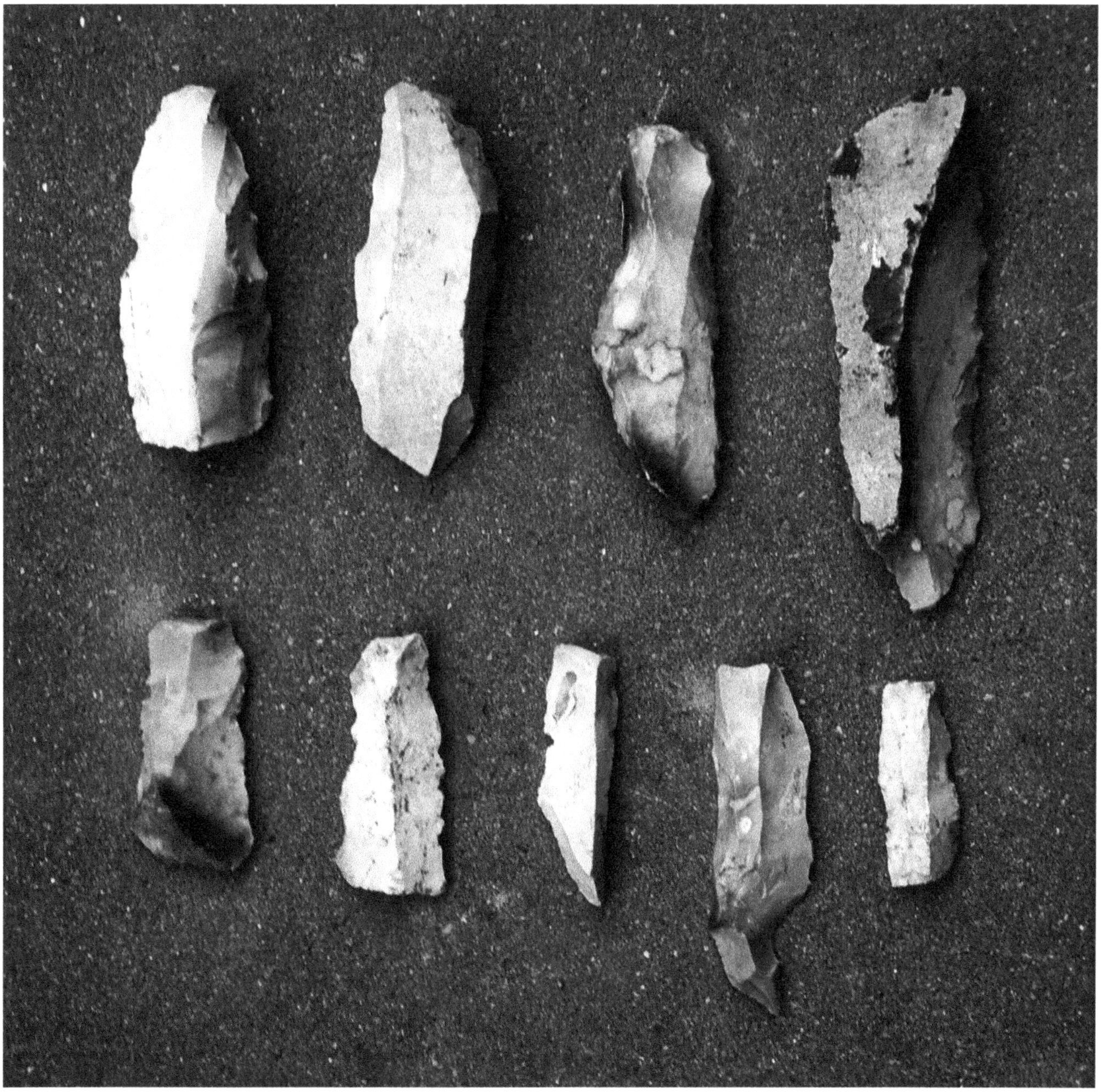

Figure 2.154b. Blade-like flakes from Reidervitz, nine from inventory number ALM 2002/39,1.

for example, flakes, cores, preforms, or failures; places where things were consumed are those where used objects are found together with waste from settlement site activities. This way of thinking is far too static. If a place must be either/or, it rules out the possibility that it could be both/and (Speal 2006). Södra Sallerup is a site that shows the importance of thinking of both possibilities.

Hanaskog was previously unknown as a manufacturing site. Since the area has only been surface-surveyed without the finds being collected, it is not possible to say anything definite about the character of the site. What is clear is that there was large-scale manufacture of large blade knives of Kristianstad flint here.

All the sites except Hanaskog have yielded retouched preforms for large blade knives which broke during the retouching of the preform for a blade knife. If none of these have been found at Hanaskog it is probably because only a fraction of the material from there has been collected. The occurrence of preforms that broke during retouching shows that finished large blade knives were produced on the sites. This is interesting in view of the fact that the analysis of the assemblages showed that both unretouched preforms and blade-like flakes were distributed from the manufacturing sites. It is thus obvious that people at these special manufacturing sites made and distributed finished large blade knives, preforms for large blade knives, and blade-like flakes from the production of these tools. Or in other words, both large blade knives, preforms and blade-like flakes should be perceived as products from these places.

Table 2.78. How the flint was available or made available at manufacturing sites for large blade knives.

Place	How was the flint available or made available?	Was the flint in limestone or chalk layers on the site?
Tommestrup	Chalk cliffs and on the shore	Yes
Wierzbica "Zele"	Mines	Yes
Södra Sallerup	Mines? Open pits? Dumps?	Yes
Hanaskog	Mines? Open pits?	Yes
Reidervitz	Quarried or picked from the till	No
Ruschvitz	Quarried or picked from the till	No

At all of the places there is a plentiful supply of high-quality flint. It will be evident from table 2.78, however, that there are both differences and similarities as regards the question of how flint was obtained during the Late Bronze Age. The common factor is that the flint was in some way extracted at the sites. This was done by quarrying it from visible layers of limestone and chalk, picking or quarrying it out of the till, or by mining. At Tommestrup the flint may also have been picked up on the shore. As regards Södra Sallerup, we do not know how the extraction was done.

It is difficult to discuss the scope of manufacture at each place based on the analyses presented here. Although experimental studies have demonstrated the possibility of using flakes resulting from core reduction to calculate the extent of manufacture (Carr & Bradbury 2001), these studies have not yet been developed to the extent that they are applicable to the assemblages discussed here. It is clear that thousands of large blade knives were made at places like Tommestrup, Reidervitz, and Ruschvitz. The enormous amounts of flakes in the collections from these places are testimony to this.

If the scope of production is difficult to ascertain, the intensity is even harder. One blade knife per month for 400 years – roughly the length of time large blade knives of flint were used – gives a total of 4,800 tools. But if two people each make ten blade knives every day and work five days a week, it does not take more than just over a year to produce the same number. The example demonstrates the difficulty of discussing the intensity of production based on analyses of debitage, as has also been pointed out in several other studies (see e.g. Whittaker & Kaldahl 2001). It is important to stress, however, that the craft as such requires a certain continuity in practice if it is to be maintained. There must therefore have been some intensity in the manufacture at the sites. Whether this intensity was sporadic, seasonal, dictated by needs, or in short-term bouts of high intensity is a question I must leave unanswered.

Huge numbers of large blade knives were thus made at these places. But how should we understand the sites? How might they have been perceived?

When Stone Age flint-knapping places in flint-rich parts of southern Scandinavia are discussed, it often concerns sites where small-scale or short-lived production took place. From the Late Palaeolithic and the Mesolithic, traces of flint craft are usually found on sites where other activities were performed as well: in complex areas for settlement and work, with features such as houses, hearths, activity areas, and graves (Karsten & Knarrström 2003), beside a hut (Fischer 1990), or in a small settlement and activity area with remains of hunting and fishing (Jonsson 2005), figure 2.155. From the Neolithic there are vestiges of the production of small tools, usually within small, demarcated areas where a specific task was performed (Högberg 2006) or in complex areas with settlement site remains, activity areas, and burial grounds (Sarnäs & Nord Paulsson 2001; Öijeberg 2004; Gidlöf, Hammarstrand Dehman & Johansson 2006).

What these forms of manufacture have in common is that they were performed at places where other activities were also pursued. Manufacturing was done, for instance, beside dwellings, ritual activities, and craft work, and the majority of what was made on the site was also used there. And although Neolithic tools such as axes, daggers, or sickles were made at special production places (Högberg 1999a; Apel 2001) they are virtually always limited in time in such a way that they represent a few days' work (Hansen & Madsen 1983), or are found together with other types of remains such as dwelling houses (Jensen 1973).

The manufacturing sites for Late Bronze Age large blade knives, preforms for blade knives, and blade-like flakes discussed here differ from the types of Stone Age manufacturing sites described above. Apart from Södra Sallerup, none of them have traces of settlement site remains and the blade knives that were made on the sites were not used there. To find analogies in order to illustrate what the manufacturing sites might have looked like, I have chosen to go forward rather than backward in time. I do so with two examples.

The first comes from the nineteenth-century gun-flint knappers at Brandon in England (Forrest 1983:98, see also

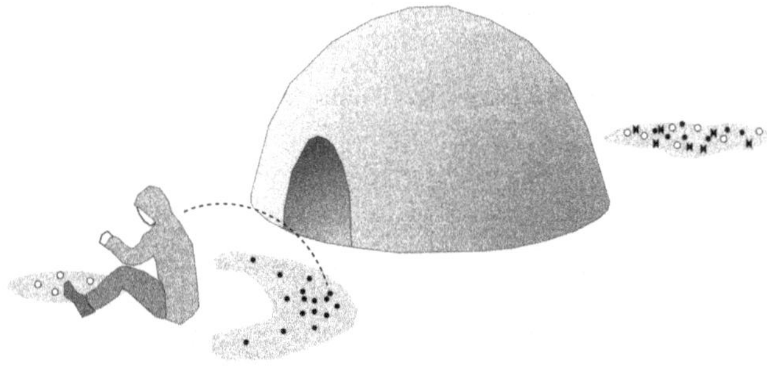

Figure 2.155. An interpretation of what a flint-knapping site from the Stone Age might have looked like (from Eriksen 2000:159)

Figure 2.156. A workshop for the making of gun flints, 1876. The picture comes from Forrest (1983:98).

e.g. Mason 1978:22ff). Figure 2.156 shows a snapshot of a workshop from 1876. The room is totally dominated by the craft. There are flint nodules and flakes everywhere. The tools are specialized and ready at hand. The workplace is organized to suit the craft and thus provides the infrastructure required by the job. The work that is done is standardized and the focus in the room is on the production. On tables and spare surfaces behind and in front of the flint-knappers there are containers of different sizes where the standardized gun flints are sorted according to shape and size. The room is a workplace, to which a person repeatedly goes to work, leaving it when the work is done.

The other example comes from the first half of the twentieth century. At several places in northern Europe, large, brick-like flint blocks were made during the twentieth century. These blocks were exported all over the world, to brick, clay, and cement industries where they were used to line the walls of the cylinder room in ball mills (Slotta 1980; Högberg 1999b, 2004b). On the photograph in figure 2.157, flint-knappers pose in front of a pallet of newly made blocks ready for sale. To the far left can be seen one of the knappers behind a barrel that was used as a table for flint knapping. He is holding a finished block in one hand and the flint-knapping hammer in the other. Around the flint-knappers there are copious nodules and flakes. The manufacture is standardized, those who work there knap flint and nothing else, and the place is specially intended for the production.

It is important to stress that the examples come from a time when workshop crafts functioned within a modern system of capital and industry. I am not implying that such a system was active in the Late Bronze Age. The examples have been brought up here to illustrate what the places may have looked like and how they may have been perceived, and to emphasize the differences between a flint-knapping place from the Stone Age, as interpreted for example in figure 2.155, and special manufacturing sites for the manufacture of large blade knives from the Late Bronze Age, not to illustrate any form of socio-economic system that the places may have been a part of.

The result of the study shows that the manufacturing sites for large blade knives of flint were places where the craft dominated and few or no other activities were performed. They were organized for production. They were strewn with nodules, cores, tools, blade knives, preforms for blade knives, and waste material.

As mentioned in the introduction to this chapter, it is important to stress that this analysis concerns special manufacturing sites for large blade knives. Later on I will analyse places where large blade knives were made in other contexts.

Discussion and conclusion – large blade knives of flint

The analyses of large blade knives of flint have been performed in five chapters, each dealing with a separate aspect of morphology and manufacture, functions and uses, dating, contexts, and geographical distribution, assemblages, and manufacturing sites. The results of the analyses show that large blade

Figure 2.157. Flint-knappers from the early twentieth century posing at their workplace in the limestone quarry at Klagshamn, south of Malmö in Scania, Sweden. In the foreground we see the knappers' products, the flint blocks, stacked up. The flint blocks were cut to a standardized thickness. They were stacked for sale on square-metre pallets and sold per unit area. The photograph was first published in Högberg (1999b). With the permission of Föreningen Klagshamns och Västra Klagstorps Historia.

knives of flint should be regarded as artefacts that display not only great similarities but also significant differences.

Based on attribute analyses, four groups of large blade knives have been defined. The groups are similar yet there are differences. They were produced by different manufacturing processes and partly in different places. The groups are:

- Type I large blade knives with a well-prepared platform.
- Type I large blade knives without platform preparation.
- Type II large blade knives.
- Rügen blade knives.

The most common are type I large blade knives with a well-prepared platform. They are normally 11 to 15 centimetres long, weigh 60 to 120 gram, are three to four centimetres wide and one to one and a half centimetres thick. They usually have one or two ridges and two or three negative removals on the dorsal side where they lack cortex. The majority of them have a straight and well-formed edge and an exterior platform angle of 70–80 degrees. The platform is flat and rhombic with a ring crack around the impact point, usually 1.5 to 3 centimetres wide and 0.6 to 1.5 centimetres thick. Platform preparation in the form of trimming along the platform edge and thin preparation flakes running down over the dorsal side are frequent features. As part of the preparation, quite a few of them also have preparation flakes with hinged and feathered termination.

Type I large blade knives without platform preparation are morphologically similar to those with a well-prepared platform. What distinguishes the two groups is the platform preparation. On type I large blade knives without platform preparation, the platform edge instead consists of a ridge formed by two negative removals running from the platform down over the surface of the dorsal side. They often have cortex on the dorsal side. They are not as common as type I large blade knives with a well-prepared platform.

Type II large blade knives show partly the same attributes as type I large blade knives, but they are shorter and the retouch at the distal part gives the blade knife a rounded end, unlike type I large blade knives which have a more angular end. Attributes that distinguish type II large blade knives from type I are that none of the type II large blade knives have any great occurrence of cortex on the dorsal side, a much larger proportion of type II than type I blade knives have bidirectional dorsal flake scars, and there is a difference between the number of ridges, as a larger share of type II large blade knives have two ridges while the majority of type I blade knives have one ridge. Type II large blade knives have greater density than type I. Type II large blade knives make up a small but distinct part of my material, with 5% of the studied blade knives.

Rügen blade knives have technological attributes that resemble those of type I large blade knives, but they differ in size. They are smaller. The majority are between 8 and 10 centimetres long, weigh 20–60 grams, and are slightly narrower and thinner than other type I blade knives.

The majority of the large blade knives, irrespective of which group they are defined as, have a marked and well-formed bulb of percussion with a bulbar scar, moustache, and fissures and continuous, regular, and direct retouch. Roughly half of them have the retouch on the left side, the other half have it on the right side. The majority of the blade knives are made of Senonian flint.

The blade knives in my material that come from Schleswig-Holstein in Germany can be grouped with the type I large blade knives, not with the Rügen blade knives. There are also blade knives from excavations in south-western Denmark which resemble Rügen blade knives. There is thus a difference in the appearance of blade knives from northern Germany and south-western Denmark. It is interesting to note that the three blade knives of Zele type from Poland in my material have platform angles of 80, 85, and 90 degrees. They thus differ from the majority of the other large blade knives.

The experimental analysis shows that large blade knives were made according to standardized production and a systematic chaîne opératoire. The manufacture requires a good knowledge of lithic technology and a supply of high-quality raw material. Although type I large blade knives with and without platform preparation have a great deal in common, the chaîne opératoire for their manufacture is different. The main difference is in the choice of raw material. Type I large blade knives with a well-prepared platform are usually made of block flint. Type I large blade knives without platform preparation are usually made of tabular flint. This difference in the choice of raw material is directly related to the differences in the morphology of the blade knives and their mode of manufacture.

Manufacture generates diagnostic flakes in the form of blade-like flakes and preparation flakes. Even if the blade-like flakes are fragmented or have been reworked into tools, the majority of them are still diagnostic.

Blade knives of type I and type II have previously been described by Björhem and Säfvestad (1993). The difference that I have defined between large blade knives of type I with a well-prepared platform and large blade knives of type I without platform preparation fits within the framework of Björhem and Säfvestad's type I. This means that, although the two groups of type I blade knives that I defined can be grouped in Björhem and Säfvestad's definition, they are different kinds of blade knives as regards the way they were made. Rügen blade knives have not been described previously. The different chaîne opératoire for making blade knives and the waste material from their manufacture has not hitherto been described either.

The chapter about functional analyses began with a survey of previous research on the function of large blade knives. Through a detailed study of how polish from silicaceous plants has been described, and through experimental trials, a foundation was established for analyses of prehistoric large blade knives from excavations in the area around Malmö in south-western Scania, Sweden. The result shows that a blade knife should be regarded as a Late Bronze Age artefact that was part of a composite tool consisting of a knife blade of flint and a shaft of wood. The tool was used for cutting cereals or reeds. The experiments showed that the implement functioned very well for this task. It was also found that occasional fragments of large blade knives which broke were used for scraping hides. Previous studies have suggested that blade knives were used unshafted, and that they were used for other functions, chiefly cutting leaves. There is nothing in the results of my studies to confirm this.

In the chapter about the dating, context, and geographical distribution of large blade knives I demonstrated that large blade knives occur in Late Bronze Age contexts together with phase B pottery. They do not occur in the early part of the Late Bronze Age, in what is called Phase A, nor in later periods, that is, the Pre-Roman Iron Age. This result eliminates the former uncertainty as to the terminus post quem of large blade knives.

Large blade knives are usually found in settlement site material all over southern Scandinavia. Most blade knives from excavations and in museum collections are found in Scania and Halland in Sweden, in Denmark, and along the Baltic coast of northern Germany. In Sweden they also occur in smaller numbers along the west coast, in the east along the coast as far north as the Kalmar district, and at a few places in central Sweden. From Norway we have a small number of finds of blade knives. In Poland they have been found in settlement site excavations and from a flint-mining area.

The analysis of assemblages of large blade knives shows that this category of find should not be linked to an ancient Neolithic tradition of deposits in the ground, as previous studies have suggested, but should instead be viewed in relation to other deposits in the Late Bronze Age. The assemblages of blade knives are not in any way a uniform find category. The result of the analyses shows that assemblages of large blade knives display a complex variation in the manner of deposition and in the composition of objects with different biographies. Earlier studies of hoards from the Late Bronze Age have only studied finds of bronzes. The twenty-one finds that I have studied considerably increase the number of known deposits from periods V and VI and also qualify our picture of eastern Sweden as an area with more deposits than the rest of southern Scandinavia.

In the last chapter of the analyses of large blade knives I examined whether there were special manufacturing sites for blade knives and if so where. The result shows that these places existed. Six have been located. They are Wierzbica "Zele" in Poland, Tommestrup in Denmark, Södra Sallerup and Hanaskog, both in Sweden, and Reidervitz and Ruschvitz in Germany. All of these places are at supplies of plentiful raw material, and huge amounts of large blade knives were manufactured there in the Late Bronze Age. Södra Sallerup is the only one of these manufacturing sites where used large blade knives were found together with remains from the production of blade knives. It is also the only one of the sites that has yielded extensive settlement site remains from the Late Bronze Age. The other places can be regarded as purely manufacturing sites.

The result of the analyses shows that the manufacturing sites for large blade knives were places where craft dominated and few or no other activities were performed. They were organized for production. Finished large blade knives were made on the sites and distributed from them, together with preforms for large blade knives and blade-like flakes from the production of large blade knives. Large blade knives, preforms for large blade knives and blade-like flakes should thus all be perceived as products from these places.

Large blade knives, a specialized artefact

An important issue to discuss is whether large blade knives should be regarded as specialized. But the question is what any such specialization and standardization looked like. One point of departure for a discussion of these concepts is the understanding that tools of flint can be manufactured in different ways:

> The amount of effort expended in stone tool production is critical for understanding tool production processes. Some stone tools can be produced with very little effort and others require a great deal of production effort (Andrefsky 1998:30).

Although I am not primarily interested in effort, this quotation puts its finger on something important. Some flint objects are easy to make, others are not. And to link up with the discussion of chaîne opératoire in the first part of this book, certain types of manufacture require extensive knowledge and know-how, while others call for only knowledge (Bamforth & Finlay 2008:1ff). As Pelegrin has shown, the difference between what he calls ordinary knapping and elaborate production is large. Hypothetically, with proper raw material, tools, and some practice virtually anyone can do ordinary knapping, such as the production of flakes from a simple core. However, proper raw material, tools, and practice are not enough for the production of something like Late Neolithic Danish daggers (Pelegrin 1990:123; Apel 2001:34ff; Olausson 2008:30ff). An important aspect for defining manufacture as specialized or not is thus the type of knowledge and know-how required by the production.

But specialization involves more than knowledge and know-how. It is also about what is manufactured, what it looks like, and whether there are similarities or variations in the production. It concerns the site of the manufacture, whether it was a specially intended production place or just any place (Costin 1991). It is also about social aspects of production and specialization (Flad & Hruby 2007). Use is important in this content too, whether an object was used for a wide range of things in several different ways, or used for specific tasks and in just a few ways (Costin 1991). The question of specialized manufacture is thus multifaceted, and the subject have also been studied from a large number of aspects (see e.g. Pelegrin 1990; Roux 1990; Costin 1991; Olausson 1998; 2008; Hirth 2003; Yerkes 2003; Hruby & Flad 2007). Here I content myself with a discussion of manufacture and use. I return to other questions in Part III of the book.

The attribute analyses have shown that large blade knives were made with a technology that is uniform for a large proportion of the blade knives in this study. They also show that blade knives in each type are similar in appearance. The majority of the technological and morphological attributes are found on virtually all the blade knives, and the variations in each attribute exist within limits. Few large blade knives differ from the mass. The functional analyses reveal a uniformity in the manner and purpose of their use. The experimental manufacture shows that the blade knives were made in a complex chaîne opératoire that cannot be mastered by just anyone. Although Sørensen and I managed to make large blade knives through several series of experiments, we did not do this from scratch. I have studied large blade knives for more than 10 years, and Sørensen began to knap flint before he reached his teens. Making large blade knives requires both knowledge and know-how. This shows that large blade knives of flint are specialized and standardized in the sense that their production requires specialized skills, they are standardized in both manufacture and appearance, and they were used in specific ways and for specific purposes.

In previous studies of specialization there is one criterion above all that has been emphasized when a manufacturing site has been defined as specialized – that specialized manufacture went on there (see e.g. Brumfiel & Earle 1987). This can concern manufacturing sites for Neolithic axes (Hansen & Madsen 1983; Nordqvist 1992; Högberg 1999; 2002), places for the manufacture of Neolithic daggers (Apel 2001), or sites for the making of smaller composite tools such as awls and scrapers (Yerkes 2003). What the majority of studies like these have in common is that they are based more on analyses of what was manufactured on the site than on the site itself. Specialized manufacture on a site thus automatically gives a specialized manufacturing site.

Yet this criterion is not easy to handle either. Specialized manufacture may have been done on sites that are not specialized for manufacture, and a specialized manufacturing site may correspondingly have been used for the manufacture of non-specialized objects. In other words, manufacture may be specialized or not regardless of whether it was done in a place specialized for manufacture or not, and places can be specialized regardless of whether the manufacture performed there was specialized or not (Yerkes 2003:17ff). It therefore takes more than an identification of what was made on the site to determine whether it was specialized.

Several authors have stressed the importance of a workshop character for distinguishing manufacturing sites. They have thus been defined as places where only specialized production is represented, that is, the same type of tool was repeatedly made there. This is contrasted to non-specialized places where a variety of activities were performed (Roux 1990:145; Rosen 1997:113; Apel 2001:189; Odell 2004:204). Others have expressed this as a difference between household production and non-domestic workshops (Pastrana & Hirth 2003:198).

Although few studies explicitly state the quantity of what is made as a criterion for designating a place as a specialized manufacturing site, it is often implicitly understood that places where mass production of same type of artefact went on are also to be regarded as specialized (Brumfiel & Earle 1987).

Another criterion which has been used is the availability of raw material. If the production requires large quantities of high-quality raw material, as the experimental study has shown to apply to the making of large blade knives, it is not uncommon that places where the raw material is easily accessible are also specialized manufacturing sites (Högberg 2002; Odell 2004:204).

Based on this survey of previous definitions, it can be said that special manufacturing sites for large blade knives are specialized in so far as only one kind of tool was made on the site, and that no household production was done there. The material can be interpreted as the result of standardized mass production on sites where there was an ample supply of high-quality raw material.

The absence of settlement site material at the place is a criterion that does not fit Södra Sallerup, however. This means that this particular site should perhaps not be perceived as specialized in the same way as the others. The Södra Sallerup site is complex, as has also been noticed in studies of other archaeological remains from there (Björhem & Magnusson Staaf 2006). I will return to this question later on.

Household production of flint tools in the Late Bronze Age

The results of the analyses of large blade knives of flint presented above show that these were specialized and standardized tools, manufactured with specialized technology at specialized production sites, and used for standardized purposes as sickles or cutting tools for specific materials, and as parts of assembled finds. To be able to discuss flint technology in the Late Bronze Age, it is essential also to investigate other parts of this than the large blade knives. This is what I intend to do in this chapter. In the same way as I worked with the large blade knives above, the study here is based on attribute analyses of flakes, cores, and retouched tools, raw material analyses, and a dedicated use-wear analysis chiefly of unmodified flakes.

Although the occurrence of knapped flint from later periods than the Stone Age has long been known in archaeology, this find material has frequently been regarded as irrelevant for interpretations of prehistory. The following quotation from Müller (1919:50) is an example of this:

> In order that the excavation [of Voldtofte in Denmark] could leave a specific and lasting idea about the occurrence of flint waste, for most of the excavation it was collected and counted, but afterwards discarded, although particularly characteristic pieces were preserved.

Another example is the excavations in the 1950s and 1960s of the extensive settlement site remains at Fragtrup in northern Jutland. Here a great deal of the flint was dumped after it was registered (Draiby 1985:205). This treatment of flint has also occurred in other parts of the world (Shott 1994:70). Steven A. Rosen, in a study of Metal Age flint technology in the Levant, has defined this as "the pretty piece syndrome". Archaeologists have kept what they thought was beautiful (Rosen 1997:37). Yet another example is a debate in Britain where knapped flint from the Late Bronze Age and the Iron Age has been seriously questioned and even called "a pure fiction" (Saville 1981), which provoked controversy (Humphrey & Young 1999). In Britain this debate has helped to bring the subject up for discussion in a way that is lacking in other parts of Europe (McLaren 2008) and stimulated new studies with good results (Humphrey 2003, 2004).

A debate from the opposite point of view was carried on in Sweden during the 1920s. The historians Lauritz Weibull (1923) and Curt Weibull (1923) argued, without much nuance, that flint objects were so important in the Bronze Age that the period should really be perceived as a Stone Age culture. Archaeologists argued against this, claiming that the image of the Bronze Age as a Stone Age culture was not as clear as this (Ekholm 1923) or downright wrong (Nordén 1926). The debate did not lead to any new studies to speak of.

In southern Scandinavia in recent years, a number of studies of flint material from the Late Bronze Age have shown that, contemporary with the large blade knife technology there was also another flint-knapping technology (Björhem & Säfvestad 1993; Högberg 2001a; Knarrström 2000, 2001; Olesen & Eriksen 2007). Studies of so-called settlement site material, that is, knapped flint discovered together with waste material such as pottery and bone close to settlement site remains, have shown that flint was used as raw material for making simple *ad hoc* tools for household use. This technology has been called household production (see e.g. Aperlo 1994) and is characterized by an *ad hoc* technology whereby flakes were struck from simple cores with a hard technique that gave a high degree of fragmentation, and a small proportion of formal tools in relation to the amount of flakes in each assemblage (Björhem & Säfvestad 1993; Högberg 2001a; Knarrström 2001; Olesen & Eriksen 2007).

It is interesting that household production is a phenomenon with a wide geographical distribution. Studies from Scandinavia (Högberg 2001a; Knarrström 2001; Olesen & Eriksen 2007), Poland (Migal 2004), Germany (Brandt & Jantzen 1994), the Netherlands (van Gijn & Nickus 2001), Britain (Humphrey 2003), and France (Prost 2002) show that, while other parts of the material culture such as metal objects, pottery, and houses display a complex image of both variations and similarities over the large geographical areas covered by these studies, the household treatment of knapped flint is very similar in the studies from these different countries. These studies raise many important questions to work with as regards settlement site flint from the Late Bronze Age. What does the material look like? How was it made? What were the everyday knapped household flints used for? These are all important questions for this study. Others include how household production was related to the production of large blade knives. Are there any points of contact? Were large blade knives manufactured in the same contexts as household production?

Attribute analyses

To work with these questions I have done attribute analyses of knapped flint from three excavations of settlement sites in the Malmö area: Lockarp 7E, Tygelsjö, and Lockarp CT 7, tables 2.79 and 2.80. These three areas have been selected on the criteria that they have recently been excavated, which makes it easy for me to survey the contexts, the find material is uniform as regards dating and find contexts, and the features in the selected areas lack any admixture of finds from earlier or later periods. It should be added that, besides the material from these sites I have studied several other Bronze Age flint assemblages (Högberg 2001a, 2001b), although they will not be considered here.

The attribute analysis is based on 1,090 flakes, 114 cores, 15 pieces of retouched flint, and six formal tools, figure 2.158. The flakes have been registered in ten attribute categories according to a total of twenty-five variables. The attributes and the variables are partly the same as those used above in the analyses of the large blade knives, and partly different. The definitions

Table 2.79. The three excavation areas from which the flint in the analyses was taken.

Excavation area	Reference	Description
Lockarp 7E	Nord & Sarnäs 2005	The area is one of the larger excavation areas in the Öresund Fixed Link Project. It was excavated in 1997. The investigation revealed a settlement site area with a large number of features, dated to the Late Bronze Age and Early Iron Age: extensive house remains, a couple of large areas with post-holes, several large and small pits and pit systems, wells, collections of stones, two inhumation graves from the Early Iron Age and two damp hollows containing finds. The excavation yielded extensive settlement site material in the form of various structures, pottery, flint objects, bones, metals, etc.
Tygelsjö	Grehn 2006	The area is located south of Malmö. It was totally excavated in 2003, revealing extensive settlement site remains from the Late Bronze Age. These consisted of different types of pits, hearths, and remains of a house. Also found in the area were house remains from the Early and Late Iron Age. The excavation uncovered settlement site material from the Late Bronze Age in the form of various structures, pottery, flint objects, bones, and groundstone objects.
Lockarp CT 7	Celin 2006	The area is one of the smaller sites excavated by the Citytunnelprojektet. It is located a few kilometres north-east of Lockarp 7E. The area was excavated in 2001, revealing a number of pits and hearths from the Late Bronze Age, remains of four houses from the Early Iron Age and one from the Viking Age. The excavation yielded a small amount of settlement site material in the form of different structures, pottery, flint objects, bones, etc.

Table 2.80. Features from the excavation areas Lockarp 7E, Tygelsjö, and Lockarp CT 7 included in the study, the amount of flint from each feature, and the basis for the dating of the features. In these excavations the ploughed topsoil and the finds of knapped flint from it have been analysed. I thus know that the flint assemblages from the features presented here are representative in that they look like the knapped flint that was found in the topsoil (see Sarnäs 2008 for discussion). Lockarp 7E has catalogue number MHM 8546, Tygelsjö has catalogue number MK 141, and Lockarp CT 7 has catalogue number MHM 12881. The finds are stored at Malmö Museums.

Excavation area	Feature number	Number of registered flint objects	Basis for dating
Lockarp 7E	11513	198	Pottery, phase B
	32278	35	Pottery, phase B
Tygelsjö	241	54	Pottery, phase B
	310	15	Pottery, phase B, find of a large blade knife
	329	93	Pottery, phase B
	351	71	Pottery, phase B
	1696	58	Pottery, phase B, radiocarbon analysis
	2157	2	Pottery, phase B, find of a large blade knife
	2182	17	Pottery, phase B
	2201	32	Pottery, phase B
	2215	143	Pottery, phase B, find of a large blade knife
	2253	57	Pottery, phase B, find of a large blade knife
	18629	62	Pottery, phase B
	18678	67	Pottery, phase B
	18694	31	Pottery, phase B
	18768	77	Pottery, phase B
	26038	14	Pottery, phase B
	26050	64	Pottery, phase B
Lockarp CT 7	154	15	Find of a large blade knife
	450	36	Pottery, phase B
	451	68	Pottery, phase B, find of a large blade knife

Figure 2.158. Examples of flakes, cores, and retouched flint included in the attribute analyses.

Table 2.81. Survey of attributes and variables used for the analysis of the flakes.

Attribute	Variables
Appearance of the flake	A – whole flake
	B – flake without distal end
	C – flake without proximal end
	D – proximal part of flake
	E – medial part of flake
	F – distal part of flake
	G – flake with radial break
	U – undefined flake
Size of the flake	Length/width ratio, actual size of the flake
	Index (width/length = index)
The number of negative removals on the dorsal side	The number of previous removals from the part of the core from which the flake was detached
Morphology of the platform	S – small platform
	L – large platform
	C – crushed platform
	B – broken platform
Platform type	Fl – flat platform
	Fa – faceted platform
	U – unidentifiable platform
Platform shape	Rhombic platform
	Wing-shaped platform
Platform preparation	Occurrence of preparation at the platform edge
Hertz cone in the bulb of percussion	Distinct cone fracture in the bulb of percussion
Ring crack	Ring crack around the point of impact
	Ring crack on the platform beside the point of impact
Well-formed ventral side	Clean and well-formed ventral side

and terms used are the same as before. The variables for each attribute are specified in table 2.81. Cores and retouched flint have been registered according to morphological attributes.

The comparisons of the different attributes are based on the number and the percentage. Percentages are rounded to whole numbers. Only in exceptional cases are they stated with halves. The percentages should be regarded as relational values, not absolute values. This means, for example, that the occurrence of individual attributes in relation to several hundred other attributes can be stated as zero per cent.

The morphology and size of the flakes

The morphology of the flakes has been analysed according to the degree of fragmentation and fracture patterns and registered in eight groups, A–G, plus a group of undefinable flakes, table 2.82 and figure 2.159 (Madsen 1992:105).

A total of 1,090 flakes have been analysed. Table 2.83 sums up the distribution of the flakes in the different morphological groups.

Figure 2.160 presents the data from table 2.83 in visual form. Whole flakes, A-flakes, dominate. What is also evident is that, although A-flakes account for the largest percentage at each site, there is considerable variation in the appearance of the flakes. Between 40 and 45 per cent of the flakes are not intact. They are divided into flakes in groups B–G. It is worth noting that there is a larger share of U-flakes in the material from Lockarp 7E compared to the other two sites.

Different flint-knapping techniques and methods, combined with the use of varying raw material, give flakes of different sizes. It is therefore essential to register not only the degree of fragmentation but also size variables (Shott 1994:77; Andrefsky 1998:96f). The size of the flakes has been registered in

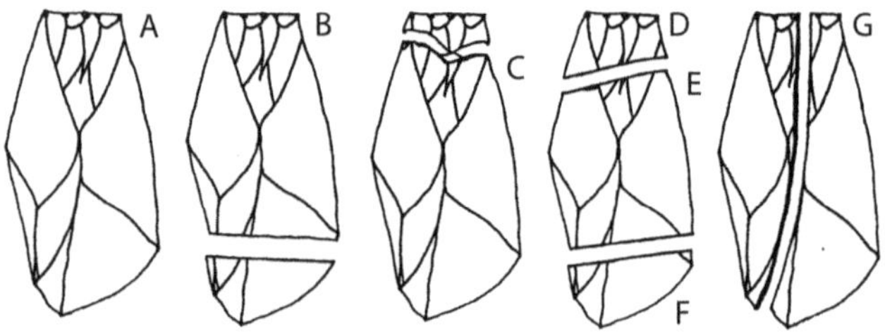

Figure 2.159. Flake groups A–G.

Table 2.82. The eight groups into which the flakes have been classified to determine the degree of fragmentation and fracture patterns.

Group	Definition
A	Whole flake
B	Flake without distal end
C	Flake without proximal end
D	Proximal part of flake
E	Medial part of flake
F	Distal part of flake
G	Flake with radial fracture in the form of a "siret", i.e. a break of a flake following the debitage axis
U	Undefinable flake which cannot be assigned to any of the other groups

terms of a length/width ratio and a length/width index (width divided by length = index). The maximum width of the flake is measured in millimetres at right angles to the direction of percussion. The maximum length is measured in millimetres along the direction of percussion (Andrefsky 1998:98: figure 5.8a & c). Flakes in groups D, E, F, G, and U have not been measured.

Table 2.84 shows the flakes divided by size in intervals of five millimetres. It is obvious that the majority of the flakes are in and around the interval 11–35 millimetres. Few flakes are bigger than 35 millimetres. This applies to both length and width.

The index (width/length) states the length of the flake in relation to its width. Only A-flakes have been registered. An index < 100 shows that the length of the flake is greater than the width, while an index > 100 shows that the length of the flake is less than the width, see figure 2.161.

The index for the A-flakes is presented in table 2.85, which shows that just over half of the flakes have an index < 100, meaning that more A-flakes are longer than they are wide.

Number of negative removals on the dorsal side

The occurrence of a ridge demarcating the negative removal is a necessary requirement if it is to be counted, and this ridge should be ≥ 10 mm. The value ≥ 10 mm has been chosen as a standard measure for debris (Andersson, Rex-Svensson & Wigforss 1978; Högberg et al. 2000; Larson & Kornfeld 1997:7). Negative removals from platform preparation, if any, are not counted.

Around 55–60 per cent of the flakes from the three sites have one or more negative removals on the dorsal side. This means that between 40 and 45 per cent of the flakes lack negative removals, table 2.86.

Of the flakes with negative removals on the dorsal side, more than 90 per cent have only one or two negative removals. A few per cent have three or four negative removals. No flake has more than four, table 2.87.

Figure 2.162 presents the data from table 2.87 in visual form. It is obvious that the majority of the flakes have no, one, or two negative removals on the dorsal side. Very few flakes have more than two.

The platform

The platform is defined here according to two variables: platform morphology and platform type. The morphology of the platform has been registered in four different groups, see table 2.88 and figure 2.163.

Table 2.83. Flakes from the three sites, number and percentage of morphological groups A–G and U.

	Lockarp 7E		Tygelsjö		Lockarp CT 7	
Morphology	**No.**	**%**	**No.**	**%**	**No.**	**%**
A	129	60	413	54	66	59
B	32	15	122	16	19	17
C	8	4	22	3	5	5
D	1	0.5	37	5	2	2
E	10	5	40	5	4	4
F	6	3	51	6	8	7
G	1	0.5	59	8	6	5
U	25	12	23	3	1	1
Total	**212**	**100**	**767**	**100**	**111**	**100**

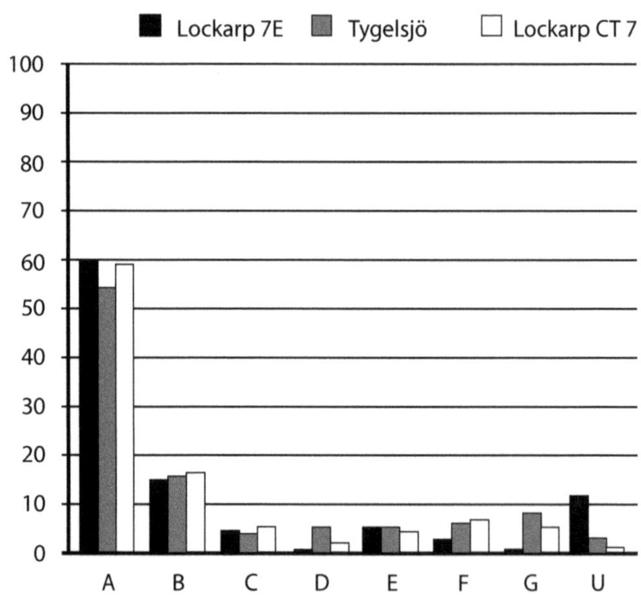

Figure 2.160. Percentages of flake registered in the different morphological groups.

Table 2.84. Width/length ratio of the flakes. The table states the width and length of flakes from each site together with the proportion of flakes of each dimension as a percentage of the total number of flakes from each site. The dimension in the column to the far left is in millimetres.

	Lockarp 7E				Tygelsjö				Lockarp CT 7			
Width/length ratio	**Width no.**	**Width %**	**Length no.**	**Length %**	**Width no.**	**Width %**	**Length no.**	**Length %**	**Width no.**	**Width %**	**Length no.**	**Length %**
–10	7	5	2	1	6	1	2	0.5	4	4.5	1	1
11–15	23	15.5	17	11	52	10	29	5.5	16	18	8	9
16–20	30	20	28	19	104	19.5	88	16.5	17	18.5	23	26
21–25	24	16	22	15	113	21	105	19.5	25	28	17	19
26–30	23	15.5	28	19	96	18	102	19.5	14	16	19	21
31–35	14	9.5	17	11.5	63	12	69	13	1	1	6	7
36–40	10	7	15	10	42	8	54	10	5	5.5	8	9
41–45	10	7	10	7	25	5	36	7	1	1	1	1
46–50	1	0.5	7	5	18	3	16	3	1	1	2	2
51–55	3	2	1	0.5	7	1	17	3	4	4.5	3	3
56–60	1	0.5	0	0	4	0.5	11	2	1	1	0	0
61–65	1	0.5	2	1	1	0	3	0.5	0	0	0	0
66–70	2	1	0	0	2	0.5	1	0	0	0	2	2
71–75	0	0	0	0	0	0	1	0	1	1	0	0
76–80	0	0	0	0	2	0.5	1	0	0	0	0	0
Total	**149**	**100**	**149**	**100**	**535**	**100**	**535**	**100**	**90**	**100**	**90**	**100**

Figure 2.161. Length/width index of the flakes.

Table 2.85. Index for A-flakes, stated in number and percentage.

	Lockarp 7E		Tygelsjö		Lockarp CT 7	
Index for A flakes	**No.**	**%**	**No.**	**%**	**No.**	**%**
< 100	68	53	231	56	36	55
100	3	2	13	3	4	6
> 100	58	45	169	41	26	39
Total	**129**	**100**	**413**	**100**	**66**	**100**

Table 2.86. The number of flakes with and without negative removals on the dorsal side, stated in number and percentage.

	Lockarp 7E		Tygelsjö		Lockarp CT 7	
Occurrence of negative removals on the dorsal side	**No.**	**%**	**No.**	**%**	**No.**	**%**
With	126	59	411	54	62	56
Without	86	41	356	46	49	44
Total	**212**	**100**	**767**	**100**	**111**	**100**

Table 2.87. Number of negative removals on the dorsal side of individual flakes, stated in number and percentage.

	Lockarp 7E		Tygelsjö		Lockarp CT 7	
No. of negative removals on the dorsal side	**No. of flakes**	**%**	**No. of flakes**	**%**	**No. of flakes**	**%**
1	82	65	292	71	46	74
2	36	29	93	22.5	13	21
3	6	5	24	6	2	3
4	2	1	2	0.5	1	2
Total	**126**	**100**	**411**	**100**	**62**	**100**

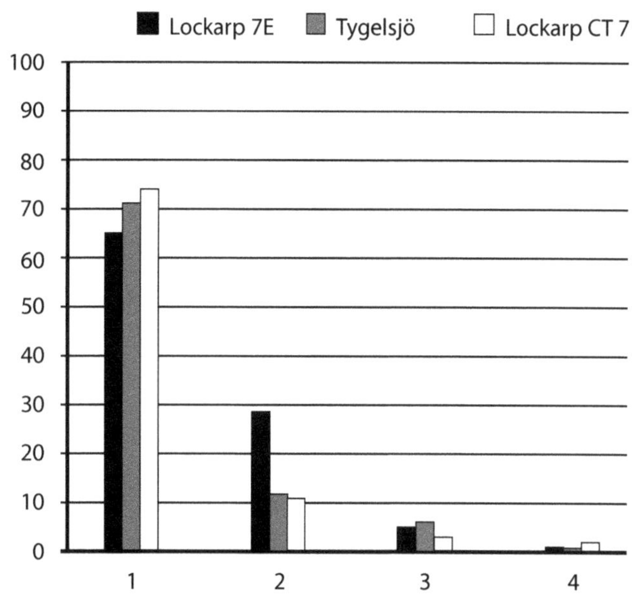

Figure 2.162. The number of negative removals on the dorsal side, percentage.

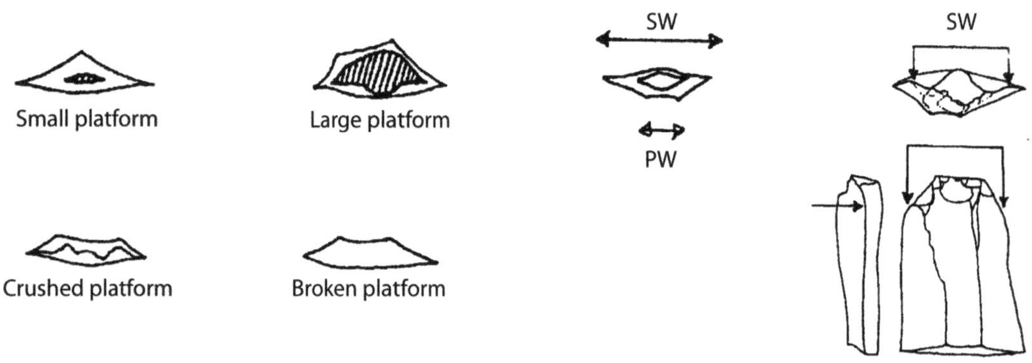

Figure 2.163. The size and morphology of the platform (after Madsen 1992:105).

Table 2.88. Definitions of platform morphology (after Madsen 1992:105f).

Group	Definition
Small platform	Platform width (pw) <2/3 of shoulder width (sw)
Large platform	Platform width (pw) >2/3 of shoulder width (sw)
Crushed platform	Platform crushed during flaking
Broken platform	Platform missing because it was broken off during flaking

A total of 891 flakes have been registered. Table 2.89 shows that flakes with large platforms dominate. Few flakes have a small, crushed, or broken platform.

Figure 2.164 presents the data from table 2.89 in visual form. Flakes with a large platform show a distinct dominance. The figure reveals that there is a difference in the material from Lockarp 7E and that from Tygelsjö and Lockarp CT 7. From Lockarp 7E the proportion of flakes with a large platform is slightly smaller and those with a small or crushed platform slightly larger.

Platform type has been registered in three groups: flat platform – the platform consists of a single negative removal and thus lacks ridges; faceted platform – the platform has one or more ridges; and an undefinable or irregular platform – the platform cannot be defined as flat or faceted, figure 2.165.

A total of 834 flakes have been registered. Table 2.90 shows that a flat platform is most common. Between 76 and 87 per cent of the flakes have a flat platform. Flakes with an irregular platform account for a small share of the registered flakes. Few flakes have a faceted platform.

Figure 2.166 presents the data from table 2.90 in visual form. Flakes with a flat platform clearly dominate. The figure also shows that there is a difference between the sites.

Table 2.89. Platform morphology, stated in number and percentage.

	Lockarp 7E		Tygelsjö		Lockarp CT 7	
Platform morphology	**No.**	**%**	**No.**	**%**	**No.**	**%**
Large	132	77	568	91	85	89
Small	22	13	25	4	5	5
Crushed	16	9.5	25	4	6	6
Broken	1	0.5	6	1	0	0
Total	**171**	**100**	**624**	**100**	**96**	**100**

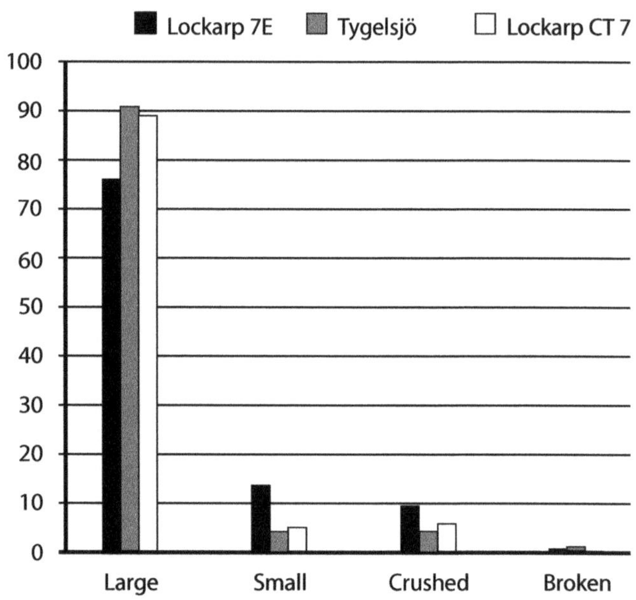

Figure 2.164. Platform morphology, stated in percentages.

Platform shape

The shape of the platform has been determined according to which type of impactor may have been used for the knapping. A rhombic platform has a marked transition between ring crack and platform edge on the ventral side of the flake, indicating that the flake was detached by direct technique with a tool of hard material. A wing platform has a soft but slightly marked transition between ring crack and platform edge, along with a wing shape, indicating that the flake was detached by direct technique with a tool of soft material (Ohnuma & Bergman 1982:161ff; Inizan, Roche & Tixier 1992:81; Knarrström & Wrentner 1996:85), figure 2.167.

A total of 846 flake have been registered under this heading. Table 2.91 shows the percentage of hard and soft technique for these.

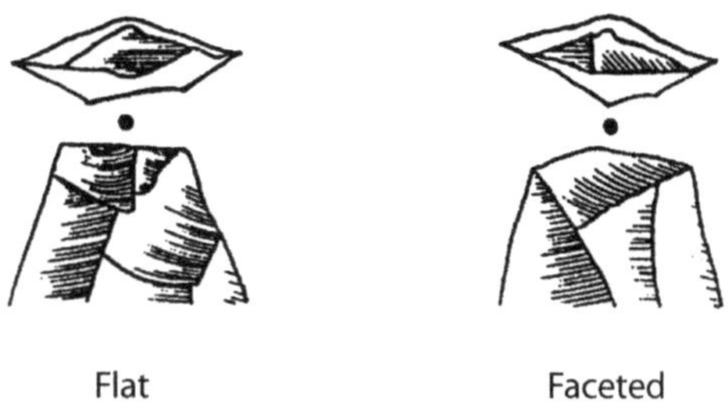

Figure 2.165. Platform type (after Inizan, Roche & Tixier 1992:81).

Figure 2.167. The shape of the platform. A rhombic platform indicates hard technique, a wing platform indicates soft technique (after Knarrström & Wrentner 1996).

Table 2.90. Platform type, stated in number and percentage.

	Lockarp 7E		**Tygelsjö**		**Lockarp CT 7**	
Platform type	**No.**	**%**	**No.**	**%**	**No.**	**%**
Flat	113	76	520	87	78	87
Faceted	12	8	12	2	0	0
Irregular	23	16	64	11	12	13
Total	**148**	**100**	**596**	**100**	**90**	**100**

Table 2.91. The distribution of rhombic and wing platforms, stated in number and percentage.

	Lockarp 7E		**Tygelsjö**		**Lockarp CT 7**	
Platform shape	**No.**	**%**	**No.**	**%**	**No.**	**%**
Rhombic platform	107	79	610	99	92	99
Wing platform	29	21	7	1	1	1
Total	**136**	**100**	**617**	**100**	**93**	**100**

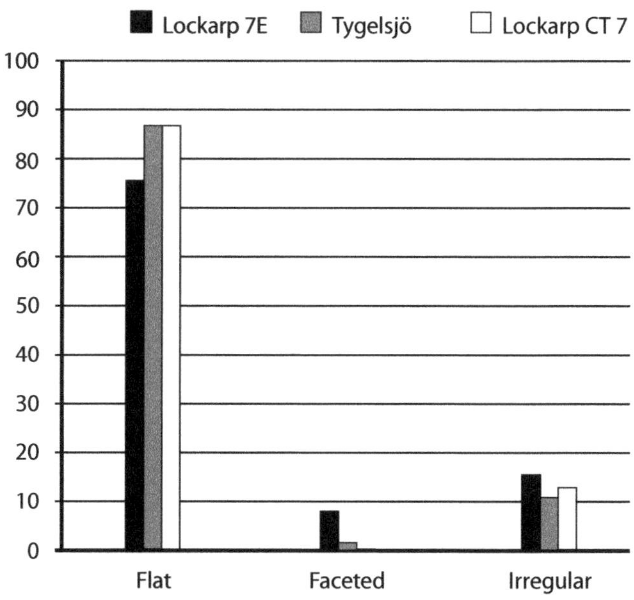

Figure 2.166. Platform type, stated in percentages.

The vast majority of the flakes have a rhombic platform, which indicates that they were detached using a hard technique. For the Tygelsjö and Lockarp CT 7 sites, the number of flakes with a wing-shaped platform is extremely small. There is, however, a difference between these two sites and Lockarp 7E. At the latter site one fifth of the flakes have a wing-shaped platform.

Platform preparation

Platform preparation is unusual in the material, table 2.92. Of altogether 821 flakes of A, B, and D type, which could hypothetically display traces of platform preparation, only 25, or 3 per cent, are prepared along the platform edge. It is thus evident that only a few flakes have platform preparation.

Hertz cone in the bulb of percussion

A Hertz cone is the shape of the fracture that arises when the flint is subjected to force in the form of percussion or pressure. This cone fracture is usually not visible in the bulb of percussion. However, excess force in the blow, in combination with a steep angle of percussion, can lead to the formation of a distinct Hertz cone in the bulb of percussion. The point of the cone is then located at exactly the point of impact and the cone stands out from the bulb of percussion and is separated from it by distinct cracks, figure 2.168. A Hertz cone in the bulb of percussion is usually associated with hard technique (Ohnuma & Bergman 1982:164f; Cotterell & Kamminga 1987:686).

A total of 231 flakes have been registered with a Hertz cone in the bulb of percussion. Table 2.93 shows that between 9 and 22 per cent of the flakes have a visible Hertz cone in the bulb of percussion.

Figure 2.169 presents the data form table 2.93 in visual form. The occurrence of a Hertz cone in the bulb of percussion on A- and B-flakes from the three sites is compared. This shows that there is a difference between Lockarp 7E and the other two sites, Tygelsjö and Lockarp CT 7.

Table 2.92. The number of flakes with platform preparation.

Lockarp 7E	12 flakes have platform preparation, out of a total 162 A, B and D flakes	7%
Tygelsjö	11 flakes have platform preparation, out of a total 572 A, B and D flakes	2%
Lockarp CT 7	2 flakes have platform preparation, out of a total 87 A, B and D flakes	2%

Figure 2.168. Hertz cone in the bulb of percussion.

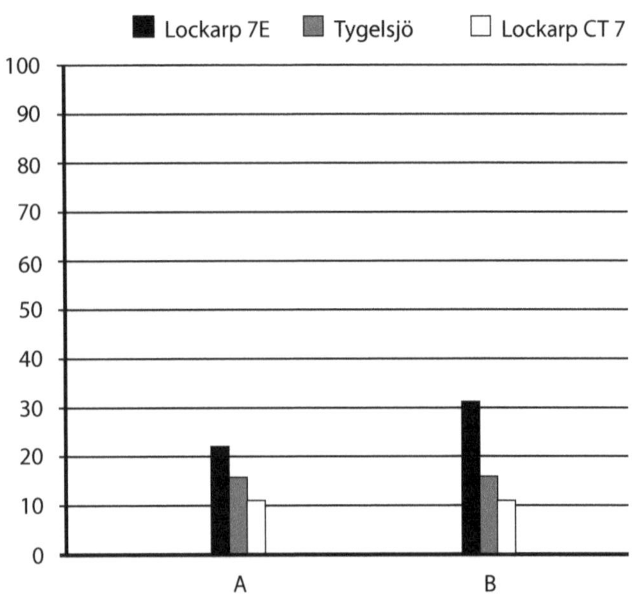

Figure 2.169. Occurrence of Hertz cone in the bulb of percussion on A- and B-flakes, percentage.

Ring crack

Ring crack has been registered in two variants, ring crack around the point of impact and ring crack on the platform beside the point of impact. Ring crack around the point of impact is formed at the point where the force of the blow that detached the flake was transferred to the platform. Ring crack around the point of impact is characterized by a scar in the form of a ring around the point of impact. In cases where a Hertz cone can be seen in the bulb of percussion, the ring crack is visible as the tip of the cone. Ring crack on the platform beside the point of impact consists of rings on the platform of the flake representing repeated blows on the core, that is, what is intended to become the flake platform, before the flake was detached, figure 2.170.

A total of 460 flakes with ring crack around the point of impact have been registered. From the Tygelsjö and Lockarp CT 7 sites, more than half of the A- and B-flakes have ring crack around the point of impact, table 2.94. Notice that there is a difference between Lockarp 7E and the other two sites.

Table 2.93. Flakes with a Hertz cone in the bulb of percussion, stated in number and percentage. The percentage is calculated in relation to the total number of flakes from each site and in each morphological group. Example: 129 A-flakes have been registered from Lockarp 7E, at the top of column one. Of these, 29 have a Hertz cone in the bulb of percussion, column two. This corresponds to a percentage of 22%, column three.

	Lockarp 7E			Tygelsjö			Lockarp CT 7		
Flakes with Hertz cone in the bulb of percussion	**Total number of flakes**	**No. with cone**	**% with cone of the total number**	**Total number of flakes**	**No. with cone**	**% with cone of the total number**	**Total number of flakes**	**No. with cone**	**% with cone of the total number**
A	129	29	22	413	68	16	66	7	11
B	32	10	31	122	21	17	19	2	11
C	–	–	–	22	–	–	5	–	–
D	–	–	–	37	3	8	2	–	–
G	–	–	–	59	1	2	6	–	–
U	25	1	4	–	–	–	–	–	–
Total	**186**	**40**	**22**	**653**	**93**	**14**	**98**	**9**	**9**

Figure 2.170. Platform with ring crack around the point of impact and ring crack beside the point of impact. Drawing: Anders Högberg.

A total of 246 flakes have been registered with ring crack on the platform beside the point of impact. Of these, the majority are A- and B-flakes. As table 2.95 shows, many have ring crack beside the point of impact. From the Tygelsjö and Lockarp CT 7 sites these flakes account for roughly one third of the A- and B-flakes. One difference between the sites should be noted: Tygelsjö and Lockarp CT 7 have a similar distribution of ring crack around and beside the point of impact, while the flakes from Lockarp 7E have a slightly smaller proportion of these attributes.

Well-formed ventral side

A well-formed ventral side is characterized by a flaking surface with a clearly formed and fully developed bulb of percussion, the absence of cracks, and clean, even waves of percussion emanating from the point of impact and the bulb of percussion, figure 2.171. A well-formed ventral side is an intuitive, non-metric variable that I use to define the visual impression made by the flake and thus to enable different reduction strategies to be distinguished.

Of altogether 1,090 registered flakes, only 20, or just under 2 per cent, have a well-formed ventral side, table 2.96. Once

Table 2.94. Occurrence of ring crack around the point of impact on flake platforms. For each site the table states first the number of flakes with ring crack, then the percentage of each morphological type with ring crack, then the total number of flakes of each morphological type, taken from table 2.83, and finally the percentage of flakes in each morphological group showing ring crack around the point of impact. Example: under the heading Lockarp 7E the first column shows how many flakes with ring crack around the point of impact occur in the material (for A-flakes, there are 43), the second column shows the percentage of these flakes in each morphological group (for A-flakes I have thus divided 43 by 58 to obtain 74%), the third column shows the total number of flakes in each morphological group from each site if this group contains any flakes with ring crack around the point of impact (for A-flakes there are 129), and the fourth column shows the percentage of flakes with ring crack around the point of impact in relation to the total number of flakes in each morphological group (for A-flakes I have thus divided 43 by 129 to obtain 33%).

	Lockarp 7E				Tygelsjö				Lockarp CT 7			
Ring crack around point of impact	No. with ring crack	% of type with ring crack	Total no. of flakes	% with ring crack of the total	No. with ring crack	% of type with ring crack	Total no. of flakes	% with ring crack of the total	No. with ring crack	% of type with ring crack	Total no. of flakes	% with ring crack of of the total
A	43	74	129	33	245	71	413	59	41	72	66	62
B	13	22	32	41	74	21	122	61	15	26	19	79
D	1	2	1	100	11	3.5	37	30	–	–	–	–
G	–	–	–	–	15	4.5	59	25	1	2	6	17
U	1	2	25	4	–	–	–	–	–	–	–	–
Total	**58**	**100**	**187**	**31**	**345**	**100**	**631**	**55**	**57**	**100**	**91**	**63**

Table 2.95. Occurrence of ring crack beside the point of impact on flake platforms. For each site the table states first the number of flakes with ring crack, then the percentage of each morphological type with ring crack, then the total number of flakes of each morphological type, taken from table 2.83, and finally the percentage of flakes in each morphological group showing ring crack beside the point of impact. Example: under the heading Lockarp 7E the first column shows how many flakes with ring crack beside the point of impact occur in the material (for A-flakes, there are 15), the second column shows the percentage of these flakes in each morphological group (for A-flakes I have thus divided 15 by 20 to obtain 75%), the third column shows the total number of flakes in each morphological group from each site if this group contains any flakes with ring crack beside the point of impact (for A-flakes there are 129), and the fourth column shows the percentage of flakes with ring crack beside the point of impact in relation to the total number of flakes in each morphological group (for A-flakes I have thus divided 15 by 129 to obtain 12%).

	Lockarp 7E				Tygelsjö				Lockarp CT 7			
Ring crack beside the point of impact	No. with ring crack	% of type with ring crack	Total no. of flakes	% with ring crack of the total	No. with ring crack	% of type with ring crack	Total no. of flakes	% with ring crack of the total	No. with ring crack	% of type with ring crack	Total no. of flakes	% with ring crack of of the total
A	15	75	129	12	150	73	413	36	11	55	66	17
B	4	20	32	12	36	17	122	30	8	40	19	42
D	–	–	–	–	9	5	37	24	–	–	–	–
G	–	–	–	–	11	5	59	19	1	5	6	17
U	1	5	25	4	–	–	–	–	–	–	–	–
Total	**20**	**100**	**186**	**11**	**206**	**100**	**631**	**33**	**20**	**100**	**91**	**22**

Figure 2.171. A flake with a well-formed ventral side, left, and a flake with a ventral side that is not well-formed, right.

Table 2.96. Flakes with a well-formed ventral side, stated in number and percentage.

Site	Number of flakes with well-formed ventral side	%
Lockarp 7E	Of 212 flakes, only 15 show a well-formed ventral side	7
Tygelsjö	Of 767 flakes, only 4 show a well-formed ventral side	0.5
Lockarp CT 7	Of 111 flakes, only 1 shows a well-formed ventral side	Just under 1

again it is worth noting a difference between Lockarp 7E and the other two sites.

Choice of raw material

Having now analysed the character of the flakes, I will briefly look at the raw material, to investigate which type of flint was used and where it may have been obtained. In a study performed as part of the Öresund Fixed Link Project I evaluated the occurrence of till flint in the topsoil on an excavation site just south of Malmö (Högberg 2001a:31). By till flint I mean here secondarily deposited flint transported by ice or till, which was broken out of the layer of limestone or chalk in which the flint was originally formed and subsequently deposited in the till when the ice melted (Knarrström 2000:25ff). The aim of the study was to ascertain what proportion of the till flint was of such a quality that it could be used as raw material for toolmaking. All the flint was collected from two topsoil squares, 50 × 50 centimetres. The squares together contained about 3.5 kilos of unworked flint, with everything from large nodules measuring approximately 10 × 5 × 4 cm to small pieces of one to two centimetres. A large share of the material was downright unsuitable as raw material because of the poor quality of the flint in the form of metamorphosed structure, thick cortex, small size, or a large admixture of fossils and other contaminants. But the topsoil also contained suitable flint nodules for making cores. Moreover, the squares contained about twenty frost-weathered bits which are suitable preforms for scrapers. This study was a simple spot check of a site which proved that the topsoil contained flint of different qualities which could be used to make cores and small tools. But the question is whether this picture is representative of other parts of the area around Malmö.

To explore this I got the people working on the Citytunnelprojektet to gather all the flint from a number of topsoil squares on different excavation sites. These are located in four places around Malmö. In this way I have obtained a representative picture of the occurrence of flint in the topsoil in the studied area, table 2.97.

As table 2.97 shows, there is plenty of flint in the topsoil. Even though this flint is of varying quality, there are many small and large nodules that can be used for making cores, figure 2.172. It is thus evident that earlier findings from studies of raw material deposits are still valid (Högberg 2001a:31).

Table 2.97. The number of small and large possible preforms for flake cores in the topsoil at six excavation sites around Malmö. Site Elinelund 2A–B was part of the Öresund Fixed Link Project (Sarnäs & Nord Paulsson 2001). The data in the table from this site come from an earlier study (Högberg 2001a). The Almhov (Gidlöf, Hammarstrand Dehman & Johansson 2006), Bunkeflo (Lövgren et al. 2007), Palissaden (Brink & Hydén 2006), Lockarp (Hansson & Celin 2006), and Hotelltomten (Hadevik et al. 2006) sites are all part of the Citytunnelprojektet. Note that the Öresund Fixed Link Project excavated topsoil in squares of 50 × 50 centimetres and the Citytunnelprojektet in volumes of 400 litres per square, which means that the volume of the squares is not comparable. It should be added that the quantity stated in the table cannot be said to correspond to the situation in prehistoric times. The flint in the topsoil was retrieved and used by prehistoric tool makers, which must have reduced the number of useful flint nodules in the topsoil; flint may have been added through raising by frost, soil improvement with the addition of lime containing pieces of flint, and modern deep ploughing bringing useful new flint nodules up into the topsoil from the underlying soil (Högberg 2001a:31). The study should primarily be regarded as a qualitative rather than a quantitative assessment.

Site name	Number of squares from which flint was collected	Number of possible preforms for flake cores, large	Number of possible preforms for flake cores, small
Elinelund 2A–B	2	8	20
Almhov	4	6	39
Bunkeflo	4	2	25
Palissaden	4	2	18
Lockarp	7	1	42
Hotelltomten	4	4	32
Total	**25**	**23**	**176**

Figure 2.172. Some small and large possible preforms for flake cores, from the Almhov area (Gidlöf, Hammarstrand Dehman & Johansson 2006). The nodule at the bottom right is approximately eight centimetres long.

The Citytunnelprojektet was a large-scale excavation project which stripped areas of several square kilometres of topsoil to expose till layers. This gave me an opportunity to investigate the occurrence of flint in large continuous areas, not just in the topsoil, but also in the till. At several places around Malmö there are streaks of limestone or chalk mixed with clay, sand, and gravel. These contain occasional large nodules of Scandinavian Senonian flint, Grey Band Danian flint, and Matte Danian flint (Högberg & Olausson 2007). Although they have been pulled out of their original contexts in limestone or chalk layers and deposited in the till, these flint nodules are often of high quality, figure 2.173.

The result of the raw material analysis shows that there is a plentiful supply of flint in topsoil and till in the Malmö area. On the basis of the appearance of the debitage as regards cortex, structure, and colour, together with the experience gained from the raw material studies presented here, it can be stated that the flint that was used for knapping at Lockarp 7E, Tygelsjö, and Lockarp CT 7 came from the topsoil, picked directly from the ground, in the form of flint with frost cracking, flint with metamorphosed structure, and flint heavily contaminated with fossil-like elements or thick cortex formation. The type of nodules of high quality that occur in till deposits were not used as raw material for household production.

The technological and morphological character of the debitage

The overall impression gained from the analysis of household production gives a picture of a heterogeneous body of material. True characteristic features are few, and it is in fact the absence

Figure 2.173. A number of large nodules of high quality flint picked from the till at the following places, from left to right: Lockarp, Bunkeflo, Lockarp, Lockarp again, and Palissaden. For references to each area see the heading of table 2.97. The boxes in which the flints are lying are 50 centimetres long, 30 centimetres wide, and 12 and 25 centimetres high.

of distinctive features that is the most striking characteristic. It is, however, possible to single out certain features that describe and define the character of the debitage more exactly.

- The degree of fragmentation is high. Just over half of the flakes are whole, and just under half of the flakes are broken in one way or another.
- The length and width of the flakes is on average within the interval 11–35 mm. Of the whole flakes, slightly more are longer than they are wide.
- The majority of flakes lack negative removals on the dorsal side. If they do occur, there are mostly just one or two.
- The platform, on flakes where this is preserved, is an irregular flat platform. Platform preparation occurs very rarely.
- The predominant technique is a hard technique resulting in a rhombic platform.
- About a fifth of the flakes have a distinct Hertz cone in the bulb of percussion.
- Ring crack around the point of impact and ring crack on the platform beside the point of impact are common.
- The ventral side of the flake is virtually never well-formed.
- The raw material was local flint picked directly from the ground on or near the sites.

A small number of flakes with a flat or faceted platform, platform preparation, wing platform, and well-formed ventral side show that, besides the predominant appearance of the flakes described above, there are simultaneously flakes with other characteristics, but they make up a very small proportion of the material.

It is important here to deal briefly with the relationship between the character of the raw material and the appearance of the flakes. Several studies have demonstrated the relationship between a flake's breakage patterns and the type of raw material that was used (see e.g. Amick & Mauldin 1997:25; Prentiss 1998:648). This raises the question whether the results of the debitage analyses presented above actually have more to do with the choice of flint from the till as raw material than with the flint-knapping technique and the method used.

This question is easily studied through a comparison with the use of the same raw material in other periods. In a study of Middle Neolithic debitage from an excavation just beside the site called Palissaden in table 2.97, I have examined raw material use as part of the technological treatment of flint tools. The study shows that nodules picked directly from the ground were used as raw material for systematic production of flakes made from prepared platform cores (Högberg 2008b). The same type of raw material that was used in the Late Bronze Age for unsystematic production was thus used in the Middle Neolithic for systematic production, figure 2.174.

The raw material affects production in that it sets limits. For example, nodules that are eight centimetres in diameter cannot be used for making flakes that are nine centimetres long. In this way there is a relationship between the use of small nodules of flint picked from the till and the size of the debitage I analysed. But it is possible to use flint from the ground as raw material for making whole, well-prepared flakes in systematic production. This shows that the character of the debitage from the Late Bronze Age is not due to the choice of raw material, but to the flint-knapping method that was practised.

There is a difference between Lockarp 7E and the other two sites, Tygelsjö and Lockarp CT 7. This difference is not large but it recurs for several of the attributes. The difference does not affect the overall impression of the flint material, but it is still worth noting that, despite the great similarities in several of the attributes, there are also differences. Subtle differences of this kind between different sites have been noted in other contexts where household production of flint tools has been studied. It has then been explained in terms of differences in the economies of the sites (Aperlo 1994). The differences in the material analysed here, however, are so small that they could be explained by the way individuals managed the same technology (Shott 1994:73). The differences that are expressed, for example, in producing a whole or a fragmented flake are differences that have repeatedly been shown in experimental studies to be variations between different persons' ways of knapping flint within a technology which gives an overall impression of similarity more than difference (see e.g. Ferguson 2003; Finlay 2008). It should also be noted that the size of the registered differences between the sites is within the framework of the variation over time in measurement errors that

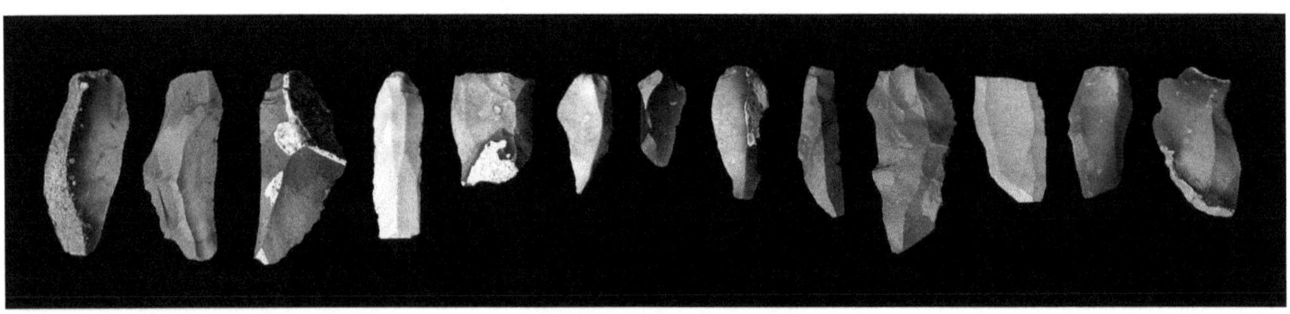

Figure 2.174. Some flakes from a Middle Neolithic flint assemblage from an excavation beside the area called Palissaden in table 2.97. The flakes are made of the same type of flint picked from the ground as was used in the Late Bronze Age. The flake to the far left is 7.5 centimetres long. From Högberg (2008b).

has been observed in experimental tests for one and the same observer. Studies of how different people's ways of registering attributes can change over time have demonstrated that it is not uncommon to find a variation or deviation of about 10 per cent between registrations of the same material performed by the same persons but on different occasions. The same studies show that a variation of 20 per cent or more is uncommon (Shott 1994:74f). Since several years elapsed between the time when I analysed Lockarp 7E on the one hand and Tygelsjö and Lockarp CT 7 on the other hand, this is a fairly reasonable explanation for the differences. However the differences between the sites are perceived, this does not affect the overall impression of the debitage.

Cores

Having analysed the debitage, I now turn to what the flakes were detached from: the cores. The appearance of the cores is intimately associated with what is produced. The core dictates the shape of the flakes that are removed from it (Dibble 1997:156ff; Andrefsky 1998:137ff). It is therefore important here to determine whether the characteristics revealed by the analysis of the flakes can also be found among the cores. This is in order to investigate whether the flakes that I analysed come from cores from the site or if there are cores in the material from which flakes were produced that are not found in the debitage, that is to say, they were taken from the site or handled somehow differently from the rest of the debitage. These analyses are particularly important for deciding whether remains of the manufacture of large blade knives can be found in the material or not.

To get some picture of the morphology of the cores, they have been classified in two groups: platform cores and other cores. A platform core is a core with one or more platforms from which more than one negative removal runs down over the front of the core, while other cores are those which have been knapped all around, with an irregular form. Other cores lack a clearly shaped platform and front (Andersson, Rex-Svensson & Wigforss 1978; Högberg et al. 2000). As cores I count all flint nodules with one or more negative removals (Inizan, Roche & Tixier 1992:84). This differs from Scandinavian practice, where something that is classified as a core usually has to have a surface dominated by three or more negative removals (Andersson, Rex-Svensson & Wigforss 1978:242ff; Ballin 1996:12; Högberg et al. 2000:8ff).

Of the total 114 cores, 112 have been registered as other cores and two as platform cores, table 2.98. A couple of round cores, as they have been defined by Björhem & Säfvestad (1993:70ff), occur in the material from Tygelsjö. From the same place, 28 cores are registered as having been struck when supported against an anvil (Knarrström 2001), a technique that should not be confused with hand-held direct percussion against a stationary anvil (Holdaway & Stern 2004:13). Cores knapped supported against an anvil have also been noted by Knarrström (2000:76f). Two such cores occur among the finds from Lockarp CT 7. In the material from Lockarp 7E, round cores and anvil technique were not registered. As table 2.99

Table 2.98. Type of cores, stated in numbers.

Type of core	Lockarp 7E	Tygelsjö	Lockarp CT 7	Total
Other cores	18	87	7	112
Platform cores	0	1	1	2
Total	**18**	**88**	**8**	**114**

Table 2.99. The number of negative removals on other cores, stated in number and percentage.

	Lockarp 7E		Tygelsjö		Lockarp CT 7	
Number of negative removals on other cores	**No.**	**%**	**No.**	**%**	**No.**	**%**
1	5	28	40	46	1	14
2	4	22	21	24	3	43
3	6	33	10	11	2	29
4	–	–	7	8	1	14
5	1	6	3	4	–	–
6	–	–	3	4	–	–
7	2	11	–	–	–	–
8	–	–	2	2	–	–
9	–	–	–	–	–	–
10	–	–	1	1	–	–
Total	**18**	**100**	**87**	**100**	**7**	**100**

shows, the majority have only one or a few negative removals. Extremely few cores have four or more negative removals.

It is worth stressing that more than half of the cores from all the sites show only one or two negative removals and would therefore, if they were classified according to current Scandinavian nomenclature (Andersson, Rex-Svensson & Wigforss 1978; Ballin 1996; Högberg et al. 2000), not have been defined as cores at all.

The result of the analysis of the cores agrees well with the result of the analysis of the flakes. The majority of the flakes and of the cores have only one or two negative removals on the dorsal side. The cores found in the material are of the kind from which the flakes were removed. There is thus nothing to suggest that anything manufactured at any of the sites was taken away from there. What was manufactured on a site was also deposited there.

A technological and methodological description of household production

Based on the description of the different attributes of the flint, I will now propose an interpretation of the technological and methodological character of the knapping. In this interpretation I distinguish between the terms method and technique, whereby the flint-knapping method can be explained as a sequence or series of actions performed with the ambition of making a predetermined product. The direct physical activity that is performed when flint is struck, for example, with a hammerstone, or the ways in which the core is held still during manufacture is the actual flint-knapping technique (Pelegrin 1990:116ff; Inizan, Roche & Tixier 1992:34). The method is thus the mental template for the knapping, a template that involves a goal and a predetermined idea of how to achieve this goal through different strategies. The technique is a practical implementation of the goal of the method in the form of the concrete actions performed by the flintsmith.

The use of these terms as interpretations of two different parts of flint knapping has been challenged for being an over-simplified interpretation of a complex craft (Apel 2001). The distinction here between method and technique should be regarded solely as an analytical device intended to elucidate the character of the knapping.

The occurrence and absence of platform preparation on flakes is an attribute that can be used to determine whether the flint-knapping method used for manufacture was simple or composite in form. Platform preparation is a way for the flint-knapper to control his work. It helps the flint-knapper to steer the knapping and thus remove flakes that correspond to the intentions behind the work. Platform preparation is therefore usually associated with a well thought-out and planned flint-knapping method (Inizan, Roche & Tixier 1992:46f). In the material from household production only a few flakes have platform preparation.

The number of negative removals on the dorsal side of the flake, in combination with the number of negative removals on the cores, is an attribute that also indicates the character of the method (Inizan, Roche & Tixier 1992:47). If there are no or few negative removals on the dorsal side of the flakes this shows that only a few flakes were removed from the same part of the flint nodule, as is confirmed by the analysis of the cores.

A uniform appearance of the platforms in a flint assemblage suggests structured and conscious methodological working (Dibble 1997:156ff). In contrast, it can be said that, if the appearance of the platforms varies, as is the case with the debitage analysed here, this suggests a multifaceted and unsystematic working method. This shows that no effort was made to shape cores with elaborated platforms and core fronts which could be used to produce series of well-formed and uniform flakes (Inizan, Roche & Tixier 1992:46ff).

The lack of platform preparation, the few negative removals on the dorsal side of the flakes, the variation in platform morphology and flake size – all this shows that the flint-knapping method used for household production in the Late Bronze Age was simple and unsystematic in character. Flakes were produced without the flaking being preceded by any proper preparation of the nodule. On each flaking occasion, the flint-knapper chose a place on the flint nodule to remove a flake, without proper planning and with regard only for the morphology of the nodule. The flint-knapper seems to have handled each removal more or less as it came. Several of the cores were knapped when supported against an anvil.

The high degree of fragmentation indicates a technique that caused the flakes to split and fragment during the knapping, and the majority of the flakes with a preserved platform have a rhombic platform. The degree of fragmentation, together with the occurrence of rhombic platforms, indicates that a direct and hard technique was used, presumably with a hammerstone as the tool (Ohnuma & Bergman 1982:161ff; Inizan, Roche & Tixier 1992:81; Knarrström & Wrentner 1996). This technique totally dominates the material.

A few flakes with distinctly flat or faceted platforms with wing-shape show that, parallel to the hard direct technique there was a soft direct technique whereby the flakes were detached with an antler club, a soft and light hammerstone, or the like. Platform preparation and choice of technique are usually considered to be variables that affect the appearance of a flake. Well-developed platform preparation and a soft technique are, as a rule, associated with flakes with a greater length/width ratio (Dibble 1997:150ff.). This is not the case with the debitage from Lockarp. Flakes with platform preparation using a soft technique do not differ in size from other flakes in the material. The majority of the flakes are found in the size range 11 to 35 millimetres.

In an earlier study I have discussed the possibility that metal in the form of composite tools containing copper, bronze, or iron, for example punches with bronze tips, were used for knapping flint in the Late Bronze Age (Högberg 2001a:36ff). The argument for this was an association between the attributes Hertz cone in the bulb of percussion and ring crack around the point of impact. In the study I relied on the results of a technological analysis of flint production at flint mines from the Bronze Age at Rohri Hills, Sindh in Pakistan, where these and other attributes were used to demonstrate the use of copper-tipped implements for blade production (Briois et al. 2006). The experimental study of the manufacture of large blade knives presented above, however, shows that the association between, on the one hand, Hertz cone in the bulb of percussion and ring crack around the point of impact, and on the other hand a flint-knapping tool of metal, is not as clear as I formerly claimed (Högberg 2001a:37). There is no evidence in the analyses, as I see it, that metal was used for knapping flint in the south Scandinavian Late Bronze Age.

To sum up, it can be observed that the flint shows signs of impulsiveness and the constant use of *ad hoc* solutions. It seems to lack any trace of elaborate and generally accepted strategies for flint-working. But the very lack of such strategies is in itself an attitude that can in fact be said to be representative of the flint analysed here. A similar situation as regards strategies characterized by an absence of strategies has also been demonstrated in a study of lithic technology after the introduction of metal in the Levant (Rosen 1997:103ff).

The result of the analyses shows that the manufacture represented by household production is totally unlike the manufacture of large blade knives. Household production and large blade knife production represent two completely different manufacturing technologies.

The different techniques that have been demonstrated in the material from household production are evidence that flint-working on settlement sites from the Late Bronze Age in the Malmö area nevertheless contains more than one dimension, and despite the variation in the material there are traces of different manufacturing techniques. On the basis of method and technique, the flint knapping can be summed up as follows:

- The method was simple in character. The lack of long-term step-by-step thinking is striking. Each stage in the flint knapping seems to have been performed with no thought for the consequences that the action would have for the continued working of the objective piece.

Two different techniques can be detected in the material:

- A hard direct technique with a hammerstone is most common.
- Less common is a soft direct technique with a soft hammerstone or antler club.

What was made?

After this analysis of the methodological and technological character of the debitage it is time to ask the question: What was made? What types of tools are found in the material? What were they used for?

I have worked in two ways with these questions. I have studied the appearance and form of retouched flakes, tools, and frost-cracked pieces, and I have performed focused use-wear analyses of retouched and unmodified flakes and frost-cracked pieces, as well as tools.

Retouched flakes and frost-cracked pieces with retouch

Of the total 15 retouched flakes and frost-cracked pieces with retouch, one third are frost-cracked pieces, see table 2.100. A frost-cracked piece arises when water in cracks in the flint freezes and expands, thus causing a fracture, detaching a flake. This fracture is characterized by the way waves, similar to waves of percussion, emanate from a point in the middle of the piece towards the edges. The detachment of a flake by the effect of frost always lacks a platform. A frost-cracked piece with retouch is one that has been retouched after having been cracked by frost.

The appearance of the retouch and form

The retouched flint is characterized in terms of the appearance and form of the retouch in two parts, retouch morphology and retouch position. These are the same attributes as used in the analyses of the retouch on the large blade knives, see above. The analysis was performed in order to gain an idea of how uniform the retouch is and whether there is any form of

Table 2.100. Retouched flakes and frost-cracked pieces from the Lockarp 7E and Tygelsjö sites, stated in numbers. The Lockarp CT 7 site yielded no retouched flakes or frost-cracked pieces.

	Lockarp 7E	Tygelsjö
Flakes with retouch	6	4
Frost-cracked pieces with retouch	3	2

Table 2.101. Retouch morphology, stated in number and percentage. There are no retouched flakes from the Lockarp CT 7 site.

	Lockarp 7E		Tygelsjö	
Retouch morphology	**No.**	**%**	**No.**	**%**
Continuous and regular	6	33	6	55
Partial and regular	1	6	–	–
Continuous and irregular	3	17	5	45
Partial and irregular	8	44	–	–
Total	**18**	**100**	**11**	**100**

Table 2.102. Retouch position, stated in number. There are no retouched flakes in the material from Lockarp CT 7.

	Direct	Opposed	Alternate	Alternating
Lockarp 7E	8	3	3	–
Tygelsjö	7	2	2	–

standardization in the look of the retouch, despite the non-standardized character of the debitage.

As table 2.100 shows, few flakes and tools are retouched. Of those that show retouch, it is irregular in the majority of cases, table 2.101. A characteristic feature is that the retouches vary in appearance.

The position of the retouch on the flakes varies, table 2.102. In this context it is not relevant to register retouch position on frost-cracked pieces with retouch, since these lack what is traditionally perceived as a ventral side. It is evident that the majority of the flakes have a direct retouch position. A smaller share are retouched from the dorsal side, and some of the flakes have alternate retouch. No flakes have alternating retouch.

The results of the analysis show that the retouches on the flakes are mostly irregular and direct, that is, consisting of a retouch with the negative removals of varying height and width, where the retouch has been struck from the ventral side of the flake and is thus visible on the dorsal side of the flake. Frost-cracked pieces were used for retouching and make up about a third of the retouched material. The non-standardized character revealed by the debitage analysis above is thus a pattern that recurs in the morphology of the retouch.

Formal tools

There are few formal tools, defined as such according to nomenclature based on Stone Age material (Andersson, Rex-Svensson & Wigforss 1978; Ballin 1996; Högberg et al. 2000), table 2.103. There are no tools from the Lockarp 7E and Lockarp CT 7 sites.

Six formal tools have been registered from Tygelsjö. Again, these tools have been defined according to nomenclature systems that are all more or less based on knowledge derived from studies of Stone Age material and therefore adjusted to suit that material (Andersson, Rex-Svensson & Wigforss 1978; Ballin 1996; Högberg et al. 2000). Of the six formal tools that

Table 2.103. Formal tools from the Tygelsjö site, stated in number. There are no tools from the Lockarp 7E and Lockarp CT 7 sites.

	Tygelsjö
Awls	3
Scrapers	2
Strike-a-light	1
Total	**6**

have been registered, three are awls, two are scrapers, and one is a strike-a-light.

It can be observed that the total number of formal tools, just six, in relation to the total of 767 flakes from all the features at Tygelsjö is very small. Tools account for only 0.8 per cent of the total flints. This is characteristic (Knarrström 2000:93; Högberg 2001a:47).

The ratio of formal tools to flakes is interesting. If only six tools have been identified in the material, what production did the 767 flakes come from? There are about 130 flakes for each formal tool, which seems like a very high proportion of flakes per produced tool. There may be many reasons for this. The material was in all probability selected, in the manufacturing situation, in the use, in the deposition, and in the excavation (see e.g. LaMotta & Schiffer 1999). But it is also likely that the material actually does contain many more tools than I have been able to define in terms of the nomenclature used (Knarrström 2000, 2001).

One category of flint object that is not usually regarded as a tool in Scandinavian archaeology is the unmodified flake. This is not so strange in view of the strong typological tradition to which Scandinavian archaeology belongs (see e.g. Fuglevik 2007/08), a tradition that views unmodified flakes exclusively as waste from manufacture, which often was not even collected as part of the finds from excavations, see the discussion above (exceptions include Stjernquist 1969; Knutsson 1988; Knarrström 2000, 2001; Högberg 2001a). From another angle, however, it is strange, because there are several archaeological and

Table 2.104. Examples of operations and contact materials in my reference collection. Flint types used are Scandinavian Senonian flint, Common Kristianstad flint, and Matte Danian flint, Östra Torp variety (Högberg & Olausson 2007).

Tool	Type of flint	Working time in minutes	Operation	Contact material
Burin	Senonian	30	Scrape	Elk antler
Burin	Senonian	30	Scrape	Roe deer bone, dried, clean
Unmodified blade	Senonian	30	Cut	Fresh oak
Unmodified blade	Senonian	3 × 15	Cut	Elk antler, dry, clean roe deer bone, and fresh hazel
Unmodified blade	Senonian	30	Cut	Dried oak
Unmodified blade	Kristianstad	30	Cut	Raw ox hide
Unmodified blade	Kristianstad	30	Cut	Dried ash
Unmodified blade	Kristianstad	30	Cut	Fresh hazel
Unmodified blade	Kristianstad	30	Cut	Fresh reeds
Unmodified blade	Kristianstad	30	Cut	Fresh rib from wild boar
Unmodified blade	Senonian	30	Cut	Dried ash
Unmodified blade	Senonian	30	Cut	Fresh hazel
Unmodified blade	Senonian	30	Cut	Tall grass
Unmodified blade	Senonian	30	Cut	Fresh reeds
Unmodified blade	Senonian	30	Cut	Raw ox hide
Unmodified blade	Senonian	30	Cut	Fresh rib from wild boar
Unmodified blade	Senonian	30	Cut	Roe deer bone, dried, clean
Unmodified blade	Senonian	30	Cut	Elk antler
Unmodified blade	Senonian	30	Cut	Raw ox hide
Scraper	Senonian	30	Scrape	Fresh wild boar bone
Scraper	Senonian	30	Scrape	Dried elm
Scraper	Senonian	30	Scrape	Roe deer bone, dried, clean
Unmodified blade	Senonian	30	Scrape	Elk antler
Unmodified flake	Senonian	30	Scrape	Raw ox hide
Unmodified blade	Senonian	30	Scrape	Fresh elm
Unmodified flake	Senonian	30	Scrape	Fresh ox meat
Unmodified flake	Danian	30	Cut	Dried oak
Unmodified flake	Danian	30	Cut	Raw ox hide
Unmodified flake	Danian	30	Cut	Fresh rib from wild boar
Unmodified flake	Danian	30	Cut	Fresh reeds
Unmodified flake	Danian	30	Cut	Fresh hazel

ethnographic studies which clearly show that unmodified flakes make up a considerable share of the lithic tools used above all in household production, for cutting, scraping, and planing (Knutsson 1988;143ff; Sillitoe & Hardy 2003; Shott & Sillitoe 2005). An important issue to investigate is therefore whether there are, concealed among the unmodified flakes, tools that were used for various purposes. To find out, I have performed focused use-wear analyses.

Use-wear analyses of unmodified flakes from household production

Unlike the use-wear analyses of large blade knives described above, the analyses presented here were undertaken solely to determine the occurrence or non-occurrence of polish and to identify the contact material in cases where this was possible. The polish has been described in terms of its position and appearance, that is, where it occurs on the edge of the tool, and whether it looks as if it could have come from the use of the tool (see Levi Sala; 1986, 1996; van Gijn 1989; Jensen 1994). Identification of the contact material was done with the aid of published references (Keeley 1980; Vaughan 1985; van Gijn 1989; Jensen 1994), together with an experimentally produced reference collection, table 2.104.

This means that, when I identify polish as coming from, say, contact with meat, it is because I observed polish on the prehistoric tool which resembles previously published studies of polish from contact with meat and also looks like the polish on tools in my reference collection which were used for working with meat. This does not mean that the prehistoric tool which I analysed was necessarily used for working with meat. The problems of source criticism in simple, context-free identifications of contact material have been studied in detail by Levi Sala (1996). Her posthumously published study shows how polish that commonly arises from work with a specific contact material can also arise in several other ways. Use-wear analyses are thus employed here chiefly to investigate whether the flakes and the tools could or could not have been used, not specifically to determine what kind of contact material they were used on.

A total of 110 tools and unmodified or retouched flakes and frost-cracked pieces have been analysed from the Tygelsjö and Lockarp CT 7 sites, table 2.105. I judged the debitage from Lockarp 7E as having suffered too much post-deposition damage to be worth analysing. The unmodified flakes in the study were selected according to the criterion that they should have a continuous edge of at least two centimetres that

Table 2.105. List of features and use-wear analyses performed on the flint from Tygelsjö and Lockarp CT 7.

Site and feature number	No. of flakes from the feature	Flakes and tools included in the analysis	Polish
Tygelsjö A18629	59	10 unmodified flakes	Two flakes have polish, one generic polish, figure 2.175a, and one from contact with wood, figure 2.175b.
Tygelsjö A2182	16	6 unmodified flakes	Two flakes have polish, one from contact with wood and one from contact with meat.
Tygelsjö A1696	51	14 unmodified flakes	Two flakes have generic polish, figures 2.175c and 2.175d.
Tygelsjö A351	61	7 unmodified flakes, 1 unmodified frost-cracked piece, and 1 scraper	One flake has generic polish. The scraper has polish from contact with wood.
Tygelsjö A329	89	13 unmodified flakes	One of the flakes has polish from contact with silicaceous plants, figure 2.175e.
Tygelsjö A241	48	3 unmodified flakes, 1 unmodified frost-cracked piece, and 1 retouched flake	The retouched flake has polish along the retouched edge, contact material wood.
Tygelsjö A18678	58	5 unmodified flakes and 1 awl	One flake has generic polish.
Tygelsjö A18694	29 flakes, 1 frost-cracked retouched bit	2 unmodified flakes and 1 retouched frost-cracked piece	The frost-cracked retouched piece has an all-over post-deposition patina rendering HPA difficult. Along the retouched edge there is a very glossy and domed polish extending in over the surface 1–2 mm. The contact material cannot be determined because of the patina.
Tygelsjö A18768	75	4 unmodified flakes and 1 retouched flake	One flake has generic polish.
Tygelsjö A26050	58	6 unmodified flakes	No polish could be observed.
Tygelsjö A310	14	1 flake with retouch	No polish could be observed.
Tygelsjö A26038	13	2 unmodified flakes	One flake has polish from contact with wood, figure 2.175f.
Tygelsjö A2157	2 flakes and 1 awl	1 awl	No polish could be observed.
Tygelsjö A2215	118 flakes, 1 awl (flake), and 1 scraper (frost-cracked piece re-touched into a scraper)	11 unmodified flakes, 2 retouched flakes, 1 unmodified frost-cracked piece, 1 awl, and 1 scraper	No polish could be observed.
Tygelsjö A2253	49	6 unmodified flakes and 1 retouched flake	The retouched flake has polish from contact with hides, figure 2.175g.
Lockarp CT 7 A451	65	5 unmodified flakes	One flake has a polish which I cannot interpret, figure 2.175h.
Lockarp CT 7 A450	31	1 unmodified flakes	No polish could be observed.

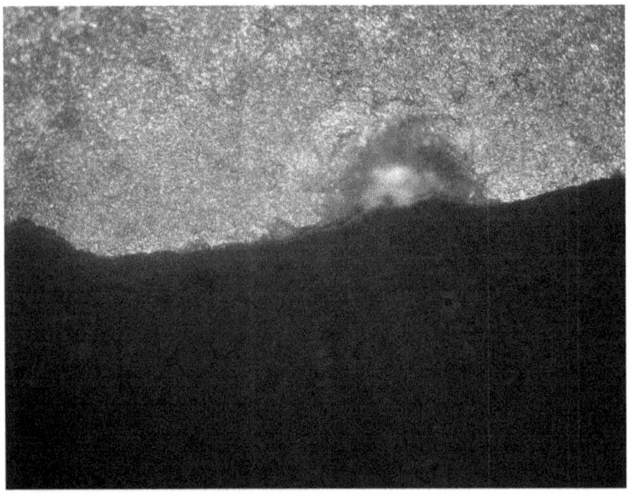

Figure 2.175a. An unmodified flake with microwear and generic polish. Original magnification 100×. Find number MK 141:79.

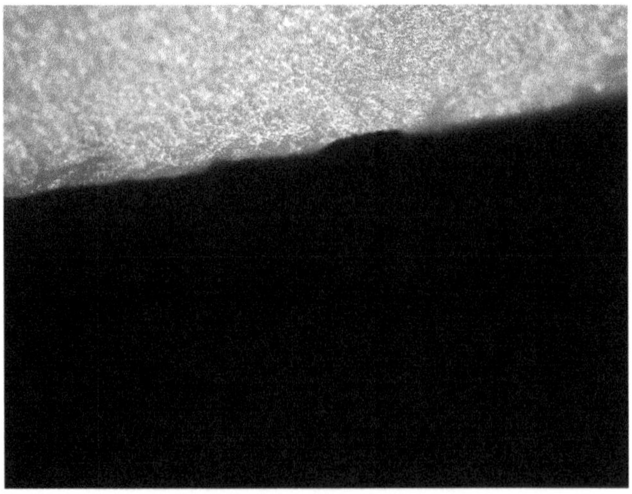

Figure 2.175b. An unmodified flake with polish interpreted as coming from contact with wood. Original magnification 200×. Find number MK 141:37.

Figure 2.175c. An unmodified flake with microwear and generic polish. Original magnification 100×. Find number MK141:442.

Figure 2.175d. An unmodified flake with generic polish. Original magnification 100×. Find number MK 141:439.

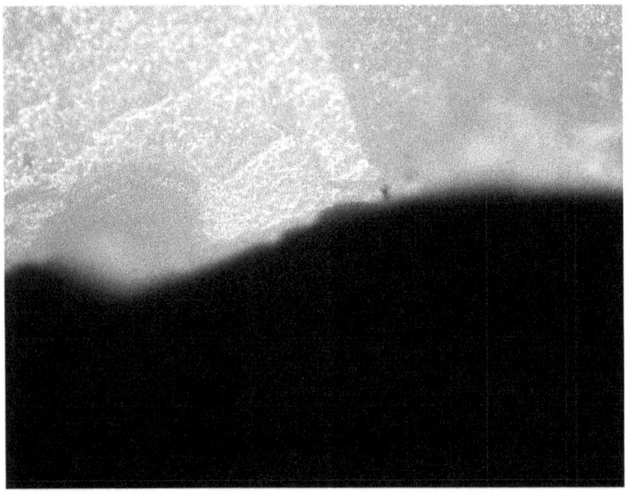

Figure 2.175e. An unmodified flake with polish interpreted as coming from contact with siliceous plants. Original magnification. 200×. Find number MK 141:470.

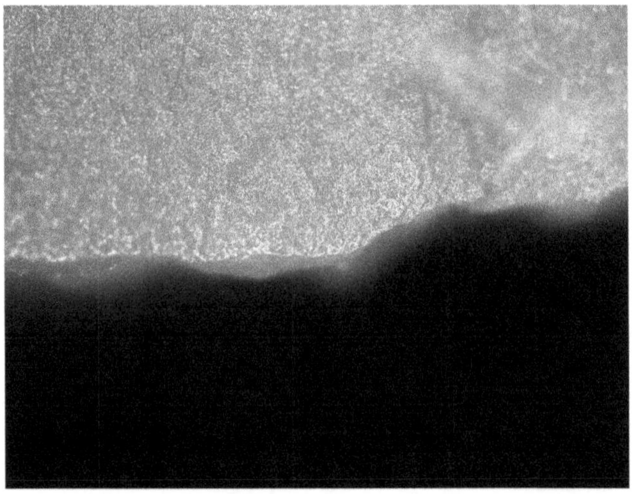

Figure 2.175f. An unmodified flake with polish interpreted as coming from contact with wood. Original magnification 200×. Find number MK141:70.

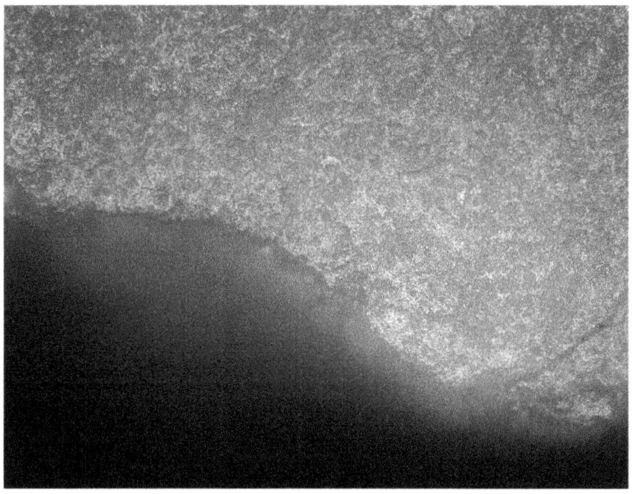

Figure 2.175g. A retouched flake with polish interpreted as coming from contact with hides. Original magnification 200×. EFI two images. Find number MK141:305.

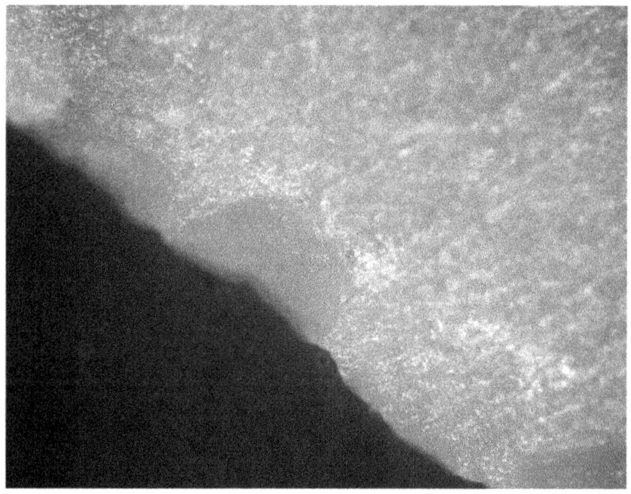

Figure 2.175h. An unmodified flake with a polish that I cannot interpret. Original magnification 100×. Find number 12881:943.

Table 2.106. Types of polish observed, number, taken from table 2.105.

Generic polish	Wood	Silicaceous plants	Hides	Meat	Unidentifiable	Total
6	5	1	1	1	2	**16**

may have been used for cutting or scraping with (Knarrström 2000:32). Of the 110 analysed flints, 16 display polish, figure 2.175a–h, tables 2.106 and 2.107.

As table 2.107 shows, polish has been observed on unmodified flakes, retouched flakes, retouched frost-cracked pieces, and a scraper. The strike-a-light from Tygelsjö has not been analysed.

The result of the analysis shows that the most commonly used flint tool in household production from the Late Bronze Age is the unmodified flake. These flakes were used on a great many different contact materials.

Table 2.107. The sixteen observed cases of polish and the contact materials, listed by the group of analysed tools.

Unmodified flake	Retouched flake	Retouched frost-cracked pieces	Scraper
Generic polish	Wood	Unidentified	Wood
Wood	Hides		
Wood			
Meat			
Generic polish			
Generic polish			
Generic polish			
Silicaceous plants			
Generic polish			
Generic polish			
Wood			
Unidentified			

It should be stressed here that I do not know anything about whether the tools that have no polish were used or not. The situational, *ad hoc* character of the flint technology within which the flakes were made and used on the sites, together with the number of flakes in my analysis showing generic polish, indicates that flakes may have been used as tools for such a short time that no polish developed, a possibility that is also supported by ethnographic studies on the curation of unmodified used flakes (Shott & Sillitoe 2005:656). If, for example, a flake was struck from a core and was then used for a few minutes to cut meat with, this was not enough to cause any polish on the flake that I can recognize in my microscope.

Comparisons with flint from earlier and later periods

Since this book is about sociotechnical change and persistence, it is essential to investigate whether there are similarities and differences over time as regards household production. Diachronic studies of flint have demonstrated that both similarities and differences can occur over long time spans (Rosen 1997; Knarrström 2000, 2001). Scandinavian diachronic studies, however, have been geared to change over a very long time – from the Mesolithic to the Viking Age, or about 8,000 years (Knarrström 2000, 2001).

My studies focus specifically on Phase B of the Late Bronze Age, and an analysis of change and similarities in household production should consider what flint looks like from the period before Phase B of the Late Bronze Age, that is, Phase A of the Late Bronze Age, and from the period after phase B, the Pre-Roman Iron Age. This entails focusing on a time span

Table 2.108. The flint assemblages in the comparative analysis of household production, total 757 flakes.

Period	Excavation	Reference	Description
Middle Neolithic	Burlöv 20B	Friman 2005	The analysis includes 180 flakes from feature A300, square D6. The feature has been interpreted as an activity area where flakes and tools of flint were made. The feature is dated by radiocarbon and pottery. The excavation has catalogue number MHM 9158. The finds are stored at Malmö Museums.
Late Bronze Age Phase A	Svågertorp 8B–C	v. Rostoványi & Hydén 2002	The analysis includes 186 flakes from feature A41121. The feature has been interpreted as a clay extraction pit with a filling of waste material and is dated with the aid of pottery. The excavation has catalogue number MHM 8541. The finds are stored at Malmö Museums.
Pre-Roman Iron Age	Lockarp 7E	Nord & Sarnäs 2005	The analysis includes 391 flakes from feature A2071. The feature has been interpreted as a clay extraction pit with a filling of waste material and is dated with the aid of pottery. The excavation has catalogue number MHM 8546. The finds are stored at Malmö Museums.

from *c.* 1000 BC to *c.* 300 BC. In the analysis I have also added a flint assemblage from the Middle Neolithic in order to show the great difference between knapped flint from the Stone Age and knapped flint from the Late Bronze Age and the Early Iron Age (Högberg 2001). The flint assemblages used in the comparative analysis come from excavations as part of the Öresund Fixed Link Project, table 2.108. They were selected according to the same criteria as the debitage analysed above.

The analysis is based on similarities and differences. When the attribute analysis reveals similar things occurring at different times, the assemblages are similar. Differences are harder to define. How large do differences have to be if the attribute analysis is to consider it worth attaching any importance to them? As discussed above, there are small variations in all flint assemblages. These may be due to factors such as individual preferences. As we also saw above, there are variations when the same assemblage is registered by the same person but at different times. These variations are often as high as 10 per cent but rarely more than 20 per cent (Shott 1994:74ff). I have therefore elected to regard differences of up to 10 per cent in the comparative attribute analysis as small and falling within the margin of error, and differences of 20 per cent and more as large and resulting from actual differences in the flint assemblage.

The result of the comparative attribute analysis of flakes from different periods shows that the Middle Neolithic debitage looks completely different from that of the Late Bronze Age and Early Iron Age, figure 2.176a–h. The differences are large. This finding is not new; it confirms similar results from previous comparisons (Knarrström 2000; Högberg 2001).

As regards the assemblages from the Bronze Age, there are differences between phase A and phase B, and, as noted above, between the different sites dated to phase B. These differences are small and lie within the interval I chose to define above as the margin of error. The result of the analysis thus shows that

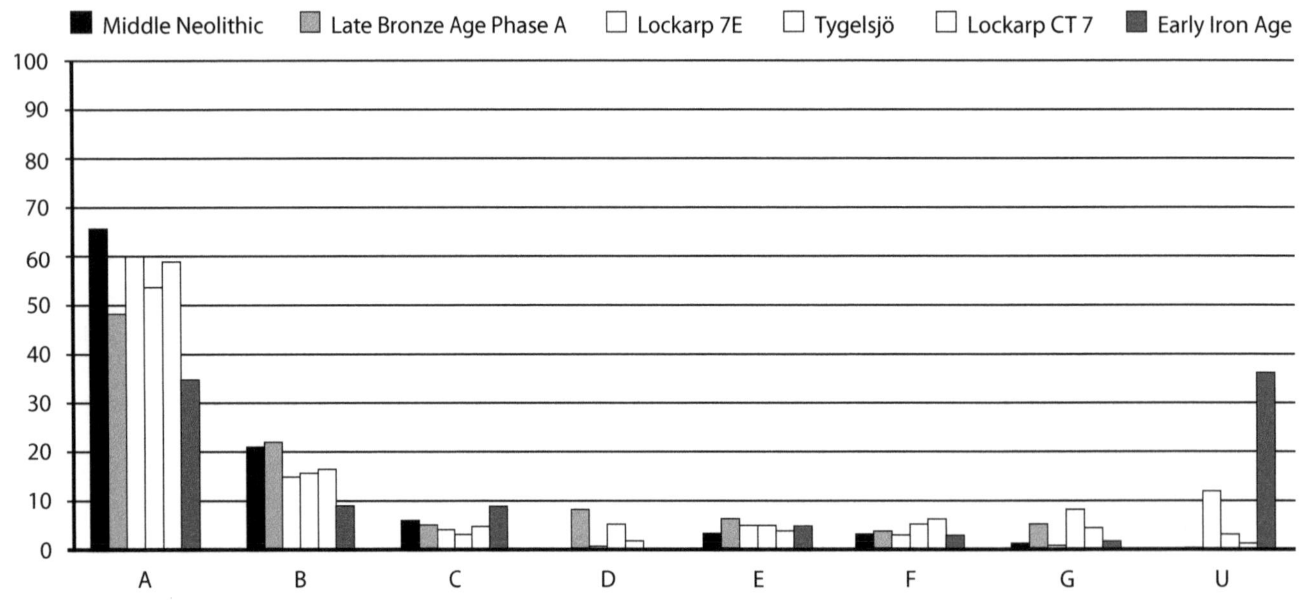

Figure 2.176a. Diachronic comparison of flake morphology.

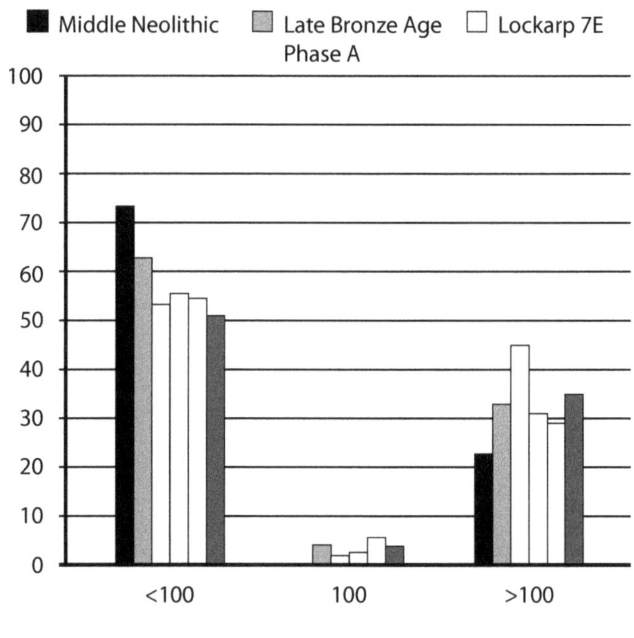

Figure 2.176b. Comparison of index figures.

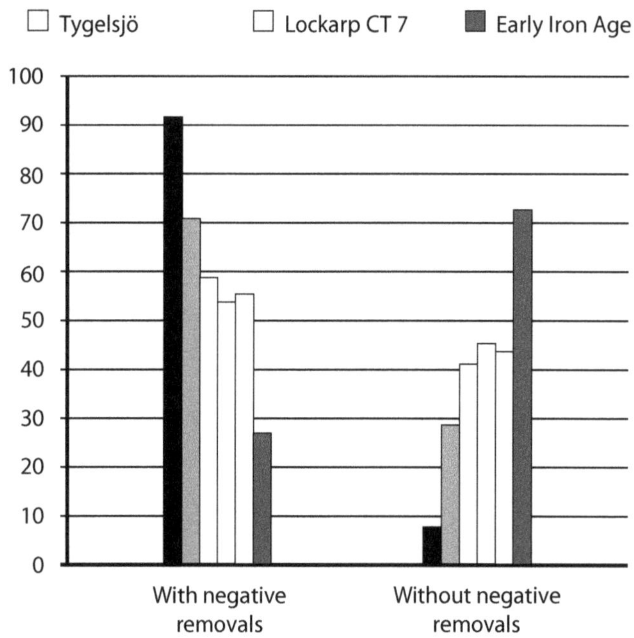

Figure 2.176c. Comparison of the occurrence of negative removals.

household production did not change to speak of during the Late Bronze Age.

When it comes to the flint assemblage from the Early Iron Age there are both similarities and differences. The similarities consist in the fact that the Late Bronze Age household production that I defined as simple and unsystematic also occurs in the Early Iron Age. The difference is that in the Iron Age there are more fragmented and crushed flakes which lack negative removals on the dorsal side. These differences are large. They are probably due to the fact that flint knapping in the Early Iron Age was characterized by an even greater absence of a strategic attitude than in the Bronze Age.

Studies of household production in the Early Bronze Age at the site of Bjerre in Thy in north-western Jutland have shown that flint tools in the Early Bronze Age were used for a variety of tasks. Flakes and cores show similar features of situational and unstructured flint knapping to the material from the Late Bronze Age (Aperlo 1994). What distinguishes the Early from the Late Bronze Age is the slightly larger proportion of formal tools in the material from the Early Bronze Age. It is thus clear that an everyday *ad hoc* technology occurred throughout the Bronze Age, but that there are differences in that the Early Bronze Age has a slightly larger share of formal tools.

Figure 2.176d. Comparison of the number of negative removals.

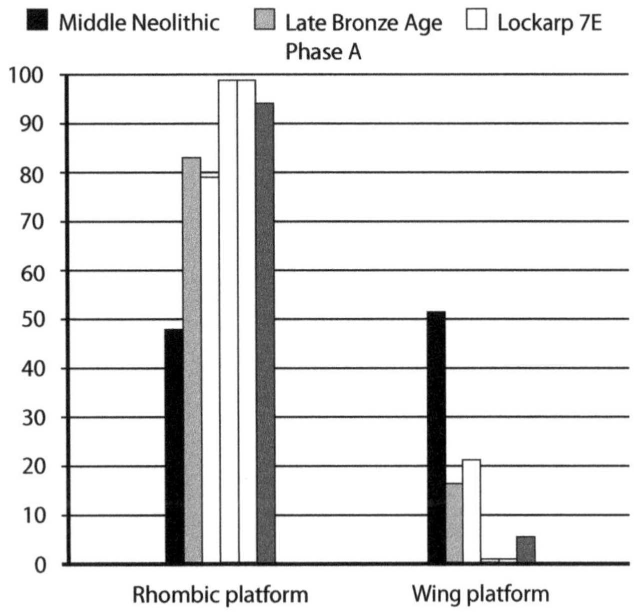

Figure 2.176e. Comparison of impactors.

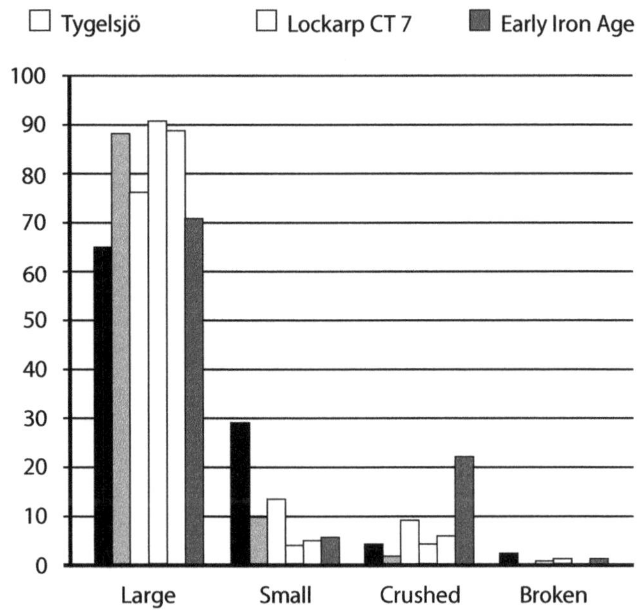

Figure 2.176f. Comparison of platform morphology.

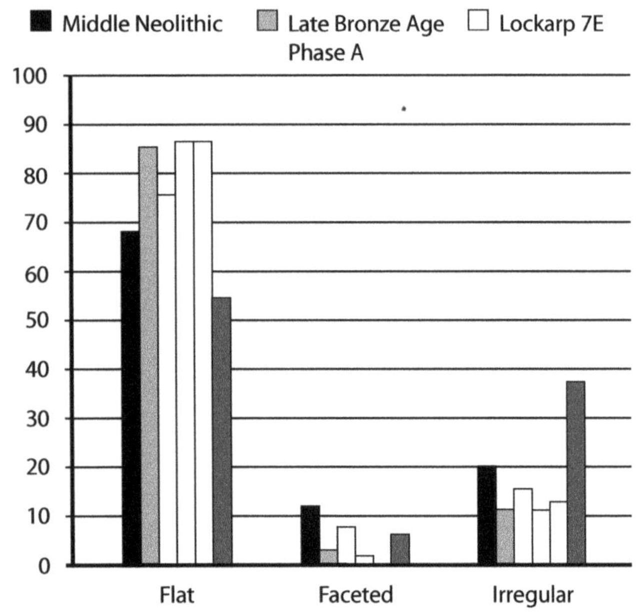

Figure 2.176g. Comparison of platform type.

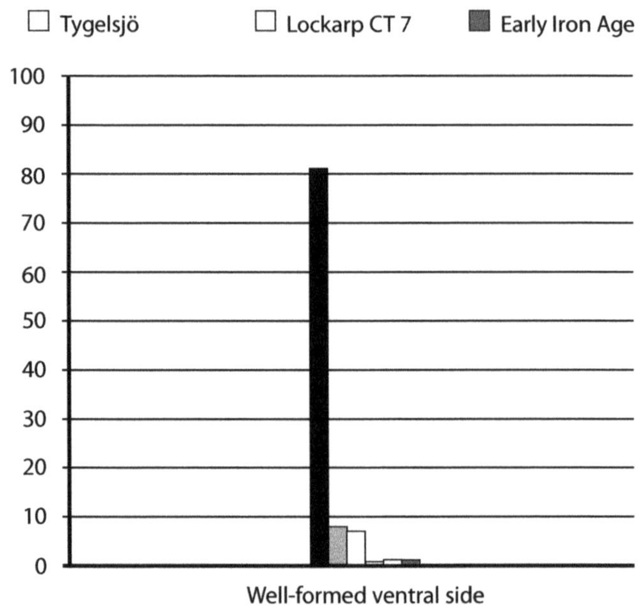

Figure 2.176h. Comparison of the occurrence of a well-formed ventral side, percentage.

Results of the analyses of household production

It is appropriate here to go back to the questions formulated in the introduction to this chapter about what the flints from household production look like, how they were made, and what they were used for. The findings of the analyses of 1,090 flakes, 114 cores, 15 pieces of retouched flint, and six tools are that the flint assemblage can be characterized by impulsiveness and constant use of *ad hoc* solutions. The material seems to lack any trace of elaborate and generally agreed and accepted strategies for the attitude to flint-working, even if the lack of a strategy can be said to be a strategy in itself. The result also shows that the use of tools in settlement site technology in the Late Bronze Age was mostly based on informal tools, that is, tools that archaeologists do not perceive as tools according to their established typologies, even though the shape may have been deliberately produced (Speal 2006:9). Unmodified flakes were used for working a great many different materials. In figure 2.177 this finding is expressed in a chaîne opératoire for household production.

There are studies which have emphasized an association between raw material quality on the one hand and formal and informal tool design on the other. The occurrence of low-quality raw material is linked in these studies to a high proportion of informal tools (see Brantingham et al. 2000; Wenzel & Shelley 2001 for a discussion). At first glance this association may seem to apply to the flints from the Late Bronze Age. Low-quality raw material was used for making almost exclusively informal tools. But the analysis of the occurrence of flint in the ground shows that there is plenty of raw material of high quality available in the till, yet this flint was not used for household production. This shows that the association between the quality of raw material and tool design is more complex than a simple relationship between low-quality raw material and a high percentage of informal tools in an assemblage.

The analyses of knapped flint from the three settlement sites also shows that large blade knives of flint were not manufactured at these places.

A comparison of household production over time shows that the way in which knapped flint was handled during Phase B of the Late Bronze Age was the same as in earlier periods of the Bronze Age and that it did not change in the Early Iron Age. From the Early Iron Age, however, there are more fragmented and crushed flakes lacking negative removals on the dorsal side.

Two flint-knapping strategies and a hybrid

The large blade knives of flint are found in archaeological assemblages from settlement remains, together with flint waste material from the everyday household production and use of flint. Here the products of two different flint-knapping strategies in the Late Bronze Age are displayed. They are found together, the everyday household production in the form of both production waste and used tools, and the specialized large blade knife production as used tools. What is interesting is the fact that these two strategies are in use at the same time in Late Bronze Age society, but so far in the material analysed and presented above, the two technological strategies never meet. They do not merge into a hybrid technology bearing traces of both strategies. The large blade knives are blade knives, produced with a large blade knife technology. The everyday household production is a household manufacture, produced with a household production technology.

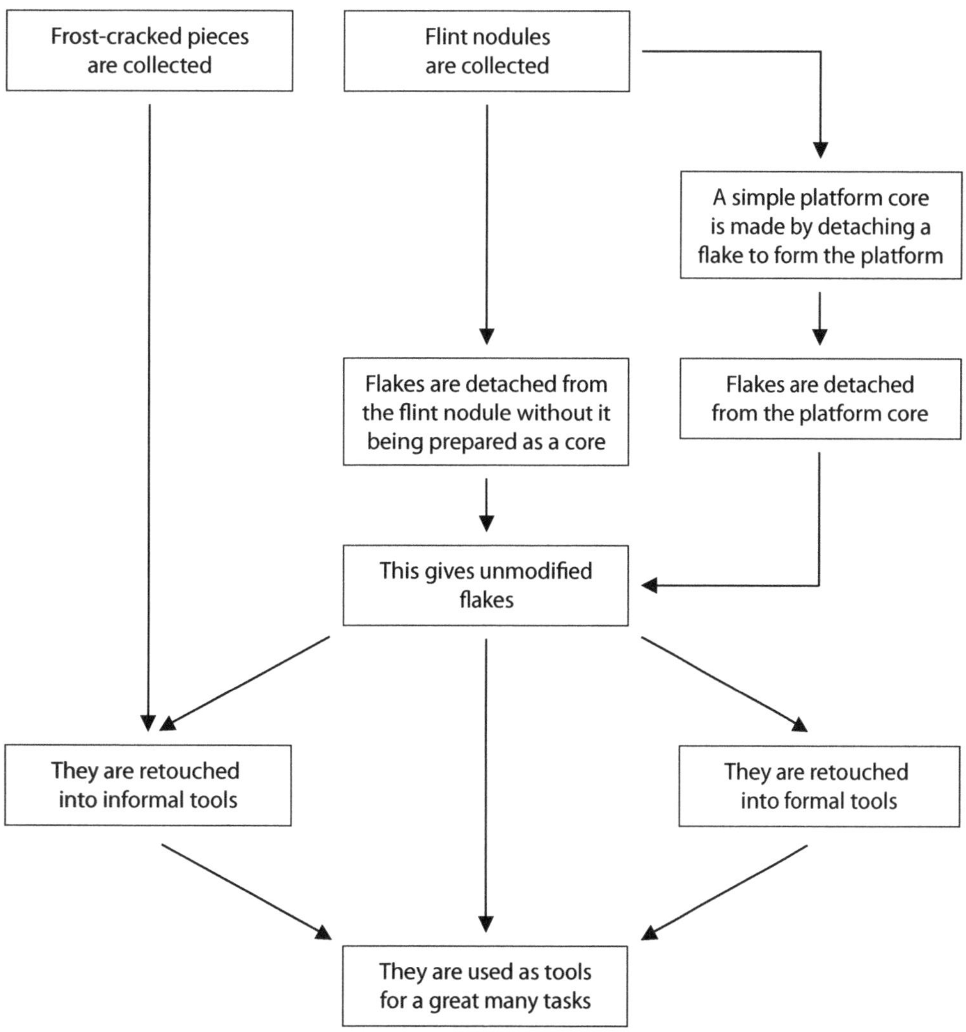

Figure 2.177. Chaîne opératoire for household production, based on the analysis of Late Bronze Age flint assemblages in this chapter.

But are these two flint-knapping strategies completely separated? Did they never meet in a technological expression, always remaining totally unconnected in all contexts of flint production? I have proposed this in previous studies (Högberg 2004a), but that suggestion has been challenged by the Danish researcher Berit Valentin Eriksen, who has discussed the results of detailed spatial, functional, and technological analysis of the flint assemblages from the well-preserved remains of a house at Bjerre in north-west Jutland, Denmark (Eriksen, in press). Eriksen has convincingly shown that the two technological strategies are not that well separated in the assemblages she analysed. In the archaeological material from Bjerre there are large blade knives and household production made with the different flint-knapping strategies I have identified. But there are also large blade knives of flint produced with a technological strategy which cannot be classified as either of the strategies I defined before. These large blade knives look like the specialized blade knives I have previously analysed, but technologically they are knapped with a strategy which more resembles the household production technology than the large blade knife technology.

These results raise questions about my previously published picture of two separate technological strategies. Maybe it is a bit too simple? Is it so that even if these flint-knapping strategies are clearly separated in several contexts, they meet in others?

An important question to discuss is obviously: Do traces of production and use exist where these two strategies merge? An evident question following this is: How does this merger take place and what is the result of it? Hypothetically it can take place in the following ways:

- *Technologically less elaborately manufactured large blade knives.* A situation where someone who has the knowledge but lacks the know-how takes up blade knife production and through trial and error creates something which looks like a blade knife, but is not manufactured with a specialized blade knife technology. The result is a tool which looks like or almost like a large blade knife, but lacks all or some of the technological attributes diagnostic for specialized production of large blade knives.
- *Technologically more elaborately knapped household production flint.* A situation where someone who is well-acquainted with large blade knife production and who has the knowledge and know-how produces *ad hoc* tools in household production. The result is domestic *ad hoc* tools which are produced with a systematic and well developed technology and in which the knowledge and know-how from the specialized large blade knife production are passed on to household production.
- *Waste material from large blade knife production which has been modified with an* ad hoc *household production technology.* A situation where someone without experience of specialized large blade knife production uses the waste material from large blade knife production and knaps it into *ad hoc* household tools. The results are tools and debitage products from large blade knife production which have been secondarily worked with a household technology.
- *Waste material from large blade knife production which has been modified with a well-developed technology.* A situation where someone who is well-acquainted with large blade knife production and who has the right knowledge and know-how uses the waste material from large blade knife production and knaps it into tools. The result is tools manufactured from waste material from large blade knife production, knapped with a systematic and well developed technology.

Hypothetically it is also possible to imagine a scenario where someone with the right knowledge and know-how for large blade knife production omits this knowledge and know-how and produces *ad hoc* household flint with the technology normally used for producing *ad hoc* household flint. In an analysis the result would be diagnostic as typical *ad hoc* household flint and would therefore be impossible to recognize. But it is not this kind of merging of technologies which is the topic of this part of the study.

The four hypotheses formulated above have been tested on a recently excavated flint assemblage. At the farm of Kristineberg a few kilometres south-west of Malmö city an archaeological excavation took place a few years ago (Rostoványi 2005).

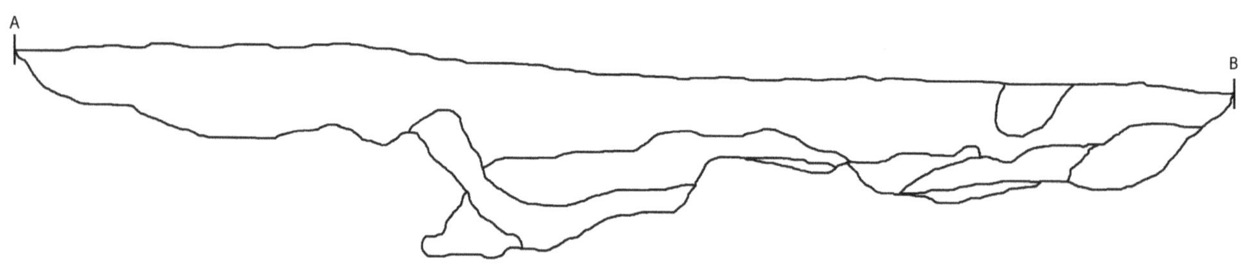

Figure 2.178. Section, feature A1751 from the excavation at Kristineberg. The feature was interpreted during excavation as a pit for extraction of clay and sand, with a secondary filling of domestic household waste material. Radiocarbon dates and finds of ceramics as well as the large blade knives of flint date the filling in the feature to the Late Bronze Age, phase B. A total of 1,160 pieces of knapped flint were found in the fillings with a total weight of seven and a half kilograms. The section is c. 4.5 meter long.

The archaeological material from this site dating to the Late Bronze Age is rich, including a number of features containing flint of both household production and several large blade knives. (These knives are included in the HPA analysis presented above.) This site makes it possible to analyse an assemblage of large blade knives and household flint together in a context. One of the features from the site (called A1751) has been chosen for this purpose, figure 2.178.

Apart from the large blade knives, the flint material from the feature consists for the greater part of unmodified flakes, mostly of the Scandinavian Senonian flint type. Tools are few. The flint is heavily fragmented.

The flint assemblage from this feature will be analysed according to the four hypotheses specified above. The purpose of the analysis is to find out whether large blade knives were produced at the site, which technology was used, whether there are large blade knives in the assemblage that were produced at the kind of highly specialized production places mentioned above, and whether there are traces of household production technology in the assemblage. All is done with the aim of looking for a potential merging of technologies. This will be done by first looking for traces of production of large blade knives, next looking for traces of household technology, and finally with a detailed analysis of the seven type I large blade knives found at the site.

Waste material from the production of large blade knives

A technological analysis of the flint assemblage from the feature reveals sixteen flakes interpreted as diagnostic for large blade knife production, table 2.109. These consist of blade-like flakes, fragments of blade-like flakes, and examples of the two kinds of preparation flakes, i.e. with a hinged or a feathered termination, all referring to the experimental study presented earlier. These flakes show that large blade knives of flint have been produced at the site. This means that this site is a different production place for large blade knives from the special production sites analysed in a previous chapter, figure 2.179.

The fragments of the blade-like flakes were all, except for one burnt piece, fragmented subsequent to the actual production event, i.e. they have once been complete blade-like flakes which have been cut up into pieces. They did not break during the production of blade knives. This is an important statement, because it shows that the waste material from large blade knife production was used as a raw material for other production. The technology used to break the blade-like flakes into pieces is a non-systematic technology. Two of the flakes have polish showing they were used as tools after the breakage, figure 2.180.

The results so far of the analysis of a flint assemblage from the feature at the Kristineberg site show that large blade knives of flint were produced at the site. Consequently, this is a small production site for large blade knives, not located at a supply of raw flint and, in contrast to the production on the previously analysed special manufacturing sites, not mass production. The results also show that the waste materials from this production were re-used as raw material in the non-systematic household production and use of unmodified flake tools.

Table 2.109. The type of flakes from feature A1751 interpreted as coming from the production of large blade knives.

	Number
Complete blade-like flake	6
Proximal part of blade-like flake	2
Medial part of blade-like flake	2
Distal part of blade-like flake	1
Preparation flake with a hinged termination	1
Preparation flake with a feathered termination	4
Total	**16**

Other production – an analysis of cores

A crucial question is whether or not there are traces of household production in the assemblage. To some extent this has already been discussed. The re-use of the waste material from the large blade knife manufacture discussed above is a tangible proof that the production manifested in the flint material not only comes from the actual making of large blade knives, but also from another kind of production. But the question here about this other production applies to more than that. It is formulated bearing in mind the household production and use discussed earlier.

But the question is how household production can best be analysed in the flint assemblage from the feature at Kristineberg. The task is not as easy as one might first think. There is a methodological problem to deal with. It concerns the documented production of large blade knives. As discussed earlier, the production of large blade knives results in flakes which are diagnostic for that kind of production. However, it also results in flakes which are not diagnostic, or rather are diagnostic for many more technologies than just the production of large blade knives. Since the household production is based completely on the making of flakes, this complicates the analysis. Thus if I want to find out whether there are any traces of household production within the flake assemblage, the flakes from the large blade knife production which are not diagnostic for just that particular production will be hard to separate from the household production.

Therefore I have to use other research material than the flakes. This is why the cores from the assemblage have been chosen for this analysis. Cores from large blade knife production look totally different from cores from household tool production. Consequently, if any household production is present in the assemblage, this will show in an analysis of the cores.

Twelve cores from the flint assemblage excavated from the feature at Kristineberg have been analysed for technological attributes. The result shows that the cores in the assemblage all have traces that are typical of household production. They are irregular, have few flake scars and numerous traces of points of impact on platforms and other parts of the core, figure 2.181.

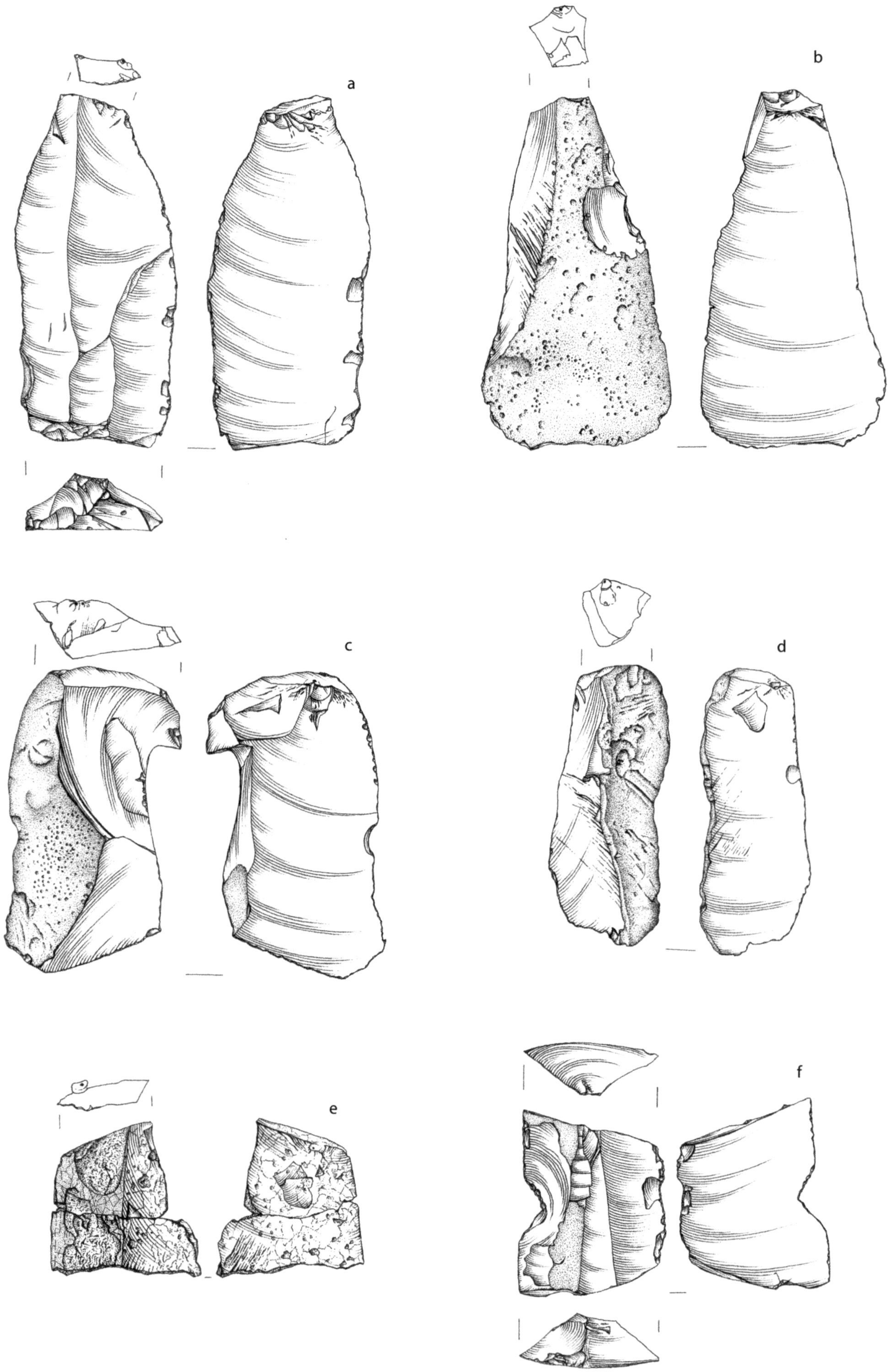
a
b
c
d
e
f

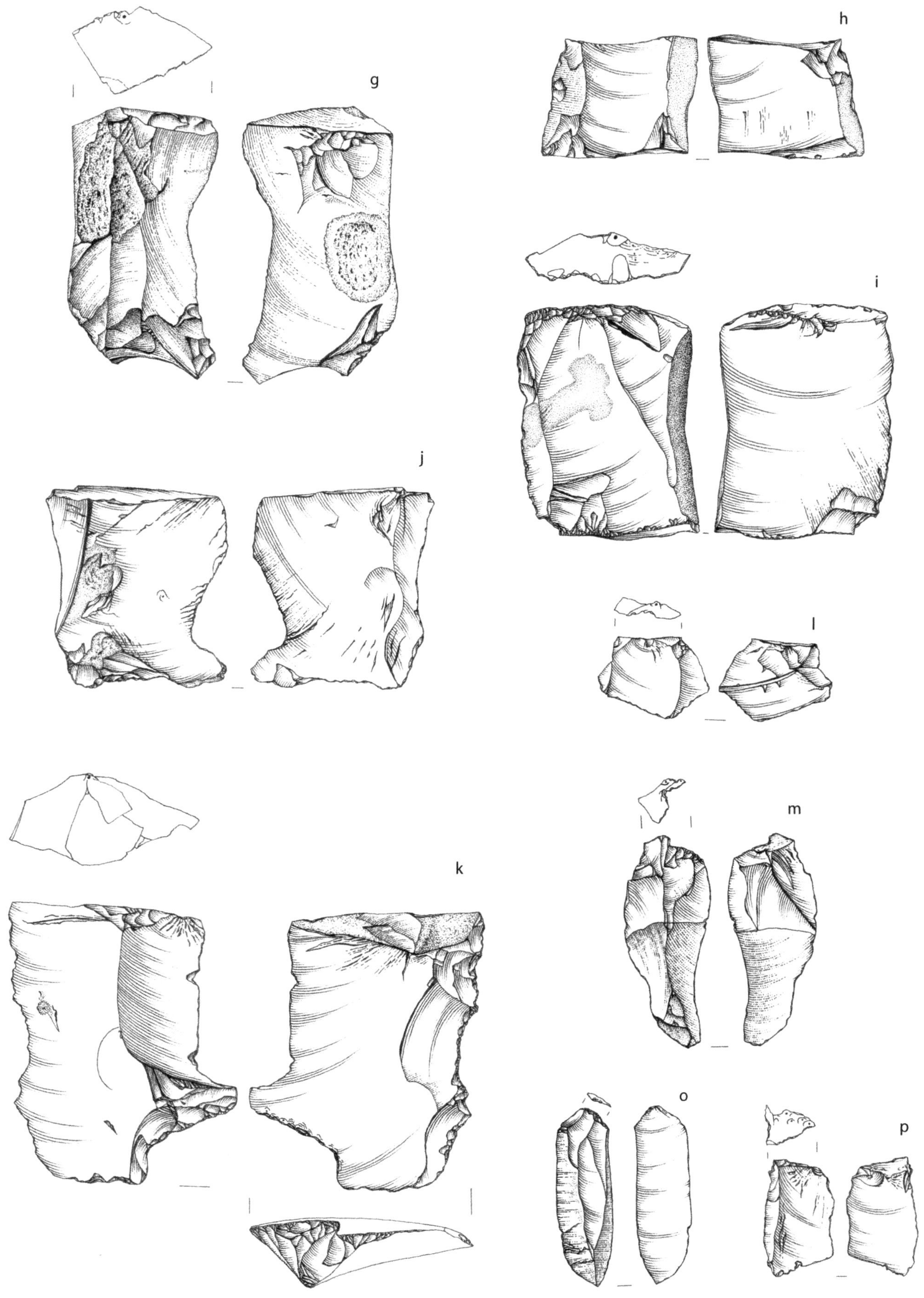

Figure 2.179. Flakes diagnostic for large blade knife production from feature A1751 at Kristineberg. 2.179a–d – complete blade-like flakes, 2.179e – a fire-cracked proximal part of a blade-like flake, 2.179f–j – different proximal, medial and distal parts of blade-like flakes all fragmented subsequent to the actual production event, 2.179k – a proximal part of a blade-like flake used as a core for household production (compare figure 2.79), 2.179l – a preparation flake with a hinged termination, and 2.179m–p – preparation flakes with a feathered termination. Blade-like flake 2.179a is 8.5 centimetres long. Drawings: Björn Nilsson.

Figure 2.180. Polish on a medial part of a blade-like flake in figure 2.179f. It has been broken off a blade-like flake with two different impacts, using the dorsal side of the flake as platform. On the proximal ventral left side a polish is visible, a polish I have interpreted as coming from working on wood. This tool is an example that waste material from the production of large blade knives was used as raw material for ad hoc household production tools. Original enlargement 200×. EFI four photos. Drawings: Björn Nilsson.

The analysis shows that there are traces of another production than the large blade knife production in the assemblage. It also shows that the systematic technology used to produce the large blade knives did not influence the technology with which the cores were reduced.

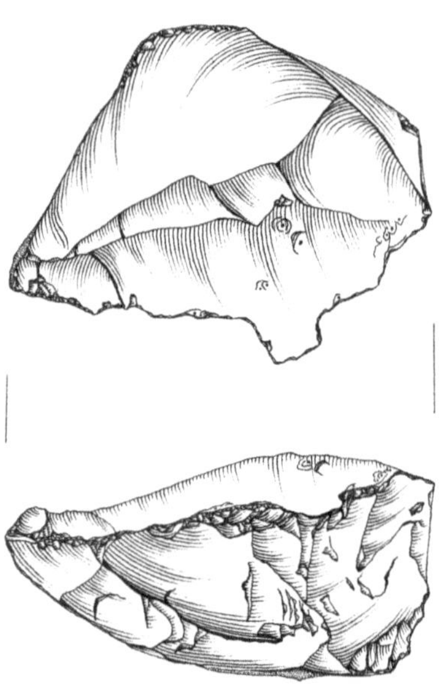

Figure 2.181. A core from household production at the Kristineberg site. Note the impact points on one of the platforms, typical of Late Bronze Age household production cores. No cores from the production of large blade knives have been found in the flint assemblage from Kristineberg. The core is 6.5 centimetres long. Drawing: Björn Nilsson.

Other production – function

Another issue vital to this study deals with function and the question of what the material from household production was used for. This question has been analysed by means of use-wear analysis of flakes with retouched edges and a selection of unmodified flakes.

The assemblage contains three retouched flakes. On two of these no use-wear can be observed. The third retouched flake, however, has a well-developed polish along all the edges, figure 2.182. The extension of the polish, the intensity, the topography, the light reflection, and the extent of linkage show that the tool was used intensively. I have looked for hafting traces but have not found any.

The use-wear on the retouched flake is the same kind as is typical for large blade knives. This is an interesting result. It shows that the tool was used to work the same material as large blade knives of flint. In this specific case I have interpreted the polish as coming from working reed. This implies that reed obviously was handled in a complex production context where different kinds of tools were required for different purposes and for different parts of the production sequences – one purpose for which the large blade knife was regarded as a suitable tool, another purpose for which a small retouched flake was regarded as a suitable tool.

A selection of the unmodified flakes have been examined for use-wear in order to investigate whether these were used as tools. A total of 201 flakes were analysed, out of a total 1,160 pieces of knapped flint. Of these flakes twenty-one have an edge which is longer than two centimetres and have been judged as suitable for cutting or scraping. None of these flakes show any sign of use, figure 2.183.

The results of the analysis of cores, retouched flakes, and unmodified flakes show that together with the production of large blade knives there was also household production on the site. The analysis also shows that the large blade knives and other tools were involved in a complex working procedure in which different kinds of tools were used for different purposes in the production sequences.

Large blade knives of flint from the site

A total of seventeen large blade knives of flint have been found at the site. Three of these, two complete blade knives and one fragment, come from the same feature as the flint assemblage analysed above, figure 2.184.

The seven complete large blade knives of type I were selected for a detailed analysis. The purpose was to compare these large blade knives with the general picture of the large blade knives presented above, as well as to examine the possibility that some of the large blade knives may have been produced at the site. This was done with a technological and morphological analysis according to the results presented earlier. A raw material analysis including all of the seventeen blade knives from the site was also performed.

The result shows that six of the seven large blade knives of type I are uniform as regards technological and morphological

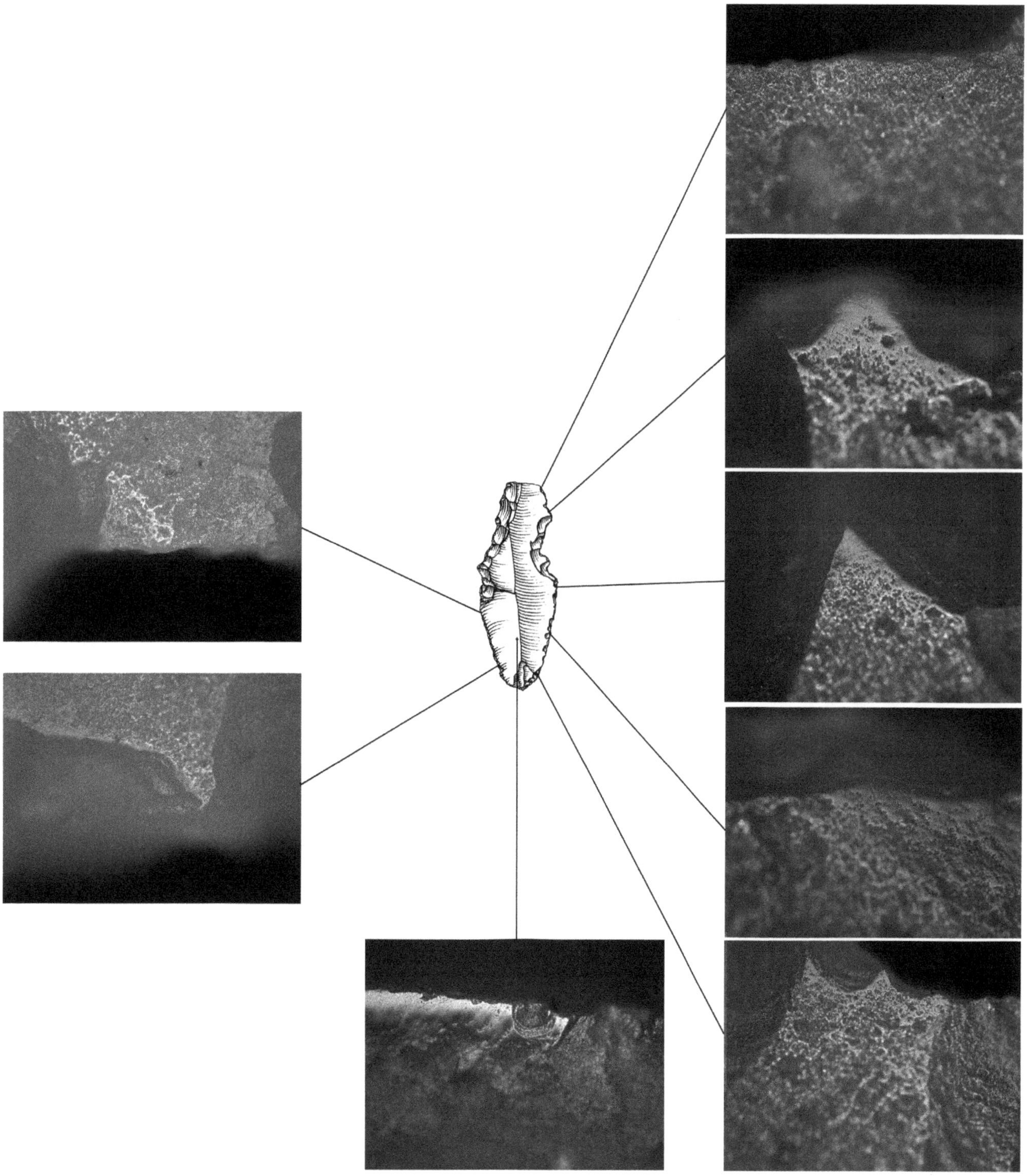

Figure 2.182. A retouched flake (MK20:201712) from feature A1751, Kristineberg. The figure shows a series of HPA photos of the polish along the edges of the tool. The polish is present on all of the tool edges and is a typical blade knife polish. The worked material is interpreted as reed. The retouched edge to the left in the figure was knapped during prehistory, the one to the right is caused by damage during excavation. Original magnification 200×. The flake is 4 centimetres long. Drawing: Björn Nilsson.

attributes such as platform angle, edge outline, platform size, and length. One blade knife, number 4, differs from the others regarding platform angle, edge outline, and platform size, table 2.110.

If we look at the raw material and different flint types used to produce all of the seventeen large blade knives from the site we can see that fifteen are made of Scandinavian Senonian flint, one of Falster flint, and one of a Senonian till flint. The Senonian till flint is classified as such from the status of the cortex. It is weathered and patinated, bearing the typical traces of once being deposited in glacial tills. A field investigation at the excavation site showed that many flint nodules of this

Figure 2.183. From feature A1751 a selection of 201 flakes was selected from a find context. Out of these flakes twenty-one have an edge which is longer than two centimetres and have been judged as suitable for cutting or scraping. I have done use-wear analysis on these twenty-one flakes. None of them exhibit any sign of use. The photograph shows the selected flakes with the twenty-one tool-like flakes to the right. Photo: Jenny Thornell, Malmö Museums.

and other types can be found in the glacial till there, see also the results from the analysis of raw material access in previous chapter. The large blade knife made of till flint is the one which also differs regarding the technological and morphological attributes, i.e. number four in table 2.110 and figure 2.184.

The results of the technological and morphological attribute analysis as well as the raw material analysis of the large blade knives from the site give a picture corresponding to the general view of blade knives as being highly standardized products, produced at special production sites with access to suitable high-quality flint.

However, one of the blade knives, number four in table 2.110 and figure 2.184, differs from this picture. It is a blade knife which does not show the typical standardized attributes and which is made of till flint. This blade knife is not a highly standardized product made of high-quality raw flint at specialized production sites. Instead it bears traces of a less standardized and less elaborate technological approach. It looks roughly like the standardized large blade knives, but it is not one, figure 2.185.

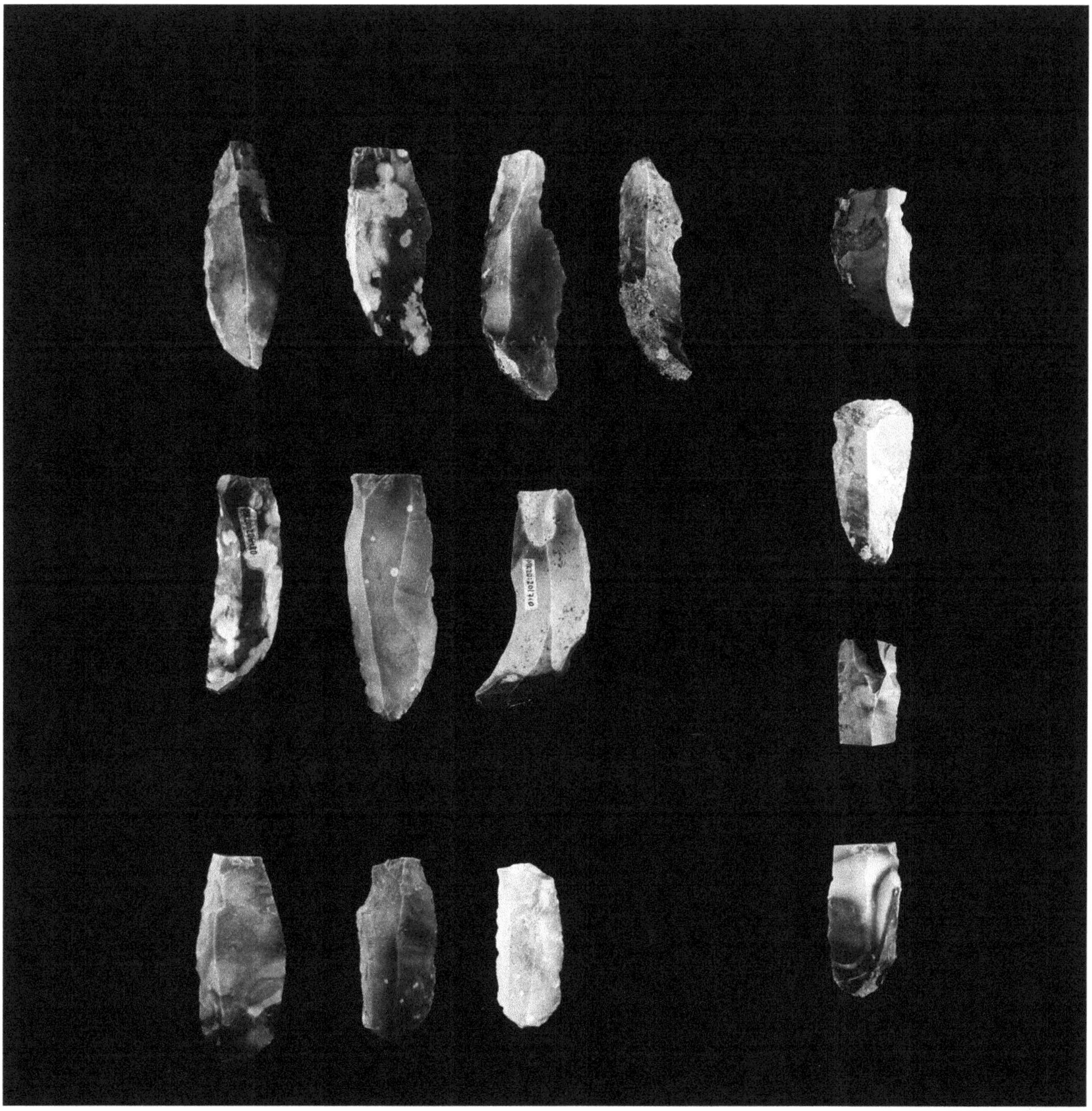

Figure 2.184. Fourteen of the large blade knives of flint from the Kristineberg site, seven of type I, three of type II, and four fragments of blade knives. Three knives from the site are missing in the picture. The four blade knives in the top row are numbers 1–4 and the three in the row in the middle are numbers 5–7 according to table 2.110. The knife to the lower left is approximately 11 cm long. Photo: Jenny Thornell, Malmö Museum.

Table 2.110. Technological and morphological attributes for the seven large blade knives of type I. As is evident, the value for blade knife number 4 differs regarding platform angle, edge outline, and platform size.

Knife number	Platform angle		Edge outline		Length in cm			Platform size	
	65	75–80	Curved	Straight to slightly curved	11	12	13	Small	Medium
1		X		X	X				X
2		X		X	X				X
3		X		X			X		X
4	X		X			X		X	
5		X		X	X				X
6		X		X			X		X
7		X		X		X			X

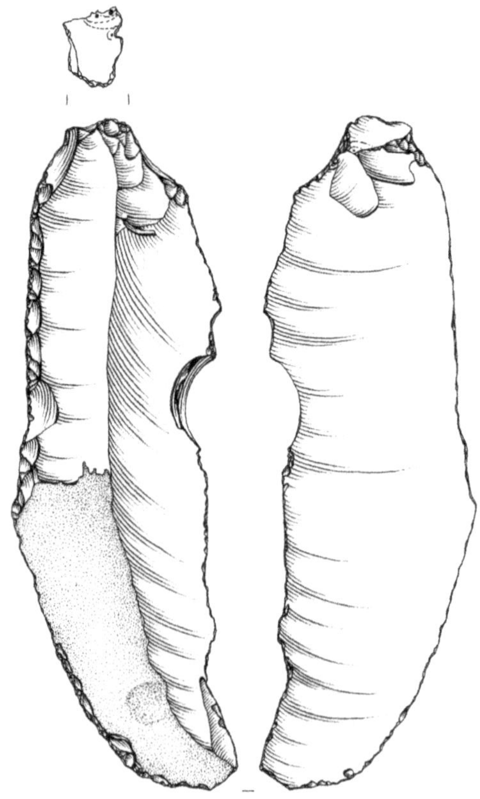

Figure 2.185. The large blade knife number four in table 2.110 and figure 2.184 from the Kristineberg site which does not show the typical attributes of the standardized large blade knives of flint and which is made of till flint. The blade is 12 centimetres long. Drawing: Björn Nilsson.

Discussion

During the Scandinavian Late Bronze Age two clearly distinguished flint-knapping strategies were in use. One was the household production of *ad hoc* tools and one was the highly specialized production of large blade knives. As discussed earlier, the household production was a situational technology, satisfying the needs on a domestic everyday basis. The production of large blade knives was concentrated in a few specialized production places from which standardized tools were distributed, satisfying the needs of a large geographical area. Even though these two technologies in many situations were clearly separated, situations existed where the two technologies met and merged into what I have considered a hybrid. Consequently, there are two technologies and a hybrid.

This result needs some more discussion, however, because there is one more question to ask: Is it really suitable to categorize the merging of the two technologies as a hybrid? That it actually is a hybrid, from the point of view of an archaeological classification, is clear. The only way I have been able to pinpoint this merging of technologies is to first define the "obvious", i.e. the two clearly separated flint-knapping strategies, and then formulate the questions about how different kinds of encounters of these technologies hypothetically took place. The result of this archaeological approach is obviously something that is suitably called a hybrid. But is it a suitable categorization?

Technologies are born out of ideas, ideas arise from different kinds of needs. The needs can be a demand for something existing, a wish for something new, or a demand for something existing which results in something new. As Bijker (1995) has stated, technology is created not solely from prevailing circumstances, as for example social structures or power relations, but also out of creativity and the emotional commitments of individuals.

The flint assemblages analysed in this study show the stability of the Late Bronze Age technological system, i.e. the two clearly separated technological strategies manifested in the household production and the specialized production of large blade knives. But what we also see is an example of technological change where this stability was challenged. This was happening in a place where these two flint-knapping strategies were under negotiation and redefinition, a negotiation and redefinition which took place in an interaction between, on the one hand, the technological traditions and, on the other hand, the needs created by the community and the creativity of the flint knapper. The result of this is what has been classified here as the hybrid.

However, from this it is impossible to maintain this label for the meeting of the two technologies. The meeting is actually a technology of its own. It is not a hybrid. It is a technology which stands for itself. It is situated in, inspired by, and dependent on the two prevailing Late Bronze Age technological flint-knapping strategies. But still it is a technology of its own.

That the two technologies are different and separated is demonstrated in two different ways:

- in the evidence of *ad hoc* domestic household production and use of simple tools
- in the evidence of large blade knives of flint showing the technological attributes of specialized production and use.

The indications that a merging or meeting of technologies took place are demonstrated in four different ways:

- in the evidence of the production of large blade knives on settlement sites
- in the fact that the waste material from the production of large blade knives was used as raw material for household tool production
- in the use-wear which shows that tools from the different technologies were used in different parts of the same production sequences
- in the presence of large blade knives produced with a less elaborate technology.

The study shows that the two technological flint-knapping strategies in the Late Bronze Age were stable. Even though there are examples of technological change in which this stability is challenged, the two strategies are still evident in the Late Bronze Age flint industry.

Figure 2.186. Flakes and a large blade knife of type II made from Common Kristianstad flint from the Årup site in north-east Scania, Sweden. The blade knife is the one to the left. The flakes, in the top row, are three proximal parts of blade-like flakes, to the lower left four preparation flakes with a hinged termination and one with a feathered termination, and to the lower right two distal parts of blade-like flakes. The blade knife is 8 centimetres long (Knarrström 2007).

Other places

As we have seen, the sites Lockarp 7E, Tygelsjö and Lockarp CT 7 which I have examined in previous chapter and which display traces of household production were not used for the manufacture of large blade knives, and at all but one of the specialised manufacturing sites for large blade knives which I have analysed, nothing but large blade knives was made.

However, based on the detailed analyses of the flint assemblages from Kristineberg presented here, I have found a further three places where household production and the manufacture of large blade knives went on together. These places are Årup in north-east Skåne, Sweden (Björk 2003; Knarrström 2007), Høbjerg and Holmehusbakke at Roskilde on Zealand, Denmark (Sørensen 1980, 1984-85), and Marlow (Auler 1999) on Rügen, Germany. From all these sites there is a large amount of settlement site material dated to the Late Bronze Age. It includes flint assemblages containing household production, large blade knives, and diagnostic flakes from the production of large blade knives, figures 2.186, 2.187, and 2.188. Since sites like these were previously unknown – I have found them through studies of the literature and by searching through archaeological material in museum stores – it is likely that more such places will be discovered when more find assemblages are analysed. Yet it should be emphasized once again that the many Late Bronze Age flint assemblages with household production which I have studied (Högberg 2001a, 2001b) do not include any trace of the manufacture of large blade knives.

Since large blade knives were made on these sites, they differ from others which I have studied where large blade knives occur in the flint assemblages but where the production on the site was only for household use, i.e. household production. But the question is whether anything else distinguishes these places.

An analysis of what is otherwise found on the sites makes it obvious that the places where large blade knives were manufactured together with household production have a heavy touch of craft work, table 2.111.

These sites were thus used not only for the production of large blade knives but also for crafts such as pottery making, stone work, bronze casting, and goldsmithing. This shows that these places differ in several ways from other contemporary settlement site remains.

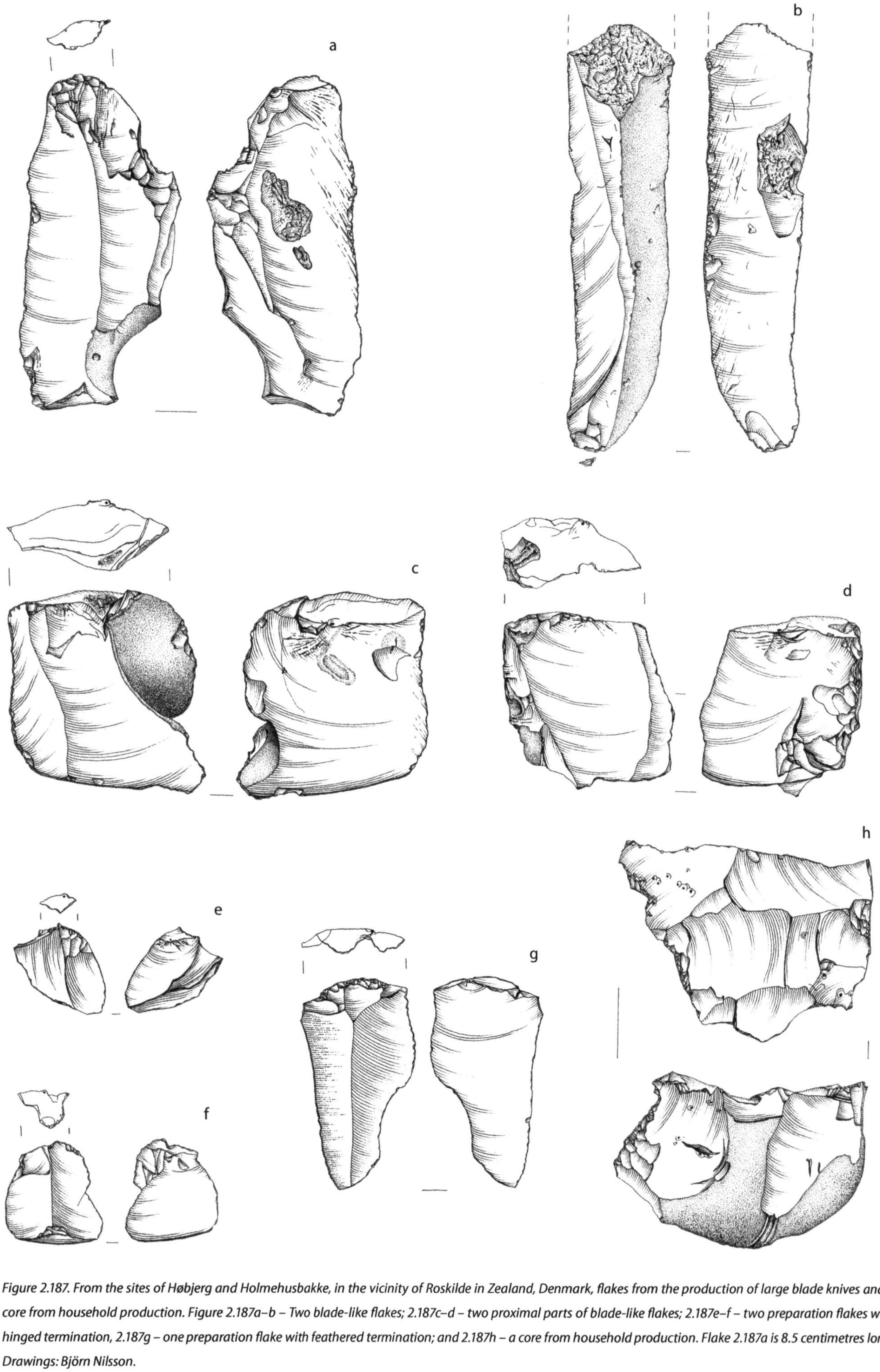

Figure 2.187. From the sites of Høbjerg and Holmehusbakke, in the vicinity of Roskilde in Zealand, Denmark, flakes from the production of large blade knives and a core from household production. Figure 2.187a–b – Two blade-like flakes; 2.187c–d – two proximal parts of blade-like flakes; 2.187e–f – two preparation flakes with hinged termination, 2.187g – one preparation flake with feathered termination; and 2.187h – a core from household production. Flake 2.187a is 8.5 centimetres long. Drawings: Björn Nilsson.

Figure 2.188. From the Marlow site on Rügen in Germany (Auler 1999). Flakes and cores from household production, to the left, and two proximal fragments and a distal fragment of blade-like flakes and a preparation flake with feathered termination from large blade knife production, to the right.

Table 2.111. Craft finds from the sites where the manufacture of large blade knives occurred together with settlement site material.

Site	Finds
Kristineberg	Site used for making pottery, casting bronze, and working gold.
Årup	Site used for casting bronze.
Høbjerg and Holmehusbakke	Finds of smoothing stones, a bone needle, and crucible fragments
Marlow	An unusual stone object which has been interpreted as having been used in very special manufacture whereby soft groundstone was sawn in slices as preforms for manufacture.

Apalle and Hallunda – a previously unknown flint-knapping technology

As mentioned before, I have studied the flint assemblages from the well-known sites of Apalle (Ullén 2003) and Hallunda (Jaanusson 1981) in order to look for large blade knives. The search resulted in no finds of blade knives from Hallunda and a fragment of a blade knife from Apalle. On both these sites, remains from the Late Bronze Age have been excavated. The discovery of house remains, occupation layers, bronze-casting places, and other craft areas, together with rich and varied finds, have made these sites among the most frequently cited in Scandinavian settlement site archaeology (see e.g. Goldhahn 2007).

In my study of the flint assemblages from Apalle and Hallunda I found something that I had not expected. On both sites there was a flint-knapping technology that I did not previously know of and that I had not found anywhere else. The flints that drew my attention to this technology consist of small, well-prepared platform cores and small flakes struck from these cores, figure 2.189 and table 2.112. Some space should be devoted to a description of this technology.

Table 2.112. Finds and contexts from Apalle. The content of the middle column, "Contexts", is of concern only to those who are especially interested in Apalle. What is interesting for the reader is the phases stated in the right-hand column. All the flakes come from phases 4, 4,5 or 5. These phases are the latest settlement layers from the Bronze Age on the site, and their dating lies between 900 and 600 BC. The data for Apalle come from Inga Ullén, Museum of National Antiquities (e-mail, Ullén 2007). From Hallunda the data on context are uncertain, so I have not included data on this for the two finds from Hallunda that are included in my analyses.

Find number	Contexts	Phase
Apalle F55558	Level 1, in non-preserved floor. K24.	Phase 4
Apalle F55560	Level 2, in floor/wall house K2.	Phase 5
Apalle F55533	Level 1, in occupation layer over K16.	Phase 4,5
Apalle F55539	Level 2, in occupation layer over well K94.	Phase 4,5
Apalle F28191	A2272, late filling in pit.	Phase 4,5
Apalle F28401	Level 1, in occupation layer outside house K8/K6.	Phase 5
Apalle F24056	Level 2, in house K29.	Phase 5
Apalle F55524	Level 1, in occupation layer over well K94.	Phase 4,5
Apalle F55552	Level 4, probably fell down into post-hole 1871, K26.	Phase 5

Figure 2.189. Flakes and a core from Apalle and Hallunda. The technology represented by these finds has not been described before and was previously unknown to me. The flake to the upper left is 3.5 cm in width.

Table 2.113. Results of an attribute analysis of ten unmodified flakes from Apalle and Hallunda. Weight in grams and length and width in centimetres.

Find no.	Weight	Length	Width	Number of negatives on the dorsal side	Angle de chasse	Miscellaneous
Apalle F55558	>5	2.0	2.4	1	50	Crushed platform
Apalle F55560	>5	1.8	2.8	2	55	Prepared platform
Apalle F55533	5	2.6	2.6	3	55	Prepared platform, cortex 25%
Apalle F55539	5–10	3.4	3.7	1	65	No platform preparation, cortex 80%
Apalle F28191	>5	1.2	2.1	2	55	Prepared platform
Apalle F28401	5–10	1.8	3.3	2	–	Crushed platform, distal part missing, cortex 25%
Apalle F24056	10	4.2	2.5	2	50	Cortex 33%
Apalle F55524	>5	2.2	2.1	3	50	Prepared platform, cortex 25%
Apalle F55552	5–10	2.3	2.3	3	50	Prepared platform, cortex 10%
Hallunda F131	5–10	2.8	3.5	5	50	Prepared platform, cortex 10%

The flakes and cores were made of Scandinavian Senonian flint. It is not possible to pick flint from the ground in this part of Sweden. All the flint that is found in the archaeological material was brought to the site. Scandinavian Senonian flint is a type that is found at several different places in southern Scandinavia. It is thus impossible to determine the source of the raw material used at Apalle and Hallunda. The lithic technology of flaked stone tools otherwise consists chiefly of manufacture based on quartz and quartzite (Ullén 2003:147ff).

The technology at both sites is very similar. Small flakes which are mostly wider than they are long and never weigh more than ten grams were detached with a hard technique, presumably a hammerstone, from small, well-prepared, and well-worked platform cores. Virtually all the flakes have a well-prepared flat rhombic platform. They were detached with a platform angle (angle de chasse) of 50 to 55 degrees, table 2.113.

From Hallunda there is one core in the assemblage. It is a platform core with a well-prepared platform edge with nine negative flaking scars on the front. The length of the front, that is, the dimension of the longest flake that could be removed from the core as it looks now, is 2.7 centimetres. The core has cortex on the base and about 10 per cent on one of the sides. The platform angle is around 55 degrees.

The result of this small-scale attribute analysis shows that the flakes and the core are technologically very similar. The flakes in the material were removed from cores which look like the flake core in the assemblage. The flakes are almost square, but slightly wider than they are long. The technology is standardized, with flakes of very uniform appearance.

The question is what this technology was used for, or rather what the resulting flakes were used for. To study this, all the flakes in this analysis underwent functional analysis by HPA. Table 2.114 shows that several flakes have polish indicating that they were used. However, I have not been able to identify the contact material on any of the flakes. Two flakes have polish along the edge from a hard contact material which I cannot interpret. Two flakes have generic polish. Five of ten flakes show post-deposition damage rendering analysis impossible. One flake lacks polish. All the flakes have post-excavation damage in the form of striations and metal gloss. All the flakes in the material from Hallunda are marked with a find number which is varnished over. This prevents functional analyses by HPA.

It is frustrating not to be able to analyse the function better than this. With the discovery of a new, previously unknown technology it would of course have been interesting to specify the function more precisely. However, this is not possible, but there is a task here for future research to tackle when new assemblages like those from Apalle and Hallunda are uncovered by excavations.

From both Apalle and Hallunda there is knapped flint which was not made using the technology just described, but with a technology that has been defined through attribute analysis in a previous chapter as household production, figure 2.192. The same relationship between systematic and well-defined flint knapping and more *ad hoc* household production, as described above, is thus found in the assemblages from Apalle

Table 2.114. Result of functional analyses (HPA) of the ten flakes from Apalle and Hallunda.

Find number	Observations
Apalle F55558	All-over post-deposition patina renders analysis impossible.
Apalle F55560	Generic weak polish.
Apalle F55533	The flake has polish along one of the edges. The polish does not extend over the surface of the flint. From this extent I interpret the polish as coming from a hard contact material. I cannot further interpret the polish.
Apalle F55539	No polish could be observed.
Apalle F28191	All-over post-deposition patina together with serious post-excavation damage renders analysis impossible, figure 2.190.
Apalle F28401	All-over post-deposition patina renders analysis impossible.
Apalle F24056	Generic weak polish.
Apalle F55524	All-over post-deposition patina renders analysis impossible.
Apalle F55552	The flake has polish along one of the edges, not extending over the surface of the flint. From this I interpret the polish as coming from a hard contact material. I cannot further interpret the polish, figure 2.191.
Hallunda F131	The flake is marked with find number which is varnished over. This makes analysis impossible.

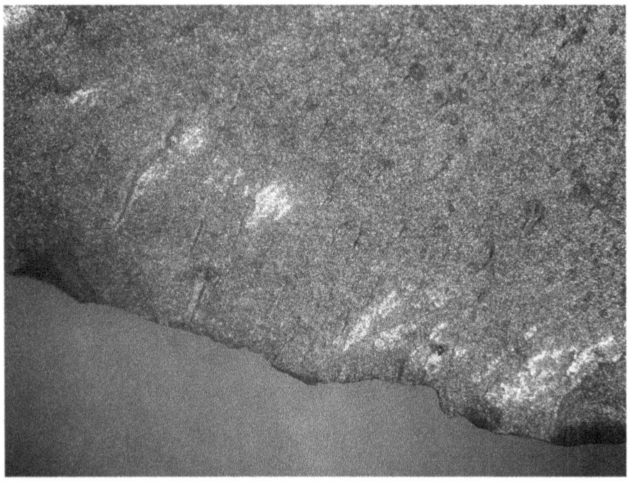

Figure 2.190. A photograph from the use-wear analysis of a flake from Apalle (find number F28191) showing an all-over post-deposition patina together with serious post-excavation damage rendering analysis impossible. Original magnification 50×.

Figure 2.191. A photograph from the use-wear analysis of a flake from Apalle (find number F55552). The flake has polish along one of the edges, not extending over the surface of the flint. I thus interpret this polish as coming from a hard contact material. I cannot further interpret the polish. Original magnification 200×.

Figure 2.192. A bipolar core from Apalle (find number F24045). The core has a maximum length of 3 centimetres.

and Hallunda. It is interesting in this context that, just like the sites I analysed above, where both household production and the manufacture of large blade knives was evidenced in the finds, both Apalle and Hallunda have extensive remains of other crafts (Ullén 2003:129ff; Goldhahn 2007:293ff). To avoid misunderstanding, however, it must be stressed that these crafts have nothing to do with the production of large blade knives of flint.

Results – Apalle and Hallunda, a previously unknown flint-knapping technology

The result of this brief analysis of flint from Apalle and Hallunda shows that flint was worked on these sites in the latter part of the Late Bronze Age, in a way that I do not know from any other sites. Small flakes were manufactured from well-prepared platform cores. Some of these flakes were used on a hard, unidentifiable, contact material. Simultaneously with this systematic and standardized technology, flint was also knapped in the way that I defined in a previous chapter as household production.

Flint objects in cremation graves

Finds of flint in graves from the Late Bronze Age are not uncommon. In the large-scale excavations of the burial cairns at Sannarp in Halland, Sweden, for example, huge amounts of burnt flint were found together with crushed quartz in the filling in several of the graves (Streiffert & Strömberg 1998), and an excavation of a ritual and burial place from 1000–400 BC at Ringeby in Östergötland, Sweden, unearthed finds of flint (Kaliff 1996:78). From the well-known excavations of the cemetery at Simris in south-east Scania, Sweden, by Berta Stjernquist, there are several examples of finds of flint in cremation graves (Stjernquist 1961:88). At the cemeteries of Svarte and Piledal, also in south-east Scania, flints were found in cremation graves (Olausson 1987). When Knut Kjellmark reported the results of the excavation of a cemetery with small barrows and cremation graves at Kämpinge in south-western Scania, he was careful to emphasize the finds of flint in the graves as part of the mortuary material from the Late Bronze Age (Kjellmark 1925:150ff). Examples of the opposite also exist, where knapped flint found in graves has been interpreted as coming from older occupation layers on the site (Rudebeck & Ödman 2000:129).

It is thus not unknown that both knapped and non-knapped flint occurs in mortuary contexts. Yet interpretations of these finds are rare. I know of one Scandinavian study which uses finds of knapped flint in cremation graves from the Late Bronze Age as a basis for interpretations (Goldhahn 2007), but the number of studies is small in relation to the published finds. In the elaborate publication of the results from the excavations by the West Coast Line Project in south-western Scania, for example, Caroline Arcini, of the National Heritage Board in Sweden, and Fredrik Svanberg, of the Museum of National Antiquities in Sweden, reported finds of flint in cremation graves:

> Flint was retrieved from at least 27 graves, but it has not been studied in detail (Arcini & Svanberg 2005:314).

This quotation is symptomatic of Swedish archaeology. In Arcini and Svanberg's study, the pottery and bones from the graves underwent detailed analysis by specialists. As for the other finds, bronze was found in five graves, six graves contained worked groundstone, and a few graves had finds of resin caulking and iron fragments (Arcini & Svanberg 2005:314). Although flint evidently accounted for the largest category of find after the burial urns and the cremated bones, it was not analysed. This should not be regarded as a criticism of Arcini and Svanberg's study. It is thorough and well done. Instead it should be viewed as one of many examples of how knapped flint found in cremation graves is not considered worth worrying about.

This is not unique for Swedish archaeology. In Poland, for example, large urn cemeteries have been excavated. At several places, tools and flakes of flint were found in cremation graves. Some scholars think that these finds should be regarded as older material ending up in the graves by mistake, while others collected the flint objects as part of the find material but did not use it in the interpretations of individual graves, rituals, or the burial ground (for a discussion see Piotrowska 2000:327ff).

In a study of flint objects in urn graves from Poland, Danuta Piotrowska, has clearly shown that these artefacts are contemporary with the graves. She interprets flakes and tools of flint as ceremonially important objects that were used in the protracted preparations, the cremations and burial ceremonies that took place in and near the burial grounds. Since several of the finds are not burnt, the flint objects were presumably used chiefly in connection with the actual burial (Piotrowska 2000:329). The flint as a material may have been associated with fire, since it was used to make fire. The symbolic meanings of the material as something containing fire may have been a significant element to emphasize in rituals. Piotrowska cites another suggested interpretation from Yaxchilan in Mexico, where obsidian knives were used for blood-letting in funeral rituals. Piotrowska puts forward the hypothesis that the unburnt flint objects found in cremation graves might have been used in a similar way in the Late Bronze Age (Piotrowska 2000:316).

This is interesting and it raises some questions for further investigation. Is the flint found in Scandinavia cremation graves burnt or not? Is it knapped flint or just pieces of the raw material flint that are found in cremation graves? Is it possible to interpret the finds in relation to the grave or the funeral ceremonies? Are there finds of large blade knives of flint from mortuary contexts? These are questions that I will consider based on a selection of published studies of cremation cemeteries. It should be underlined that I am unable here to carry out an exhaustive analysis of flint from cremation graves. It would be too huge a task to search the museum stores for finds from excavations conducted up to a hundred years ago, to make them accessible and to ascertain the chronology and context. I have therefore chosen instead to highlight a few assemblages with finds of flint in order to demonstrate the potential of analyses of this archaeological material. For a historical survey of excavations and interpretations of cremation cemeteries from Scania, see Arcini and Svanberg (2005).

Analyses of cremation graves with finds of knapped flint

In 1977 extensive archaeological remains were excavated at Kristineberg, just outside Malmö in south-western Scania (Rudebeck & Ödman 2000). This excavation was near the place discussed above as a site where large blade knives were made together with household production. The 1977 excavation uncovered a number of cremation graves from the Late Bronze Age (Rudebeck & Ödman 2000:129ff). Nine of the graves are dated to Montelius periods V–VI. Two of these contain flint, table 2.115 and figure 2.193. This I have analysed.

The flakes from feature 190 were made of till flint following the same chaîne opératoire as that defined above as household production. The flakes had not been detached from the core found in the grave. Of the five flakes, four have a continuous edge longer than two centimetres and ending in a point. I have performed use-wear analysis on all of the flakes. They all have a post-deposition patina that makes it impossible to interpret the contact material, figure 2.194.

The number of flakes with edges is interesting. In the analyses of household production presented above, for example, from the Tygelsjö excavations, there are 767 flakes. Of these, 104, or 13.5 per cent, have a continuous edge longer than two centimetres. In the cremation grave, feature 190, from Kristineberg, four of the five flakes have a continuous edge longer than two centimetres. I interpret this as a sign that specially selected unmodified flakes were placed in the grave. If just any flakes had been placed in the grave, the number with continuous edges would not have been so large.

In the 1920s Knut Kjellmark excavated a cemetery with small barrows and cremation graves at Kämpinge in south-western Scania, Sweden (Kjellmark 1925). Two of the cremation graves, which were dated to the Late Bronze Age, contain

Table 2.115. Data on two cremation graves from the excavations at Kristineberg.

Designation	Find	Description of the flint objects	Reference
Urn pit, feature 190	Urn with burnt bones and a razor of bronze.	One burnt flake, an unburnt core knapped using anvil technology, and five unburnt flakes.	Rudebeck & Ödman 2000:129
Urn cremation pit, feature 151	Urn with burnt bones.	A piece of burnt flint.	Rudebeck & Ödman 2000:136f

Figure 2.193. The urn pit, feature 190, from the excavations at Kristineberg (Rudebeck & Ödman 2000:130, figure 97).

finds of knapped flint, table 2.116. I have not had been able to analyse the finds.

At Klåhög just outside Helsingborg, northwestern Scania, a number of burial mounds and a cremation cemetery were excavated in 1967 and 1968 (archival material, Lund University Historical Museum, inv. no. LUHM 31875). The dating of the graves ranges from the Early to the Late Bronze Age. Two graves which have been dated to the Late Bronze Age contain finds of knapped flint, table 2.117. This I have analysed.

The flake from the grave called "Mårtenssons hög 16" is made of till flint following the chaîne opératoire defined above as household production. It was struck using bipolar technology and has a continuous edge longer than two centimetres and ending in a tip. The flake is patinated in a way that renders

Table 2.116. Finds from two cremation graves from Kjellmark's excavations at Kämpinge in Scania.

Designation	Find	Description of the flint objects	Reference
Barrow 5	Urn with burnt bones and lid, double button of bronze.	"Three fairly large unburnt flint sherds."	Kjellmark 1925:156
Barrow 7	Urn with burnt bones and lid, double button of bronze, awl of bronze, ring of bronze.	"Three pieces of unburnt flint, one of them a piece of a core-like nodule."	Kjellmark 1925:158

Figure 2.194. Flakes from feature 190. The flake to the left is 7.7 cm long.

use-wear analysis impossible. Nor have I been able to analyse the three burnt flakes for use-wear.

In the years around 1950 some seventy cremation graves and similar features dated to the Late Bronze Age were excavated at Simris in south-east Scania, Sweden. Seven cremation graves contain finds of flint, and three of these are dated to Montelius periods V and VI (Stjernquist 1961:88), table 2.118.

I have not been able to examine the finds myself, but in the publication they are depicted with photographs which show that both flakes, that is, those from graves 50 C and 98, have a continuous and well-formed edge more than two centimetres long (Stjernquist 1961, Pl. LV:2 & 4).

The results of combining the finds from four excavations show that knapped flint found in cremation graves from Montelius periods V and VI in Scania mostly consists of unburnt, unmodified flakes with continuous and well-formed edges more than two centimetres long. These flakes were made according to the same chaîne opératoire as that defined for household production. A comparison of the proportion of unmodified flakes with edges in the graves in relation to the proportion of the same type of flakes in settlement site material shows that the flakes in the graves were specially chosen. They were not randomly collected flakes. My interpretation is that they were selected to meet the criterion of having a continuous and well-formed edge more than two centimetres long. The fact that the flakes are unburnt shows that they were not included in the cremations. Apart from these flakes, there are also finds of burnt flakes and tools, as well as flint which was burnt so severely that it cannot be identified. This flint was probably included in the cremation (Stjernquist 1961:88).

Table 2.117. Data on two cremation graves with finds of flint from excavations at Klåhög, Helsingborg, north-western Scania, Sweden.

Designation	Find	Description of the flint objects	Reference
Mårtenssons hög 16. Grave 5, urn grave.	Urn with burnt bones, lid.	An unburnt flake.	Archival material LUHM 31875.
Barrow VI. Mårtenssons hög 52. Feature 1.	Pottery fragment.	Three burnt flakes.	Archival material LUHM 31875.

Table 2.118. Data on three cremation graves from the excavations at Simris. Details of the finds of flint objects are taken from the publication (Stjernquist 1961). I have not been able to examine the objects myself.

Designation	Find	Description of the flint objects	Reference
No. 6	Razor of bronze, resin caulking from a wooden box, burnt fragments of bone with dot-and-circle ornaments, piece of iron, potsherds and cremated bone.	A burnt strike-a-light.	Stjernquist 1961:11, 88
No. 50 C	Potsherds and cremated bone.	An unburnt flake.	Stjernquist 1961:17, 88
No. 98	Double button of bronze, bronze hook, and cremated bone.	An unburnt flake.	Stjernquist 1961:24, 88

A burnt large blade knife of flint from a cremation grave

Just north-west of Greifswald, directly across the Strelasund from Rügen, Germany, at a place named Ladebow, a cremation cemetery has been excavated. This was done repeatedly during the 1990s by a committed and knowledgeable private archaeologist (personal communication Michael Schirrin). The cemetery is unpublished. It has been dated through finds of pottery and metal to the period from the Late Bronze Age to the Early Iron Age.

In one of the graves, number 23, several fragments of burnt flint, including a large blade knife, were found together with the burnt bones. This is the only find, to my knowledge, of a large blade knife of flint from a mortuary context, figure 2.195.

The grave was a 30 × 20 cm pit in the ground which was filled with burnt bones. No pottery was found in the pit. The flint was discovered in the filling with the burnt bones. The flint is burnt, leaving it white-calcinated and cracked. It consists of 26 pieces in all. One of these pieces is not knapped, the rest are. Burnt flint is difficult to identify since it is often badly fragmented. To obtain some idea of the minimum number of flakes in the find material, I sorted out flakes with the bulb of percussion wholly or partly preserved. Eight such pieces were found.

Among the pieces was a fragment of a large blade knife. The proximal part of the blade knife is preserved but the platform was broken off by the heat. The knife is retouched on the right side. It weighs 40 grams. Since it is burnt, the large blade knife presumably accompanied the body on the pyre.

Conclusion – flint objects in cremation graves

Recent years' excavations of burial places from the Late Bronze Age have highlighted the complexity of the sites and the actions performed on them. Extensive rituals and ceremonies were enacted, and the cremations and burials should be perceived as complex actions lasting a long time (Arcini & Svanberg 2005; Arcini 2007; Arcini, Höst & Svanberg 2007; Svanberg 2007; Goldhahn 2007; Kaliff 2007).

Flint in cremation graves from the Late Bronze Age is a category of find that has not previously been used for interpretations of grave goods, grave structures, and funeral ceremonies. The result of the studies performed here show that flint flakes and other flint tools from cremation graves should be regarded as part of the find material from cremation graves. The finds of knapped flint from the graves are dominated by unmodified flakes with sharp, continuous edges suitable for cutting with. These were evidently specially selected parts of the household production that I analysed above. To interpret the unmodified

Figure 2.195. Fragment of a burnt large blade knife of flint and burnt flakes from a cremation grave dated to the Late Bronze Age. The find comes from Ladebow, just north-west of Greifswald in northern Germany. The find is the only one to my knowledge of a blade knife from a mortuary context. The finds are in the stores of the Landesamt für Bodendenkmalpflege in Mecklenburg-Vorpommern and have inventory number ALM 1999/1014,59. The fragment of the blade knife is 8 centimetres long.

unburnt flakes from the cremation graves as objects used in these rituals and ceremonies, in the way Piotrowska (2000) does, seems reasonable. I know of one instance of a large blade knife of flint found in a cremation grave. It is burnt. It is a unique find.

Large blade knives and other flint assemblages from the Late Bronze Age – conclusions

During the years since flint assemblages from the Late Bronze Age became known in archaeological research, they have been evaluated in different ways. By far the most common perception is that flint was a substitute for metal. This view is found in both early and recent publications, and it still exerts some dominance even today. It has also been common to make comparisons with Stone Age flint technology. This approach made its breakthrough in the 1930s and was to set the standard for studies of Late Bronze Age flint tools and technology for several decades to come. The comparative assessments still survive to some extent today, rating the flint technology of the Late Bronze Age as a coarse, primitive simplification of the elaborate flint craft of the Stone Age, as a relic of a vanished age, a late echo of the high quality of an earlier period which had almost – but only almost – become extinct. It has been less common for flint assemblages from the Late Bronze Age to be ascribed a value of their own, worth studying on their own terms. These values have affected the way in which research on flint assemblages from the Late Bronze Age has developed. Negative judgements together with a shortage of good analyses have had the result that Late Bronze Age flint assemblages, even though they often account for the largest share of excavation finds by volume and weight, have not been used to any great extent in interpretations.

What is known about large blade knives of flint has changed during the more than a hundred years that have elapsed since the implement from Stenild was published (Blinkenberg 1898). A survey of previous studies shows that archaeological research on blade knives has not led to any accumulation of knowledge. Instead the studies have displayed a mixture of new questions, elaborate hypotheses, new methods, good analyses, and new knowledge on the one hand, and on the other hand misunderstandings, incorrect assumptions, reproduction of old claims, unsubstantiated statements, and carelessness.

At an early stage Müller (1919) established that large blade knives and other flint objects are found in flint assemblages from the Late Bronze Age. Since then opinions have differed as to which periods blade knives should be assigned to. It was not really until the contract excavations in the late 1970s and early 1980s that the question of dating was seriously considered again and blade knives were dated to the Late Bronze Age. Despite this, there has been some doubt among scholars as to whether the large blade knife should be dated to Montelius periods IV, V, and VI or only to periods V and VI – or Phase A and phase B, or only phase B to put it in other terms. The studies performed here clearly show that large blade knives of flint occur together with pottery of phase B type. The large blade knives were introduced some time just after 900 BC and went out of use some time just before 500 BC.

I have come across large blade knives in museum collections and excavation finds, and seen them mentioned in publications from Sweden, Denmark, Norway, Germany, and Poland. This corresponds to the areas of influence defined in other works on the basis of studies of bronzes (Jensen 2002:497ff). Most of the blade knives occur in collections and assemblages from Denmark, from Scania and Halland in Sweden, and from the island of Rügen in northern Germany. The further away from these areas we come, the scarcer large blade knives of flint become in archaeological finds. As regards Poland, however, I have not studied the numbers or the geographical extent.

Large blade knives were not made by just anybody. Knowledge and know-how were required for successful manufacture. The experimental production series conducted for this study demonstrate that blade knives were a standardized and specialized product which followed well-defined stages of manufacture. Although modern experimental manufacture can never reflect a prehistoric situation, the experiments were performed in such a way that the chaîne opératoire defined above is, in my opinion, close to the prehistoric manufacture in that the defined stages of manufacture are stages that the prehistoric production also went through.

Large blade knives of flint are standardized. Their appearance is not random. The variation in the size and technological attributes of blade knives falls within set limits. I have defined four groups – type I large blade knives with a well-prepared platform, type I large blade knives without platform preparation, type II large blade knives, and Rügen blade knives. These definitions are partly new, partly based on earlier studies (Björhem & Säfvestad 1993). The different groups of blade knives that have been defined have a great deal in common in appearance and manufacture, but minor differences reveal that they are morphologically different and that manufacture took place within different chaînes opératoires.

There are regional similarities and differences. One obvious difference is that Rügen blade knives are smaller than type I large blade knives from other areas. The blade knives in these assemblages represent a separate manufacturing tradition. Assemblages from Rügen contain only this type of blade knife. But this manufacturing tradition is also found in other places. In the flint assemblages from excavations in south-western Denmark, Ethelberg noticed that some sites lacking large blade knives have knives that are equivalent to them. It is difficult to interpret flint artefacts solely on the basis of drawings in a publication, but the smaller knives that Ethelberg describes look to me like fragments of Rügen blade knives (Ethelberg 1982:122). The same applies to the excavations at Fragtrup in Jutland, Denmark. None of the finds from there were defined as large blade knives, but judging by the drawings of the finds, there is a number of small knives. These resemble Rügen blade knives (Draiby 1985:206). I have not found Rügen blade knives in eastern Denmark or in Sweden. It is important to point out that the finds from western Denmark also include plenty of the other types of large blade knives (see e.g. Berthelsen et al. 1996; Eriksen, in press). Large blade knives from Schleswig-Holstein are more like blade knives from eastern Denmark than those from Rügen.

If we look eastwards we find that Rügen blade knives are like the published blade knives from settlement sites in Poland (Moskwa 1976; Kruk 1994). But large blade knives of Zele type from the flint-mining area of Wierzbica in Poland (Lech & Lech 1997) do not look like Rügen blade knives. They are more like type I large blade knives with a well-prepared platform. Yet the three blade knives of Zele type in my analysis differ from other type I large blade knives in my material as regards some of the technological attributes.

Another difference concerning the areas in western Denmark and north-western Germany is the presence of bifacial technology in assemblages that are contemporary with large blade knives. From Sweden and eastern Denmark I have not found any example of finds of bifacial technology in contexts with blade knives from the Late Bronze Age, as Sørensen (1984-85) also noted. Both in the assemblages from Fragtrup in Jutland (Draiby 1985) and in those from south-western Denmark published by Ethelberg (1982) there are finds of bifacial technology. If we look further west to the Netherlands, we see that thick crescent-shaped bifacials have been found in contexts dated to the Late Bronze Age. These areas show no finds of large blade knives (van Gijn 1988:197ff; Woltering 2001:364). It is obvious that the flint technology in these areas in the latter part of the Late Bronze Age was geared to bifacial production. Ethelberg has discussed the problem that these bifacials have previously been dated as Late Neolithic or Early Bronze Age. He thinks that parts of the finds from northern Germany and Jutland should probably be dated to the Late Bronze Age, just like the finds from the Netherlands and north-west Germany (Ethelberg et al. 2000:237ff).

Looking at the margins of the areas within which I have found large blade knives, in Central Sweden there is a specialized technology consisting of the production of small flakes, a technology that I have not seen anywhere else. This is from a site where a large blade knife has also been found in the assemblage.

It is clear that south-western Denmark and northern Germany constitute an area where several different technologies met. These are technologies manifested in blade knives of the type found in the rest of Scandinavia, blade knives of the type found on Rügen and to the east in Poland and the thick crescent-shaped bifacials found in north-western Germany and the Netherlands. Further south in Poland, towards Wierzbica, yet another technology is represented in large blade knives of Zele type. I have not studied the encounter of the technologies in Poland. To the north in central Sweden yet another encounter between different flint technologies took place, between the large blade knife as an artefact and the specialized manufacture of small flakes.

The function of large blade knives has been discussed for a long time, and several different uses have been proposed for the tool. Up until the end of the 1970s these discussions were based on the look of the knives, the occurrence of gloss visible to the naked eye on the edges, and practical experiments. Since the end of the 1970s the discussions have mainly been based on functional analyses. Several studies claimed that large blade knives of flint were multi-tools. It has been suggested that they were used for reaping cereals, cutting reeds, harvesting leaves, doing woodwork, and preparing hides. It has also been suggested that large blade knives were shafted or that they were held in the hand unshafted when they were used.

The studies conducted here show that the large blade knives in my functional analyses were mainly used for cutting cereals or reeds. There is nothing to suggest that large blade knives of flint were used for cutting leaves, the function indicated in the Swedish name of the tool, *lövkniv*. In all the cases where analyses could be performed, they show that the large blade knives were shafted in a way identical or similar to the tool from Stenild; the knife blade was wedged in a wooden shaft. A large blade knife was thus a blade in a composite cutting and reaping tool.

The implement was an everyday tool used close to places of work and habitation. Since it was primarily employed for cutting large amounts of cereals and reeds, it was a seasonal tool. Grain was harvested in the late summer, reeds in the period between summer and late winter. This means that the large blade knife was not a spring tool.

The knife blades, that is to say the artefacts I call large blade knives, were also used for other purposes. If they broke, the pieces could be reused. On one site which I have examined, three fragments were used for working hides. Often, however, the broken blade knives were discarded on rubbish dumps and in pits.

Another way to reuse large blade knives was to deposit two or more of them together in the ground. These hoards have previously been interpreted by archaeologists as a uniform group of finds and associated with a Neolithic tradition of depositing flint tools in the ground. There is nothing in the studies conducted here to support this idea. Assemblages of large blade knives have not previously been interpreted together with contemporary hoards of bronze. This has seriously limited the interpretations of the assemblages.

The deposition of large blade knives in the Late Bronze Age was a complex procedure. Sometimes forty or more objects could be deposited together. The majority of these had not been shafted and used as tools. They came straight from the place of manufacture and consisted of newly made large blade knives, preforms for large blade knives, and blade-like flakes. Several of them had a shared biography in the sense that they met before they were deposited. At other times a small number of large blade knives were deposited together. These consisted of knife blades which had been shafted and used as cutting or reaping implements in the same way as other large blade knives. These large blade knives all had different life cycles in the sense that they were not manufactured and distributed together. They did not meet until they were deposited. Further examples show that large blade knives were deposited on different occasions but in the same hoard, in a pit in a house beside a roof-bearing post. Knife blades were repeatedly placed in this pit together with other objects. These large blade knives all had different biographies, in the same way as in the hoards just described, but they did not meet until after they were deposited, that is, not in the deposition but in the deposit. There is nothing to suggest that whole tools – knife blade plus shaft – were deposited together in these hoards. It must be added that single finds of large blade knives have not been analysed here.

The result of the analyses of the assemblages of large blade knives shows that previous knowledge about Late Bronze Age hoards must be revised as regards the number of deposits and their geographical distribution.

The experimental studies gave the information needed to look for sites where large blade knives were manufactured. Blade-like flakes and two kinds of preparation flakes proved to be diagnostic, and with the aid of these I have gone through a large number of flint assemblages from archaeological excavations in search of traces of blade knife production. One of the clearest results is that large blade knives were not manufactured at many places. On the majority of the sites where people lived in the Late Bronze Age, no large blade knives were produced. They were instead manufactured at a number of special manufacturing sites located beside plentiful supplies of raw material. I have analysed six such sites where blade knife knappers produced standardized large blade knives, preforms for blade knives, and blade-like flakes in huge numbers. These sites were specialized for this manufacture, and the raw material was extracted on the site. The composition of the assemblages shows that there were other places of this kind, but they remain to be located.

One site, Södra Sallerup in Sweden, differs from the other five in that it was not just a special manufacturing site located beside a rich supply of raw material, but also a place where people lived and performed other craft work.

Large blade knives, preforms for blade knives, and blade-like flakes were the products of the specialized manufacture on these sites. It is thus evident that these sites were used for the production of other objects than the knife blades which the blade knives were used as when they were shafted to give cutting and reaping implements. Large blade knives, preforms for blade knives, and blade-like flakes were distributed locally over large areas to the north in Sweden and Norway and to the south-west in Denmark, to places where flint was not available in the ground. Several of the assemblages show that these large blade knives, preforms for blade knives, and blade-like flakes could be packed on a manufacturing site and distributed together over large areas. Large blade knives from the settlement sites indicate that the blade knives which were used there could come from different manufacturing sites and have different biographies. From this we see that the distribution of large blade knives took place in different ways.

Contemporary with the large blade knives, a completely different flint-knapping technology was used. This consisted of *ad hoc* household production in the places where people normally lived and worked, whereby flakes were made from simple cores. The most common tool in this technology was the unmodified flake, which was used for many different things. This *ad hoc* household production looked the same over very large geographical areas. The similarity in this production also extends over time. In the centuries both before and after the introduction, use, and abandonment of the large blade knife, household production persisted relatively unchanged.

In the Late Bronze Age there were thus two different flint technologies side by side. Although these technologies were present in everyday life, manifested in different tools, they usually did not influence each other but were kept separate. This is most clearly seen in all the large blade knives that are found together with household production in settlement site material. It is also evident in the specialized flake production that is found in Central Sweden together with household production. The simple, everyday flake tools must have been so suitable for their purposes that it was not considered essential to transform this household technology according to the model of the large blade knife or specialized flake production. Alternatively, large blade knives were so radically different from household production that people did not associate the technologies at all.

There are, however, a number of sites where the two technologies met. These are places where large-scale craft activities such as pottery making, manufacture of stone tools, bronze casting, and other metalwork was performed. On these sites the specialized large blade knives were used as knife blades for the tools of which they were a part. Household production was also pursued there. But what is distinctive is that someone also made large blade knives on these sites. The blade knives manufactured on-site have technological features of both specialized large blade knife production and household production. At these places, then, several different crafts and their practitioners met, and someone challenged the distinction into two different flint technologies that were so clearly separated at other places.

Knapped flint was a part of Late Bronze Age mortuary rituals. In what is today northern Germany, a person was buried with a large blade knife. This blade knife was cremated, just like the person's body. On other sites, unburnt, unmodified flint flakes were deposited in graves; they were thus not placed on the funeral pyre. These flint flakes are distinguished by having a continuous sharp edge. They were made using the same technology as flakes from household production. They were employed in funeral ceremonies during or after the placing of the urn, the cremated bones, or the remains of the pyre in the ground.

To conclude the second part of this book, it can be said that I am now beginning to gain a grasp of the large blade knives and other flint assemblages from the south Scandinavian Late Bronze Age. The questions asked in the introduction to this part of the book have been considered, table 2.119, and there is a good foundation on which to stand as I now return, in the last part of the book, to focus on the why-question that was posed in the introduction.

Table 2.119. Results in brief, listed as answers to questions formulated in the chapter "A history of previous archaeological research".

The placing of the retouch has been associated with right- and left-handedness. Is there any justification for this interpretation? Could there be other reasons for left and right retouch?	The experimental analysis shows that the choice of right or left retouch on large blade knives depends on the way in which they were made. Preforms for blade knives which were removed from the right side of the core have the edge to the left and preforms removed from the left side of the core have the edge to the right. I have not examined whether this can be related to right- and left-handedness during the Late Bronze Age.
Are there functional and technological types that Sørensen (1984-85) or Björhem & Säfvestad (1993) have not defined because their methods did not consider function and technology?	The attribute analysis shows that four different blade knives can be defined in terms of technology. Of these, three were included in the functional analysis, the results of which show that blade knives were shafted in the same way and were used on the same kind of material. So even though I defined different technological types manufactured in different ways, they were used in the same way.
From which periods are there large blade knives in the archaeological material and which periods have none? When was the large blade knife introduced? How long was it used? When did it cease to be used?	The analyses of dated contexts shows that large blade knives occur together with other archaeological material which can be pinpointed to Phase B of the Late Bronze Age, that is, Montelius periods V and VI. The large blade knife was introduced some time after 900 BC and fell out of use some time just before 500 BC. This means a period of use of at most 400 years.
How were blade knives used, for what purpose and in what contexts?	The large blade knife was used as a knife blade in a composite tool. It was shafted with the aid of wedges in a wooden shaft. It was used as a cutting or reaping tool for cereals and reeds. The contexts in which blade knives were used are close to the settlement site. Knife blades were also used as objects in deposits.
Are all assemblages of large blade knives similar? Are there different kinds of deposits? Is it right to place hoards of blade knives within the same interpretative framework as Stone Age hoards?	There is variation between different assemblages. They contain different artefacts, artefacts which had been handled in different ways and deposited in different ways. There is nothing in my results to support the claim that large blade knives belong within the same interpretative framework as Stone Age hoards.
The studies that have been published have concerned large blade knives in settlement site contexts or hoards. Are blade knives found in other contexts?	Large blade knives occur in settlement site finds, in assemblages, as "hoards", and as stray finds, for example in ploughed soil. I know of one find of a blade knife from a mortuary context. Large blade knives also occur on special manufacturing sites. Otherwise I have not found blade knives in any other context.
It is obvious that large blade knives are found in southern Sweden, Denmark, Poland, and northern Germany. What remains to investigate is where the geographical boundary runs. Is it in Norway, in central Sweden, or in other parts of northern Europe? What do flint assemblages look like in different parts of the area where the blade knife occurs? Are they similar, different, or does this vary?	Large blade knives are chiefly found in the southern parts of Scandinavia, that is, in Denmark, along the Baltic coast of northern Germany, and in Scania and Halland in Sweden. Blade knives are found to a lesser extent along the west and east coasts of Sweden and in provinces north of Scania towards the Mälaren valley. From Norway I have found blade knives in museums in Oslo and in Trondheim. It seems to be an unknown artefact type in Norway, so there may be many blade knives that have not been recognized as such. In Poland there are large blade knives from manufacturing sites and in settlement site material. I have not studied the occurrence of blade knives in Poland in detail. From other parts of northern Europe than those mentioned here, I have not found any large blade knives. The analyses show that blade knives from Rügen in northern Germany look different from those from Sweden and Denmark. There also seems to be a difference between Scania and eastern Denmark on the one hand and western Denmark and the western parts of northern Germany on the other hand. From the latter areas there are finds of bifacial technology together with large blade knives, which are not found in the eastern parts of the studied area.
How were large blade knives made? Where were they made? Is it right to look for blade cores to demonstrate production? Is there other material which could indicate the production of blade knives?	Large blade knives were made in systematic, standardized, and specialized production. They were made on special manufacturing sites with rich natural supplies of high-quality flint. On the majority of the sites with settlement site material there are no traces of blade knife production. On some sites there are traces of blade knife production together with other settlement site material and remains of other crafts. Blade cores are unsuitable as indications of manufacturing sites. Two types of flakes – blade-like flakes and preparation flakes – are diagnostic.
How were blade knives distributed? What channels or networks of contacts could they have belonged to? Who acted in the handling of blade knives? Is it possible to detect influences between different parts of the studied area? Can studies of large blade knives contribute anything to our knowledge of topics such as contacts, economy, social relations, power, trade, or ritual?	These are questions which will be discussed to varying extents in the next part of the book.

Part III: A sociotechnical construction of the large blade knife of flint

Part II of this book may be regarded as the long task of establishing and renegotiating the large blade knife as an artefact in the discourse of archaeology. This work has sought to deconstruct the archaeological pre-understanding of this artefact by discussing what has previously been said or not said about Late Bronze Age large blade knives of flint. It has also aimed to build new knowledge, that is, to contextualize and highlight the large blade knife as a multifaceted and ambiguous archaeological source material. This means that the large blade knives of flint presented in photographs and described in words in the introduction to this book have been deconstructed. They represent different types, they were used, unused, manufactured, distributed, mass-produced, made as individual objects, discarded as waste, and deposited. They are cutting tools for reeds, reaping tools for cereals, signs of specialized manufacture, objects in assemblages, testimony to long-distance contacts and indications of geographical boundaries in material culture. It is no longer possible to perceive large blade knives of flint as just one category.

In this third part of the book I will continue deconstructing blade knives by discussing them in relation to actors, actants, and networks. I do this to be able to focus on the question posed in the introduction, about why large blade knives were introduced, were used, and disappeared in the latter part of the Late Bronze Age. By discussing contexts to which the blade knives belonged, it is my intention to demarcate their technological framework. As a result of my analyses, I conclude by discussing how what we call the Bronze Age became what we call the Iron Age.

I will not cover all aspects of flint craft in the Late Bronze Age that I have analysed. I have chosen to focus on one of the encounters between technologies that I have highlighted, and in this I will analyse sociotechnical aspects of change and persistence. In this analysis I once again concentrate on the main source material in the book, the large blade knives of flint.

Actors, actants and networks

The question is, which networks were large blade knives a part of? As we noted in the first part of the book, a discussion of this question is not a matter of analysing subject and object or persons and artefacts. It is a matter of trying to identify and discuss human as well as non-human actors and actants with potentially equal properties and abilities to shape the networks to which the blade knives belonged (Latour 2007). Objects, technologies, and persons participate in the sociotechnical, they shape it. Human beings form social networks not because they interact with other human beings but because they interact with human beings and endless other materials (Law 1992). Actors in a network are thus determined by their relations to others in the network.

I shall devote a few pages to putting names on possible actors and actants on the basis of the analyses. The aim is to clarify the networks to which the large blade knives may have belonged – although not with the ambition of listing every conceivable one – so that I can then discuss how different networks may have interacted in a time of sociotechnical change and persistence.

The tool-maker at the special manufacturing sites that I have analysed was a craftsman with knowledge and know-how that others did not have. The person kept this knowledge and know-how to him- or herself. It did not overflow into other flint knapping. The craftsmen mass-produced large blade knives. There were sites like this at various places in southern Scandinavia. The crucial thing for the establishment of such a site was a good supply of high-quality flint. The raw material was important for mass production. Someone quarried, dug out, or picked up flint for manufacturing on these sites. This called for a knowledge of the raw material. When large blade knives were manufactured on these sites, there were ideas about how they were supposed to look like. These ideas were in the head of the craftsmen. In this sense the large blade knife was a concept. It was standardized with a well-defined chaîne opératoire. It was a knife blade that had to fit the shafts that were used. The shaft thus dictated the dimensions of the blade knives. When someone acquired a new large blade knife of flint to replace an old one, it was essential that the new one looked like the old one and thus fitted in the shaft. Together with large blade knives, preforms for large blade knives and blade-like flakes were also made at the manufacturing sites.

The large blade knives were taken from the manufacturing sites and distributed over distances of varying length. This was done by people who traded, exchanged, bought, or sold. The distribution required an infrastructure. This was physical in the form of roads, paths, shipping routes, and rivers. It was also social in the form of contact networks that people established, maintained, terminated, and renewed. The distributors were people who knew this infrastructure and who moved within areas of varying extent. These people went to the manufacturing sites and there selected blade knives, preforms for blade knives, and blade-like flakes from the manufacture. Often this selection was made from one and the same production session.

The manufacturer on smaller sites, those which were not special manufacturing sites, was a craftsman who had the right knowledge but not the full know-how possessed by the makers of large blade knives at the special manufacturing sites. On these smaller manufacturing sites people made large blade knives in contexts where other crafts were performed as well, for example bronze casting, pottery making, goldsmithing, or the making of stone objects. These large blade knives were used near the place where they were made.

The large blade knife represented a preset pattern for what it was to be used for and how it was to be used. The task which involved using the tool to which the blade knife belonged must thus have been determined in advance. It cannot have been a task that arose spontaneously and was solved *ad hoc*. The regularity and consensus in the management of the task shows that it was established as a working concept and thus was probably well planned and well synchronized with other tasks. When someone used a large blade knife it was as a knife blade in a cutting and reaping implement with which that person was familiar.

Large blade knives were deposited in different ways. They ended up in waste heaps, in small assemblages, and in large assembled deposits.

At dwellings and workplaces there were stores of flint nodules. When someone needed a flake to cut or scrape with, that person picked up a nodule, knapped off a few flakes, used them, and then dumped them on the site in a waste heap, sometimes in the same pile in which used large blade knives were thrown. There was never anyone who collected these flakes to deposit them elsewhere. However, flakes of this type were employed at funeral ceremonies. The people who used the flint in these ways in household production did not have the knowledge and know-how possessed by those who manufactured large blade knives.

At funeral ceremonies, small unmodified flakes with a well-formed and continuous edge were used. They were made following the same chaîne opératoire as household production. They were deposited together with the cremated bones.

In this account a great many different actors and actants have been described. Some are relatively clearly formulated in my study, for example the maker or the large blade knives, while others more diffusely hinted at and not studied, for example, funeral ceremonies or footpaths as channels of distribution. The aim of the account is to clarify how the large blade knives and other flint assemblages were a part of Late Bronze Age society in a multifaceted way, and that actors and actants in various constellations formed networks in which the blade knife was constituted in different ways. Let me look at this in greater depth.

If we return briefly to the studies of the source material in Part II of the book, take the actors and actants defined above and follow the blade knives in, say, the assemblage from Spjald, a complex pattern emerges of actors who in different ways shaped and were shaped by the object that was the large blade knife. The six blade knives from the find were made on at least three different sites. On each of these sites, different flint-knappers manufactured a product. A large blade knife made at one of these places, which would later end up in a pit beside a roof-bearing post in a house at Spjald, was one of a great many large blade knives produced during a few hours' work. Together with other blade knives, a distributor took charge of them and transformed the large blade knife from a product into one commodity among all the others. A user selected this particular blade knife as suitable from among these goods. Through this choice the commodity was transformed into a part of a tool, a knife blade in a wooden shaft. This tool was used for cutting reeds. When the knife blade had become blunt from use it was removed from the shaft and deposited in the pit beside one of the roof-bearing posts in a dwelling house. Here it became a part of the house, and when a further five large blade knives and other objects were deposited in the pit it became part of a new context.

If we instead follow the objects in the assemblage from Stora Hammar in the same way, a different picture emerges. Of the forty large blade knives, preforms for large blade knives, and blade-like flakes in the find, thirty-nine were made on the same site. Here a flint-knapper did not solely mass-produce blade knives, but made a package of mutually related objects. These objects were not just a few in the crowd but a specific package of objects handled together. The distributor did not transform this package into an ordinary commodity. The package remained a package. The user chose this package, not because a knife blade fitted into a shaft but because it was a package. The objects were never shafted, they were never anywhere near to being used for cutting or reaping. The whole package was deposited in a pit together with another used large blade knife.

In these example the objects were not constituted in the same way. The objects in the find from Stora Hammar seem to follow a direct line from manufacture to deposition, while the objects from Spjald have a more varied life story from manufacture to deposition. The networks to which the large blade knives belonged were shaped by – and shaped – people, actions, conventions, and objects in a variety of ways.

On the premises used in these two example, then, there are great differences in the way the actors made the large blade knives into their own knife blades and artefacts. In none of these examples do I know whether it was different people or the same people who performed the roles of the actors I have sketched. That is not the point. It is not individuals but actors that I have sought to define.

To relate this to the discussion in Part I of the book about ambiguity and pluralism, we may observe that, just as in my example of a car crash which is perceived from different points of view, Late Bronze Age large blade knives were also perceived in different ways. For the maker at the special manufacturing site, for example, blade knives were primarily the result of work. When flint-knappers talked about large blade knives the conversation may have been about the supply of flint, about a beautiful fossil that someone found in a nodule of flint when it was split, about the craft, about the tools, about how someone had hit his thumb with the hammerstone, and so on. When people were in the field using the tool of which the large blade knife was a part to harvest barley, perhaps they spoke about how the knife blade fitted the shaft, how quickly or slowly it became blunt, about how they would later take charge of the barley, or about the blisters on their hand caused by working with the implement. When used large blade knives were deposited beside a stone near where the people lived, they may have spoken about things to do with the actual deposition.

These examples are of course hypothetical. I have no idea what people in the Late Bronze Age spoke about. The examples are also banal. But it is precisely these banal things that I can use to show that the large blade knives and the tools of which they were a part could be perceived in different ways depending on which situation they arose in. From different points of view the large blade knife became a part of the networks to which it belonged. The large blade knife was thus an object that was firmly rooted in Late Bronze Age society. It existed, it was handled, it was used, it was given meaning and gave meaning in different ways in various contexts.

I ask the reader to note here that the terms and concepts used, for example goods and products, are employed only for the purpose of illustration. I am fully aware that these concepts are analytical categories, but I have no ambition to assign any interpretative value to any category. When I write the word "commodity", for example, it is to illustrate that the large blade knife hypothetically went from one stage to another in the analysis I perform, nothing else.

The construction of large blade knives

With these examples I have further deconstructed the large blade knife. Different actors have been pointed out in sociotechnical networks, all with the large blade knives or other knapped flint as the common denominator. It is now time to interpret these deconstructed blade knives, to construct my image of them. Or as Bijker has put it: "once an artifact has been deconstructed into different artifacts, what has to be investigated is how these different artifacts developed" (Bijker 1995:77). I will do this by putting the different large blade knives into contexts. The contexts I use consist of previous archaeological research; i.e. other scolars′ archaeological interpretation of the Late Bronze Age on different aspects of the relationship between flint, bronze, and iron, ritual specialists, local, regional, and supraregional, changes in the landscape, ritual depositions, and bronze sickles. On this basis I formulate my interpretation of why the large blade knife was introduced, was used, and then disappeared, and about how what we call the Bronze Age became what we call the Iron Age.

Flint as a substitute for bronze

In the few examples of previous research where flint has been given a role in interpretations of the Late Bronze Age, it has been as a surrogate for bronze:

> In connection with the evident shortage of metal that began with the crisis at the start of the Late Bronze Age, flint had a renaissance as a material for tools (Ethelberg et al. 2000:237).

This quotation from a study of the history of agriculture in southern Jutland is one of many examples where this interpretation is stated (see e.g. Glob 1938; Broholm 1949; Brøndsted 1958). It is an interpretation that describes a course of cause and effect as follows: until the Late Bronze Age, bronze was available and there was no crisis. In the Late Bronze Age, bronze disappeared and there was a crisis. Bronze was then replaced with flint. The logic in this interpreted course is thus: bronze existed, disappeared, and was replaced. This logic is based on the assumption that the technology manifested in the bronze is the same over time but that the raw material changed; in other words flint was substituted for bronze.

The result of my studies in the history of archaeology show that this logic is a construction within the discipline, a taken-for-granted assumption that many endorse but few have examined. It is based on the view that flint tools in the Late Bronze Age represented a primitive substitute material and as such were a relic from the past (Glob 1938; Broholm 1949; Brøndsted 1958).

But if we instead use the results that I obtained in the analyses of Late Bronze Age flint technology and view the large blade knives as representatives of a new technology that stood on its own legs and was not a surrogate material for tools of bronze, what happens to the logic of the interpretation that bronze existed, disappeared, and was then replaced with flint?

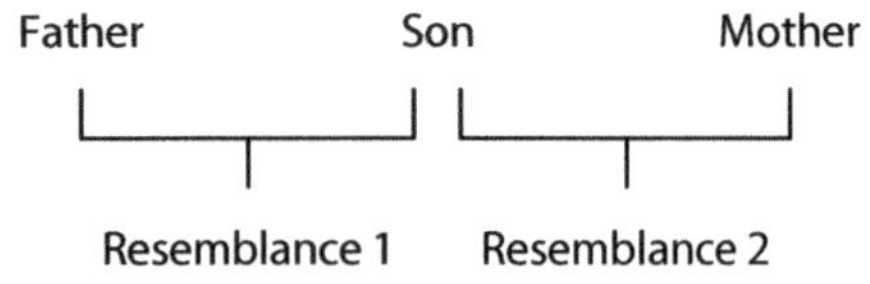

Figure 3.1. An illustration of a chain of relations according to Wittgenstein's (1953) idea of family resemblance.

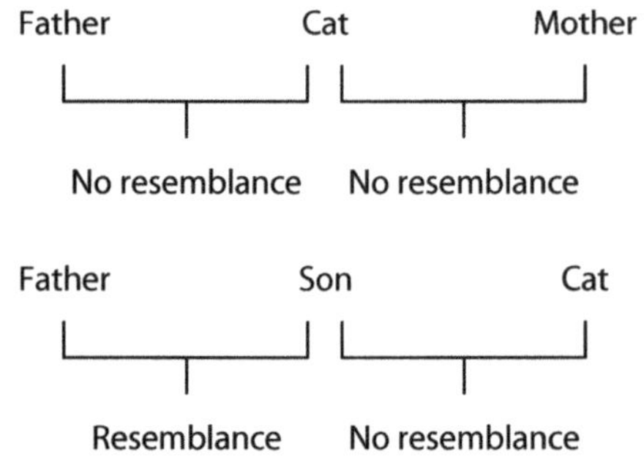

Figure 3.3. Alternative chains of relationship which do not display any family resemblance between father, son, mother and the cat.

I will discuss this with the aid of some thoughts about concepts, relations, and processes. To assist me I use Ludwig Wittgenstein's well-known idea of family resemblance or family likeness (Wittgenstein 1953). The idea itself takes its name from Wittgenstein's metaphorical description of a type of relationship he argued was exhibited by language. To demonstrate how seemingly different things can belong together, he built up analogies consisting of logical chains of relationship. One such concerns the relationships between father, son, and mother.

The members of a family resemble each other, but not in every respect. The father and the son are alike in that they have, say, the same build, curly hair, and pointed nose. The son and the mother are alike in that they have the same colour of hair, the same colour of eyes, and both get freckles on their noses in the summer. This chain of relationships is illustrated in figure 3.1.

In this chain of relationship the father and mother lack similarities, but they are linked to each other through a continuous chain of similarities and thus have an undeniable relationship. The concept of family resemblance illustrates this (Wittgenstein 1953).

If we now return to the discussion of bronze and flint in the Late Bronze Age and replace the terms father, son, and mother with the concepts "bronze existed", "bronze disappeared", and "bronze was replaced by flint", and replace resemblance one and two with cause/effect one and two, we obtain a chain that concerns interpretations of the relationship between the raw materials bronze and flint. Bronze existed, it disappeared and it was replaced with flint as a raw material. This interpretative model can be illustrated in a chain of relationships, figure 3.2.

That bronze first existed is a precondition without which this chain of relationships could not have been created, just as the assumption that there is a father is a necessary condition in Wittgenstein's example. The knowledge that bronze disappeared is linked to the knowledge that it had previously existed through a cause-and-effect relationship. The third assumption, that bronze was replaced with flint, is in turn linked to the knowledge that bronze disappeared. This is done through yet another cause-and-effect relationship. Together these relationships of cause and effect create a foundation for interpretations of what happened during the Late Bronze Age. Just as the similarities between father and son and between son and mother create a relationship between father and mother, in the example of bronze and flint a relationship is created between the existence of bronze and its replacement with flint.

If we now go further in this discussion and break out some of the parts of the chain of relationships of father, son, and mother, for example, replacing the son or the mother with the cat, then the similarities that created the family concept disappear and the whole definition of parts of the chain of relationships being linked thus collapses. Father and mother are no longer tied by a family resemblance, figure 3.3.

In the same way we can work with the chain of relationships: bronze existed, bronze disappeared, and it was replaced with flint. If, for example, we put in the word "not" in the chain of relationships, it becomes: bronze existed, bronze disappeared, and bronze was not replaced with flint. In this example the three parts of the chain are no longer linked through a family resemblance, figure 3.4.

The consequence of this discussion of concepts, relationships, and processes is that, if one part of the chain of relationships is changed, the interpretation of what happened during the Late Bronze Age is no longer valid. The interpretation that bronze disappeared and that it was replaced with flint is based

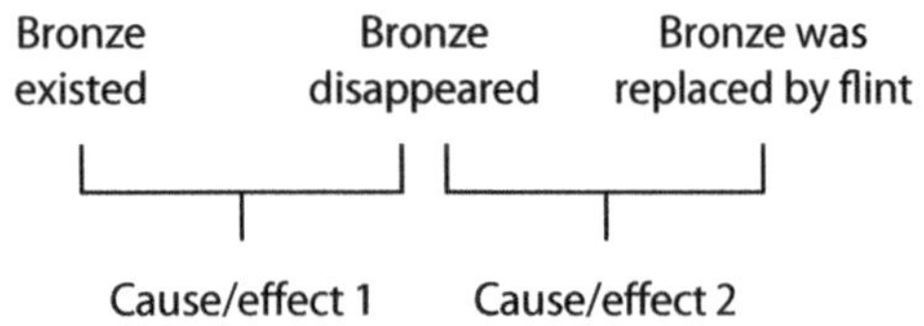

Figure 3.2. An illustration of a chain of relations according to Wittgenstein's (1953) idea of family resemblance but with bronze and flint as examples. Bronze existed, it disappeared, and it was replaced with flint. This relation is linked by the two assumptions about cause and effect.

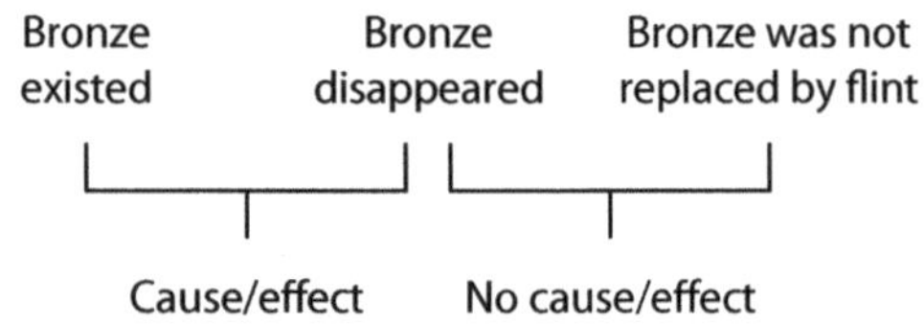

Figure 3.4. Alternative chain of relationships which does not show any family resemblance between the existence of bronze and its replacement with flint.

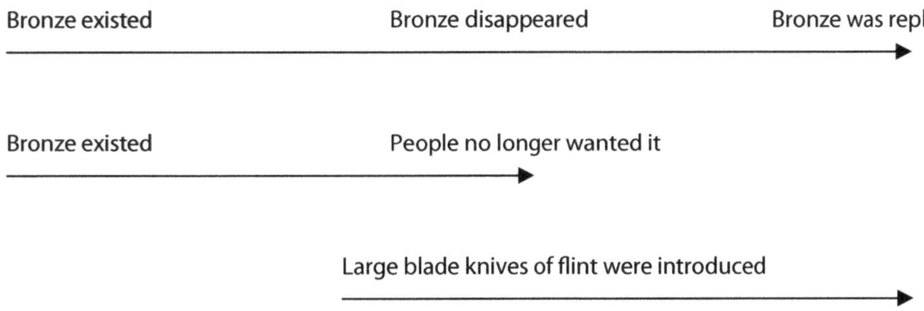

Figure 3.5. Different ways of describing the course of events in the Late Bronze Age. At the top is the interpretative model established in archaeological research. Bronze existed, it disappeared, and it was replaced with flint. This model, in my opinion, is a false linear explanation. At the bottom I illustrate a hypothetical alternative. Bronze existed but was no longer wanted. Flint as a raw material for the manufacture of large blade knives was introduced as something new.

on a logic which requires that all the parts of the chain of relationships must be included in the course of events. If any of the parts is removed, it is no longer a course of events. In this way it can be compared to what I called, in Part I of the book, a false linear explanation, where something is explained in terms of what is known in retrospect, not based on an analysis of what happened.

This means that step one is stable, it is the point of departure for the entire logic. But since we do not know whether step three, the replacement of bronze with flint, is correct, we cannot know whether step two is correct either. Perhaps bronze did not disappear in the sense that it was not available. It might have disappeared in the sense that people no longer wanted it, figure 3.5.

To sum up so far, if we regard large blade knives of flint as representatives of a new technology in the Late Bronze Age, not as a substitute material for tools made of bronze, then it is not possible to sustain an interpretation based on the explanatory chain of bronze being replaced with flint. Flint was not a surrogate. The flint was itself part of a new technology. The assumption that bronze disappeared because it was no longer available is untenable. It is a claim that cannot be taken for granted without investigation. Perhaps bronze disappeared because people thought they did not need it any more? I will pursue this question by adding yet another raw material to the discussion: iron.

From one metal to another

There is an archaeological interpretation that has totally dominated the narrative about technological change in the Late Bronze Age. It concerns societies which stopped using bronze and started to use iron as a raw material (see e.g. several articles in Sørensen & Thomas 1989; Karlenby 2003; Karlsson 2003). This is explicitly expressed in the terms Bronze Age and Iron Age. In general discussions this technological change is often held up as one of the clearest examples from prehistoric time of how one technology replaced another. The arguments for this are that bog iron ore and red earth occur at more places, in contrast to ores of copper, tin, lead, and other metals which were alloyed into bronze, and that the objects made of these metals were produced in different ways and looked different. These are arguments that these two metal technologies were different and are therefore an example of a technological change that is easy to identify (see e.g. Greene 2004:160).

Consequently, archaeologists discussing technological change in the south Scandinavian Late Bronze Age and Early Iron Age have mainly been occupied with the shift from a bronze technology to an iron technology (see e.g. Hjärthner-Holdar 1993; Nørbach 1998; Sperber 1998; Harding 2000; Greene 2004; Vandkilde 2007). It has also been taken for granted that iron was the major causal factor behind this change (for a critical discussion of this see Levinsen 1989; Sørensen 1989a). The following quotation is a textbook example:

> Given the constraints imposed by problems in the supply of copper and bronze and the deterioration of quality in such metals as were available, the use of iron now became more attractive from both a technical and an economic standpoint. The first swords and spear points made of iron or soft steel were produced – with an astounding command of iron technology – during the later half of the 9th century BC. And from that point on the use of iron increased steadily until, in the early 8th century BC, bronze was abandoned completely as a material for the production of weapons, tools and cutting implements (Sperber 1998:51).

This approach to change has had consequences for the interpretation of this period. It has been looked upon as a time of transition, when one metal technology was replaced by another. And this is of course true. What we call the Bronze Age was ending and what we call the Iron Age started, society changed, and so did technology. Yet if we interpret and explain technological change in this way, as a linear process from one thing to another over a period of 400 years – i.e. radiocarbon years, not calendar years – we obviously risk omitting to discuss what actually took place during that transition. People do not live in transitions. They live right here and right now, whether it is in south Scandinavia in the year AD 2008 (when this text is being written) or in south Scandinavia during the Late Bronze Age. This is about a difference between what is an explanation and what needs to be explained. To say that bronze disappeared and iron took over as a raw material for tool production is not an explanation. It is something that needs to be explained (Levinsen 1989:440ff; Bijker 1995:270).

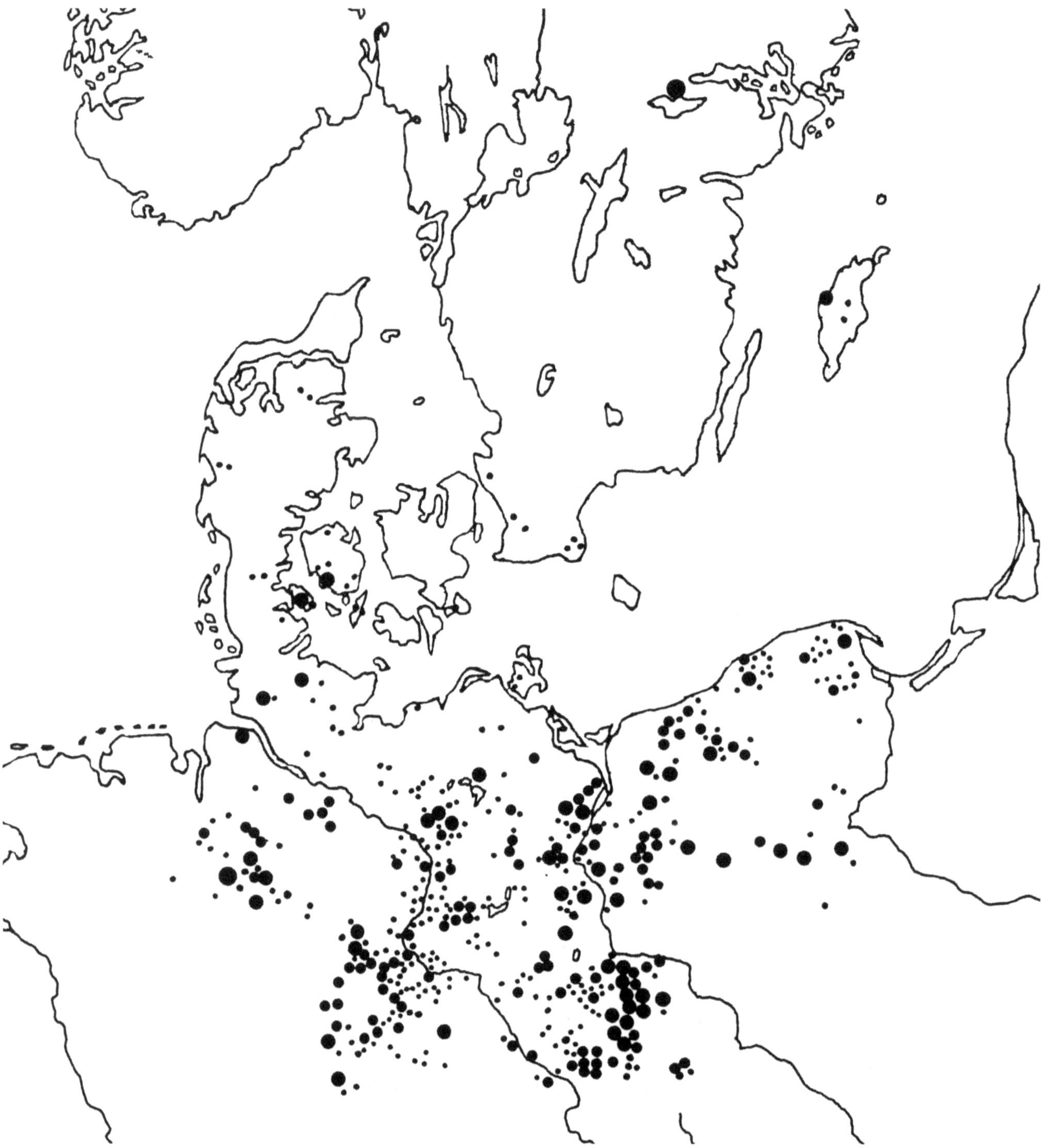

Figure 3.6. On a distribution map like this showing iron finds in period VI north of the Hallstatt culture, the differences between areas with and those without manufacture of large blade knives become clear. It can be seen that south-western Scandinavia, the Baltic coast of northern Germany, and central Poland have no dots or hardly any dots marking finds of iron. These empty spaces are exactly where large blade knives of flint were made. Notice that the map shows finds of iron objects, not production places for iron. The map comes from Thrane (1994:108), an updated and redrawn version of a map originally published by Horst (1971).

It is obvious, as we have seen, that radical changes over time occurred at many places in today's Europe at the time defined as the transition between the Bronze Age and the Iron Age (Vandkilde 2007). Iron production was developed, iron objects appear among archaeological finds from this time, and the amount and type of bronze objects found by excavations looks different if we compare excavations of Bronze Age remains from earlier periods with those from the time around 800 BC.

But the results of my studies show that it is not so easy to explain technological change at this time as a transition from bronze technology to iron technology. In both southern Scandinavia and northern Europe, artefacts were also made of flint. There is thus one more technology to take into consideration in discussions of technological change at this time (see Levinsen 1989 for a similar discussion).

Iron production

Objects of iron occur early in Bronze Age find contexts from southern Scandinavia (see e.g. Stjernquist 1961:77ff; Nørbach 1998:53ff), but it is not until the Late Bronze Age that there are extensive traces of iron production. Eva Hjärthner-Holdar, in an exhaustive study of the introduction of iron metallurgy into what is today Sweden, has shown that iron was made in the eastern and western parts of this area from around 1000 BC. In Denmark iron manufacture was not introduced until 500 BC (Hjärthner-Holdar 1993; see also Risberg & Hjärthner-Holdar 2004; Damell 2008). It is thus evident that iron was not produced to any great extent in the areas that are the central object of this study, that is, south-western Scandinavia (Sørensen 1989a:183f; Hjärthner-Holdar 1993:36). It is also clear that in the areas where there are special manufacturing sites for large blade knives, no production places for iron dated to the Late Bronze Age have been found (Hjärthner-Holdar 1998:38ff), figure 3.6.

Iron was mainly used for making utility objects such as small edged tools in the form of knives and sickles (Sørensen 1989a:191; Hjärthner-Holdar 1993:121ff). Hjärthner-Holdar shows that the introduction of iron metallurgy took about five hundred years, from the first evidence of indigenous (that is, in today's Sweden) iron production in the Late Bronze Age to a mainly iron-based economy in the early part of the Pre-Roman Iron Age (Hjärthner-Holdar 1993:185). It is striking that this time, which Hjärthner-Holdar calls an introductory period, coincides so well with the introduction, use, and disappearance of large blade knives of flint.

This is interesting. In the Late Bronze Age two technologies were introduced. In areas which had a natural supply of the raw material high quality flint, a technology was introduced which involved the specialized manufacture of large blade knives of flint. Other areas saw the introduction of a technology that involved the production of iron, which was used, among other things, for forging knife blades. If we disregard the raw material and the manufacturing technology, there is reason to view these technologies as similar. They were both new. They were specialized, meaning that not just anyone could perform the craft. The result of both technologies was tools with cutting edges. Neither of them was used to manufacture unique or special artefacts, unlike bronze (see e.g. Levy 1982:83f). The products of the technologies occurred contemporaneously in the same areas, but to differing extents.

Interpretations in contrast

Marie Louise Stig Sørensen has discussed how research on the transition between bronze and iron technology has been burdened by take-for-granted notions of the superiority of iron, which have steered interpretations. She thinks that iron and iron technology can be regarded as having very little importance for changes in southern Scandinavia during the Late Bronze Age, and that the concept of a transition from Bronze Age to Iron Age is in fact entirely based on the bronzes (Sørensen 1989a:186). Sørensen demonstrates how the use of objects in the Late Bronze Age was prescribed and constricted through the formalization and standardization of the production, use, and reproduction of bronze objects (Sørensen 1987:91, 1989a:190). She interprets this to mean that south Scandinavian Bronze Age society regarded iron as an alien element. Society was so strictly formalized and static that it was incapable of absorbing innovations (Sørensen 1989a:196). This delayed the introduction of iron to Scandinavia.

There is a paradox in Sørensen's (1989a) reasoning. While she puts her finger on a previously unquestioned assumption of the superiority of iron, and conducts an exemplary discussion stressing the importance of seeing an internal cultural driving force rather than external pressure as an explanation for tradition and change, she fails to analyse tradition and change as anything but a transition from one metal to another.

But what if tradition and change were not locked into a raw material, but into ideas that were translated into materiality in different ways at different places? What if the introduction of large blade knives of flint was a way to create, manage, or confirm the same needs and wishes that were created, managed, or confirmed with the introduction of iron production in eastern and western Sweden? This would mean that we are not dealing with a society in which the way bronze as materiality was managed – in an intricate interaction between being shaped by people and simultaneously shaping people – contributed to a conservative tradition incapable of carrying out essential changes or adapting to altered conditions (Sørensen 1987:91, 1989a:190). It is instead an innovative society where flint and iron were used in changed technologies.

Sørensen's interpretations are in stark contrast to those presented by Hjärthner-Holdar. Hjärthner-Holdar clearly shows how iron metallurgy was introduced in the time that Sørensen defines as being unable to absorb innovations. My study of large blade knives also reveals the improbability of Sørensen's argument.

At the same time, I think Sørensen makes an important point. For if I take her results seriously and do not dismiss the interpretations just because both my findings and those of Hjärthner-Holdar seem at first glance to refute them, I must then ask myself whether it is possible to combine the interpretations which show that society was conservative and not disposed to change with the interpretations which indicate radical changes and innovations. The important question then is not whether it is Hjärthner-Holdar and I or Sørensen who is right, but how these two seemingly conflicting results – both conservatism and innovation – can be contained in interpretations of Late Bronze Age society.

This is a crucial question since it brings up a perspective that emphasizes the dynamics between old and new, between conservative and innovative forces in society. This is a perspective that can help me to interpret why large blade knives of flint were introduced, were used, and disappeared. I will proceed with this by focusing on the people who made the blade knives and the places where they were made, and by analysing the actions and networks manifested in the manufacture and the manufacturing sites.

Ritual specialists and flint craft

Joakim Goldhahn has devoted a comprehensive study to ritual specialists in the Scandinavian Bronze Age (Goldhahn 2007; Østigård & Goldhahn 2007). Goldhahn's work is impressive. He covers a body of archaeological material taken from more than a thousand years of the Bronze Age in a large part of southern Scandinavia, and he puts this into a broad interpretative framework of ritual and craft. The focus in the study is on ritualized acts, a focus that is channelled through the concept of specialists, that is, people with a skill in the form of knowledge and know-how that other people lack. The concept of specialists is in turn personified in the smith as craftsman and exemplified in the rock-carving artist, the bronze caster, and the stone-knapper.

Goldhahn thinks that Bronze Age smiths, or rather the people who belonged to a social and ideological institution that can be called smithwork, were ritual specialists who, through their practice and knowledge of how different materials could be transformed from one state to another – from slab to rock art, from soapstone to mould, from bronze to bronzes, and from flint to tools – constituted and maintained the cosmology of the time. The smiths were a part of explicit knowledge systems with access to esoteric learning (Goldhahn 2007:93).

The ritual spheres of the Bronze Age, according to Goldhahn, were interwoven in symbolic semantic fields. This meant that crafts such as pottery making, carving motifs in rock art, bronze casting, quarrying quartz, and flint-knapping, were performed by these ritual specialists and that they worked in the same ritual institution. This institution was based on Bronze Age ritualized traditions through which power and ritual interacted in the same cosmology (Kristiansen & Larsson 2005). It did not necessarily mean that one and the same person did everything. A better way to view it is that, for example, rock carvings, bronze objects, pottery, and flint artefacts were produced by special actors with special insight into specific ritual spheres, but that this was united under the leadership of the same ritual specialist (Østigård & Goldhahn 2007:190). As an institution of this kind, the smith was a transformer, an all-embracing cosmologist:

> The bronze craft was performed by ritual specialists and involved a wider spectrum of society's transformation rituals, whereby copper and tin were transformed into bronze and bronze into objects with both symbolic and cosmological meanings. The smith was at the spot where pictures from ritual life were fixed in stone or where clay was transformed into pottery. The smith was also active where dead people were given a helping hand on the journey after this life (Østigård & Goldhahn 2007:188).

Goldhahn's study is important in that it clearly shows ritualized connections between different activities during the Bronze Age, activities that at first sight may seem separate but which can be linked through analyses and interpretations. In this way, an extensive and complex network of actors, human and non-human, is created. Goldhahn demonstrates through his results that the work of the transformer was charged with symbolic and cosmological metaphors:

> The "technological spheres" of the Bronze Age thus merge in the fact that they were all mastered by the smith – i.e., society's ritual specialist – and that all these media were instrumental in representing and staging the mythology and cosmology of societies. This phenomenon, the boundary-crossing of symbols, should be understand as a metaphorical image of the ritual specialists' boundary-crossing practice. The smith was the great transformer in the Bronze Age (Goldhahn 2007:321).

If one accepts Goldhahn's interpretations, as I do, this means that the transformer was not an itinerant tradesman practising the art of smithing. The transformer was the spider in the web of all the actors who were needed to stage the crafts represented by the institution – the metallurgist, the fire tender, the clay shaper, the bellows man, the flint collector, the knapper, the iconographer, and the director, to name just a few examples.

The variation in flint craft revealed by the results of my studies nevertheless indicates that an interpretation of craft should be made from a great many angles. If flint craft was varied, it is reasonable to hypothesize that there was also variation in the references to symbolism and cosmology that the craft constituted and maintained. It should be possible to ascribe different meanings to different flint crafts. It should thus be possible to arrive at a more profound picture of the transformer portrayed by Goldhahn. This should not be viewed as a critique of his study. It should be viewed as an elaboration of questions raised by the study. Below I will qualify the image of the transformer by discussing places where large blade knives of flint were made and the flint-knapping technologies manifested by different manufactures.

About places where specialists worked

The transformer – that is, the people who worked within his/her institution – was active at some of the places I analysed above. At sites like Kristineberg and Årup in Scania, Sweden, Høbjerg/Holmehusbakke on Zealand, Denmark, and Marlow on Rügen, Germany, a great many of the crafts singled out by Goldhahn were performed. Here large blade knives of flint were made together with household production. On these sites the institution of the transformer worked right beside settlement sites. This was also the site of what many scholars have defined as power, the elite, or the aristocracy, manifested in rich finds, for example, gold objects, remains of goldsmithing, and the proximity of burial places with long continuity (Jensen 1981, 1997; Thrane 1993, 1994; Kristiansen 1998; Kristiansen & Larsson 2005). On sites like these the transformer and the aristocracy worked in symbiosis. They were a unit. They manifested the site and the site manifested them (Kristiansen 2005, 2006). The manufacture of large blade knives of flint on these sites shows that large blade knives were a part of the transformer's repertoire.

But the question is how these places are related to the special manufacturing sites for large blade knives that I analysed earlier. At Wierzbica in Poland, Tommestrup in Denmark, Södra Sallerup and Hanaskog, both in Sweden, Reidervitz and Ruschvitz in Germany, and at several other places, large-scale manufacture of blade knives took place. With the exception of Södra Sallerup, the only things made at these special manufacturing sites were large blade knives of flint, preforms for large blade knives, and blade-like flakes. There are no traces of household production here. Nor are there any traces of other crafts such as bronze casting. The transformer's arena was where the elite manifested their claims to power through buildings, cemeteries, and a diversity of crafts. The special manufacturing sites lack this. They are places for mass production, with no traces of other crafts or other activities. The special manufacturing sites thus differ from the transformer's arenas. Södra Sallerup, however, is an exception to this. I will return to that site later on.

Different ways of making large blade knives

In Part I of this book I discussed the chaîne opératoire as a basis for reconstructing the actions that are expressed within a technology. In a chaîne opératoire the use of different materials can be analysed by placing each artefact into its technological context (Inizan, Roche & Tixier 1992). The chaîne opératoire is thus, as discussed above, a chain of action, gestures, and processes in production sequences which lead to something being manufactured. The concepts of knowledge and know-how are important. Knowledge is theoretical. It is communicative, something that can be passed on from one person to another through conversation or instructions. Know-how is muscular memory. It is something you acquire through practical experience. These concepts are central in a chaîne opératoire analysis since they are linked to what a flint knapper knows and does not know, what he can and cannot do – in other words, ideas about things like the method of manufacture, forms, and materials (Pelegrin 1990). A chaîne opératoire analysis manifests the knowledge and know-how in manufacturing. I do this by analysing the manufacturer, not just what was manufactured, that is, production and intervention as well as representation (Amsterdamska 2008:206).

The experimental analyses I performed show that large blade knives of flint were made within a structured and systematic chaîne opératoire. The stages of manufacture that were identified are by and large the stages of manufacture that large blade knives in the Late Bronze Age underwent. This gives me a possibility to assess how different manufacturers handled these stages, whether this was done in similar or different ways.

The results of the analyses in Part II of the book show a difference in flint knapping between the special manufacturing sites and the places where the transformer acted. The flint knapping performed at the special manufacturing sites was a systematic, specialized mass production that required a high degree of knowledge and know-how to perform. The analyses of the flint knapping at the places where the transformer worked show that this production was not systematic and specialized in the same way, see chapter *Two flint-knapping strategies and a hybrid*. It has features of both the specialized manufacture and the household production. It does not display the same knowledge and above all not the same well-developed know-how as the flint knapping on the special manufacturing sites.

I interpret this to mean that it was not the same flint knapper working on these different sites. Both kinds of manufacture proceeded from the same mental templates but configured it in different ways. The sites were structured in different ways, and it was different actors who were active at the different sites. This means that the sites were also different as actors. The transformer was not present on both sites.

The consequence of this is that there were at least two distinct contexts for the manufacture of large blade knives of flint in the Late Bronze Age, and two different social groups of manufacturers. The transformer and his/her institution was one of these. His/her social group was what has been defined in the discussions above as a conservative elite or an aristocracy. But what was the other social group, the other institution?

The innovator

The practical knowledge and understanding that is manifested in and derives from material encounters with the world serve as personal and sociotechnical epistemologies for knowing the way the world works and the place of the self within it (Dobres 2000:136f). Technology is not solely a system of ideas or products that were produced by different social groups, but also a set of conceptualized practices (Amsterdamska 2008:205). Marie-Claude Mahias, for example, in a study of pottery techniques in India, has shown how subtle variations in manufacturing techniques within the framework of the same type of production function as identity-creating codes, subtle variations which created, confirmed, and demarcated specific social groups. These identity-creating codes were both inclusive and exclusive. The general design idiom was shared by everyone. They all shared an understanding of the technology and the function. But the subtle variations were used to distinguish and separate these shared pre-understandings from each other. This means that, even if everyone shared the same overall idiom, the variations were due to and contributed to a division into social groups (Mahias 1993:177f).

Several studies have demonstrated how control over technology is also a way to control the preferential right of social interpretation (see e.g. Mahias 1993; Bijker 1995). For example, the families that made flint daggers in the Late Neolithic in south-western Scandinavia based their production on a system of displaying their skill in the parts of the technology that others could not copy, that is to say, their know-how, and conceal the parts of the technology that others could hypothetically copy, that is, their knowledge (Apel 2001).

In the same way, the dynamics between, on the one hand, the transformer and, on the other hand, the production at special manufacturing sites, can help to clarify differences in the way that different social groups in the Late Bronze Age perceived the technology. Institutions are important in this:

> Institutions are the building blocks of society … to trace them is therefore a central interpretative task. Although institutions are invisible, they are often materialised in specific and recurring ways that allow us to infer the cultural and institutional significance of the evidence. The interpretative task is to single out, from among the different objects and contexts available to us, those which formed a specific relationship defined by a set of symbolic meanings, actions and transactions that once linked them together in an institution (Kristiansen 2006:172f).

The transformer was, to use Sørensen's terminology, a conservative institution that referred to the power over production, cosmology, and ritual that his/her institution had previously possessed in different forms. One cannot overemphasize the conservative power that this institution must have developed to be able to maintain itself in different forms during the thousand years that the Bronze Age, generally speaking, lasted (Kristiansen 2006; Goldhahn 2007). The transformer built his/her institution on esoteric knowledge (Goldhahn 2007). When this knowledge was challenged by a new technology – that is, the manufacture of large blade knives at special manufacturing sites – a technology that was part of the transformer's repertoire but was not controlled within the old order, a dynamic arose between old and new. This dynamic can be said to manifest socially meaningful interactions between tradition and innovation (Lemonnier 1993:5).

The production at the special manufacturing sites was a challenge to the transformer. These places saw manufacture on a considerable scale of a knife blade for a tool that was widely used within the same geographical areas as the bronzes and other craft products of the transformer. Production at the special manufacturing sites was done within a social group that was not the same as the social group of which the transformer was a part. Since this social group made an object of flint which was new in the sense that it had no similarity to any type of earlier flint object, this social group was evidently not conservative, unlike the transformer. It was a social group that was not steered by conventions in the same way as the transformer's institution was.

If we accept the argument that the differences in the structures of the manufacturing sites, the differences in the links of the places to the transformer's power and cosmology, and the differences in the technological knowledge and know-how manifested in the different manufacturing settings show that the places and the people who were active there were different social actors, then we can say that the makers at the special manufacturing sites can be called innovators as opposed to transformers. The dynamic between different social groups illustrated by the manufacture of large blade knives of flint can then be described as a dynamic between transformers and innovators.

Having come this far in the discussion, it can be observed that the analyses of different manufacturers and manufacturing sites have contextualized the large blade knife and pinpointed a sociotechnical dynamic in the Late Bronze Age. It was a dynamic between conservative transformers and innovative craftsmen. In this way Sørensen's (1987, 1989a) interpretation of south Scandinavian society in the Late Bronze Age as conservative and not disposed to change has been combined with interpretations of a society receptive to changes, as the innovations in iron metallurgy (Hjärthner-Holdar 1993) and flint technology show. This combination of two seemingly contradictory interpretations has been enabled by highlighting different social groups in the Late Bronze Age.

If we return to the discussion above about Wittgenstein's concept of family resemblance and about the logic of interpretations based on the chain of relationships that bronze existed, disappeared, and was replaced with flint, the results thus far indicate that this chain of relationships limits interpretations of the Late Bronze Age. For the parts of society that were grounded in social relations which referred to the transformer and his/her institution, it was perhaps the case that this social group felt that bronze disappeared and was replaced with something else. But for the parts of society that were engaged in new social relations and referred to the innovator's institution, it was not that bronze disappeared. This social group might not have been interested in using bronze.

I will now proceed in the analysis and further contextualize the dynamic between these groups, between the institutions of the transformer and the innovator. I do so through a discussion of local, regional, and supraregional, changes of the landscape, ritual depositions, and bronze sickles.

Local, regional, and supraregional

During the roughly 400 radiocarbon years covered by this study, there were significant changes in central Europe (see e.g. Vandkilde 2007:121ff). Areas which had previously had contacts directed northwards were now oriented southwards towards regions around the Mediterranean. The Danish archaeologist Jørgen Jensen, in a study of this period, has discussed how these profound changes are reflected in, for example, social patterns, communication ties, religious ideas, and iconographic forms of expression (Jensen 1997:193ff, 2002:497ff). In today's northern Europe this meant that the areas in eastern Poland, northern Germany, and southern Scandinavia developed contacts with each other but had sparse contacts with areas south of the Alps:

> From the start of period VI one can certainly see how, for example, pin forms from the north-eastern Hallstatt areas from HaC are adopted with great consistency over the whole north German-Polish-south Scandinavian area, which thereby bears the stamp of a greater uniformity than before (Jensen 1997:136).

> Of proper tools from the closing phase of the Bronze Age we know very few. In votive deposits, however, a type of small axe, known as the celt, appears relatively frequently. But as in the case of the jewellery and the costume accessories, the celts are also designed in the same way over the whole north German-Polish lowlands (Jensen 2002:515).

In Bronze Age research, studies of contacts, trade, connections, and influences are common, to say the least (see e.g. Thrane 1975, 1994; Larsson 1986; Hjärthner-Holdar 1993, 1998; Kristiansen 1994, 1998; Jensen 1997). Travel and exchange over large geographical areas where the power elite, the aristocracy, in northern Europe were influenced by Mediterranean cultural expressions and their cosmology, ritual, iconography, and design have been held up as characteristic of the Bronze Age. This has been formulated most explicitly by Kristian Kristiansen and Thomas B. Larsson in the book *The Rise of Bronze Age Society: Travels, Transmissions and Transformations* (Kristiansen & Larsson 2005). Their interpretations, however, have been criticized for lacking method and for having a limited theoretical perspective (Nordquist 2005; Sjögren 2005), and several studies have highlighted distinctive local and regional characteristics to qualify the supraregional perspective (Skoglund 2005), and to nuance the view of power, elite and aristocracy (Thedéen 2005; Vandkilde 2007; Thrane 2008).

If the Scandinavian Bronze Age culture is dynamically defined as a heterogeneous phenomenon which does not necessarily find expression in uniform distribution patterns (Bolin 2005:218), the similarities I discovered in the Late Bronze Age use of a tool consisting of a knife blade of flint and a wooden shaft in several places over a large geographical area can be discussed in terms of supraregional contacts. If the amount of archaeological finds is to determine the intensity of this use, then the intensity within this area differs. But if each archaeological find is regarded as an indication of the use of this tool, it can be stated that this happened within an area that extends today from the west coast of Norway, down over the western parts of southern Sweden, taking in the whole of Denmark, the North Sea coast of Germany towards the Netherlands, along the Baltic coast of Germany, down into central Poland, up along the eastern parts of southern Sweden to the Mälaren valley and across the lakes Vättern and Vänern towards Oslo Fjord. This is a large area within which a technology was managed with the same type of tool. This technology is manifested in finds of large blade knives of flint from Sweden, Denmark, Norway, Poland, and northern Germany, and in finds of thick bifacials from south-western Denmark, north-western Germany, and the Netherlands. Yet while this reveals a supraregional similarity, that is, a society that introduced, within the same period, new tools with knife blades made of flint, the technology nevertheless shows regional variations. Large blade knives from parts of Poland do not resemble other blade knives. Other blade knives from Poland are like those from northern Germany and western Denmark. From Denmark, Sweden, Norway, and northern Germany there are other large blade knives which are similar, and from south-western Denmark, north-western Germany, and the Netherlands there are bifacial sickles. Although the tool existed over large areas, there are differences in regional – and in some places also local – ways of handling it. The concept of the implement is supraregional, the solutions are regional and sometimes local.

To link up with the discussions that are so popular in Bronze Age research, concerning travel, influence, and exchange over large areas, it may be noted that large blade knives of flint were a part of the material culture that was shared by large areas, while there were simultaneously regional variations in the way societies managed this supraregionality. This is a finding that agrees well with the results of studies based on other source material (see e.g. Thrane 1994; Berglund 2005; Skoglund 2005). The dynamic between the transformer and the innovator was thus something that existed over large geographical areas. But just as there were regional variations in the appearance of a tool with a shaft of wood and a knife blade of flint, it is reasonable to assume that, even if the dynamic between the transformer and the innovator was supraregional, it could presumably be expressed differently both regionally and locally. I will now go on to look at the use of large blade knives of flint in relation to changes in the landscape.

Changes in the landscape

The period around 800 BC is characterized by an expansion of the open landscape in northern Europe (Berglund 2005:52). In Denmark, southern Sweden, northern Germany, and parts of Poland this expansion occurred slightly earlier than in the areas further north (Harding 1989:174ff; Ostoja-Zagórski 1989:393ff; Lagerås 2002:385ff; Petersson 2006:24; Regnell & Sjögren 2006a:79). The expansion is generally characterized by the opening of the landscape, the creation of grasslands, and the introduction of new crops. Studies from recent years have demonstrated that these general changes had distinctive regional features and took place on different scales, differing in duration in different places (Skoglund 1999:287; Regnell & Sjögren 2006b:168).

In a synthesis of the results of the extensive settlement site excavations in the Öresund Fixed Link Project in the Malmö area, Nils Björhem and Björn Magnusson Staaf (2006) have shown how settlement sites underwent changes at the time of the transition between Late Bronze Age Phase A and Phase B. Whereas this change had previously been interpreted as having been caused by population growth and subsequent expansion of the number of farms (see e.g. Welinder 1998:192), it was now viewed as a redistribution of the landscape, that is, a division of the existing settlement and increased utilization of the lands around the settlement sites. Formerly uninhabited or less intensively used parts of the landscape, such as areas beside wetlands and in the inner hummocky landscape, were now subject to settlement and land use (Björhem & Magnusson Staaf 2006:176). Yet even if this redistribution entailed increased exploitation and new settlement in the landscape, the individual farms should be regarded as parts of a larger general concentrated organization. Terms such as villages and central places have been suggested for discussing this organization (Björhem & Magnusson Staaf 2006:172ff). Similar results have been presented from studies in northern Scania, Sweden, where increased breaking of new land through stone clearance and fires has been demonstrated from the same period (Lagerås 2002:385ff). These are important findings. They demonstrate a radical change in the way houses, lands, and resource areas were used.

Other changes connected to agriculture took place at this time. Peter Skoglund has shown how the composite ard was introduced some time around 800 BC. He thinks that it was introduced in a package together with the paddle spade and new cutting tools. This package made it possible to clear and sow in a new way, and Skoglund associated the introduction of these tools with increased land use (Skoglund 2008:324ff). Stefan Gustafsson, in his comprehensive studies of carbonized seed finds from excavations in south and central Sweden, has shown that Early Bronze Age agriculture was based on speltoid wheat and naked barley, while during the Late Bronze Age hulled barley took over as the most important crop (Gustafsson 1998). This finding has been confirmed in subsequent studies (Regnell & Sjögren 2006b:155f).

The changes in the landscape that occurred in present-day southern Sweden during the Late Bronze Age are clearly connected to the introduction of the tool of which the large blade knife of flint was a part. In the areas where large blade knives are found in flint assemblages from excavations and museum collections, this change in the landscape has also been demonstrated. In the parts of Sweden where large blade knives have not been found, changes in the landscape have not been found to the same extent (Gaillard et al. 1994; Berglund 2005).

In this picture of a dynamic change in the way the landscape was used, it is not difficult to set in knife blades of flint and iron, but they should not be classified as parts of a tool for collecting fodder for animals stabled in the winter, i.e. leaf fodder or hay mowing, as has previously been suggested (Björhem & Säfvestad 1993:76f; Gaillard et al. 1994:60; Welinder 1998:92; Skoglund 2008:332). Instead they should be viewed as a tool that was introduced together with a great many other changes, and used as a reaping tool on newly cleared arable land and as an implement for collecting reeds for all the purposes for which reeds can be used when establishing new farms and habitation areas. When, for example, one of the Late Neolithic houses found by the Fosie IV excavations was reconstructed in full scale, many kilos of reeds were used for thatching, to make mats for walls and floors, screens and other things (Björhem & Säfvestad 1987; see also Draiby 1991). Modern thatching shows that a roof of about 35 square metres requires roughly 600 kilos of thatching material (*Halmtak* 1998:35).

The possibility that the tool was used for collecting fodder for stabled animals is still, however, a current hypothesis. In a study of carbonized plant remains from Fosie IV, Roger Engelmark shows that a change in harvesting strategy occurred over time (Engelmark 1993). In the Early Bronze Age only the ears of grain were harvested, whereas during the Late Bronze Age the grain was cut closer to the root. Engelmark hypothesized that a more effective harvesting strategy and a desire to harvest barley straw suitable for animal fodder lay behind this change (Engelmark 1993:407). These results have been used by Gaillard et al. (1994:61) in a discussion of archaeological evidence for winter fodder production:

> The absence of any artefact resembling the scythe in southern Scandinavia prior to the Roman Iron Age has been taken as proof that hay mowing did not occur earlier. ... [H]ere [i.e. in Engelmark's analysis] is a clear evidence that mowing, however not necessarily hay mowing, was in fact practised by the Late Bronze Age, apparently with something other than a scythe (Gaillard et al. 1994:61).

This "something other than a scythe" is the tool of which large blade knives of flint were a part. But the link to mowing is still not at all clear. In a comprehensive study of animal husbandry in western Östergötland, Sweden, during the Late Bronze Age, Maria Peterson has demonstrated how pasturing was a part of animal husbandry. Pasturing was regulated and controlled in a grazing system with outdoor pasturing the whole year round and with places to which the livestock were recurrently driven (Petersson 2006:249f). The studies show a situation in which the livestock were not kept indoors during the winter and there was no great need to collect winter fodder. An analysis of pollen data from western Scania, Sweden, found that much of the landscape in the Late Bronze Age consisted of open grassland, which could confirm Peterson's results (Regnell & Sjögren 2006b:147f). On the other hand, the cornfields may have been used to give fodder as a supplement during the winter.

To sum up, the discussion about land use shows that the tool of which the large blade knife of flint was a part can be said to have been introduced to meet new needs. New lands were cultivated in a new rotation system and in a context where new agricultural tools, new crops, and pasturing were introduced. The tool was used for reaping, mainly hulled barley. It was also used for cutting reeds. At a time when new areas were brought into use and farms were established, reeds were important building material (Becker 1990).

The idea that Bronze Age cosmology played a vital role in the organization of the physical landscape and the way it was inhabited and used, and vice versa, that the organization of the landscape had an impact on cosmology, has been well researched (see e.g. Barrett 1989; Olausson 1999; Bradley 2000; Gröhn 2004; Goldhahn 2007). The cosmos was everywhere and the landscape and the world were seen as contiguous and spatially reiterated (Kaul 1998; Kristiansen 2005).

But the Late Bronze Age was a time of change. If, for example, the placing of settlements, barrows, cemeteries, sanctuaries, or rock art is to be understood as meaningful in relation to the cosmological order of society (Kristiansen 2005:135), then the changes in land use discussed above, manifested in the reorganization and redistribution of places where people lived and farmed the land as demonstrated by Björhem and Magnusson Staaf (2006), are meaningful in one way or another in relation to the cosmology.

A changing landscape need not in itself entail a revolutionary change in the cosmology. But the large blade knives of flint are an example of a challenge to repetition. They were a novelty. Things were not as they had always been. The cosmos was no longer everything, and the landscape and the world was no longer contiguous and spatially reiterated, in the way it had been before.

In the dialectic between tradition and challenge, the large blade knives of flint are a reminder of new and old. Whether or not this was perceived by the Late Bronze Age people who used the tool including the large blade knife is something I cannot determine. The fact that I as an archaeologist can discuss change in the Late Bronze Age need not mean that the people then saw it as a change. With my long-term perspective I know that what happened in the Early Bronze Age was not the same as what happened in the Late Bronze Age. I also know that there would subsequently be change in the Early Iron Age (Hedeager 1992; Kristiansen 2006; Vandkilde 2007). Here the large blade knives are a lucid example of how material culture with close connections to changes in agriculture and land use also helped to challenge tradition and were thus a part of a coming change. Material culture was created and shaped while simultaneously shaping the person doing the creating.

The tool has a given place in agriculture, as part of the day-to-day toil performed by the large share of the population who worked in the fields and built homes to live in. At the same time, everyday life in the Bronze Age was permeated by ritual. It was part of the daily life through the concept of ritualization (Barrett 1989; Bradley 2003; Kristiansen 2005). Here the tool also has a given place, as small hoards in houses and on dwelling sites and as larger deposits besides stones and at other places. I will now look more closely at this.

Ritual depositions

In a survey of archaeological history examining different terms used in Scandinavian archaeology to discuss the type of find material that I have hitherto called assemblages of large blade knives, Åsa Berggren has held up "ritual depositions" as a useful term. She thinks that if we use this term we shift our focus from the motives to the actions of the prehistoric people without predefining them as sacrifices or offerings (Berggren 2006:306). This is a conceptual discussion that is in line with a shift of focus from studies of ritual to ritualization. Ritual is a kind of practice, a performance which is defined on its own. If we reject the idea that the only function of ritual is to communicate religious beliefs, it becomes unnecessary to separate the practice of ritualization from the patterns of daily life. They form a continuum (Bradley 2003:12; compare also with Levy 1982:5 for a similar thesis).

This means that the practice expressed was a special social strategy which can be studied in terms of the action of ritualization. This makes it possible to discuss what contexts different rituals were created and performed in, and what consequences these actions may have had. Or, as Richard Bradley has put it:

> Ritualization is both a way of acting which reveals some of the dominant concerns of society, and a process by which certain parts of life are selected and provided with an added emphasis. Again that process is essentially historical, for it is unlikely to develop instantaneously. In principle, that means that it can be traced over time and studied in its wider setting. By following the development of rituals in this way it should be possible to identify a few of the ideas that they were meant to express (Bradley 2003:12).

As we saw in the analyses above of the assemblages, the majority of earlier interpretations have not included large blade knives of flint in analyses of Bronze Age ritual depositions. As a result, interpretations have been presented which stress differences between Early and Late Bronze Age ritual depositions as regards the proportions of tools and personal objects:

> This [the difference] signals a clear shift in emphasis from artefacts possibly associated with labour and production to non-utilitarian symbols with a reproductive social function (Larsson 1986:169).

My analyses show that this is not correct. The many finds of large blade knives of flint are evidence that the tool was a part of the Late Bronze Age ritual depositions. What is interesting instead is the difference expressed in the choice of raw material to make the artefacts in the ritual depositions. In the Early Bronze Age bronze was used in ritual depositions for "artefacts possibly associated with labour and production as well as for non-utilitarian symbols with a reproductive social function", to use the terms from the quotation above. In the Late Bronze Age this was not the case. Bronze was then used mainly for ritual depositions for "non-utilitarian symbols with a reproductive social function" while flint was used for "artefacts possibly associated with labour", table 3.1.

This shows how the innovator, through the large blade knife of flint, was part of the ritualization. Another example is the assemblages I discussed earlier of packaged large blade knives, preforms for large blade knives, and blade-like flakes which were taken directly from a special manufacturing site and deposited together. This agrees well with studies showing that bronze objects were also manufactured for immediate deposition (Sørensen 1989a:190). What was new, i.e. a new flint technology, was thus accepted in the old, or rather, the innovation was negotiated along with the old to give something that was valid for the time:

> The large group of hoards ... relate to, and produce, different forms of social identity ... these different spheres can be considered fields of interaction in which power-relations between actors were acted out, maintained and expanded (Vandkilde 2007:21).

Sørensen (1989a:190) thinks that during the Late Bronze Age a single group of objects, i.e. bronzes, dominates symbolic

Table 3.1. Differences between raw material in Early and Late Bronze Age ritual depositions. Note that the terms "non-utilitarian symbols" and "labour" come from Larsson (1986:26ff). Here they should be viewed solely as terms to illustrate differences in the choice of raw material, not as descriptive or interpretative terms. See also Sørensen (1987:96) for a similar discussion. It should be stressed here that tools of flint were also deposited in the Early Bronze Age. This is demonstrated by the large number of assembled finds of bifacial sickles.

	Non-utilitarian symbols	Labour
Early Bronze Age	Bronze	Flint and bronze
Late Bronze Age	Bronze	Flint

communication and ritual activities. My studies show that this should be viewed as something that applied to the social groups to which the transformer and his/her institution belonged. Other social groups used flint and large blade knives as part of their symbolic communication and ritual activities. In Late Bronze Age ritual depositions, then, we see the same dynamic between the transformer and the innovator as that exhibited by the technological analyses of different reduction strategies for the manufacture of large blade knives from different places. The actors, actants, and networks I revealed above acted in different ways and in different constellations.

And if we link this dynamic to the changes in the landscape discussed above, the ritual depositions of large blade knives of flint can be regarded as examples of how the transformer lost his/her cosmological power over the social groups that changed the landscape and introduced new agricultural tools. This is something that Sørensen has also stressed, albeit in a different interpretative framework from mine:

> The archaeological concept of a transition from Bronze Age to Iron Age in Scandinavia seems therefore to be a reflection of a society shifting from a previously rigidly formalised and ritualistic expression of its superstructure to an integration of rituals and the means of subsistence production (Sørensen 1989b:482).

From the Late Bronze Age there are ritual depositions of bronzes which have been interpreted as signs of an exclusive elite with all-embracing influence and power, that is, the institution within which the transformer worked:

> The distribution of special occupational, religious, and political items in the hoards suggests the greatest social complexity in periods IV and V. An elite, combining symbols of economic, political, and religious influence … is most visible at this time (Levy 1982:84).

These ritual depositions contain objects which are visually striking, rare, extremely valuable in metal weight, and are either imports from central Europe or unique Nordic manufactures (Levy 1982:83). The transformer displayed all his/her force and splendour in these ritual deposits. The innovator's institution posed a challenge in other arenas, through the special manufacturing sites, through the big ritual deposits of large blade knives, preforms for large blade knives, and blade-like flakes, and through land use that made new claims. I will look more closely at this dynamic by briefly touching on the issue of objects in graves, burial places, and rock carvings.

Graves, burial places, and rock carvings

In his study of transformers Goldhahn considers the absence of smiths – that is, the transformer according to his definition – in graves:

> Although we have found tens of thousands of bronze objects and rock carvings it is virtually impossible to identify unambiguously any smiths in the thousands of graves that have been excavated by archaeologists (Goldhahn 2007:319).

He singles out one grave, from Galgehøj on Funen in Denmark, with a content that can be interpreted as the burial of a bronze smith/caster. It is dated to Montelius period III. It may be mentioned here parenthetically that this grave contained the same type of highly unusual tool (Randsborg 1986) as found at Marlow on Rügen, that is, in a place that I interpreted as one where the transformer made large blade knives of flint (Auler 1999). Based on the lack of definite smith graves pointed out by Goldhahn, the cremation grave from Ladebow in northern Germany discussed earlier stands out as unique. This large urn field cemetery had a pit with cremated bone and burnt flint objects. One of the objects was a burnt large blade knife.

Yet even if we do not find the transformer in any graves, this institution was present at burial ceremonies, as Goldhahn has also emphasized (Goldhahn 2007). My analyses of unmodified flakes from cremation graves showed that the technology represented by household production was also manifested at burial ceremonies. Unmodified flakes were used when urns containing cremated bone were deposited in the ground. There are no signs here of the innovator being present.

The extensive analyses of burial places in Scania, Sweden, conducted by Caroline Arcini and Fredrik Svanberg revealed variations in the appearance and structure of burial places in the Late Bronze Age (Arcini & Svanberg 2005). Some cemeteries consist of small groups of graves which had no connection to older burial mounds, while at other cemeteries a large number of graves were dug over a long time, from the Early to the Late Bronze Age and with clear associations with Early Bronze Age mounds (see also Jennbert 1993). Arcini and Svanberg have interpreted the smaller burial places as having been newly established in the Late Bronze Age (Arcini & Svanberg 2005:333). Perhaps we see here a difference between the institutions of the innovator and the transformer? It would be interesting in the future to investigate whether it is the case that the worked flint found in graves and following the same chaîne opératoire as household production, which I have interpreted here as the presence of the transformer at burial ceremonies, can be found at cemeteries with long continuity but is lacking at the new cemeteries established in the Late Bronze Age. All the flakes analysed above, however, come from cemeteries displaying long continuity.

In times of change the past is regarded as a social resource with the ability to legitimize the order of society (Bradley 2000). Ann-Mari Hållans Stenholm, for example, in a study of graves from the Scandinavian Viking Age, has shown how these were constructed in such a way that they partly overlie graves that were five hundred to a thousand years older. Hållans Stenholm believes that this is evidence of how a society in change, where older institutions were challenged by new ones, referred to a mythic past in order to create legitimacy through a ritualized construction of memory (Hållans Stenholm 2006:341ff). She thinks that these references to the past took place at the same time as other changes in society and that special structures from the past in the cemeteries were selected to link the present with the past (Hållans Stenholm 2006:342).

Burials in the Late Bronze Age usually took place at cemeteries in cremation graves without barrows or in small barrows and cairns (see e.g. Andersson 1999; Karlenby 1999; Arcini & Svanberg 2005; Arcini, Höst & Svanberg 2007). Large barrows were mainly built in the Early Bronze Age (Jennbert 1993). Yet in the Late Bronze Age too, a few large barrows were built in eastern Denmark and south-western Sweden (Thrane 1984, 1994; Andersson 1997, 1999; Jensen 1997; Kristiansen 2006). This has been interpreted to mean that there was social stratification on such a scale that leading families could maintain a tradition of building large barrows (Thrane 1993:79ff; Jensen 1997:201).

If we add the Late Bronze Age large barrows to our earlier discussion of a dynamic between the transformer and the innovator and simultaneously allow ourselves to be inspired by Hållans Stenholm's example from the Viking Age, these large barrows can perhaps be regarded as an attempt to uphold old power positions through references to barrows from bygone times (cf. Andersson 1999:19ff). By not only using burial places with Early Bronze Age barrows to mark references backwards in time, "family memory" as it has been called (Jennbert 1993:77), but also taking over the architectural form of the barrow, in itself from a mythic past, and building large barrows over cremation places and graves, the transformer manifested his/her claims in a vigorous way.

In a publication of the results from excavations of Lusehøj at Voltofte on Funen, Denmark, Thrane has discussed various aspects of the functions of Late Bronze Age large barrows. One interpretation he puts forward is that the barrows can be regarded as a manifestation by those in power of their preferential right of interpretation in a time of social and economic change (Thrane 1984:174). Thrane advances the hypothesis that the large barrows can be viewed as an expression of the emerging power structure that would become fully visible in the Pre-Roman Iron Age (Thrane 1984:175). The hypothesis I present here is the reverse, that the large barrows are manifestations of and references to what had been, not what was to come.

Another example of references to the past is that, in finds from ritual depositions there are objects that had been in circulation for several hundred years before they were deposited in the Late Bronze Age together with bronzes which were not as old (Karlenby 2007). This phenomenon has been interpreted as proof of the shortage of bronze, as a way of economizing on a valuable raw material (Jensen 2002:516). Perhaps it should instead be viewed as an attempt by a conservative elite or aristocracy to manifest its power through references more to the past than the present. Things that endure may have been associated with memories and histories (Hodder 2004:50).

I found yet another example in Katherine Hauptman Wahlgren's work (Wahlgren 2002:240). She has shown how the thousand-year-old tradition represented by rock carvings changed during the Late Bronze Age, becoming less restrictive and attaining a greater extent:

> Through more numerous and more accessible places, the rock carvings seem to have become a more open concern, and perhaps the carving sessions also took place at a faster rhythm, or else they simply involved more people ... instead of the concentration on boundaries and transcending them, the emphasis is on presence and belonging in the place and its vicinity (Wahlgren 2002:240, 243).

It is clear that more groups were made visible in the ritual practices, at the same time as there was a greater focus on what was near and known, the home and the place. Perhaps we can see here once again the dynamic between the new and the old, between the innovator and the transformer?

These brief glances at ritual depositions, burials, and rock carvings has had the intention of raising questions, hypotheses, and examples where the transformer's institution may have extended its sphere of interpretation to counter the claims of the innovator. I will discuss this further on the basis of an artefact that may resemble the large blade knife of flint – the bronze sickle.

Bronze sickles

Sickles of bronze were manufactured during the Late Bronze Age within the framework of the transformer's institution and within the geographical areas where large blade knives of flint were used. What the bronze sickles were used for has not been ascertained (Sprockhof 1956:129ff; Baudou 1960:45; Larsson 1986:64f). Eva Weiler has shown that the kind of sickles known as *Rückenzapfensichel* (Baudou 1960:45ff) were made in moulds intended for mass production. She has calculated that the finds of moulds from southern Sweden could together have been used for the manufacture of more than 4,000 sickles (Weiler 1994:124, 1997:25). Weiler states that this number is in sharp contrast to the roughly hundred finds of sickles from southern Sweden (Baudou 1960:23ff; Larsson 1986:64f). She thus believes that far fewer sickles were made than the estimated capacity would have sufficed for. Weiler has also shown that the majority of the sickles that were made were never used for cutting or reaping (Weiler 1994:140ff):

> In almost all scrap finds, both in Sweden and on the continent, there are broken sickles. Many of them were not even used, but come straight from the mould. This can be seen in the way the burrs are still there, the metal that leaked out of the mould because it did not hold tight (Weiler 1997:25).

Although Weiler's calculation of the number of manufactured bronze sickles is difficult to accept – a bronze sickle can be recast in the same mould in which it was once cast – her discussion about scrap finds and production is interesting. Weiler's interpretation is that there were plans for the mass production of bronze sickles in southern Sweden but that this never happened. She thinks that the broken sickles are evidence of attempts to get this mass production started, but evidently they did not work out. The sickles were not used as cutting or reaping tools. Instead they were cut into pieces and ended up in "scrap finds" as raw materials (Weiler 1997:24f).

The bronze sickles have previously been described as difficult to interpret (Larsson 1986:64f). Perhaps the moulds with the capacity for mass production, together with the quantities of unused bronze sickles, should be regarded as attempts by the transformer and his/her institution to regain the initiative and make something to match the large blade knives of flint. Yet another observation speaks in favour of this. The moulds for the sickles are richly designed and decorated in a way that makes them differ from other objects (Weiler 1994:124f). Weiler has summed this up, more in a reflection than in an interpretation:

> The sickles of bronze seem at any rate to represent production of quite a specific kind ... it was not shortage of metal that lay behind this, but completely different causes (Weiler 1994:126).

I interpret these causes as a part of the dynamic of the time, between the transformer and the innovator. It was not just mass production within the transformer's institution that had to be done. It was to be done with a multifaceted symbolism carved in the moulds.

An age of flint, iron, and bronze

Having come this far in the discussion, it is time to tie it all together. I have proceeded from a dialectic between research which has viewed Late Bronze Age society as conservative and undisposed to change (Sørensen 1989a) and other research, including my own, that has been presented in this study, which has seen signs of major technological changes in this period (Hjärthner-Holdar 1993). Both Sørensen (1989a:194) and Hjärthner-Holdar (1993:188) have interpreted the early iron technology as a part of the thousand-year-old Bronze Age technological tradition of bronze casting. Iron was accepted by the bronze casters as part of their craft, and it was not until the Early Iron Age that iron became a technology in itself. I have shown how this interpretative framework has been steered by an assumption that one metal technology replaced another and that the change from a bronze technology to an iron technology happened within the same social group.

I find this interpretative framework insufficient. If we analyse the sociotechnical change more as a dynamic between different social groups – between those defined by Goldhahn as the institution of the transformer and what I have defined here as the institution of the innovator – and less as something that happened within a social group, other interpretations can be made of what happened during the time when what we call the Bronze Age became what we call the Iron Age.

In the latter part of the Late Bronze Age the uses of bronze became increasingly fixed, with the result that bronzes became strikingly similar and standardized, which also resulted in the formalization of ritual depositions (Levy 1982; Larsson 1986; Sørensen 1987, 1989a). Sørensen believes that a society had been shaped at this time where the use of the objects, the materiality, was determined and limited by a preset formalization and standardization of the production, use, and reproduction of bronze objects and "a cultural system in which a single material group (bronzes) dominates symbolic communication and ritual activities" (Sørensen 1989a:190).

At the same time, two new technologies were introduced – the manufacture of large blade knives of flint for use as knife blades in wooden shafts, and the manufacture of iron to forge into objects such as knife blades which could be shafted. These technologies were innovations with a wide distribution and use in the same areas of contact and influence to which the bronzes belonged. The large blade knives of flint acquired a clear place in ritual depositions. The codes manifested in the bronzes through their iconography and design were not recreated in objects of iron or flint. This displays a dynamic in Late Bronze Age society (see also Larsson 1986:187).

Sørensen has shown that iron in the Bronze Age was used as a kind of inferior metallic substitute and was treated as an uninteresting and even an unattractive material (Sørensen 1989a:195). But I would argue that this did not apply to all social groups. It applied to the aristocracy which, through a symbiosis of different claims to power, manifested through the institution of the transformer, had shaped Bronze Age society for a long time. For the social groups that adopted innovations such as iron production and flint technology, the novelties were anything but uninterested and unattractive. The new things were fully used as part of their ritualized everyday life. In the fields, at other work, at manufacturing sites, and in ritual depositions.

As Sørensen has shown, the control of the things that the transformer and his/her institution made claims to created an exclusive aristocracy who rejected any changes that threatened their preferential right of interpretation. The consequence was that this social group increasingly became detached from its original economic and social basis (Sørensen 1989a:190). The transformer was thus part of an institution and a social group that gradually became increasingly separated from the rest of society. In this the manufacture of large blade knives of flint on sites where the transformer was active, the manufacture of bronze sickles, and perhaps also the burial where a large blade knife accompanied the body on the funeral pyre, can be regarded as attempts to regain the preferential right of interpretation, a strategy of copying to incorporate the new in the old.

The chronological resolution with which I am forced to work gives me no scope to analyse how this dynamic of sociotechnical change and persistence was expressed over time, or to ascertain its length. Perhaps it meant that, for several generations, the transformer increased his/her power and influence while at other times it was the innovator who prevailed, or perhaps it was a short process that lasted a generation or two. It is clear, at any rate, that when the roughly four hundred radiocarbon years covered by my study were over, the transformer had disappeared (Østigård 2007).

The reason for the change that took place during these four centuries at the close of the Late Bronze Age cannot be interpreted in terms of one material being more efficient or more favoured than another. Instead the change must be viewed as a result of how networks of actors – humans and non-humans – interacted in a dynamic of introduction, stabilization, destabi-

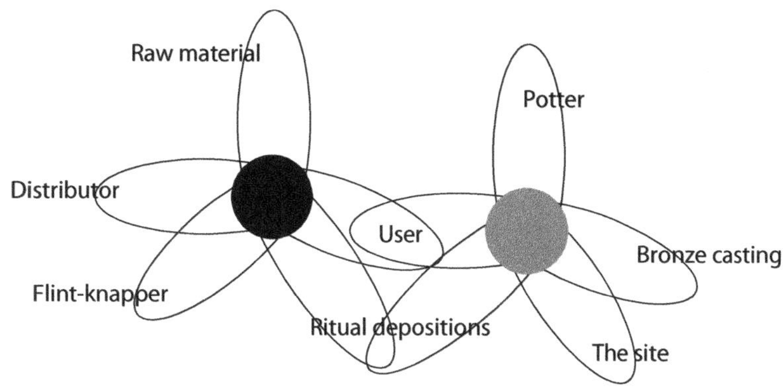

Figure 3.7. An illustration of differences in the structure of networks and their points of contact for large blade knives made by the innovator, to the left, black spot, and large blade knives made by the transformer, to the right, grey spot. Notice that I do not claim that this figure illustrates the complexity of the networks to which the transformer and the innovator belonged. The figure is solely intended to show that certain parts of the transformer's and innovator's networks overlapped.

lization, and restabilization. I will discuss this by going back to the concept of technological frame from Part I of the book.

Change and persistence

Innovation has been described as a complex process which is difficult to fully understand (Torrence & van der Leeuw 1989). I have not been able to detect any innovation or acceptance in my material. With the chronological resolution at my disposal, the closure process seems to be immediate. Even if the chronological resolution does not permit detailed discussions, it is obvious that the closure, by archaeological standards, was direct. There are signs of failure, that is to say, artefacts or technologies that were introduced but did not catch on. As discussed above, I regard bronze sickles as an artefact that was rejected.

The large blade knife of flint was negotiated in the dynamic between the transformer and the innovator. If we view the innovator as the one who introduced the technology of the large blade knives and the transformer as the one who copied it, then the transformer's action can be regarded as a correlation, an attempt to negotiate the technology within the conservative interpretative framework where the transformer worked. The innovator, however, developed the knife blade in a way that entailed a distinct closure for the technology. The similarity in the way it was made, used, and handled over large geographical areas is evidence of a clear stabilization.

In this perspective, the special manufacturing site of Södra Sallerup is interesting. Large-scale production of large blade knives of flint was pursued at this place with its rich supply of flint. The excavation here also uncovered extensive settlement site remains and the residue of other crafts, including bronze casting. Here, then, the transformer carried on systematic and specialized manufacture similar to that of the innovator. Södra Sallerup could thus be a place where the aristocracy, through systematic manufacture of large blade knives of flint within the institution of the transformer, managed to transform the novelty into something within the framework of the old.

Together with the technology represented by the large blade knives of flint, there was household production that did not change much over time. This household production seems to have been persistent, unaffected by the changes over a long time that I as an archaeologist can see.

The different types of manufacture constituted networks which partly interlocked with each other. The complexity as regards actors and actants that has been discussed above shows that these networks were multifaceted. I can only scratch the surface of this complexity. The social groups that I have pointed out – the institutions of the transformer and the innovator – should therefore not be viewed as exclusive categories. They participated in different networks, but these networks also had points of contact, for example, through the use and ritual depositions of large blade knives. It is thus not a matter of reducing all aspects of the sociotechnical to a few variables (see Hendon 2004 for discussion). The transformer and the innovator should be envisaged here as generalized actors who manifest persons, places, action, and objects as actants in chains of connections and intricate interaction within networks of humans and non-humans, figure 3.7.

In a study of changes in a long temporal perspective, Helle Vandkilde has discussed aspects of innovation processes:

> Inventions ... became integrated in day-to-day life, thereby contributing to social change at different points in time. This was not only due to practical advantages related to their use, but indeed also due to the social functions and symbolic meanings they carried. They thus became true innovations, which were further developed and hence instigated further change in the course of time (Vandkilde 2007:13).

If we look at the large blade knife of flint from this point of view, we can say that it was not only because the technology was efficient that it caught on and was accepted by many people. It became efficient because it was accepted by many people, in other words, the social function and symbolic meaning were part of the efficiency of the technology, together with its practical advantages (Bijker 1995:85). This is evidence of a technological culture where social interaction is mediated

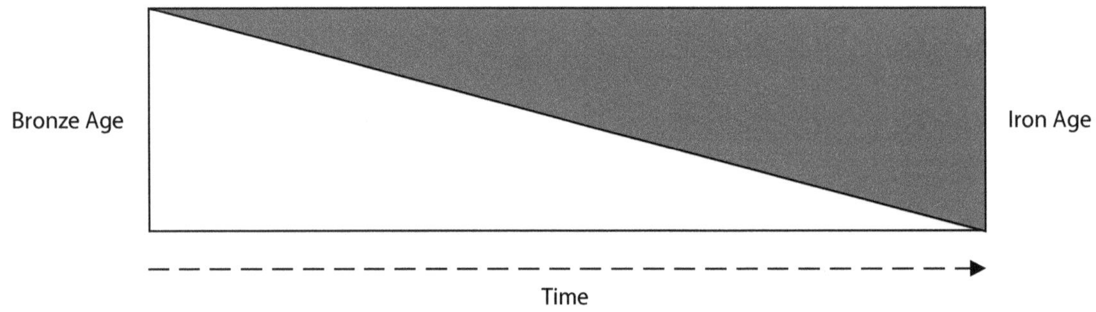

Figure 3.8. An illustration of a non-dynamic view of change, where something gradually proceeds from one thing to another, in this case the bronze technology of the Bronze Age, the white field, which gradually gives way to the iron technology of the Iron Age, the grey field. This is a way of viewing change which my work has shown to be limited for interpretations of what happened during the Late Bronze Age, and it is a view of change that I do not share.

through technologies and the technologies are only functional when embedded in social institutions.

Conservatism was manifested in the way the transformer and his/her institutions maintained their preferential right of interpretation. This was challenged by the innovator and his/her institution; manufacturing and distributing mass-produced large blade knives of flint over large areas in today's southern Sweden, Denmark, northern Germany, and Poland, and perhaps introducing iron technology in today's western and eastern Central Sweden.

I will now go on to discuss how this dynamic can be fitted into interpretations of what is traditionally called the transition between the Bronze Age and the Iron Age.

From Bronze Age to something else and from something else to Iron Age

During the Late Bronze Age there was a dynamic in society between what can be perceived as old and new. With the long-term perspective of archaeology we can see that this dynamic was something that gradually led to the change that what we

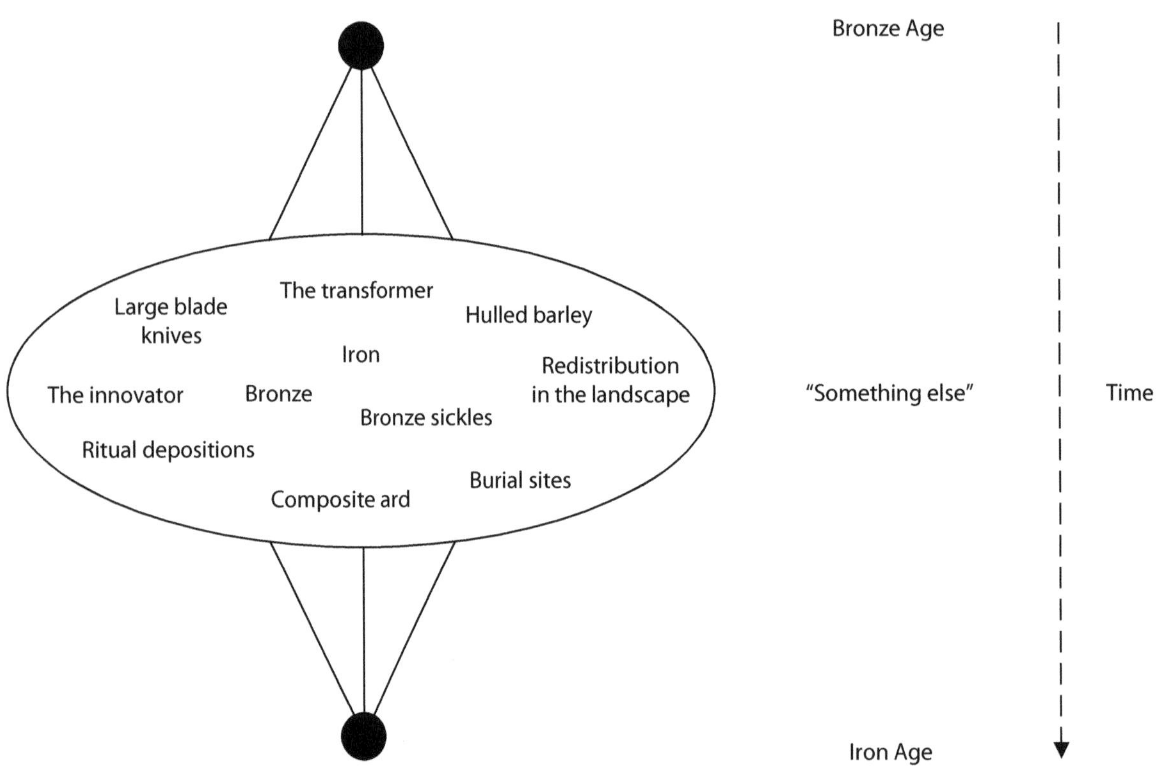

Figure 3.9. An illustration of how I perceive the dynamic in the changes occurring in the later part of the Late Bronze Age. It was not a bronze technology changing to an iron technology. What we call the Bronze Age did not become the Iron Age. It was a change that consisted of a dynamic between old and new. This dynamic in turn created something new: what we call the Iron Age. Notice that the aim of the figure is to illustrate the complexity of the time that I call "something else". The Bronze Age and the Iron Age have therefore each been reduced to a dot. Everyone understands that these dots were also complex. I have nevertheless chosen the dots so that the time between them will be visually highlighted in the figure.

call the Bronze Age became what we call the Iron Age. The results of my studies show that this change should not be viewed as a steadily intensified development from one technology to another, figure 3.8.

Instead the change should be perceived as a dynamic between old and new where negotiation, interaction, innovation, conservatism, inclusion, exclusion, ignoring, idealizing, etc. created something that had not previously existed – not out of the old but out of the dynamic of the time. Events do not change the course of history. Events make the course of history (Beck et al. 2007:835).

The Bronze Age did not become the Iron Age. The Bronze Age became something else which in turn became the Iron Age. There is thus an essential difference here between effect and process (Bijker 1995). The people who were the bearers of the change that we as archaeologists define as the Bronze Age becoming the Iron Age were not just the innovators who were disposed towards change. The conservative institution of the transformer also had a part in this change. Large blade knives of flint, the transformer, hulled barley, iron, the innovator, the shift in the landscape, bronze sickles, the composite ard, the bronze, burial places, ritual depositions, and much more were actors and actants in complex and interacting networks of sociotechnical change and persistence, figure 3.9.

In a study from 1989, Karin Tweddell Levinsen put forward the hypothesis that "iron replaced materials such as stone, flint, bone, and antler rather than bronzes as a force of production" (Levinsen 1989:452). The studies I have presented here show that Levinsen's hypothesis is correct. Bronze was bronze until roughly period IV. When a flint technology and an iron technology were introduced some time just before period V, bronze was transformed. It was then no longer the bronze that had existed before, but bronze in relation to what was new, that is, bronze-flint-iron. In the Pre-Roman Iron Age this technology of bronze-flint-iron was transformed. Flint in the form of large blade knives ceased to be used, the use of bronze was restricted to other things than before, and the iron technology then was not a technology in relation to flint and bronze, but a technology in itself. This means that the iron that existed before should not be regarded as the iron that existed in the Iron Age.

Kristiansen and Larsson argue that the changes that took place in the Late Bronze Age were a change whereby the transformer's all-embracing preferential right of interpretation when it came to power and religion could be questioned, in that power and religion were no longer the same thing. This means that if someone questioned power, they did not simultaneously question the gods and the divine order (Kristiansen & Larsson 2005:367). What I have tried to pinpoint is the dynamic of sociotechnical change and persistence that opened the door for this questioning.

Vandkilde has observed that certain periods in prehistory are more eventful than others (Vandkilde 2007:12). The time from just after 900 until just before 500 BC can be regarded as one such period. In that time it was not obvious which path society would take. We call this time the latter part of the Late Bronze Age. This has steered archaeological interpretations:

> In central Europe the Bronze Age came to a terminological conclusion around 700 BC and was succeeded by the Hallstatt period, which is also the early Iron Age. In southern Scandinavia, by contrast, the Bronze Age continued in some form (Vandkilde 2007:171).

Perceptions of what is and is not the Bronze Age have led to an outlook which more or less takes it for granted that it was within the social system of the Bronze Age that changes took place (for a discussion see Kristiansen 1998:359ff). I have shown that at this time there was a dynamic between the raw materials bronze, flint, and iron, a dynamic that changed Bronze Age society into something else before it became what we know as the Iron Age. Different raw materials were essential in different ways for different groups in society. This means that the Bronze Age as we know it ended some time in period V, and just before 500 BC the dynamic in society had the result that the institution which had enjoyed power during the Bronze Age, the transformer, no longer existed. Iron technology now caught on throughout southern Scandinavia.

To return, finally, to the discussion in Part I where I cited Latour's (1992) study of room keys in hotels, we may note that intentions and consequences cannot be followed in a straight line through the Late Bronze Age. The intentions were renegotiated during the process. The dynamic between the transformer and the innovator contributed to a shift between the original intentions and the consequences. Material culture as an actor was an effective part of this shift. In this way the large blade knives of flint reflect not just a social relationship, they were an active part of the manner in which these relationships were created and how people chose to act and were steered into acting. When the large blade knives were introduced it was scarcely with the ambition that the consequence would be a change to a time of thousand-year-old traditions, or that this change would gradually lead to the disappearance of the large blade knives. But that is what happened. The paradox is thus that the reason why large blade knives of flint disappeared was that they were introduced.

Conclusion – the sociotechnical construction of the large blade knife of flint

In this third part of the book I have discussed how interaction between actors and actants during the Late Bronze Age created sociotechnical networks of change and persistence. Flint technology was a palpable part of this, contributing to a technical shaping of society. At the same time, there was a social shaping of technology. People used technology to create things, and things in use created people.

By focusing on manufacturing sites and different ways of making large blade knives of flint, I have brought out a dynamic between different claims in society, between two social groups – the institution of the transformer and the institution of the innovator. Large blade knives of flint were a point of reference to ideas about new technology in the form of the use of flint and iron, an actor in the dynamic between old and new, between the transformer's preferential right of interpretation and the innovator's claim to this right.

This was a dynamic that gradually marginalized older power positions, and over a long time it had the effect that society took a different shape. My findings show that this was not a change between Bronze Age and Iron Age. There was something else in between. This "something else" has not been formulated before.

The result shows how intentions and consequences cannot be followed in straight lines. Some time just after 900 BC a tool was introduced with a shaft of wood and a knife blade of flint. It was manufactured and used for cutting and reaping over a large geographical area. It was included in the ritual depositions of the age. Through time the original intention of making knife blades for a composite tool was renegotiated. The tool became part of a dynamic between old and new, for example, through manufacturing sites, use, and deposits. The consequence by the time just before 500 BC was that society had changed. Iron attained widespread distribution and the large blade knives of flint disappeared.

References

Adamsen, C. 2002. Tagdækning. *Skalk*, december 2002. Nr. 6:11–12.

Althin, C.-A. 1951. The Scanian flint mines. *Meddelande från Lunds Universitets Historiska Museum*. Lund. Pp. 139–158.

Amick, D. S. & Mauldin, R. P. 1997. Effects of Raw Material on Flake Breakage Patterns. *Lithic Technology*. Volume 22, number 1, Spring 1997:18–32.

Amsterdamska, O. 2008. Practices, People, and Places. In (eds.) Hackett, E. J., Amsterdamska, O., Lynch, M. & Wajcman, J. *The Handbook of Science and Technology Studies. Third Edition*. The MIT Press, Cambridge, Massachusetts. Pp. 205–209.

Andersson, M. 1997. Tranarpshögen. En gravhög från yngre bronsålder. In (ed.) Karsten, P. *Carpe Scaniam. Axplock ur Skånes förflutna*. Riksantikvarieämbetet Arkeologiska Undersökningar. Skrifter nr 22. Lund. Pp. 59–88.

Andersson, M. 1999. Högens betydelse som socialt och religiöst monument – En studie över gravhögar från yngre bronsålder i nordvästra Skåne och södra Halland. In (ed.) Olausson, M. *Spiralens öga. Tjugo artiklar kring aktuell bronsåldersforskning*. Avdelningen för arkeologiska undersökningar. Skrifter nr. 25. Riksantikvarieämbetet. Pp. 9–25.

Andersson, S., Rex-Svensson, K. & Wigforss, J. 1978. Sorteringsschema för flinta. *Fynd rapporter 1978*. Rapporter över Göteborgs Arkeologiska Musei undersökningar. Pp. 215–252.

Andrefsky, W. Jr. 1998. *Lithics. Macroscopic approaches to analysis*. Cambridge Manuals in Archaeology. Cambridge University Press. Cambridge.

Andrefsky, W. Jr. (ed.) 2001. *Lithic Debitage. Context, Form, Meaning*. The University of Utah Press. Salt Lake City.

Andrén, A., Jennbert, K. & Raudvere, C. (eds.) 2006. *Old Norse religion in long-term perspectives. Origins, changes, and interactions*. Vägar till Midgård 8. Nordic Academic Press. Lund.

Apel, J. 2001. *Daggers, knowledge and power. The social aspects of flint-dagger technology in Scandinavia 2350–1500 cal BC*. Kust till kust-böcker 3. Uppsala.

Aperlo, P. E. 1994. *Crude but effective: Stone tools and household production in early Bronze Age Thy, Denmark*. M.A. thesis, University of California, Los Angeles.

Arcini, C. 2007. Elden utplånar inte allt. Brandgravar och bålplatser vid Gualöv. In (ed.) Artursson, M. *Vägar till Vætland. En bronsåldersbygd i nordöstra Skåne 2300–500 f.Kr.* Riksantikvarieämbetet. Regionmuseet Kristanstad. Stockholm. Pp. 169–186.

Arcini, C., Höst, E. & Svanberg, F. 2007. Gravar, bålplatser och två bronsåldersfamiljer i Gualöv. Studier av gravmiljö. In (ed.) Artursson, M. *Vägar till Vætland. En bronsåldersbygd i nordöstra Skåne 2300–500 f.Kr.* Riksantikvarieämbetet. Regionmuseet Kristianstad. Stockholm. Pp. 107–168.

Arcini, C. & Svanberg, F. 2005. Den yngre bronsålderns brandgravsmiljöer. In (eds.) Lagerås, P & Strömberg, B. *Bronsåldersbygd. 2300–500 f.Kr. Skånska spår – arkeologi längs Västkustbanan*. Riksantikvarieämbetet. Lund.

Arkeologi och naturgas i Skåne. 1985. Riksantikvarieämbetet UV Syds skriftserie nr 6.

Audouze, F. 2002. Leroi-Gourhan, a Philosopher of Technique and Evolution. *Journal of Archaeological Research*. Vol. 10, No 4:277–306.

Auler, J. 1997. Klingentechnik der Spätbronzezeit. *Archäologie in Deutschland*, 1997(4):46.

Auler, J. 1999. Die Siedlung Marlow 1. Ldkr. Rügen, aus der späten Bronze-bzw. Frühen Eisenzeit. *Ethnographisch-Archäologische Zeitschrift EAZ 40. Jahrgang Heft 1*, 1999:1–20.

Ballin, T. B. 1995. Teknologiske profiler – datering af stenaldersbopladser ved attributanalyse. *Universitetets Oldsaksamling Årbok* 1993/1994:25–46.

Ballin, T. B. 1996. *Klassifikationssystem for stenartefakter*. Varia 36. Universitetets Oldsaksamling. Oslo.

Bamforth, D. B. 1988. Investigating Microwear Polishes with Blind Tests: The Institutie Results in Context. *Journal of Archaeological Science* 1988:15:11–23.

Bamforth, D. B. & Finley, N. 2008. Introduction: Archaeological Approaches to Lithic Production Skill and Craft Learning. *Journal of Archaeological Method and Theory*. Volume 15, number 1 March 2008:1–27.

Barrett, J. C. 1989. Time and tradition: The rituals of everyday life. In (eds.) Nordström, H.-Å. & Knape, A. *Bronze Age Studies. Transactions of the British-Scandinavian Colloquium in Stockholm, May 10–11, 1985*. The Museum of National Antiquities, Stockholm Studies 6. Stockholm. Pp. 113–126.

Barret, J. 1994. *Fragments from Antiquity: An Archaeology of Social Life in Britain, 2900–1200 BC*. Blackwell, Oxford.

Barrowclough, D. A. 2004. The Secrets of the Craft Production of Scandinavian Late Neolithic Flint Daggers. *Lithic Technology*, volume 29, number 1, Spring 2004. Pp. 75–86.

Baudou, E. 1960. *Die regionale und chronologische Einteilund der jüngeren Bronzezeit im Nordischen Kreis*. Acta Universitatis Stockholmiensis. Studies in North-European Archaeology. Almqvist & Wiksell. Stockholm.

Baudou, E. 2004. *Den nordiska arkeologin - historia och tolkningar*. Kungl. Vitterhets Historie och Antikvitets Akademien. Stockholm.

Beck, R. A., Bolender, D. J., Brown, J. A. & Earle, T. K. 2007. Eventful Archaeology. The Place of Space in Structural Transformation. *Current Anthropology*, volume 48, number 6, 2007:833–860.

Becker, C. J. von 1959. Flint-mining in Neolithic Denmark. *Antiquity* 33:87–92.

Becker, C. J. von 1972. Hal og hus i yngre bronszealder. *Nationalmuseets Arbejdsmark*. Copenhagen. Pp. 5–16.

Becker, C. J. von 1980a. Hov, Gem. Sennels, Amt Thisted, Jütland. In (ed.) Weisgerber, G. *5000 Jahre Feuersteinbergbau. Die Suche nach dem Stahl der Steinzeit*. Deutsches Bergbau Museum, Bochum. Pp. 457–464.

Becker, C. J. von 1980b. Hillerslev, Amt Thisted Jütland. In (ed.) Weisgerber, G. *5000 Jahre Feuersteinbergbau. Die Suche nach dem Stahl der Steinzeit*. Deutsches Bergbau Museum, Bochum. P. 470.

Becker, C. J. von 1980c. Aalborg, Skovakken (oder Hasseris), Amt Aalborg, Jütland. In (ed.) Weisgerber, G. *5000 Jahre Feuersteinbergbau. Die Suche nach dem Stahl der Steinzeit*. Deutsches Bergbau Museum, Bochum. P. 469.

Becker, C. J. von. 1990. Ein Ungewöhnliches "Depot" der jüngeren Bronzezeit von Spjald, Westjütland. I. Archäologischer Befund. *Acta Archaeologica* vol. 60 1989. Copenhagen. Pp. 193–207.

Berggren, Å. 2006. Archaeology and sacrifice. A discussion of interpretations. In (eds.) Andrén, A., Jennbert, K. & Raudvere, C. *Old Norse religion in long-term perspectives. Origins, changes, and interactions*. Vägar till Midgård 8. Nordic Academic Press. Lund. Pp. 303–307.

Berggren, Å. 2007. *Till och från ett kärr. Den arkeologiska undersökningen av Hindbygården*. Malmöfynd nr 17. Malmö Kulturmiljö. Malmö.

Berglund, B. E. 2005. Samhälle och miljö under sex tusen år. In (eds.) Bunte, C. Berglund, B. E. & Larson, L. *Arkeologi och Naturvetenskap. Gyllenstiernska Krapperupstiftelsens Symposium Nr 6 år 2003*. Gyllenstiernska Krapperupstiftelsen. Nyhamnsläge. Pp. 49–61.

Bergström, A. & Edman, V. 2005. *Folkhemmets museum. Byggnader och rum för kulturhistoriska samlingar*. Byggförlaget AB. Stockholm.

Bertelsen, J. B., Christensen, M., Mikkelsen, M., Mikkelsen, P., Nielsen, J. & Simonsen, J. 1996. *Bronzealderens bopladser i Midt- og Nordvestjylland*. Udgivet af de Arkæologiske museer i Viborg Amt. Skive Museum, Skive.

Bijker, W. E. 1995. *Of Bicycles, Bakelites, and Bulbs. Toward a Theory of Sociotechnical Change*. The MIT Press. Cambridge.

Bijker, W. E. 2006a. The Vulnerability of Technological Culture. In. (ed.) Nowotny, H. *Cultures of Technology and the Quest for Innovation*. Berghahn Books. New York. Pp. 52–69.

Bijker, W. E. 2006b. Why and How Technology Matters. In (eds.) Goodin, R. E. & Tilly, C. *Oxford Handbook of Contextual Political Analysis*. Oxford University Press. Oxford. Pp. 681–706.

Billberg, I., Björhem, N., Magnusson Staaf, B. & Thörn, R. (eds.) 1996. *Öresundsförbindelsen och arkeologi: projektprogram och undersökningsplaner för arkeologiska slutundersökningar*. Stadsantikvariska avdelningen, Malmö Museer. Malmö.

Billberg, I. & Magnusson Staaf, B. (eds.) 1999. *Öresundsförbindelsen. II. Rapport över arkeologiska förundersökningar*. Stadsantikvariska avdelningen, Malmö Museer. Malmö.

Binford, L. R. 1962. Archaeology as Antropology. *American Antiquity* 28:217–225.

Bisson, M. S. 2000. Nineteenth Century Tools for Twenty-First Century Archaeology? Why the Middle Paleolithic Typology of François Bordes Must Be Replaced. *Journal of Archaeological Method and Theory*, vol 7, no 1:1–48.

Björhem, N. & Säfvestad, U. 1983. Fosie IV – En långdragen historia. *Ale 1/83*:3–29

Björhem, N. & Säfvestad, U. 1987. *Stenåldershus. Rekonstruktion av ett 4000 år gammalt hus.* Rapport nr 2. Malmö Museer. Stadsantikvariska avdelningen. Malmö.

Björhem, N. & Säfvestad, U. 1989. *Fosie IV. Byggnadstradition och bosättningsmönster under senneolitikum.* Malmöfynd 5. Malmö Museer. Malmö.

Björhem, N. & Säfvestad, U. 1993. *Fosie IV. Bebyggelsen under brons- och järnålder.* Malmöfynd 6. Malmö Museer. Malmö.

Björhem, N. & Magnusson Staaf, B. 2006. *Långhuslandskapet. En studie av bebyggelse och samhälle från stenålder till järnålder.* Malmöfynd 8. Malmö Kulturmiljö. Malmö.

Björk, T. 2003. Bronsgjutarna vid Årup. In (ed.) Carlie, A. *Berättelser från Vætland. En arkeologisk resa längs E22 i Skåne.* Riksantikvarieämbetet, Lund and Regionmuseet Kristianstad/Landsantikvarien i Skåne, Kristianstad. Pp. 70–73.

Bleed, P. 2001. Trees or Chains, Links or Branches: Conceptual Alternatives for Consideration of Stone Tool Production and Other Sequential Activities. *Journal of Archaeological Method and Theory*, Vol 8, No 1, 2001:101–127.

Blinkenberg, Chr. 1898. Skæftede stenalders redskaber. *Aarbøger for Nordisk Oldkyndighed og Historie.* Utgivne af Det Kongelige Nordiske Oldskrift-Selskab. 1898, II. Række. 13. Pp. 125–156.

Bolin, H. 2005. Ockuperad förhistoria – om östligt inflytande i den nordiska bronsålderskretsen. In (ed.) Goldhahn, J. *Mellan sten och järn, del 1. Rapport från det 9:e nordiska bronsålderssymposiet, Göteborg 2003-10-09/12.* Gotarc Serie C. Arkeologiska Skrifter No 59. Göteborg. Pp. 217–227.

Borna-Ahlkvist, H. 2002. *Hällristarnas hem. Gårdsbebyggelse och struktur i Pryssgården under bronsålder.* Riksantikvarieämbetet arkeologiska undersökningar skrifter 42. Riksantikvarieämbetet.

Borna-Ahlkvist, H., Lindgren-Hertz, L. & Stålbom, U. 1998. *Pryssgården. Från stenålder till medeltid. Arkeologisk slutundersökning RAÄ 166 och 167. Östra Eneby socken, Norrköpings kommun, Östergötland.* Riksantikvareämbetet Rapport UV Linköping 1998:13. Linköping.

Bowker, G. C. 2005. *Memory Practices in the Sciences.* The MIT Press. Cambridge.

Bra Böckers Lexikon 1980. Band nummer 20, entry SAAB, page 102.

Bradley, B. A. 1975. Lithic Reduction Sequences: a Glossary and Discussion. In (ed.) Swanson, E. *Lithic Technology. Making and Using Stone Tools.* Mouton Publishers. Paris. Pp. 5–13.

Bradley, R. 2000. *An Archaeology Of Natural Places.* Routledge. London.

Bradley, R. 2003. A Life Less Ordinary: the Ritualization of the Domestic Sphere in Later Prehistoric Europe. *Cambridge Archaeological Journal* 13:1, 5–23

Brandt, J. & Jantzen, D. 1995. Ein Grubenbefund der jüngeren Bronzezeit und eine Giessform aus Garvmühlen, Kreis Bad Doberan. *Bodendenkmalpflege in Mecklenburg-Vorpommern Jahrbuch 1994*:93–120.

Brantingham, P. J, Olsen, J. W., Rech, J. A. & Krivoshapkin, A. I. 2000. Raw Material Quality and Prepared Core Technologies in Northeast Asia. *Journal of Archaeological Science*, 2000, 27:255–271.

Briois, F., Negrino, F., Pelegrin, J. & Starnini, E. 2006. Flint exploitation and blade production during the Harappan period (Bronze Age): testing the evidence from the Rohri Hills flint mines (Sindh-Pakistan) throughout an experimental approach. In (eds.) Körlin, G. & Weisgerber, G. *Stone Age – Mining Age. VIII Flint Symposium. Der Anschnitt, Zeitschrift für Kunst und Kultur im Bergbau.* Beiheft 19:307–313.

Briz i Godino, I. 2006. Lithic Analysis in Spanish Archaeology. *Lithic Technology*, vol 31, no 2 Fall 2006:89–99.

Briz, I., Clemente, I., Pijoan, J., Terradas, X. & Vila, A. 2005. Stone Tools in Ethnoarchaeological Contexts: Theoretical-Methodological Inferences. In (ed.) Terrades, X. *Lithic Toolkits in Ethnoarchaeological Context.* BAR International Series 1370. Oxford. Pp. 1–8.

Broholm, H.C. 1949. *Danmarks Bronzealder. Fjerde bind. Danmarks kultur i den yngre bronzealder.* Nyt Nordisk Forlag. Arnold Busck. København.

Brumfiel, E. M. & Earle, T. K. (eds.) 1987. *Specialization, Exchange, and Complex Societies.* Cambridge University Press. Cambridge.

Brøndsted, J. 1958 (1939). *Danmarks Oldtid. II Bronzealdere.* Gyldendal. Copenhagen.

Buchli, V. 2004. Material Culture: Current Problems. (eds.) Meskell, L. & Preucel, R. W. *A Companion to Social Archaeology.* Blackwell Publishing. Oxford. Pp. 179–194.

Burenhult, G. (ed.) 1999. *Arkeologi i Norden 2.* Natur och kultur, Stockholm.

Callahan, E. 1987. *An Evaluation of the Lithic Technology in Middle Sweden during the Mesolithic and Neolithic.* Aun 8. Uppsala.

Callon, M. 1997. *Actor Network-Theory – the Market Test*, published by the Department of Sociology, Lancaster University, Lancaster LA1 4YL, UK at http://www.comp.lancs.ac.uk/sociology/papers/Callon-Market-test.pdf. Accessed on 8th April 2008.

Carlsson, T. 2007. *Mesolitiska möten. Strandvägen, en senmesolitisk boplats vid Motala ström.* Acta Archaeologica Lundensia. Series Altera in 8°, No 54. Riksantikvarieämbetet. Stockholm.

Carlstedt, C.-A. 1912. Ett märkligt depåfynd från stenåldern. *Falkenbergs–Posten*, lördagen den 18 maj 1912.

Carr, P. J. & Bradbury, A. P. 2001. Flake Debris Analysis, Levels of Production, and the Organization of Technology. In (ed.) Andrefsky, W. Jr. *Lithic Debitage. Context, Form, Meaning*. The University of Utah Press. Salt Lake City. Pp. 126–146.

Celin, U. 2006. *Citytunnelprojektet. Lockarp-delområde 7. Rapport över arkeologisk slutundersökning*. Rapport nr 44. Malmö Kulturmiljö. Malmö.

Christensen, M. 1996. Slid på flintredskapber fra bronzealderen. *Bronzealderens bopladser i Midt- og Nordvestjylland*. Udgivet af de Arkæologiske museer i Viborg Amt. Skive Museum, Skive. Pp. 76–89.

Clarkson, C., Vinicius, L. & Mirazón Lahr, M. 2006. Quantifying flake scar patterning on core using 3D recording techniques. *Journal of Archaeological Science* 33 (2006):132–142.

Coles, J. M. 1979. *Experimental archaeology*. Academic Press. London.

Copeland, L. & Moloney, N. 2003. The Mousterian Lithic Assemblages from Ras El-Kelb Cave, Lebanon. In (eds.) Moloney, N. & Shott, M. J. *Lithic Analysis at the Millennium*. Institute of Archaeology. University College. London. Pp. 17–28.

Cornell, P. & Fahlander, F. 2002. *Social praktik och stumma monument. Introduktion till mikroarkeologi*. Gotarc Serie C, No. 46. Institutionen för Arkeologi Göteborgs university. Göteborg.

Costin, C. L. 1991. Craft Specialization: Issues in Defining, Documenting, and Explaining the Organization of Production. In (ed.) Schiffer, M. B. *Archaeological Method and Theory, vol. 3*. University of Arizona Press. Tucson. Pp. 1–56.

Cotterell, B. & Kamminga, J. 1987. The Formation of Flakes. *American Antiquity*. Volume 52, number 4:675–708.

Cresswell, R. 1996. *Prométhée ou Pandore? Propos de Technologie Culturelle*. Kimé. Paris.

Damell, D. 2008. Om järnproduktion under yngre bronsålder och äldre järnålder i Linde Bergslag. In (ed.) Goldhahn, J. *Gropar & monument. En vänbok till Dag Widholm*. Kalmar Studies in Archaeology IV. Kalmar. Pp. 353–362.

Daun, Å. 1999. *Det allmänmänskliga och det kulturbundna*. Prisma. Stockholm.

Davis, Z. J. & Shea, J. J. 1998. Quantifying Lithic Curation: An Experimental Test of Dibble and Pelcin's Original Flake-Tool Mass Predictor. *Journal of Archaeological Science* 25:603–610.

Dibble, H. L. 1987. The interpretation of Middle Paleolithic scraper morphology. *American Antiquity* 52:109–117.

Dibble, H. L. 1995. Middle Paleolithic Scraper Reduction: Background, Clarification, and Review of the Evidence to Date. *Journal of Archaeological Method and Theory*, Vol. 2, No. 4:299–368.

Dibble, H. L. 1997. Platform variability and flake morphology: a comparison of experimental and archaeological data and implications for interpreting prehistoric lithic technological strategies. *Lithic Technology*. Volume 22, number 2, Fall 1997:150–170.

Dibble, H. L. & Pelicin, A. 1995. The Effect of Hammer Mass and Velocity on Flake Mass. *Journal of Archaeological Science* 22:429–439.

Dibble, H. L. & Whittaker, J. C. 1981. New Experimental Evidence on the Relation Between Percussion Flaking and Flake Variation. *Journal of Archaeological Science*. 1981, 8:283–296.

Dobres, M-A. 2000. *Technology and Social Agency. Outlining a Practice Framework for Archaeology*. Blackwell Publishers. Oxford.

Dobres, M-A. 2006. Skilled Production and Social Reproduction in prehistory and contemporary archaeology: a personal exegesis on dominant themes and their psychosocial influences. In (eds.) Apel, J. & Knutsson, K. *Skilled Production and Social Reproduction. Aspects of Traditional Stone-Tool Technologies. Proceedings of a Symposium in Uppsala*, August 20–24, 2003. SAU Stone Studies 2. Uppsala. Pp. 25–36.

Draiby, B. 1985. Fragtrup – en boplads fra yngre bronzealder i Vesthimmerland. *Aarbøger for nordisk Oldkyndighed og Historie* 1984:127–215.

Draiby, B. 1991. Studier i jernalderens husbygning. Rekonstruktion af et langhus fra ældre romersk jernalder. *Eksperimentel Arkæologi. Studier i teknologi og kultur* nr. 1. 1991:103–133.

Ebbesen, K. 1981. Bronzealderens offerfund af flintredskaber. *Fra Holbaek Amt 1981*:7–20.

Edring, A. 2000. *Snårarp – en boplats från yngre bronsålder/förromersk järnålder. Arkeologisk undersökning, 2000*. Regionmuseet, Kristianstad. Kristianstad.

Ekholm, G. 1923. Det nyaste bidraget till vår fornhistoria. Några kritiska synpunkter. *Fornvännen*. Årgång 18:60–80.

Eliasson, L. & Kishonti, I. 2003. *Öresundsförbindelsen. Lockarp 7B. Rapport över arkeologisk slutundersökning*. Rapport nr 17. Malmö Kulturmiljö. Malmö.

Engelmark, R. 1993. Makrofossilanalyser från Fosie IV. Appendix II. In Björhem, N. & Säfvestad, U. *Fosie IV. Bebyggelsen under brons- och järnålder*. Malmöfynd 6. Malmö Museer. Malmö. Pp. 406–407.

Eriksen, B. V. 2000. Chaîne opératoire – den operative proces og kunsten at tænke som en flinthugger. In (ed.) Eriksen, B. V. *Flintstudier. En håndbog i systematiske analyser af flintinventarer*. Aarhus University Press. Århus. Pp. 75–100.

Eriksen, B. in press. Bronze Age Flint Working at Bjerre, Thy. In (eds.) Bech J-H. et.al *Bronze Age Settlement Structure and Land Use in Thy, Northwest Denmark.*

Ethelberg, P. 1982. Bopladsgruber med flint og keramik fra yngre bronzealder. *Nordslesvigske Museer. Årbog for museerne i Sønderjyllands amt. 9.* Tønder. Pp. 108–132.

Ethelberg, P., Jørgensen, E. Meier, D. & Robinson D. E. (eds.) 2000. *Det sønderjyske landbrugs historie : sten- og bronzealder.* Historisk Samfund for Sønderjylland. Haderslev Museum. Haderslev.

EuroREA, 2005. (Re)construction and Experiment in Archaeology – European Platform, Volume 2 – 2005.

Exarc. 2008. The European organisation of Archaeological Open Air Museums and other facilities involved in Experimental Archaeology. http://www.exarc.eu/. Last update 17 03 2008. Accessed 27 04 2008.

Fahlander, F. & Oestigaard, T. (eds.). 2004. *Material Culture and Other Things - Post-disciplinary studies in the 21 st century.* Gotarc Series C, No. 61. Gothenburg.

Fansa, M. 2002. *Experimentelle Archäologie in Europa. Bilanz 2002.* Isensee Verlag. Oldenburg.

Fendin, T. 1999. *Boplats och härdgropsområde från bronsåldern vid Glumslöv. Skåne, Glumslövs sn, Övra Glumslöv 10:5. Västkustbanan 3:3.* Arkeologisk undersökning UV syd rapport 1999:39. Riksantikvarieämbetet. Lund.

Ferguson, J. R. 2003. An Experimental Test of the Conservation of Raw Material in Flintknapping Skill Acquisition. *Lithic Technology.* Volume 28, number 1. Spring 2003:113–131.

Finley, N. 2008. Blank Concerns: Issues of Skill and Consistency in the Replication of Scottish Later Mesolithic Blades. *Journal of Archaeological Method and Theory.* Volume 15. Number 1. March 2008:68–90.

Fischer, A. 1990. A late Palaeolithic "School" of flint-knapping at Trollesgave, Denmark. Results from refitting. *Acta Archaeologica*, 60:33–49.

Flad, R. K. & Hruby, Z. X. 2007. "Specialized" Production in Archaeological Contexts: Rethinking Specialization, the Social Value of Products, and the Practice of Production. In (eds.) Hruby, Z. X. & Flad, R. K. *Rethinking Craft Specialization in Complex Societies: Archaeological Analyses of the Social Meaning of Production. Archeological Papers of the American Anthropological Association*, Number 17:1–19.

Ford, J. A. 1952. Measurements of some prehistoric design developments in the Southeastern states. *Antropological Papers of the American Museum of Natural History* 44(3).

Ford, J. A. 1954. Comment on A. C. Spaulding, "Statistical Techniques for the Discovery of Artifact Types". *American Antiquity*, Vol. 19, No. 4, Apr., 1954:390–391.

Forrest, A. J. 1983. *Masters of Flint.* Terence Dalton Limited. Suffolk.

Forsman, C. & Victor, H. 2007. *Sommaränge Skog. Begravning, ritualer och bebyggelse från senneolitikum, bronsålder och folkvandringstid.* Arkeologi E4. SAU Skrifter 18. Societas Archaeologica Upsaliensis. Uppsala.

Friman, B. 2005. *Öresundsförbindelsen. Burlöv 20A–B. Rapport över arkeologisk slutundersökning.* Rapport nummer 35. Malmö Kulturmiljö. Malmö.

Fuglevik, L. M. 2007/08. Treperiodesystemets konstruktsjon tolket i sammen heng med sekulariseringen og vitenskapeliggjøringen av samfunnet i første halvdel av det nittende århundre. *Primitive tider.* 2007/08, 10:31–39.

Fullagar, R. (ed.) 1998. *A Closer Look. Recent Australian Studies of Stone Tools.* Sydney University Archaeological Methods Series 6. University of Sydney. Sydney.

Gaillard, M.-J., Birks, H. J. B., Emanuelsson, U., Karlsson, S., Lagerås, P. & Olausson, D. 1994. Application of modern pollen/land-use relationships to the interpretation of pollen diagrams-reconstructions of landuse history in south Sweden, 3000–0 BP. *Review of Palaeobotany and Palynology* 82 (1994):47–73.

Gidlöf, K., Hammarstrand Dehman, K. & Johansson, T. 2006. *Citytunnelprojektet. Almhov – delområde 1. Rapport över arkeologisk slutundersökning.* Rapport nr 39. Malmö Kulturmiljö. Malmö.

van Gijn, A. L. 1988. The use of Bronze age flint sickles in the Netherlands: a preliminary report. In (ed.) Beyries, S. Industries lithiques: tracéologie et technologie, BAR int. Ser. 411, vol 1, Oxford. Pp. 197–218.

van Gijn, A. L. 1989. *The wear and tear of flint. Principles of functional analysis applied to Dutch Neolithic assemblages.* Analecta Praehistorica Leidensia 22. University of Leiden, Leiden.

van Gijn, A. L. 1994. Introduction : There is more to life than butchering and harvesting. *Helenium*, XXXIV/2, 1994:176–185.

van Gijn, A. & Niekus, M. J. L. Th. 2001. Bronze Age Settlement Flint from the Netherlands: the Cinderella of Lithic Research. In (eds.) Metz, W. H., van Beek, B. L. & Steegstra, H. *Patina: Essays presented to Jay Jordan Butler on the occasion of his 80th birthday.* Groningen, Amsterdam. Pp. 305–320.

Gebel, H. G. 1980. Eine rezente Abschlagindustrie in der Region Hilvan/Südosttürkei. In (ed.) Weisgerberg, G. *5000 Jahre Feuersteinbergbau. Die Suche nach dem Stahl der Steinzeit.* Herausgegeben vom Deutschen Bergbau-Museum Bochum 1980. Pp. 396–401.

Glob, P. V. 1938. Stenredskaber fra Bronzealderen. *Winther Festskrift. Arkæologiske studier tilegnet Jens Winther.* Ejnar Munksgaard, Copenhagen. Pp. 40–74.

Glob, P. V. 1951. En flintsmedie paa Fornæs. *KUML* 1951:23–39.

Goldhahn, J. 2007. *Dödens hand – en essä om brons- och hällsmed.* Gotarc Serie C. Arkeologiska Skrifter No 65. Institutionen för Arkeologi och antikens kultur. Göteborgs universitet. Göteborg.

Graves-Brown, P. M. (ed.) 2000. *Matter, Materiality and Modern Culture*. Routledge. London and New York.

Greene, K. 2004. Archaeology and Technology. In (ed.) Bintliff, J. *A Companion to Archaeology.* Blackwell Publishing. Oxford. Pp. 155–173.

Gregersen, S. 1960. Efterlyst. *Skalk. Nyt om gammelt.* 1960. Nr. 4:28–29.

Grehn, F. 2006. *Tygelsjö 41:1. Inför planerad villabebyggelse vid Laavägen.* Arkeologisk slutundersökning 2003. Tygelsjö socken i Malmö Stad. Skåne län. Rapport 2006:074. Malmö Kulturmiljö. Malmö.

Gröhn, A. 2004. *Positioning the Bronze Age in social theory and research context.* Acta Archaeologica Lundensia Series in 8° No 47. Almqvist & Wiksell International. Stockholm.

Gurstad-Nilsson, H. 1999. En bronsålders framväxt och konsolidering – Om kontinuitet och social strategi i Möre under senneolitikum och bronsålder. In (ed.) Olausson, M. *Spiralens Öga. Tjugo artiklar kring aktuell bronsåldersforskning.* Riksantikvarieämbetet. Stockholm. Pp. 181–212.

Gustafsson, A. 1996. *Arkeologins egna historier. Reflexioner kring arkeologihistoria, dess histografi och användning.* Gotarc Serie C. Arkeologiska Skrifter No 12. Institutionen för arkeologi, Göteborgs Universitet. Göteborg.

Gustafsson, A. 2001. *Arkeologihistoria som historia och som arkeologi. Studier i arkeologins egenhistorier.* GOTARC. Serie B. Archaeological Theses 17. Institutionen för arkeologi, Göteborgs Universitet. Göteborg.

Gustafsson, A. & Karlsson, H. 2004. *Plats på scen. Kring beskrivning och förmedling av fasta fornlämningar i Bohuslän genom tiderna.* Riksantikvarieämbetet & Bohusläns museum. Stockholm.

Gustafsson, S. 1998. The Farming Economy in South and Central Sweden during the Bronze Age. A Study Based on Carbonised Botanical Evidence. *Current Swedish Archaeology.* Vol. 6, 1998:63–71.

Göteborgs museums årstryck. 1915. Göteborgs Litografiska Aktiebolag. (afd. Bonniers). Göteborg.

Göteborgs museums årstryck. 1921. Göteborgs Litografiska Aktiebolag. Göteborg.

Hadevik, C. & Gidlöf, K. 2003. *Öresundsförbindelsen. Fosie 11A–D samt Broläge Larsbovägen. Rapport över arkeologisk slutundersökning.* Rapport nr 22. Malmö Kulturmiljö. Malmö.

Hadevik, C., Jansen, J., Jansson, P., Sagrén, J. & Winkler, M. 2006. *Citytunnelprojektet. Hyllie station – delområde 3 och Hotelltomten. Rapport över arkeologisk slutundersökning.* Rapport nr 41. Malmö Kulturmiljö. Malmö.

Hagen, A. 1967 (1977). *Norges oldtid.* J.W. Cappelens Forlag AS. Oslo.

Hagner, M. & Rheinberger, H–J. 2003. Institutionalizing Science & Technology Studies in the Academy. In (eds.) Joerges, B. & Nowotny, H. *Social Studies of Science and Technology: Looking Back, Ahead.* Kluwer Academic Publishers. London.

Halmtak. Historik och framtid kring skånsk taktäckning. 1998. Länsmuseet i Kristianstad, Retzner Arén rekonstruktion & arkitektur AB, Riksantikvarieämbetet. Kristianstad.

Hansen, P. V. & Madsen, B. 1983. Flint Axe manufacture in the Neolithic. An Experimental Investigation of a Flint Axe Manufacture Site at Hastrup Vænget, East Zealand. *Journal of Danish Archaeology* vol. 2 1983. Pp. 43-59.

Hansson, K. & Celin, U. 2006. *Citytunnelprojektet. Lockarp–delområde 6. Rapport över arkeologisk slutundersökning.* Rapport nr 43. Malmö Kulturmiljö. Malmö.

Harding, A. F. 1989. Interpreting the evidence for agricultural change in the Late Bronze Age in northern Europe. In (eds.) Nordström, H.-Å. & Knape, A. *Bronze Age Studies. Transactions of the British-Scandinavian Colloquium in Stockholm, May 10–11, 1985.* The Museum of National Antiquities, Stockholm Studies 6. Stockholm. Pp. 173–181.

Harding, A. F. 2000. *European Societies in the Bronze Age.* Cambridge University Press. Cambridge.

Hastrup, K. 1999. *Viljen til Viden. En humanistisk grundbog.* Gyldendal. Copenhagen.

Hatt, G. 1937. *Landbrug i Danmarks Oldtid.* Nielsen & Lydiche. Copenhagen.

Hedeager, L. 1992. *Danmarks jernalder. Mellem stamme og stat.* Aarhus University Press. Århus.

Heimer, O. 2006. *Öresundsförbindelsen. Södra Sallerup 15B, D–E.* Rapport över arkeologisk slutundersökning. Rapport nr 27. Malmö Kulturmiljö. Malmö.

Helskog, K., Indrelid, S. & Mikkelsen, E. 1976. *Morfologisk klassifisering av slåtte steinartefakter.* Universitetets oldsaksamlings årbok 1972–1974. Oslo.

Hendon, J. A. 2004. Living and Working at Home: The Social Archaeology of Household Production and Social Relations. In (eds.) Meskell, L. & Preucel, R. W. *A Companion to Social Archaeology.* Blackwell Publishing. Oxford. Pp. 282–286.

Hinnerson Berglund, M. 2000. Style as Aesthetics – An attempt at a qualitative approach in Palaeo-Inuit research on Greenland. In (eds.) Appelt, M., Berglund, J. and Christian Gulløv, H. *Identities and Cultural Contacts in the Arctic: Proceedings from a Conference at the Danish National Museum, Copenhagen, November 30 to December 2, 1999.* The Danish National Museum & Danish Polar Center. Copenhagen.

Hirth, K. G. (ed.) 2003. *Mesoamerican Lithic Technology. Experimentation and Interpretation.* The University of Utah Press. Salt Lake City.

Hjärthner-Holdar, E. 1993. *Järnets och järnmetallurgins introduction i Sverige.* AUN 16. Societas Archaeologica Upsaliensis. Uppsala.

Hjärthner-Holdar, E. 1998. Samspel mellan olika regioner i Sverige och Ryssland under yngre bronsålder sett utifrån järnteknologins införande. In (ed.) Løken, T. *Bronsealder i Norden: regioner og interaksjon.* AmS Varia 33. Stavanger. Pp. 35–44.

Hodder, I. 1982. *Symbols in action.* Cambridge University Press. Cambridge.

Hodder, I. 1986 (1992). *Reading the Past. Current approaches to interpretation in archaeology.* Second edition. Cambridge University Press. Cambridge.

Hodder, I. (ed.) 2001. *Archaeological Theory Today.* Polity Press. Cambridge.

Hodder, I. 2004. Neo-thingness. In (eds.) Cherry, J., Scarre, C. & Shennan, S. *Explaining social change: studies in honour of Colin Renfrew.* McDonald Institute Monographs. Cambridge. Pp. 45–52.

Hodges, H. 1964. *Artifacts. An introduction to early materials and technology.* John Baker. London.

Holdaway, S. & Stern, N. 2004. *A Record in Stone. The Study of Australia's Flaked Stone Artefacts.* Victoria Museum. Aboriginal Studies Press. Melbourne & Canberra.

Holst, N.-O. 1907. Flintgrufvor och flintgräfvare i Tullstorpstrakten. *YMER. Tidskrift utgifven af svenska sällskapet för antropologi och geografi.* 26 årgången 1906:139–174.

Holm, L. 1985. Variation in Flake Material. Different Utilization of Stone Material. *Archaeology and Environment* 4, 1985:303–311.

Horst, F. 1971. Hallstattimporte und –einflüsse im Elb-Havelgebiet. *Zeitschrift für Archäologie* 5:192–214.

Hoskins, J. 1998. *Biographical Objects. How Things Tell the Stories of People's Lives.* Routledge. New York and London.

Hruby, Z. X. & Flad, R. K. (eds.) 2007. *Rethinking Craft Specialization in Complex Societies: Archaeological Analyses of the Social Meaning of Production. Archaeological Papers of the American Anthropological Association.* Number 17.

Humphrey, J. 2003. The utilization and technology of flint in the British Iron Age. In (ed.) Humphrey, J. *Re-searching the Iron Age. Selected papers from the proceedings of the Iron Age Research Student Seminars, 1999 and 2000.* Leicester Archaeology Monographs No. 11 University of Leicester. Pp. 17–23.

Humphrey, J. 2004. The use of flint in the British Iron Age: results from some recent research. In (eds.) Walker, E. A., Wenban-Smith, F. & Healy, F. *Lithics in Action. Papers from the conference Lithic Studies in the Year 2000.* Oxbow Books Lithic Studies Society Occasional Paper No. 8. Oxford. Pp. 243–251.

Humphrey, J. & Young, R. 1999. Flint Use in Later Bronze Age and Iron Age England – Still a Fiction? *Lithics. The Newsletter of the Lithic Studies Society.* No 20:57–61.

Hurcombe, L. 2007. A sense of materials and sensory perception in concepts of materiality. *World Archaeology 39:4*:532–545.

Hållans Stenholm, A-M. 2006. Past memories. Spatial returning as ritualized remembrance. In (eds.) Andrén, A., Jennbert, K. & Raudvere, C. *Old Norse religion in long-term perspectives. Origins, changes, and interactions.* Vägar till Midgård 8. Nordic Academic Press. Lund. Pp. 341–345.

Högberg, A. 1997. Flinta är flinta är flinta – eller ...? – En experimentell studie av tre skånska flinttypers frakturmönster. *Bulletin för arkeologisk forskning i Sydsverige* nr. 2/1997.

Högberg, A. 1999a. Child and Adult at a Knapping Area. A technological Flake Analysis of the Manufacture of a Neolithic Square Sectioned Axe and a Child's Flint knapping Activities on an Assemblage excavated as Part of the Öresund Fixed Link Project. *Acta Archaeologica vol 70, 1999.* Copenhagen. Pp. 79–106.

Högberg, A. 1999b. Flintsmeder i Klagshamn – flinthuggkonst och flinthuggare i början av 1900-talet. *Elbogen. Malmö fornminnesförenings årsskrift* 1998:181–192.

Högberg, A. (ed.) 2000. *Artfakter, arkeologiska ting. En bok om föremål ur ett arkeologiskt perspektiv.* University of Lund Institute of Archaeology. Report Series No. 71. Lund.

Högberg, A. 2001a. *Öresundsförbindelsen. Flinta under yngre bronsålder och äldre järnålder. Rapport över arkeologisk slutundersökning.* Rapport nr 37. Malmö Kulturmiljö. Malmö.

Högberg, A. 2001b. Flint in the Malmö area. In Knarrström, B. *Flint, a Scanian Hardware.* Skånska spår – arkeologi längs Västkustbanan. Riksantikvarieämbetet. Stockholm. Pp 172–204.

Högberg, A. 2001c. Flinta. In (eds.) Lindhé, E., Sarnäs, P. & Steineke, M. *Citytunneln och spåren i landskapet. Projektprogram och undersökningsplaner för arkeologiska slutundersökningar för Citytunnelns spårsträckningar och Hotelltomten.* Malmö Kulturmiljö. Malmö. Pp. 57–65.

Högberg, A. 2002. Production Sites on the Beach Ridge of Järavallen. Aspects of Tool Preforms, Action, Technology, Ritual and the Continuity of Place. *Current Swedish Archaeology.* Vol. 10. 2002:137–162.

Högberg, A. 2004a. The use of flint during the south Scandinavian Late Bronze Age: two technologies, two traditions. In (eds.) Walker, E. A., Wenban-Smith, F. & Healy, F. *Lithics in Action. Papers from the conference Lithic Studies in the Year 2000.* Oxbow Books Lithic Studies Society Occasional Paper No. 8. Oxford. Pp. 229–242.

Högberg, A. 2004b. Flintavslag, åskviggar och eld – några aspekter på flinta i medeltid. Elbogen. *Malmö Kulturhistoriska förenings årsskrift* 2003:29–40.

Högberg, A. 2005a. Hanaskog – en nyupptäckt produktionsplats för lövknivar. *ALE Historisk tidskrift för Skåne, Halland och Blekinge.* Nr 1 2005:1–6.

Högberg, A. 2006. Continuity of place: actions and narratives. In (eds.) Apel, J. & Knutsson, K. *Skilled Production and Social Reproduction. Aspects of Traditional Stone-Tool Technologies. Proceedings of a Symposium in Uppsala,* August 20–24, 2003. SAU Stone Studies 2. Uppsala. Pp. 187–206.

Högberg, A. 2008a. Playing with Flint: Tracing a Child's Imitation of Adult Work in a Lithic Assemblages. *Journal of Archaeological Method and Theory.* Volume 15. Number 1. March 2008:112–131.

Högberg, A. 2008b. Avslag, knivar och skrapor av flinta – en analys av råmaterial, teknologi och funktion. In (ed.) Skoglund, P. *Fest Slakt Odling Neolitikum och järnålder i Hyllie.* Malmö Kulturmiljö. Malmö. Pp. 54–73.

Högberg, A., Apel, J., Knutsson, K., Olausson, D. and Rudebeck, E. 2001. The Spread of Flint Axes and Daggers in Neolithic Scandinavia. *Památky Archeologické* 42(2):193–221.

Högberg, A., Mardell, L., Rudebeck, E., Sarnäs, P., Sheker, L. & Ödman, C. 2000. *Nomenklatur och sorteringsschema för flintregistrering. Utarbetad inom ramen för projektet Öresundsförbindelsen.* Stadsantikvariska avdelningen Kultur Malmö. Malmö.

Högberg, A. & Olausson, D. 2005. Flinta under yngre bronsåldern – några tankar om nu och då. In (ed.) Goldhahn, J. *Mellan sten och järn, del 1. Rapport från det 9:e nordiska bronsålderssymposiet, Göteborg 2003-10-09/12.* Gotarc Serie C. Arkeologiska Skrifter No 59. Göteborg. Pp. 97–122.

Högberg, A. & Olausson, D. 2007. *Scandinavian Flint – an Archaeological Perspective.* Aarhus University Press. Århus.

Ibánez, J. J. & González, J. E. 2003. Use-Wear in the 1990s in Western Europe: Potential and Limitations of a Method. In (eds.) Moloney, N. & Shott. M. J. *Lithic Analysis at the Millennium.* Institute of Archaeology. University College. London. Pp. 163–172.

Ingold, T. 1990. Society, Nature and the Concept of Technology. In (eds.) Sinclair, A. & Schlanger, N. *Technology in the humanities. Archaeological Review from Cambridge.* Volume 9:1:5–17.

Ingold, T. 2000. Making Culture and Weaving the World. In (ed.) Graves-Brown, P.M. *Matter, Materiality and Modern Culture.* Routledge. London and New York. Pp. 50–71.

Ingold, T. 2007. Materials against materiality. *Archaeological Dialogues* 14(1):1–16.

Ingvardson, G. T. 2005. Hyrdehøj-oven. Keramikproduktion i yngre bronzealder. *Aarbøger for Nordisk Oldkyndighed og Historie.* 2002. Det Kongelige Nordiske Oldskriftselskab København 2005:39–78.

Inizan, M-L., Roche, H. & Tixier, J. 1992. *Technology of Knapped Stone.* Préhistoire de la Pierre Taillée. Crep. Meudon.

Jaanusson, H. 1981. *Hallunda. A Study of Pottery from a Late Bronze Age Settlement in Central Sweden.* The Museum of National Antiquities, Stockholm Studies 1. Statens Historiska Museum. Stockholm.

Jennbert, K. 1993. Släkters hågkomst. Om bruket av bronsåldershögar. In (ed.) Larsson. L. *Bronsålderns gravhögar. Rapporter från ett symposium i Lund 15.XI–16.XI 1991.* University of Lund. Institute of Archaeology Report Series No. 48. Lund. Pp. 69–78.

Jensen, H. J. 1990. Funktionsanalyse der bronzenzeitlichen Flintmesser von Spjald und Stenild. Vorläufige Mitteilung. *Acta Archaeologica* 60:209–214.

Jensen, H. J. 1994. *Flint Tools and Plant Working. Hidden Traces of Stone Age Technology. A use wear study of some Danish Mesolithic and TRB implements.* Aarhus University Press. Århus.

Jensen, H. J. 2006. Flintknive fra yngre bronzealder. *KUML* 2006. Årbog for Jysk Arkæologisk Selskab. Pp. 101–116.

Jensen, J. 1981. Et rigdomscenter fra yngre bronzealder på Sjælland. *Aarbøger for Nordisk Oldkyndighed og Historie 1981:48–96.*

Jensen, J. 1987. Bronze Age Research in Denmark 1970–1985. *Journal of Danish Archaeology* vol. 6. 1987:155–174.

Jensen, J. 1997. *Fra Bronze- til Jernalder – en kronologisk undersøgelse.* Nordiske Fortidsminder Serie B Bind 15. Det Kongelige Nordiske Oldskriftselskab. Copenhagen.

Jensen, J. 2002. *Danmarks Oldtid. Bronzealder 2.000–500 f.Kr.* Gyldendal. Copenhagen.

Jensen, J. A. 1973. Myrhøj, 3 hustomter med klokkebægerkeramik. *KUML* 1972. Årbog for Jysk Arkæelogisk Selskab. Pp. 61–122.

Jensen, O. & Karlsson; K. (eds.) 2000. *Archaeological Conditions. Epistemological and Ontological Examples.* Gotarc Serie C, Arkeologiska Skrifter. No 40. Institutionen för arkeologi, Göteborgs Universitet. Göteborg.

Johansen, Ø. K. 1993. *Norske depotfunn fra bronsealderen.* Universitetets Oldsaksamlings Skrifter Ny rekke Nr. 15. Oslo.

Johnson, L. L. 1978. A History of Flint-Knapping Experimentation, 1836–1976. *Current Anthropology* 19(2):337–372.

Johnson, M. H. 2006. On the nature of theoretical archaeology and archaeological theory. *Archaeological dialogues.* Vol 13. No. 2. 2006:117–131.

Jonsson, E. 2005. *Öresundsförbindelsen. Skjutbanorna 1A. Rapport över arkeologisk slutundersökning.* Rapport nr 8. Malmö Kulturmiljö. Malmö.

Kaliff, A. 1996. *Ringeby. En kult- och gravplats från yngre bronsålder.* Arkeologisk undersökning RAÄ 6, Kvillinge socken Norrköpings kommun Östergötland. Riksantikvarieämbetet. Avdelningen för arkeologiska undersökningar UV Linköping 1995:51. Linköping.

Kaliff, A. 2007. *Fire, Water, Heaven and Earth. Ritual practice and cosmology in ancient Scandinavia: An Indo-European perspective.* Riksantikvarieämbetet. Stockholm.

Karlenby, L. 1999. Deposition i skärvstenshögar – En studie kring avfallshantering och religion under äldre och yngre bronsåldern i sydvästra Uppland. In (ed.) Olausson, M. *Spiralens öga. Tjugo artiklar kring aktuell bronsåldersforskning.* Avdelningen för arkeologiska undersökningar. Skrifter nr. 25. Riksantikvarieämbetet. Pp. 115–125.

Karlenby, L. 2003. Till frågan om Närkes bronsålder. In (ed.) Karlenby, L. *Mittens Rike. Arkeologiska berättelser från Närke.* Riksantikvarieämbetet. Arkeologiska undersökningar. Skrifter No 50. Stockholm. Pp. 133–156.

Karlenby, L. (ed.) 2007. *Om makt och offer. Röster om centralmaktens utveckling i tiden före historien.* Riksantikvarieämbetet. Stockholm.

Karlin, C. & Julien, M. 1994. Prehistoric technology: a cognitive science? In (eds.) Renfrew, C. & Zubrow, E. B. *The Ancient Mind. Elements of cognitive archaeology.* New Directions in Archaeology. Cambridge University Press. Cambridge. Pp. 152–164.

Karlsson, C. 2003. Mellan brons och järn. In (ed.) Karlenby, L. *Mittens Rike. Arkeologiska berättelser från Närke.* Riksantikvarieämbetet. Arkeologiska undersökningar. Skrifter No 50. Stockholm. Pp. 363–376.

Karlsson, H. 2000. Swedish Archaeology in the Twenty-First Century. The Necessity of a (self-) Critical Dialogue. *Current Swedish Archaeology* Vol 8:143–155.

Karsten, P. 1994. *Att kasta yxan i sjön. En studie över rituell tradition och förändring utifrån skånska neolitiska offerfynd.* Acta Archaeologica Lundensia. Series in 8° 23. Almqvist & Wiksell. Stockholm.

Karsten, P. 2005. Västskånska offerfynd från 2300–500 f.Kr. In (eds.) Lagerås, P. & Strömberg, B. *Bronsåldersbygd. 2300–500 f.Kr.* Skånska spår – arkeologi längs Västkustbanan. Riksantikvarieämbetet. Stockholm. Pp. 420–443.

Karsten, P. & Knarrström, B. 2003. *The Tågerup Excavations.* Skånska spår – arkeologi längs Västkustbanan. Riksantikvarieämbetet. Stockholm.

Kaul, F. 1998. *Ships on Bronzes. A Study in Bronze Age Religion and Iconography.* PNM Studies in Archaeology & Historiy 3:1. Copenhagen.

Keeley, L. H. 1980. *Experimental Determination of Stone Tool Uses. A Microwear Analysis.* University of Chicago Press. Chicago.

Keeley, L. H. 1982. Hafting and Retooling: Effects on the Archaeological Record. *American Antiquity*, Vol. 47, No. 4, (Oct., 1982), Pp. 798–809.

Kelterborn, P. 2002. Measurable Flintknapping. Experimentelle Archäologie in Europa. *Bilanz* 2002:35–49.

Kelterborn, P. 2005a. What Widening Gap? *Lithic Technology.* Vol 30, no 2, Fall 2005:86–88.

Kelterborn, P. 2005b. Principles of experimental research in archaeology. *EuroRea* 2/2005:120–124.

Kishonti, I. 2006. *Öresundsförbindelsen. Södra Sallerup 15 H. Rapport över arkeologisk slutundersökning.* Rapport nr 29. Malmö Kulturmiljö. Malmö.

Kjellmark, K. 1925. Boplats och gravar från bronsåldern vid Kämpinge i Skåne. *Fornvännen*, årgång 20:138–161.

Kjellmark, K. 1944. Stenåldersboplatserna inom Mörrumsåns vattenområde. *Fornvännen* 1944. Häfte 1. Pp. 257–279.

Knarrström, B. 2000. *Flinta i sydvästra Skåne. En diakron studie av råmaterial, produktion och funktion med fokus på boplatsteknologi och metalltida flintutnyttjande.* Acta Archaeologica Lundensia Series in 8°, No. 33. Almqvist & Wiksell International. Stockholm.

Knarrström, B. 2001. *Flint a Scanian Hardware.* Skånska spår – arkeologi längs Västkustbanan. Riksantikvarieämbetet. Stockholm.

Knarrström, B. 2007. Boplatsaktiviteter belysta genom flintstudier. Råmaterial, teknologi och funktion. In (ed.) Artursson, M. *Vägar till Vætland. En bronsåldersbygd i nordöstra Skåne 2300–500 f.Kr.* Riksantikvarieämbetet. Regionmuseet Kristianstad. Stockholm. Pp. 97–106.

Knarrström, B. & Wrentner, R. 1996. Diagnostik av plattformar. *Bulletin för arkeologisk forskning i Sydsverige.* Nr2/97:80–87.

Knutsson, K. 1988. *Making and using stone tools. The analysis of the lithic assemblages from Middle Neolithic sites with flint in Västerbotten, northern Sweden.* Aun 11. Societas Archaeologica Uppsaliensis. Uppsala.

Knutsson, K. 2006. A genealogy of reflexivity: The skilled lithic craftsman as "scientist". In (eds.) Apel, J. & Knutsson, K. *Skilled Production and Social Reproduction. Aspects of Traditional Stone-Tool Technologies. Proceedings of a Symposium in Uppsala*, August 20–24, 2003. SAU Stone Studies 2. Uppsala. Pp. 153–186.

Koch, E. 1998. *Neolithic Bog Pots from Zealand, Møn, Lolland and Falster.* Det Kongelige Nordiske Oldskriftselskab. Copenhagen.

Kohring, S. 2006. Let's NOT talk technology? Bringing production into a discussion of technological knowledge. *Archaeological Review from Cambridge 21:1*:98–116.

Kooyman, B. P. 2001. *Understanding Stone Tools and Archaeological Sites.* University of Calgary Press. University of New Mexico Press, Albuquerque.

Kragh, A. 1964. *Mand and Flint.* Rhodos. Copenhagen.

Kristiansen, K. 1994. The Emergence of the European World System in the Bronze Age: Divergence, Convergence and Social Evolution during the First and Second Millennia BC in Europe. In (eds.) Kristiansen, K. & Jensen, J. *Europe in the First Millennium B.C.* Sheffield Archaeologicl Monographs 6. J.R. Collis Publications. Sheffield. Pp. 7–30.

Kristiansen, K. 1998. *Europe Before History*. Cambridge University Press. Cambridge.

Kristiansen, K. 2004. Genes versus agents. A discussion of the widening theoretical gap in archaeology. *Archaeological dialogues*. Vol. 11. No. 2. 2004:77–98.

Kristiansen, K. 2005. Cosmology and consumption in the Bronze Age. In (ed.) Goldhahn, J. *Mellan sten och järn, del 1. Rapport från det 9:e nordiska bronsålderssymposiet, Göteborg 2003-10-09/12*. Gotarc Serie C. Arkeologiska Skrifter No 59. Göteborg. Pp. 135–149.

Kristiansen, K. 2006. Cosmology, economy and long-term change in the Bronze Age Northern Europe. In (ed.) Sjögren, K-G. *Ecology and Economy in Stone Age and Bronze Age Scania*. Skånska Spår – arkeologi längs Västkustbanan. Riksantikvarieämbetet. Lund. Pp. 171–193.

Kristiansen, K. Cornell, P. & Larsson, L. (eds.) 2004. *Arkeologins referensvärld. Analyser av referenskonventioner inom arkeologin och kulturmiljövård*. Gotarc Series C, Arkeologiska Skrifter No 53. Institutionen för Arkeologi. Göteborgs Universitet. Göteborg.

Kristiansen, K. & Larsson, T. B. 2005. *The Rise of Bronze Age Society. Travels, Transmissions and Transformations*. Cambridge University Press. Cambridge.

Kruk, K. 1994. Przemysl krzemienny ludnosci grupy tarnobreskiej kultury luzyckiej. *Wolinski Informator Muzealny 1*. Pp. 2–226.

Kärrholm, M. 2004. *Arkitekturens territorialitet. Till en diskussion om territoriell makt och gestaltning i stadens offentliga rum*. Institutionen för arkitektur, LTH. Lunds Universitet. Lund.

Lagerås, P. 2002. Skog, slåtter och stenröjning. Paleoekologiska undersökningar i trakten av Stoby i norra Skåne. In (ed.) Carlie, A. *Arbetets landskap. Tre bidrag till Stobytraktens bebyggelse- och odlingshistoria*. Riksantikvarieämbetet. Lund. Pp. 363–411.

Lagerås, P. & Regnell, M. 1999. Agrar förändring under sydsvensk bronsålder. En diskussion om skenbara samband och olösta gåtor. In (ed.) Olausson, M. *Spiralens Öga. Tjugo artiklar kring aktuell bronsåldersforskning*. Riksantikvarieämbetet. Stockholm. Pp. 263–276.

LaMotta, V. M. & Schiffer, M. B. 1999. Formation Processes of House Floor Assemblages. In (ed.) Allison, P. M. *The Archaeology of Household Activities*. Routledge. London. Pp. 19–29.

Lampe, W. 1982. Zur endbronzezeitlichen Besiedlung der Insel Usedom. *Bodendenkmalpflege in Mecklenburg, Jahrbuch* 1981:85–114.

Larson, M. L. & Kornfeld, M. 1997. Chipped Stone Nodules: Theory, Method, and Examples. *Lithic Technology*. Volume 22, number 1, Spring 1997:1–18.

Larsson, L. 1975. The Fogdarp find. A hoard from the Late Bronze Age. *Meddelande från Lunds Universitets Historiska Museum 1973–1974*. Pp. 169–238.

Larsson, L., Callmer, J. & Stjerquist, B. (eds.) 1992. *The archaeology of the cultural landscape. Field work and research in a south Swedish rural region*. Acta archaeologica Lundensia. Series in 4. Almqvist & Wiksell International. Stockholm.

Larsson, T. B. 1986. *The Bronze Age Metalwork in Southern Sweden. Aspects of Social and Spatial Organization 1800–500 B.C.* Archaeology and Environment 6. University of Umeå. Department of Archaeology. Umeå.

Latour, B. 1992. Technology is society made durable. In (ed.) Law, J. *A Sociology of Monsters: Essays on Power, Technology and Domination*. Routledge. London. Pp. 103–130.

Latour, B. 1998. *Artefaktens återkomst. Ett möte mellan organisationsteori och tingens sociologi*. Studier i företagsekonomi 5. Nerenius & Santérus förlag. Göteborg.

Latour, B. 2000. The Berliner key or how to do words with things. In (ed.) Graves-Brown, P. M. *Matter, Materiality and Modern Culture*. Routledge. London and New York. Pp. 10–21.

Latour, B. 2007. *Reassembling the Social. An Introduction to Actor-Network-Theory*. Oxford University Press. Oxford.

Law, J. 1992. *Notes on the Theory of the Actor Network: Ordering, Strategy and Heterogeneity*, published by the Centre for Science Studies, Lancaster University, Lancaster LA1 4YN, at http://www.comp.lancs.ac.uk/sociology/papers/Law-Notes-on-ANT.pdf. Last revised on 30th November 2003. Accessed on 8th April 2008.

Law, J. & Bijker, W. E. 1992. Postscript: Technology, Stability and Social Theory. In (eds.) Bijker, W. E. & Law, J. *Shaping Technology/Building Society. Studies in Sociotechnical Change*. The MIT Press. Cambridge. Pp. 290–308.

Lech, H. & Lech, J. 1984. The prehistoric flint mine at Wierzbica 'Zele': a case study from Poland. *World Archaeology volume 16 No. 2. Mines and quarries*. Pp. 186–203.

Lech, H. & Lech, J. 1997. Flintmining in the Bronze and Early Iron Ages. Research or the 'Zele' site at Wierzbica, Radom voiv. In (eds.) Lech, J. & Piotrowska, D. *Studies of Flint-Mining and Flint-Working in the Bronze and Early Iron Ages. Proceedings of the flint symposium held in Warsaw 20–22 October 1994*. Polish Scientific Publishers PWN Warszawa. Pp. 95–114.

Lech, J. & Leligdowicz, A. 2003. Studien zum mitteleuropäischen Feuersteinbergbau in der Bronzezeit. In (eds.) Stöllner, Th., Körlin, G., Steffens, G. & Cierny, J. *Man and Mining – Mensch und Bergbau. Studies in honour of Gerb Weisgerber on occasion of his 65th birthday* Der Anschnitt, Beiheft 16, Bochum. Pp. 285–300.

Lejre Forsøgcenter, 2008. http://www.lejrecenter.dk/ Accessed in 15 april 2008.

Lemonnier, P. (ed.) 1993. *Technological Choices. Transformation in material culture since the Neolithic.* Routledge. London.

Leroi-Gourhan, A. 1993. *Gesture and Speech.* Mass. MIT Press. Cambridge.

Levinsen, K. T. 1989. The introduction of Iron in Denmark. In (eds.) Sørensen, M. L. S. & Thomas, R. *The Bronze Age-Iron Age Transition in Europe. Aspects of continuity and change in European societies c. 1200 to 500 B.C.* BAR International Series 483(ii). Pp. 440–456.

Levi Sala, I. 1986. Experimental Replication of Post-Depositional Surface Modification on Flint. In (eds.) Owen, L. R. & Unrath, G. *Technical Aspects of Microwear Studies on Stone Tools.* Part I. Early Man News 9/10/11. Newsletter for Human Palecology. Tübingen. Pp. 103–109.

Levi Sala, I. 1996. *A Study of Microscopic Polish on Flint Implements.* BAR International Series 629. Oxford.

Levy, J. E. 1982. *Social and Religious Organization in Bronze Age Denmark. An Analysis of Ritual Hoard Finds.* BAR International Series 124. Oxford.

Lindhé, E., Sarnäs, P. & Steineke, M. (eds.) 2001. *Citytunneln och spåren i landskapet. Projektprogram och undersökningsplaner för arkeologiska slutundersökningar för Citytunnelns spårsträckning och Hotelltomten.* Malmö Kulturmiljö. Malmö.

Lindholm, P. 2003. Stenmaterialet i Apalle. In Ullén, I. *Bronsåldersboplatsen vid Apalle i Uppland. Arkeologi på väg – undersökningar för E18.* Riksantikvarieämbetet. UV Uppsala rapport 1997:64. Uppsala. Pp. 147–189.

Lindman, G. & Ortman, O. 1997. *Hus och odlingsspår från forntiden i Herrestad. Arkeologiska undersökningar för motorvägen Lerbo-Torp. Del 1.* Arkeologiska resultat. Uv väst rapport 1997:10. Riksantikvarieämbetet och Bohusläns museum. Uddevalla.

Lindälv, E. 1964. Depåfynd i Varbergsbygden. Offerfynd och skattgömmor i Himle och Viske härader. *Varbergs Museum årsbok 1964*:35–74.

Lindälv, E. 1967. *Fornfynd och fornminnen i norra Halland.* Varberg.

Lomborg, E. 1973. *Die Flintdolche Dänemarks. Studien über Chronologie und Kulturbeziehungen des südskandinavischen Spätneolitikums.* Nordiske Fortidsminder, Serie B – in quarto Bind 1 utgivet av Det kgl. nordiske Oldskriftselskab. Copenhagen.

Longo, L. & Skakun, N. 2005. *The Roots of Use-Wear Analysis: Selected papers of S.A. Semenov.* Published in the occasion of the Congress "Prehistoric Technology" 40 years later: Functional studies and the Russian Legacy Verona, 20–23 April 2005. Memorie del Museo Civico di Storia Naturale di Verona (2. serie). Sezione Scienze Dellúomo – N° 7 – 2005. Verona.

Luedtke, B. E. 1992. *An Archaeologist's Guide to Chert and Flint.* Archaeological Research Tools 7. Institute of Archaeology, University of California. Los Angeles.

Lundqvist, L. 1992. Fornlämning 82. Tvååkers socker. Halland. In (eds.) Artelius, T., Johansson, L. G., Jonsäter, M., Lundqvist, L., Nordqvist, B., Streiffert, J., Strömberg, B., Thorsberg, K. & Wickerts-Jensen, M. *Förhistoria i Mellersta Halland. Arkeologiska undersökningar längs väg E6.* Riksantikvarieämbetet och Statens Historiska Museer. Rapport UV 1988:15. Göteborg. Pp. 125–140.

Lucas, G. 2005. *The Archaeology of Time. Themes in Archaeology.* Routledge. New York.

Lütjens, I. 2001. Archäologische Untersuchungen zur jüngeren Bronze- und älteren Eisenzeit in Jürgenshagen, Lkr. Güstrow. *Bodendenkmalpflege in Mecklenburg-Vorpommern, Jahrbuch* 2000 (48):31–103.

Lövgren, K., Heimer, O., Ifverson, P., Ingwald, J., Koch, H. & Svensson, M. 2007. *Citytunnelprojektet. Bunkeflo – delområde 2 och Bunkeflo bytomt. Rapport över arkeologisk slutundersökning.* Rapport nr 40. Malmö Kulturmiljö. Malmö.

Mackenzie, I. (ed.) 1994. *Archaeological Theory: Progress or Posture?* Worldwide Archaeology Series; Vol. 11. Avebury.

Madsen, B. 1992. Hamburgkulturens flintteknologi i Jels. In (eds.) Holm, J. & Rieck, F. *Istidsjægere ved Jelssøerne. Hamburgkulturen i Danmark.* Skrifter fra Museumsrådet for Sønderjyllandes amt, 5. Haderslev. Pp. 93–131.

Magnusson, N. 1963. *Sveriges Geologi.* Scandinavian University Books. Svenska Bokförlaget/Nordstedts.

Magnusson Staaf, B. 2000. Ting och tanke. In (ed.) Högberg, A. *Artefakter, arkeologiska ting. En bok om föremål ur ett arkeologiskt perspektiv.* University of Lund Institute of Archaeology. Report Series No. 71. Lund. Pp. 41–49.

Mahias, M-C. 1993. Pottery Techniques in India. Technical variants and social choice. In (ed.) Lemonnier, P. *Technological Choices. Transformation in material cultures since the Neolithic.* Routledge. London and New York. Pp. 157–180.

Martin, A. 2005. Agents in Inter-Action: Bruno Latour and Agency. *Journal of Archaeological Method and Theory.* Vol 12. No 4, December 2005:283–311.

Mason, H.J. 1978. *Flint - The Versatile Stone.* Providence Press. Cambridgeshire.

Mathieu, J. R. (ed.). 2002. *Experimental Archaeology. Replicating past objects, behaviors, and processes.* BAR International Series 1035, 2002. Oxford.

McLaren, A. P. 2008. Flintworking in the British later Bronze and Iron Age: A Critical Review and Statement of Research Potential. *Lithic Technology.* Volume 33:2, Fall 2008:141–159.

Meldgaard, M. & Rasmussen, M. 1996 (eds.). *Arkæologiske eksperimenter i Lejre.* Köpenhamn.

Meskell, L. (ed.) 2005. *Archaeologies of Materiality.* Blackwell Publishing. Oxford.

Meskell, L. & Preucel, R. W. (eds.) 2004. *A Companion to Social Archaeology.* Blackwell Publishing. Oxford.

Mikkelsen, P. 1996. Gruber, fund samt aspekter af økonomien på bopladserne. *Bronzealderens bopladser i Midt- og Nordvestjylland.* Udgivet af de Arkæologiske museer i Viborg Amt. Skive Museum, Skive. Pp. 64–75.

Migal, W. 2004. Social conditions of flint-working during the Bronze Age and Early Iron Age in Poland and eastern Europe. In (eds.) Walker, E. A., Wenban-Smith, F. & Healy, F. *Lithics in Action. Papers from the conference Lithic Studies in the Year 2000.* Oxbow Books. Lithic Studies Society Occasional Paper No. 8. Oxford. Pp. 215–228.

Miller, D. 1985. *Artefacts as Categories: A Study of Ceramic Variability in Central India.* Cambridge University Press. Cambridge.

Miller, D. (ed.). 1998. *Material Cultures: Why some Things Matter.* University of Chicago Press. Chicago.

Miller, D. 2005. Afterword. In (ed.) Meskell, L. *Archaeologies of Materiality.* Blackwell Publishing. Oxford. Pp. 212–219.

Molander, B. 1996. *Kunskap i handling.* Daidalos. Göteborg.

Montelius, O. 1872. Halländska fornsaker från Hednatiden. *Hallands fornminnesförenings årsskrift.* Halmstad. Pp. 121–205.

Montelius, O. 1885. *Om tidsbestämning inom bronsåldern med särskildt afseende på Skandinavien.* Stockholm.

Montelius, O. 1917. *Minnen från vår forntid. Ordnade och beskrifna af Oscar Montelius. I Stenåldern och bronsåldern.* P. A. Norstedt & Söners Förlag, Stockholm.

Moore, L. E. 1994. The Ironies of Self-Reflection in Archaeology. In (ed.) Mackenzie, I. *Archaeological Theory: Progress or Posture?* Worldwide Archaeology Series; Vol. 11. Avebury. Pp. 43–56.

Moskwa, K. 1976. *Kultura Luzycka w poludniowo-wschodniej Polsce.* Muzeum Okregowe w rzeszowie. Rzeaszow.

Müller, S. 1884. Mindre bidrag til den forhistoriske Archæologis methode. *Aarbøger for Nordisk Oldkyndighed of Historie, udgivne af Det Kongelige Nordiske Oldskrift-Selskab.* 1884:161–216.

Müller, S. 1919. Bopladsfund fra Bronzealderen. *Aarbøger for Nordisk Oldkyndighed of Historie, udgivne af Det Kongelige Nordiske Oldskrift-Selskab.* 1919. III Række. 9. Bind:3–92.

Nicklasson, Påvel. 2001. *Strävsamma bönder och sturska stormän. Stafsinge och Halland från bronsålder till medeltid.* Acta Archaeologica Lundensia. Series in 8°, No. 35 Almqvist & Wiksell International. Stockholm.

Nielsen, B. & Rudebeck, E. 1991. Introduktion till arkeologi i Södra Sallerup. En översikt över utgrävningarna kring Ängdala gård. *Elbogen 1990. Malmö Fornminnesförenings Årsskrift,* årgång 58:64–97.

Nielsen, C. 1994. *Tillbaka till stenåldern. Handbok i stensättning.* Svenska Kommunförbundet. Stockholm.

Nielsen, F. O. 2005. Bornholm i yngre bronzealder. In (eds.) Kaul, F., Stoltze, M., Nielsen, F. O., and Milstreu, G. *Helleristninger. Billeder fra Bornholms bronzealder. Bornholms Museum.* Wormaianum. Pp. 113–133.

Nilsson, A. & Onsten Molander, A. 2004. *Öresundsförbindelsen. Södra Sallerup 15F & I. Rapport över arkeologisk slutundersökning.* Rapport nr 30. Malmö Kulturmiljö. Malmö.

Nilsson, B. 2003. *Tingens och tankarnas landskap. Försök i naturumgängets arkeologi med exempel ur Blekinges och Smålands förflutna.* Acta Archaeologica Lundensia Series in 8° No. 44. Almqvist & Wiksell International. Stockholm.

Nilsson, B. 2007. An archaeology of material stories. Dioramas as illustration and the desire of a thingless archaeology. *Archaeological Dialogues* 14(1):27–30.

Nishiaki, Y. 2000. *Lithic Technology of Neolithic Syria.* BAR International Series 840. BAR Publishing. Oxford.

Nord, J. & Sarnäs, A. 2005. *Öresundsförbindelsen. Lockarp 7D–E. Rapport över arkeologisk slutundersökning.* Rapport nr 18. Malmö Kulturmiljö. Malmö.

Nordén, A. 1926. Bronsålderns arbetsredskap och bronsen. *Fornvännen.* Årgång 21:116.

Nordin, S. 2008. *Humaniora i Sverige - Framväxt. Guldålder. Kris.* Atlantis. Stockholm.

Nordquist, G. 2005. Grekisk bronsålder – eller europeisk? In (ed.) Goldhahn, J. *Mellan sten och järn, del 1. Rapport från det 9:e nordiska bronsålderssymposiet, Göteborg 2003-10-09/12.* Gotarc Serie C. Arkeologiska Skrifter No 59. Göteborg. Pp. 125–133.

Nordqvist, B. 1992. Boplatsfunktion ur flintbearbetningsperspektiv – sett i tid och rum. In (eds.) Artelius, T., Johansson, L. G., Jonsäter, M., Lundqvist, L., Nordqvist, B., Streiffert, J., Strömberg, B., Thorsberg, K. & Wickerts-Jensen, M. *Förhistoria i Mellersta Halland. Arkeologiska undersökningar längs väg E6.* Riksantikvarieämbetet och Statens Historiska Museer. Rapport UV 1988:15. Göteborg. Pp. 7–38.

Nordqvist, B. 2005. *Kultplatsen vid Stretered.* Västergötland, Kållereds socken. Stretered 1:1, RAÄ 34, 77:2 och 85. UV väst rapport 2005:1. Arkeologisk förundersökning och undersökning. Riksantikvarieämbetet. Mölndal.

Nylén, E. 1993. Time to Separate Theory from Practice in Archaeology. In Arwidsson, G. (ed.) *Sources and resources. Studies in honour of Birgit Arrhenius.* PACT 38. Belgium, Rixenart. Pp. 165–177.

Nærøy, A. J. 2007. Review on Jan Apel & Kjel Knutsson: Skilled Production and Social Reproduction. Aspects of Traditional Stone-Tool Technologies. Proceedings of a Symposium in Uppsala, August 20–24, 2003. *Norwegian Archaeological Review,* Vol 40, No 2, 2007:208–211.

Nørbach, L. C. 1998. Ironworking in Denmark. From the Late Bronze Age to the Early Roman Iron Age. *Acta Archaeologica.* Vol. 69, 1998:53–75.

Odell, G. H. (ed.) 1996. *Stone Tools. Theoretical Insights into Human Prehistory.* Plenum Press. New York and London.

Odell, G. H. 2004. *Lithic Analysis.* Kluwer Academic/Plenium Publishers. New York.

Odell, G. H. 2005. Editor's Lair. *Lithic Technology* Volume 30, number 1, Spring 2005:3–5.

Odell, G. H. & Odell-Vereecken, F. 1980. Verifying the Reliability of Lithic Use-Wear Assessments by "Blind Tests": the Low-Power Approach. *Journal of Field Archaeology* 7:87–120.

Ohnuma, K. and Bergman, C. 1982. Experimental studies in the determination of flaking mode. *Bulletin of the Institute of Archaeology.* London No.19:161–170.

Olausson, D. 1983. *Flint and groundstone axes in the Scanian Neolithic. An evaluation of raw materials based on experiments.* Scripta Minora Regae Societatis Humaniorum Litterarum Lundensis. Studier utgivna av Kungl. Humanistiska Vetenskapssamfundet i Lund 1982–1983:2. Lund.

Olausson, D. 1987. Piledal and Svarte. A Comparison Between Two Late Bronze Age Cemeteries in Scania. *Acta Archaeologica* 57:121–152.

Olausson, D. 1992. The archaeology of the Bronze Age cultural landscape – research goals, methods, and results. In (eds.). Larsson, L., Callmer, J. & Stjernquist, B. *The Archaeology of the Cultural Landscape. Field Work and Research in a South Swedish Rural Region.* Acta Archaeologica Lundensia. Series in 4°. N° 19. Almqvist & Wiksell International, Stockholm. Pp. 251–282.

Olausson, D. 1998. Different Strokes for Different Folks. Possible reasons for variation in quality of knapping. *Lithic Technology* Vol. 23, Nr. 2. s. 90–115.

Olausson, D. 2007. On the road. Paving stones past and present. In (eds.) Hårdh, B., Jennbert, K. & Olausson, D. *On the Road. Studies in Honour of Lars Larsson.* Acta Archaeologica Lundensia in 4°, No. 26. Almqvist & Wiksell International. Stockholm. Pp. 29–34.

Olausson, D. 2008. Does Practice Make Perfect? Craft Expertise as a Factor in Aggrandizer Strategies. *Journal of Archaeological Method and Theory* Volume 15. Number 1. March 2008:28–50.

Olausson, D., Rudebeck, E. & Säfvestad, U. 1980. Die südschwedischen Feuersteingruben – Ergebnisse und Probleme. In (ed.) Weisgerberg, G. *5000 Jahre Feuersteinbergbau. Die Suche nach dem Stahl der Steinzeit.* Herausgegeben vom Deutschen Bergbau-Museum Bochum 1980. Pp. 183–204.

Olausson, M. 1999. Om vallanläggningar och boplatser i ett bronsålderslandskap. In (ed.). Olausson, M. *Spiralens öga. Tjugo artiklar kring aktuell bronsåldersforskning.* Avdelningen för arkeologiska undersökningar. Skrifter nr. 25. Riksantikvarieämbetet. Pp. 365–392.

Oldeberg, A. 1932. Some contribution to the earliest history of sickles. *Acta Archaeologica.* Vol III. Fasc. 1:209–230.

Olesen, M. W. & Eriksen, B. V. 2007. Krogstrup ved Snejbjerg. En bronzealderboplads med flintværktøj. *Herning Museum. Midtjyske fortællinger* 2007:77–92.

Olsen, B. 2003a. *Från ting till text. Teoretiska perspektiv i arkeologisk forskning.* Studentlitteratur. Lund.

Olsen, B. 2003b. Material Culture after Text: Re-Membering Things. *NAR. Norwegian Archaeological Review,* Vol. 36, No. 2:87–104.

Olsen, B. 2007. Keeping thing at arm's lengt: a genealogy of asymmetry. *World Archaeology, 39:4*: 579–588.

Ostoja-Zagórski, J. 1989. Change in the economic and social structures in northern Poland at the transition from the Bronze Age to Iron Age. In (eds.) Sørensen, M. L. S. & Thomas, R. *The Bronze Age-Iron Age Transition in Europe. Aspects of continuity and change in European societies c. 1200 to 500 B.C.* BAR International Series 483(ii). Pp. 389–407.

Owen, L. R. & Unrath, G. (eds.) 1986. *Technical Aspects of Microwear Studies on Stone Tools.* Part I. Early Man News 9/10/11. Newsletter for Human Palecology. Tübingen.

Papousek, D. A. 1989. Technological change as social rebellion. In (eds.) Leeuw, van der S.E. & Torrence, R. *What's New? A Closer Look at the Process of Innovation.* One World Archaeology 14. Unwin Hyman Ltd. London. Pp. 140–166.

Pastrana, A. & Hirth, K. G. 2003. Biface Production and Craft Specialization. A View from Sierra de las Navajas, Hidalgo. In (ed.) Hirth, K. G. *Mesoamerican Lithic Technology. Experimentation and Interpretation.* The University of Utah Press. Salt Lake City. Pp. 197–207.

Patten, B. 2005. Comments to Odell. *Lithic Technology.* Vol 30, no 2, Fall 2005:85–86.

Patterson, L. W. 1998. Raking Lithic Retouch. *Lithic Technology.* Volume 23, number 1. Spring 1998:27–30.

Pelegrin, J. 1990. Prehistoric Lithic Technology: Some Aspects of Research. In (eds.) Sinclair, A. & Schlanger, N. *Technology in the humanities. Archaeological Review from Cambridge.* Volume 9:1:116–125.

Persson, M. 2001. Yngre bronsålderskeramik i Möre. In (ed.) Magnusson, G. *Möre historien om ett småland.* Kalmar läns museum, Kalmar. Pp. 289–314.

Petersson, M. 2006. *Djurhållning och betesdrift. Djur, människor och landskap i Västra Östergötland under yngre bronsålder och äldre järnålder.* Riksantikvarieämbetet och Uppsala universitet. Stockholm.

Pinch, T., Ashmore, M. & Mulkay, M. 1992. Technology, Testing, Text: Clinical Budgeting in the U.K. National Health Service. In (eds.) Bijker, W. B. & Law, J. *Shaping Technology/Building Society. Studies in Sociotechnical Change.* The MIT Press. Cambridge. Pp. 265–288.

Piotrowska, D. 2000. Krzemienie w grobach z pól popielnicowych: przypadek czy ritual? In (eds.) Gediga, B. & Piotrowska, D. *Kultura Symboliczna Kregu pól popielnicowych epoki Brazu I wczesnej epoki Zelaze w Europie Srodkowej.* Warsawa-Wroclaw-Biskupin. Pp. 293–330.

Prentiss, W. 1998. The Reliability and Validity of a Lithic Debitage Typology: Implications for Archaeological Interpretation. *American Antiquity.* Volume 63, number 4:635–650.

Prost, D. 2002. Des silex taillés à l'Âge du Fer: une presence problématique. In (ed.) Bostyn, F. *Néolithique et protohistoire du site des Antes Rungis, val-de-Marne.* Editions Artcom, Vitry-le-François, ARPEA94. Paris. Pp. 131–181.

Randsborg, K. 1986. A Bronze Age grave on Funen containing a Metal Worker's tools. *Acta Archaeologica* 55:185–189.

Rassman, K. 2001. Die Nutzung Baltischen Feuersteins and der Schwelle zur Bronzezeit – Krise oder Konjunktur der Feuersteinverarbeitung? Vortrag zur Jahressitzung 2000 der Römisch-Germanischen Kommission. *Bericht der Römisch-Germanischen Kommisson Band 81,* 2000. Pp. 5–36.

Rathje, W. & Murphy, C. 1992. *Rubbish! The Archaeology of Garbage.* Harper Collins Publishers. New York.

Regnell, M. & Sjögren, K-G. 2006a. Vegetational development. In (ed.) Sjögren, K-G. *Ecology and Economy in Stone Age and Bronze Age Scania.* Skånska Spår – arkeologi längs Västkustbanan. Riksantikvarieämbetet. Lund. Pp. 40–79.

Regnell, M. & Sjögren, K-G. 2006b. Introduction and development of agriculture. In (ed.) Sjögren, K-G. *Ecology and Economy in Stone Age and Bronze Age Scania.* Skånska Spår – arkeologi längs Västkustbanan. Riksantikvarieämbetet. Lund. Pp. 106–169.

Renfrew, C. & Zubrow, E. B. (eds.) 1994. *The Ancient Mind. Elements of cognitive archaeology.* New Directions in Archaeology. Cambridge University Press. Cambridge.

Reybrouck, D. V. 2000. Beyond ethnoarchaeology? A critical history on the role of ethnographic analogy in contextual and post-processual archaeology. In Gramsch, A. (ed.) *Vergleichen als archäologische Metode – Analogien in den Archäologien.* BAR International Series 825. Oxford. Pp. 39–51.

Ring, C., Pedersen, E. A., Gurstad-Nilsson, H., Lindblad, J., Nilsson, M-L., Persson, M. & Svensson, I. 2001. *Kölbygärde. Bronsåldersboplatser i ett fossilt odlingslandskap. Ljungby socken, Småland.* E22-projeket, Rapport 2001:10. Kalmar Läns museum. Kalmar.

Roche, H. 2005. From Simple Flaking to Shaping: Stone-knapping Evolution among Early Hominins. In (eds.) Roux, V. & Bril, B. *Stone knapping, the necessary conditions for a uniquely hominin behaviour.* MacDonald Institute for Archaeological Research. Cambridge. Pp. 35–48.

Rosen, S. A. 1997. *Lithics After the Stone Age. A Handbook of Stone Tools from the Levant.* Altamira Press. London.

Rostholm, H. 1999. Arkæologiske undersøgelser på Brande Omfartsvej. *FRAM – Fra Rinkøbing Amts Museer* 1999:59–86.

v. Rostoványi, A. & Hydén, S. 2002. *Öresundsförbindelsen. Svågertorp 8B–C.* Rapport över arkeologisk slutundersökning. Rapport nummer 14. Malmö Kulturmiljö. Malmö.

v. Rostoványi, A. 2005. Äta bör man annars dör man – aspekter på mat och förhistoria. In (ed.) Goldhahn, J. *Mellan sten och järn, del 2. Rapport från det 9:e nordiska bronsålderssymposiet, Göteborg 2003-10-09/12.* Gotarc Serie C. Arkeologiska Skrifter No 59. Göteborg. Pp. 671–687.

Roth, W-M. 2005. Making Classifications (at) Work: Ordering Practices in Science. *Social Studies of Science 35/4*:581–621.

Rots, V 2004. Prehensile wear on flint tools. *Lithic Technology.* Volume 29, number 1, Spring 2004:7–32.

Roux, V. 1990. The Psychosocial Analysis of Technical Activities: a Contribution to the Study of Craft Specialisation. In (eds.) Sinclair, A. & Schlanger, N. *Technology in the humanities. Archaeological Review from Cambridge.* Volume 9:1:142–153.

Roux, V. 2003. A Dynamic Systems Framework for Studying Technological Change: Application to the Emergence of the Pooter's Wheel in the Southern Levant. *Journal of Archaeological Method and Theory,* Vol. 10. No 1. March 2003:1–30.

Rudebeck, E. 1986. *Ängdala. Flintgruvor från yngre stenåldern, S Sallerup. Utgrävningar 1977–81.* Rapport Nr 1. Malmö Museer, Stadsantikvariska enheten. Malmö.

Rudebeck, E. 1987. Flintmining in Sweden during the Neolithic period: new evidence from the Kvarnby-S. Sallerup area. In (eds.) Sieveking, G. de G. & Newcomer, M. *The human uses of flint and chert.* Cambridge University Press, Cambridge. Pp. 151–157.

Rudebeck, E. 1994. Ängdala och meningen med arkeologin, *Arkeologi i Sverige 3* (ny ordning). Riksantikvarieämbetet. Stockholm. Pp. 7–40.

Rudebeck, E. 1998: Flint Extraction, Axe Offering, and the Value of Cortex. In (eds.) Edmonds, M. & Richards, C. *Understanding the Neolithic of North-western Europe.* Cruithne Press, Glasgow. Pp. 312–327.

Rudebeck, E. & Ödman, C. 2000. *Kristineberg. En gravplats under 4 500 år.* Malmöfynd 7. Stadsantikvariska avdelningen. Kultur Malmö. Malmö.

Rudebeck, E., Ödman, C., Samuelsson, B-Å. & Högberg, A. 2001. *Vetenskapligt program för Malmö Kulturmiljö. Den arkeologiska verksamheten.* Malmö Kulturmiljö. Malmö.

Rundkvist, M. 2005. Arkeologer tittar på gamla saker för att ta reda på hur det var att leva för länge sedan. *META nr 1* 2005:46–52.

Russell, S. 1986. The Social Construction of Artefacts: A Response to Pinch and Bijker. *Social Studies of Science, Vol. 16, No. 2.* May, 1986:331–346.

Rybczynski, W. 1986. *Home. A short history of an idea.* Penguin. Toronto.

Rydbeck, O. 1918. Slutna mark- och mossfynd från stenåldern i Lunds universitets historiska museum, deras tidsställning och samband med religiösa föreställningar. *Från Lunds Universitets historiska Museum. Skrift utgifven med anledning af museets inflyttning i dess nya hem 1918.* Lund.

Sandklef, A. 1934. Are Scandinavian Flint Saws to be Considered as Leaf Knives? *Acta Archaeologica Vol V Fasc. 1–2*:284–290.

Sarnäs, P. 2008. *Fullåkerslandskapet. Matjord, metod, diskussion, reflektion.* Öresundsförbindelsen och arkeologin. Malmöfynd 14. Malmö Kulturmiljö. Malmö.

Sarnäs, P. & Nord Paulsson, J. 2001. *Öresundsförbindelsen. Skjutbanorna 1B & Elinelund 2A–B.* Rapport över arkeologisk slutundersökning. Rapport nr 9. Malmö Kulturmiljö. Malmö.

Saville, A. 1981. Iron Age flint working – fact or fiction? *Lithics. The Newsletter of the Lithic Studies Society.* No 2:6–9.

Scharfstein, B-A. 1989. *The Dilemma of context.* New York University Press. New York.

Schiffer, M. B. 1988. The Structure of Archaeological Theory. *American Antiquity, Vol. 53, No. 3.* July 1988:461–485.

Schiffer, M. B. 2000. Indigenous Theories, Scientific Theories and Product Histories. In (ed.) Graves-Brown, P. M. *Matter, Materiality and Modern Culture.* Routledge. London and New York. Pp. 72–96.

Schiffer, M. B. & Skibo, J. M. 1987. Theory and Experiment in the Study of Technological Change. *Current Anthropology Volume 28, Number 5,* December 1987:595–622.

Schiffer, M. B., Skibo, J. M., Boelke, T. C., Neupert, M. A. & Aronson, M. 1994. New Perspectives on Experimental Archaeology: Surface Treatments and Thermal Response of the Clay Cooking Pot. *American Antiquity.* Vol 59, No 2, Apr 1994:197–217.

Schlanger, N. 1994. Mindful technology: unleashing the chaîne opératoire for an archaeology of mind. In (eds.) Renfrew, C. & Zubrow, E. B. *The Ancient Mind. Elements of cognitive archaeology.* New Directions in Archaeology. Cambridge University Press. Cambridge. Pp. 143–151.

Schnittger, B. 1910. *Förhistoriska flintgruvor och kulturlager vid Kvarnby och S Sallerup i Skåne.* Stockholm.

Sehested, N. F. B. 1884. *Praktiske Forsøg. Archaeologiske Undersøgelser 1878–1881.* Copenhagen.

Semenov, S. A. 1964 (1957). *Prehistoric Technology. An Experimental Study of the oldest Tools and Artefacts from traces of Manufacture and Wear.* Adams & Dart. Bath.

Shanks, M. 2007. Symmetrical archaeology. *World Archaeology, 39:4*:589–596.

Shanks, M. & Hodder, I. 1995. Processual, postprocessual and interpretive archaeologies. In (eds.) Hodder, I., Shanks, M., Alexandri, A., Buchli, V., Carman, J., Last, J. & Lucas, G. *Interpreting Archaeology. Finding meaning in the past.* Routledge. London. Pp. 3–29.

Shanks, M. & Tilley, C. 1987a. *Reconstructing Archaeology.* Cambridge University Press. Cambridge.

Shanks, M. & Tilley, C. 1987b. *Social Theory and Archaeology.* Polity Press. Cambridge.

Shennan, S. 2002. *Genes, Memes and Human History. Darwinian Archaeology and Cultural Evolution.* Thames & Hudson, London.

Shott, M. J. 1994. Size and Form in the Analysis of Flake Debris: Review and Recent Approaches. *Journal of Archaeological Method and Theory,* vol. 1. No 1. 1994:69–110.

Shott, M. J. 2003. Chaine operatoire and reduction sequence. *Lithic Technology.* Volume 28, Number 2, Fall 2003:95–105.

Shott, M. J., Bradbury, A. P., Carr, P. J. & Odell, G. H. 2000. Flake Size from Platform Attributes: Predictive and Empirical Approaches. *Journal of Archaeological Science* 27:877–894.

Shott, M. J. & Sillitoe, P. 2005. Use life and curation in New Guinea experimental used flakes. *Journal of Archaeological Science* 32:653–663.

Sillitoe, P. & Hardy, K. 2003. Loving Lithics: ethnoarchaeology in Highland Papua New Guinea. *Antiquity.* Volume 77, number 297, September 2003:555–566.

Sinclair, A. & Schlanger, N. (eds.) 1990. *Technology in the humanities. Archaeological Review from Cambridge,* volume 9:1.

Sismondo, S. 2008. Science and Technology Studies and an Engaged Program. In (eds.) Hackett, E. J., Amsterdamska, O., Lynch, M. & Wajcman, J. *The Handbook of Science and Technology Studies. Third Edition.* The MIT Press, Cambridge, Massachusetts. Pp. 13–32.

Sjöberg, J. E. 1998. *Tusentals fynd från hundratusentals år. En resa i tid och rum genom stadsmuseets arkeologiska samlingar.* Göteborgs Stadsmuseum. Göteborg.

Sjögren, L. 2005. Minoiskt i norr? Om kulturella influenser från Kreta till Skandinavien. In (ed.) Goldhahn, J. *Mellan sten och järn, del 1. Rapport från det 9:e nordiska bronsålderssymposiet, Göteborg 2003-10-09/12.* Gotarc Serie C. Arkeologiska Skrifter No 59. Göteborg. Pp. 151–166.

Sjöström, A. & Nilsson, B. 2008. Rulers of southern Sweden: Technological aspects of a rediscovered tool. In (eds.) McCartan, S., Woodman, P., Schulting, R. & Warren, G. *Mesolithic Horizons: Papers presented at the Seventh International Conference on the Mesolithic in Europe, Belfast 2005.* Oxbow Books. Oxford. Pp. 786–792.

Skertchly, S. B. J. 1879. *On the Manufacture of Gun-flints, the Methods of Excavating for Flint, the Age of Paleolithic Man, and the Connection between Neolithic Art and the Gun-flint Trade.* Memoirs of the Geological Survey of England and Wales. Geological Survey. London.

Skoglund, P. 1999. De enskilda hushållens betydelse för landskapsutvecklingen under bronsåldern. In (ed.) Olausson, M. *Spiralens öga. Tjugo artiklar kring aktuell bronsåldersforskning.* Riksantikvarieämbetet. Avdelningen för arkeologiska undersökningar. Skrifter nr 25. Stockholm. Pp. 277–289.

Skoglund, P. 2005. *Vardagens Landskap – lokala perspektiv på bronsålderns materiella kultur.* Acta Archaeologica Lundensia Series in 8° No 49. Almqvist & Wiksell International. Stockholm.

Skoglund, P. 2008. The Composite Ard in Sweden – its Introduction, Geographical Distribution and Consequences for Landscape Management. In (ed.) Goldhahn, G. *Gropar & monument. En vänbok till Dag Widholm.* Kalmar Studies in Archaeology IV. Kalmar. Pp. 323–339.

Slotta, R. 1980. Die heutige Feuersteinindustrie von Eben-Emaël (Belgien). In (ed.) Weisgerberg, G. *5000 Jahre Feuersteinbergbau. Die Suche nach dem Stahl der Steinzeit.* Herausgegeben vom Deutschen Bergbau-Museum Bochum 1980. Pp. 366–374.

Smith, A. T. 2004. The end of the essential archaeological subject. *Archaeological dialogues.* Vol. 11. No. 1. 2004:1–19.

SMS dictionary 2008. http://www.smsdictionary.co.uk/ , http://www.funsms.net/sms_dictionary.htm , http://www.mobtastic.com/sms/dictionary/sms_dictionary.asp http://www.telepk.com/paksms/smsdictionary.asp, all accessed 17 April 2008.

Solli, B. 2007/2008. Fra test via tegn og tekst til ting? Eller: Pimp my site?. *Primitive tider* 2007/2008. Nr 10:155–164.

The Society of Primitive Technology. http://www.primitive.org/ Accessed 3 August 2008. Last update May 23 2008.

Spaulding, A. C. 1953. Review of "Measurements of some prehistoric design developments in the Southeastern States". *American Anthropologist* 55:588–591.

Spaulding, A. C. 1954. Reply to Ford. *American Antiquity,* Vol. 19, No. 4, Apr., 1954:391–393.

Speal, S. C. 2006. Regional Economic Intergration in the Costal Maya Lowlands: The Lithic Assemblage of Saktunha, Belize. *Lithic Technology,* volume 31, number 1. Spring 2006:3–26.

Sperber, L. 1998. Crises in Western European Metal Supply During the Late Bronze Age: From Bronze to Iron. In (eds.) Hvass, S., Jacob, W., Cachin, F., Digne, J. & Tzedakis, Y. *Gods and Heroes of the Bronze Age. Europe at the Time of Ulysses.* National Museum of Denmark / Kunst- und Ausstellungshalle der Bundesrepublik Deutschland / Réunion des musées nationaux, France / Associatio Francaise dÁction Artistique, France / Hellenic Ministry of Culture. Germany. Pp. 48–51.

Sprockhoff, E. 1956. *Jungbronzezeitliche Hortfunde. Der Südzone des Nordischen Kreises (periode V). Band I und II.* Verlag des Römisch-Germanischen Zentralmuseums. Mainz.

Stafford, M. 1998. In Search of Hindsgavl: Experiments in the Production of Neolithic Danish Flint Daggers. *Antiquity* 72:338–349.

Stamm Forssblad, E. 2001. *Lockarp 8:4. Rapport over arkeologisk slutundersökning 2001. Lockarp sn, Malmö Stad, Skåne.* Malmö Kulturmiljö. Malmö.

Steensberg, A. 1943. *Ancient Harvesting Implements. A Study in Archaeology and Human Geography.* Nationalmuseets skrifter. Arkæologisk-Historisk Række, I. Gyldendalske Boghandel, Nordisk Forlag. Copenhagen.

Steineke, M., Ekenberg, A., Hansson, K. & Ifverson, P. 2005. *Öresundsförbindelsen. Sunnanå 19A–F. Rapport över arkeologisk slutundersökning.* Rapport nr 34. Malmö Kulturmiljö. Malmö.

Stenberger, M. 1964. *Det forntida Sverige.* Almqvist & Wiksell, Stockholm.

Stjernquist, B. 1961. *Simris II. Bronze Age Problems in the Light of the Simris Excavation.* Acta Archaeologica Lundensia Series in 4°. No 5. CWK Gleerups Förlag. Lund.

Stjernquist, B. 1969. *Beiträge zum Studium von Bronzezeitlichen Siedlingen.* Acta Archaeologica Lundensia. Series in 8°. N° 8. Gleerups Förlag. Lund.

Streiffert, J. 2001. *På gården. Rumslig organisation inom bosättningsytor och byggnader under bronsålder och äldre järnålder.* Riksantikvarieämbetet arkeologiska undersökningar skrifter 35. Gotarc. serie C. Arkeologiska skrifter 37. Riksantikvarieämbetet och Göteborgs universitet.

Streiffert, J. & Strömberg, B. 1998. *Gravar och boplatslämningar vid Sannarp – Sannarp 1:1, Raä 3, Årstads socken Falkenbergs kommun, Halland.* Arkeologiska resultat UV Väst rapport 1998:8. Riksantikvarieämbetet. Stockholm.

Strömberg, M. 1952. En privatägd fornsakssamling i Trolle-Ljungby socken. *Skånes Hembygdsförbunds Årsbok.* 1952:84–98.

Strömberg, M. 1954. Bronzezeitliche Wohnplätze in Schonen. *Meddelanden från Lunds Universitets Historiska Museum* 1954:295–380.

Sullivan, A. P. III. & Rozen, K. C. 1985. Debitage Analysis and Archaeological Interpretation. *American Antiquity.* Vol. 50, No. 4. Oct. 1985:755–779.

Sundström, L. 2003. *Det hotade kollektivet. Neolitiseringsprocessen ur ett östmellansvenskt perspektiv.* Coast to coast – books – no 6. Uppsala.

Svanberg, F. & Söderberg, B. 2000. *Porten till Skåne. Löddeköpinge under järnålder och medeltid.* Riksantikvarieämbetet Arkeologiska undersökningar Skrifter No 32; Arkeologiska studier kring Borgeby och Löddeköpinge 1. Lund.

Svanberg, F. 2007. Aristokratiska husgravar under bronsåldern. Hierarki ända in i döden. In (ed.) Artursson, M. *Vägar till Vætland. En bronsåldersbygd i nordöstra Skåne 2300–500 f.Kr.* Riksantikvarieämbetet. Regionmuseet Kristianstad. Stockholm. Pp. 187–221.

Swanson, E. (ed.) 1975. *Lithic Technology. Making and Using Stone Tools*. Mouton Publishers. Paris.

Sydsvenskan 2008. http://sydsvenskan.se/inpalivet/article290566.ece. First publiced 1 January 2008. Accessed 16 April 2008.

Sørensen, M. 2006b. Teknologiske traditioner i Maglemosekulturen. En diakron analyse af Maglemosekulturens flækkeindustri. In (ed.) Eriksen, B. V. *Stenalderstudier. Tidligt mesolitiske jægere og samlere i Sydskandinavien*. Jysk Arkæologisk Selskabs Skrifter 55. Højbjerg. Pp. 19–75.

Sørensen, M. 2006c. Rethinking the lithic blade definition: towards a dynamic understanding. In (eds.) Apel, J. & Knutsson, K. *Skilled Production and Social Reproduction. Aspects of Traditional Stone-Tool Technologies. Proceedings of a Symposium in Uppsala*, August 20–24, 2003. SAU Stone Studies 2. Uppsala. Pp. 277–296.

Sørensen, M. & Desrosiers, P. (eds.) 2008. *Technology in Archaeology. Proceedings from the SILA Workshop: The study of Technology as a method for gaining insight into social and cultural aspects of Prehistory*. Publications from the National Museum Studies in Archaeology & Historiy Vol. 14. Copenhagen.

Sørensen, M. L. S. 1987. Material order and cultural classification: the role of bronze objects in the transition from Bronze Age to Iron Age in Scandinavia. In (ed.) Hodder, I. *The Archaeology of Contextual Meanings*. Cambridge University Press. Cambridge. Pp. 90–101.

Sørensen, M. L. S. 1989a. Ignoring innovation – denying change: the role of iron and the impact of external influences on the transformation of Scandinavian societies 800–500 BC. In (eds.) Leeuw, van der S. E. & Torrence, R. *What's New? A Closer Look at the Process of Innovation*. One World Archaeology 14. Unwin Hyman Ltd. London. Pp. 182–202.

Sørensen, M. L. S. 1989b. Period VI reconsidered: Continuity and change at the transition form Bronze to Iron Age in Scandinavia. In (eds.) Sørensen, M. L. S. & Thomas, R. *The Bronze Age-Iron Age Transition in Europe. Aspects of continuity and change in European societies c. 1200 to 500 B.C.* BAR International Series 483(ii). Pp. 457–492.

Sørensen, M. L. S. & Thomas, R. (eds.) 1989. *The Bronze Age-Iron Age Transition in Europe. Aspects of continuity and change in European societies c. 1200 to 500 B.C.* BAR International Series 483(ii). Oxford.

Sørensen, S. A. 1980. En omfartsvej gennem bronzealderen. *ROMU 1. Årsskrift fra Roskilde Museum 1980*:18–36.

Sørensen, S. A. 1984–85. Nye fund fra den yngre bronzealders affaldsgruber. *ROMU 3. Årsskrift fra Roskilde Museum 1984–85*:53–62.

Tel el-Far'ah Dictionary/Pictionary 2008. Last modified 13 April 2003. Accessed 18 April 2008. http://farahsouth.cgu.edu/dictionary/#C

Thedéen, S. 2005. Till frågan om hövdingadömet under bronsåldern. In (ed.) Goldhahn, J. *Mellan sten och järn, del 1. Rapport från det 9:e nordiska bronsålderssymposiet, Göteborg 2003-10-09/12*. Gotarc Serie C. Arkeologiska Skrifter No 59. Göteborg. Pp. 385–401

Thomas, J. 2004. The Great Dark Book: Archaeology, Experience, and Interpretation. In (ed.) Bintliff, J. *A Companion to Archaeology*. Blackwell Publishing. Oxford. Pp. 21–36.

Thomsen, E. 2000. Flintens geologi og mineralogi. In (ed.) Eriksen, B. V. *Flintstudier. En håndbog i systematiske analyser af flintinventarer*. Aarhus University Press. Århus. Pp. 17–36.

Thrane, H. 1971. En broncealderboplads ved Jyderup Skov i Odsherred. *Nationalmuseets Arbejdsmark* 1971:141–164.

Thrane, H. 1975. *Europæiske forbindelser. Bidrag til studiet af fremmede forbindelser i Danmarks yngre broncealder (periode IV–V)*. Nationalmuseets skrifter Arkæologisk-historisk række Bind XVI. Nationalmuseet. Copenhagen.

Thrane, H. 1980. En boplads fra broncealderens slutning fra Flædemose på Stevns. *Historisk Samfund for Præstø Amt* 1979:21–36.

Thrane, H. 1984. *Lusehøj ved Voldtofte – en sydvestfynsk storhøj fra yngre broncealder*. Fynske Studier XIII Odense bys museer. Odense.

Thrane, H. 1985. Bronze Age Settlements. In (ed.) Kristiansen, K. *Archaeological Formation Processes. The representativity of archaeological remains from Danish Prehistory*. Nationalmuseets forlag. Copenhagen. Pp. 142–151.

Thrane, H. 1993. From Mini to Maxi. Bronze Age Barrows from Funen as Illustration of Variation and Structure. In (ed.) Larsson. L. *Bronsålderns gravhögar*. Rapporter från ett symposium i Lund 15.XI–16.XI 1991. University of Lund. Institute of Archaeology Report Series No. 48. Lund. Pp. 79–90.

Thrane, H. 1994. Centres of Wealth in Northern Europe. In (eds.) Kristiansen, K. & Jensen, J. *Europe in the First Millennium B.C.* Sheffield Archaeologicl Monographs 6. J.R. Collis Publications. Sheffield. Pp. 95–110.

Thrane, H. 2008. Chiefs & Chiefdoms – høvdinge og høvdingedømmer. In (ed.) Goldhahn, G. *Gropar & monument. En vänbok till Dag Widholm*. Kalmar Studies in Archaeology IV. Kalmar. Pp. 107–116.

Tilley, C. (ed.) 1993. *Interpretative Archaeology*. Berg Publishers. Oxford.

Tilley, C. 2007. Materiality in materials. *Archaeological Dialogues* 14(1):16–20.

Tomášková, S. 2005. What is a Burin? Typology, Technology, and Interregional Comparison. *Journal of Archaeological Method and Theory*. Vol. 12, No 2, June 2005:79–115.

Torrence, R. 1989. Tools as optimal solutions. In (ed.) Torrence, R. *Time, energy and stone tools*. New Directions in Archaeology. Cambridge University Press. Cambridge. Pp. 1–6.

Torrence, R. & Leeuw, S. E. van der. 1989. Introduction: what's new about innovation? In (eds.) Leeuw, van der S. E. & Torrence, R. *What's New? A Closer Look at the Process of Innovation*. One World Archaeology 14. Unwin Hyman Ltd. London. Pp. 1–15.

Tostevin, G. B. in press. Levels of theory and social practice in the Reduction Sequence and Chaîne Opératoire methods of lithic analysis. In (ed.) Tostevisn G. B. *Reduction Sequence, Chaîne Opératoire, and Other Methods: The Epistemologies of Different Approaches to Lithic Analysis*. Monograph of the electronic symposium at the 71th Annual Meeting of the Society for American Archaeology, San Juan, Puerto Rico. Publication in preparation.

Trigger, B. G. 2006. A *History of Archaeological Thought, second edition*. Cambridge University Press. Cambridge.

Ullén, I. 2003. *Bronsåldersboplatsen vid Apalle i Uppland. Arkeologi på väg – undersökningar för E18*. Riksantikvarieämbetet. UV Uppsala rapport 1997:64. Uppsala.

Unrath, G., Owen, L. R., Gijn, A. L. van, Moss, E., Plisson, H. & Vaughan, P. 1986. An Evaluation of Microwear Studies: A Multi-Analyst Approach. In (eds.) Owen, L. R. & Unrath, G. *Technical Aspects of Microwear Studies on Stone Tools*. Part I. Early Man News 9/10/11. Newsletter for Human Palecology. Tübingen. Pp. 117–176.

Vandenberghe, F. 2002. Reconstructing Humants: Humanist Critique of Actant-Network Theory. *Theory, Culture & Society* 2002 Vol. 19(5/6): 51–67.

Vandkilde, H. 2000. Material Culture and Scandinavian Archaeology: A Review of the Concepts of Form, Function, and Context. In (eds.) Olausson, D. & Vandkilde, H. *Form, Function and Context. Material culture studies in Scandinavian archaeology*. Acta Archaeologica Lundensia. Series in 8°, No. 31. Almqvist & Wiksell International. Stockholm. Pp. 3–49.

Vandkilde, H. 2007. *Culture and Change in Central European Prehistory, 6th to 1st millennium BC*. Aarhus University Press. Århus.

Vang Petersen, P. 1993. *Flint fra Danmarks oldtid*. Høst & Søn. Copenhagen.

Vaughan, P. 1985. *Use-Wear Analysis of Flaked Stone Tools*. University of Arizona Press. Tuscon.

Wahlgren, K. H. 2002. *Bilder av betydelse. Hällristningar och bronsålderslandskap i nordöstra Östergötland*. Stockholm Studies in Archaeology 23. Bricoleur Press. Lindome.

Webmoor, T. 2007. What about one more turn after the social in archaeological reasoning? Taking things seriously. *World Archaeology, 39:4*:563–578.

Weibull, C. 1923. *Sveriges bebyggelse: bronsålderns karaktär: diskussionsinlägg*. Gleerup. Lund.

Weibull, L. 1923. Det arkeologiska treperiodsystemet. Dess uppkomst och giltighet. *Historisk tidskrift för Skåneland: Skåne, Halland, Blekinge och Bornholm*. 1923:5:247–266.

Weiler, E. 1994. *Innovationsmiljöer i bronsålderns samhälle och idévärld. Kring ny teknologi och begravningsritual i Västergötland*. Arkeologiska Institutionen Umeå Universitet. Studia Archaeologica Universitatis Umensis 5. Umeå.

Weiler, E. 1997. Livlig skrotimport på bronsåldern. *Forskning & Framsteg* nr 3 1997:20–25.

Welinder, S. 1998. Neoliticum – Bronsålder 3900–500 f.Kr. In (eds.) Welinder, S., Pedersen, E. A. & Widgren, M. *Jordbrukets första femtusen år. 4000 f Kr.–1000 e. Kr.* Natur och Kultur/Lts förlag. Stockholm. sid 11–236.

Wenban-Smith, F. F. 1989. The Use of Canonical Variates for Determination of Biface Manufacturing Technology at Boxgrove Lower Palaeolithic Site and the Behavioural Implications of this Technology. *Journal of Archaeological Science* 1989,16: 17–26.

Wenzel, K. E. & Shelley, P. H. 2001. What Put the Small in the Artic Small Tool Tradition: Raw Material Constraints on Lithic Technology at the Mosquito Lake Site, Alaska. In (ed.) Andrefsky, W. Jr. *Lithic Debitage. Context, Form, Meaning*. The University of Utah Press. Salt Lake City. Pp. 106–123.

Wheen, F. 2004. *How mumbo-jumbo conquered the world. A short history of modern delusions*. Fourth Estate. London.

Whittaker, J. C. 1994. *Flintknapping. Making & Understanding Stone Tools*. University of Texas Press. Austin.

Whittaker, J. C. 2004. *American Flintknappers. Stone Age Art in the Age of Computers*. University of Texas Press. Austin.

Whittaker, J. C. & Kaldahl, E. J. 2001. Where the Waste Went: A Knappers' Dump at Grasshopper Pueblo. In (ed.) Andrefsky, W. Jr. *Lithic Debitage. Context, Form, Meaning*. The University of Utah Press. Salt Lake City. Pp. 32–60.

Winkler, M. 2004. *Öresundsförbindelsen. Södra Sallerup 15C & A*. Rapport över arkeologisk slutundersökning. Rapport nr 28. Malmö Kulturmiljö. Malmö.

Witmore, C. L. 2007. Symmetrical archaeology: exerpts of a manifesto. *World Archaeology, 39:4*:546–562.

Wittgenstein, Ludwig 1953 (2001). *Philosophical Investigations*. Blackwell. Oxford.

Woltering, P. J. 2001. Occupation History of Texel, IV Middle Bronze Age – Late Iron Age (1500–100 BC)s. *Berichten van de Rijksdienst voor het Oudheidkunig Bodemonderzoek*. Volume 44 2000-01:9–396.

Wylie, A. 2002. *Thinking from Things. Essays in the Philosophy of Archaeology*. University of California Press. Los Angelse.

Wyszomirski, M. 1979. Svenska depåfynd med s.k. lövknivar. *Fornvännen* 1979/3:145–150.

Yerkes, R. W. 2003. Using Lithic Artifacts to Study Craft Specialization in Ancient Societies: The Hopewell Case. In (eds.) Kardulias, N. & Yerkes, R. *Written in Stone. The Multiple Dimensions of Lithic Analysis.* Lexington Books Oxford. Pp. 17–34.

Ödman, C. (ed.) 1992. *Arkeologi i Malmö. En presentation av ett antal undersökningar i Malmö under 1980-talet.* Rapport nr 4. Malmö Kulturmiljö. Malmö.

Öijeberg, J. 2004. *Öresundsförbindelsen. Vintriehemmet 3A–B.* Rapport över arkeologisk slutundersökning. Rapport nr 10. Malmö Kulturmiljö. Malmö.

Østigård, T. 2007. *Transformatøre. Ildens mester i jernalderen.* Gotarc Serie C. Arkeologiska Skrifter No 65. Institutionen för Arkeologi och antikens kultur. Göteborgs universitet. Göteborg.

Østigård, T. & Goldhahn, J. 2007. Rituella specialister i brons- och järnålder – en epilog. In Østigård, T. *Transformatøre. Ildens mester i jernalderen.* Gotarc Serie C. Arkeologiska Skrifter No 65. Institutionen för Arkeologi och antikens kultur. Göteborgs universitet. Göteborg. Pp. 187–199.

Unpublished references

Archival material, Sydsjælland Museum, inv. no. SMV 7647, material concerning the find spot of Tommestrup at Stevns, Zealand, Denmark.

Archival material, Lund University Historical Museum, inv. no. LUHM 31875. Undated and unsigned report on excavations in 1967 and 1968 at Klåhög, Scania, Raus Parish, Ättekulla (Helsingborg). The report was written by Rolf Petré.

Högberg, A. 2000b. *Lövknivar – flintteknologi i yngre bronsålder. En experimentell studie av lövknivstillverkning.* Report on experiments financed by the Historical-Archaeological Research Centre, HAF 6/00. Unpublished report. Lejre.

Högberg, A. 2003. *Lövknivar – flintteknologi i yngre bronsålder. En funktionell studie av lövknivar – riktade experiment och slitspårsanalys.* Report on experiments financed by the Historical-Archaeological Research Centre, HAF 08/03. Unpublished report. Lejre.

Högberg, A. 2004c. *Tandade flintredskap och kontext – analys av flintmaterial från ett antal neolitiska gropar.* Analysis performed as part of the sub-project "The Functional Landscape", in the Öresund Fixed Link Project. Malmö Heritage. Malmö.

Högberg, A. 2005b. *En experimentell produktion av lövknivar i Kristianstadsflinta.* Report on experiments financed by the Historical-Archaeological Research Centre, HAF 04/05. Unpublished report. Lejre.

Jonsson, E. 1986. *De arkeologiska utgrävningarna inom kvarteret Benkammen.* Unpublished brochure from Malmö Museums.

Ohlsson, Tom. n.d. *Löddeköpingeundersökningen IV. Södra delen av byområdet.* Manuscript in the archives of Lund University Historical Museum. The excavation has the provisional inventory number 80250.

Sørensen, M. 2006a. *Teknologi og Tradition i Østarktis 2500 BC – 1200 AD. En dynamisk teknologisk undersøgelse af litiske inventarer i de palæoeskimoiske traditioner. Vol. 1, tekst.* Unpublished Ph.D. dissertation, Copenhagen University, Saxo Institute, Department of Prehistoric Archaeology. Publicly defended on 19 May 2006.

Sørensen, S. A. 1986. Archive report on Højbjerg and Holmehusbakke. Dated 28/1 1986, journal number 520/82. Roskilde Museum archives.

Tostevin, G. B. 2006. Levels of theory and social practice in the reduction sequence and chaîne opératoire methods of lithic analysis. Paper submitted to the Electronic Symposium "Core Reduction, Chaîne Opératoire, and Other Methods: The Epistemologies of Different Approaches to Lithic Analysis" at the 71st Annual Meeting of the Society for American Archaeology, San Juan, Puerto Rico.

Personal communications

Cinthio, Hampus. 2008. Lund University Historical Museum. Information about the find circumstances of the assemblage from Löddeköpinge Församlingshemmet. Cinthio took part in the excavation.

Malmros, Claus. 2003. National Museum, Copenhagen. Information about the identification of types of wood on the shaft from the Stenild find.

Schirren, Michael. 2007. Landesamt für Bodendenkmalpflege in Mecklenburg-Vorpommern, Stralsund. Information about finds of large blade knives from Rügen.

Ullén, Inga. 2007. Museum of National Antiquities, Stockholm. Information about find contexts and datings of finds from Apalle, by e-mail on 8 February 2007.

www.ingramcontent.com/pod-product-compliance
Lightning Source LLC
LaVergne TN
LVHW070242220926
848482LV00036B/723